GEOMORPHOLOGY
IN THE TROPICS

FRONTISPIECE   Julius Büdel explains how etchplains form to Denys Brunsden during a visit to Scotland in 1982

GEOMORPHOLOGY IN THE TROPICS

A Study of Weathering and Denudation in Low Latitudes

MICHAEL F. THOMAS
Department of Environmental Science
University of Stirling, UK

JOHN WILEY & SONS
Chichester · New York · Brisbane · Toronto · Singapore

Copyright ©1994 by John Wiley & Sons Ltd,
Baffins Lane, Chichester,
West Sussex PO19 1UD, England

Telephone National Chichester (0243) 779777
International (+44) (243) 779777

Other Wiley Editorial Offices

John Wiley & Sons, Inc., 605 Third Avenue,
New York, NY 10158-0012, USA

Jacaranda Wiley Ltd, 33 Park Road, Milton,
Queensland 4064, Australia

John Wiley & Sons (Canada) Ltd, 22 Worcester Road,
Rexdale, Ontario M9W 1L1, Canada

John Wiley & Sons (SEA) Pte Ltd, 37 Jalan Pemimpin #05-04,
Block B, Union Industrial Building, Singapore 2057

Library of Congress Cataloging-in-Publication Data

Thomas, Michael Frederic.
    Geomorphology in the tropics : a study of weathering and
  denudation in low latitudes / Michael F. Thomas.
      p.   cm.
    Includes bibliographical references and index.
    ISBN 0-471-93035-0
    1. Geomorphology—Tropics. I. Title.
  GB446.T46 1994   93-6094
  551.4′1′0913—dc20   CIP

British Library Cataloguing in Publication Data

A catalogue record for this book is available from the British Library

ISBN 0-471-93035-0

Typeset in 10/12pt Times from author's disks by
Dobbie Typesetting Ltd, Tavistock, Devon
Printed and bound in Great Britain by Bookcraft (Bath) Ltd

In memory of Anne
and for
Gillian and Graham

Contents

PART II: DENUDATION PROCESSES

5 Surface Processes in Tropical Climates   125

PART IV: THE EVOLUTION OF TROPICAL LANDSCAPES

Preface

In deciding to write this book, I took a conscious decision to do so alone, despite the limitations this was bound to impose on the outcome. Several 'symposia' volumes with tropical themes have appeared in recent years (e.g. Goudie & Pye, 1983; Wilson, 1983; Douglas & Spencer, 1985b), but despite the merits of this format, it does not permit integration of the material, and it inhibits the consistent development of themes and arguments. At the same time, an earlier study (Thomas, 1974a), which might have been revised after 10 years had this option been available, was not only out of print but embarrassingly out of date after nearly 20 years. Despite the many advances in our understanding of tropical geomorphology that have been made in recent decades, the field remains relatively neglected. I find this puzzling in the light of widespread concerns about tropical forest (and savanna) ecosystems, and the considerable body of new literature on tropical weathering and soils, and hydrology.

It is clearly time for a new look at geomorphology in tropical (and near-tropical) environments, and in particular, it is important to shift the balance of material away from preoccupation with the geomorphology of tropical cratons, with their inselbergs and extensive plains, though these will not be neglected. To some extent this study is more concerned with the humid tropics, but no attempt is made to draw geographical or climatic boundaries around such an area, and, like Tricart (1965), I have included the savannas with the forested environments. This makes sense because of the major shifts in climate that have affected the tropical zone on timescales varying from 10^3–10^7 years. Indeed, some topics overlap with arid zone interests, currently well served by other recent publications (D. S. G. Thomas, 1989; Cooke *et al.*, 1993), and it is I think instructive to ponder how different are the approaches to geomorphological study adopted by writers, who in a real sense 'look outwards' from contrasting core environments. In this book, the perspective is as seen from the humid tropics which sharpens the contrast with arid zone studies.

The attempt here is to provide a source book for serious students of tropical environments who already have some knowledge of geomorphology and related earth sciences. It is my hope that subjects covered here will be found relevant across a wide range of special interest groups, though, inevitably many will be dissatisfied with what they find. I cannot claim to have a thorough knowledge of all tropical environments, nor of all the specialisms covered in this book, and I feel some diffidence about proffering my limited understanding. I must also ask my friends and colleagues who reside in tropical countries to forgive my presumption in writing about such matters from a base so far away. However, possibly because I go to tropical areas as a visitor, I have a heightened awareness of the differences, and sometimes the similarities, between the landforms of colder climates and those of the tropics.

The organisation of the book shows some inheritance from my earlier study, but no part of that text has been re-used. I start with an introduction on 'geomorphology from a tropical perspective', because it is my view that this perspective is often lost in the pursuit of general theory. I certainly do not wish to encourage the emergence of some unproductive, compartmentalised 'tropical' geomorphology, but, at the same time, I consider that it would be misconceived to ignore the many special characteristics of tropical environments and their impacts on the balance of forces and materials within denudation systems. Those properties hinge in my view on three driving mechanisms: (1) the radiation flux with its positive balance over the year, and all year within the inner tropics; (2) the moisture flux or water balance and the impacts of moisture surpluses and deficits within climates with a positive radiation balance, and (3) the intensity of rainfall which is distinctively greater in most tropical climates than elsewhere, with storms of > 50 mm occurring on average every 2 years, and rainfalls exceeding > 150 mm recurring often within 10 years.

These factors vary within the tropics in important ways, so that intense eluviation and leaching (*lessivage*) can produce giant podzols, perhaps even dissolve silicate rocks entirely, while alternating moisture levels can lead to the formation of ferricretes and silcretes. Rock rotting really does go to very deep levels, and comments made with some trepidation on the basis of early work in Nigeria (Thomas, 1966a) have proved to be over cautious, when viewed across a wider range of landscapes. It really is not helpful to insist on the ability of temperate climates to form laterites, not to mention podzols, as a means of downgrading the importance or distinctiveness of tropical conditions. Deep rock decay, and advanced lateritisation (ferrallitisation) are fundamentally tropical phenomena. From the humid tropical perspective their appearance in temperate latitudes is anomalous, requiring either great lengths of time or major climatic shifts, possibly both. More often the weathered mantle of temperate areas consists of *gruss* which is not common as a surface material within the more humid areas of the tropics.

The book is divided into four parts. Part I proceeds from discussions of rock weathering to a consideration of weathering products and profiles, with an extended section on laterites and bauxites (Chapters 2–4). Part II looks at erosional processes in terms of runoff and fluvial activity, and then mass movement, with an extended discussion of landsliding (Chapters 5 and 6). Part III introduces the Quaternary climate changes and their impacts on temperature and rainfall (Chapter 7), and then discusses the impact of these changes on the formation of distinctive landforms and deposits (Chapter 8). Part IV is concerned with longer term landform evolution and the emergence of residual hills, and includes an extended discussion of etchplanation (Chapters 9–11). Within each chapter, sections have been included to point to the ways in which particular aspects of tropical geomorphology and surficial geology become important in the fields of resource development and conservation. Such topics range from mineral exploration to problems of land stability and accelerated erosion.

What follows is very much a personal view of the subject matter, though I have tried to offer a balanced view of the literature. Before thanking some of the many people who have helped me over the years in my travels in the tropics, I should wish to apologise to those whose work has been unwittingly omitted, and hope that I have not misrepresented either facts or findings from the works that I have managed to read and cite in this study.

My thanks go first and foremost to my late wife, Anne, who tragically died during the preparation of this book. Her support and forbearance over many years encouraged me to continue fieldwork in distant places, and this book is dedicated to her memory.

The work of Julius Büdel has had a major influence on my thinking and I recall the excitement of sitting down with a teacher of German (Miss D. Jackson), in Ibadan in 1961, to obtain a full translation of his seminal paper from 1957, 'Die "doppelten Einebnungsflächen" in den feuchten Tropen'. Julius Büdel also had a personal magnetism and enthusiasm few can match, and his interest and encouragement over the years was invaluable. I met Joao Bigarella for the first time at a conference in Madrid in 1989, and he subsequently enabled me to visit Brazil in 1991. The time spent with Professor Bigarella and his Brazilian colleagues in the field has left a lasting impression, and I am grateful for the hospitality of his wife Iris and all his family. The late Lester King was gracious to a young disbeliever, and gave his time generously to show me the landscapes of Natal some 25 years ago.

A special word of acknowledgement is also due to Martin Thorp, whose name appears as co-author on a range of publications written from a shared experience of work in the tropics over more than a decade. During this time, many ideas have been developed jointly and a lot of work has been shared between us. I also owe his wife, Margaret, debts of friendship and hospitality, particularly during visits to Sierra Leone. In that country and in Zambia, I have also been helped by Geoffrey and Lorna Williams and by Barrie and Jean Gleave. Jack Mabbutt provided my first opportunity to visit Australia which also allowed me to see Papua New Guinea and these were formative experiences, made possible by a visit to New Zealand at the invitation of Barry Johnson. I also wish to thank all of the following for their personal interest and help in many different ways: Hanna Bremer, Mike Browne, Denys Brunsden, Des and Carol Bowden, Detlef Busche, Chris Clements, John Cooke, Barry Dalal-Clayton, Ian Douglas, Geruse Maria Duarte, Mike Everett, Adetoye Faniran, Peter Fookes, Basil Foster, Cyril Geach, Adrian Hall, David Harrison, David Hopley, Lawrence Jeje, Gear Kajoba, Owen King, David Ledger, Tom Martin, Dai Morgan, Cliff Ollier, Hélio Penha, Robert Preston-White, John Pugh, Luiz Fernando Scheibe, Stanley Schumm, Odair Gercino da Silva, Xavier de Silva, Richard and Ann Whitlow; a special word of thanks also to an engaging and enthusiastic group of young Brazilian researchers who took time at short notice to show me the Bananal area in southeast Brazil: Anice Esteves Afonso, Ceres Maria Bastos Barreto, Claudio Limeira Mello, and Maria Naisz de Oliveira Peixoto.

My thanks are also due to the following organisations for their financial and other support: the universities of Ibadan (Nigeria), Natal (South Africa), New South Wales (Australia), North Queensland (Australia), St Andrews (Scotland), Santa Caterina (Florianopolis, Brazil), Sierra Leone, Stirling (Scotland), and Zambia; also to The Carnegie Trust for the Universities of Scotland which has consistently supported my work in the tropics, enabling me to travel in Africa, Brazil, Papua New Guinea and Zambia on different occasions; the British Council has supported visits to Botswana, Liberia, Sierra Leone and Zambia; the Conshelo Nacional de Desenvolvimento Cientifico e Tecnologico (CNPq) supported my recent visit to Brazil; research in Ghana was facilitated by the National Diamond Mining Company and the UNDP. The former BP Minerals International and PT Sungai Kencana commissioned work in Sierra Leone and Indonesia and provided access to field areas and to data.

The figures for this book were painstakingly prepared by William Jameson and Mary Smith of the Cartographic Unit, Department of Environmental Science, assisted also by Ken Dockery and Tracy Grieve. The plates were prepared by Ron Stewart of Media Services, University of Stirling, from my own negatives and slides, except where separately acknowledged.

Acknowledgements

The author and publishers would like to thank the following publishers, organisations and individuals for permission to reproduce figures and tables in this book.

Academic Press Inc (London) Ltd, for Figures 3.19, 4.6, 4.7, 4.9; American Geophysical Union, Figure 7.1; American Institute for Mining and Metallurgy, Figures 2.11, 4.14; Table 2.2; American Institute of Mining Engineers, Figure 4.14; Armand Collin, Paris, Figure 2.13; A. A. Balkema (Rotterdam), Figure 6.4; Tables 2.5, 3.2, 11.2; J. J. Bigarella, Figures 8.15, 8.16; Blackwell Scientific Publications (Oxford) Ltd, Figures 2.6, 8.24; Table 2.4; Brazilian Society for Soil Mechanics, Figures 3.7, 3.8; Chapman & Hall (London) Ltd, Figures 2.4, 4.5, 8.5, 8.43, 10.1, 10.2, 10.3, 11.4; CSIRO Australia, Figure 2.10; G. H. Dury, Figure 4.2; Editions CDU et SEDES Réunies (Paris), Figure 11.13; Edward Arnold (London), Figure 5.8; Elsevier Science Publishers BV (Amsterdam), Figures 4.1, 4.3; Tables 3.3, 4.2, 5.2, 7.9; T. Feininger, Table 3.5; Gebrüder Borntraeger (Stuttgart), Figures 3.15, 8.1, 8.2, 8.10, 8.29, 8.42, 9.1, 9.6, 9.7, 9.8, 9.10, 9.11, 10.5, 10.7, 10.11, 10.16, 10.20; Tables 8.2, 9.1, 9.3; Geological Society of America, Figures 6.6, 6.7, 8.15, 8.16; Geological Society (London) Publishing House (Bristol), Figures 1.4, 2.1, 3.1, 3.3, 3.5, 3.9, 3.10, 3.12, 3.15, 6.1, 8.6; Table 3.1; Hindustan Publishing Corporation (India), Figure 4.11; IAHS Press, Institute of Hydrology (Wallingford), Figures 1.1, 5.1; Table 5.4; Institute of British Geographers (London), Figures 3.16, 3.18, 3.21; Y. Lucas, INRA (Institute National Recherche Agronomique, Paris), Figures 2.14, 2.15; R. W. Jibson, Figures 6.6, 6.7; John Wiley & Sons Ltd (Chichester), Figures 2.5, 3.20, 3.22, 4.4, 5.7, 8.7, 10.29, 10.31, 11.2; Table 6.2; John Wiley & Sons (New York), Figure 11.1; Kluwer Academic Publishers BV (Dordrecht), Figures 8.8, 11.5, 11.11; Macmillan Magazines Ltd (London), Figure 7.5; Macmillan Press Ltd (Basingstoke), Figures 1.2, 1.5, 2.8, 4.2, 9.2, 10.18, 11.7, 11.12; Tables 3.5, 3.11; Masson (Paris), Figure 1.3; Methuen Ltd (London), Figures 2.9, 9.3, 9.4; Tables 2.3, 6.3; J. R. Da Silva de Moura, Figure 8.22; J. R. da Silva de Moura and Limeira de Mello, Figure 8.19; Natural Resources Institute for permission to quote from Land Resources Division Land Resource Report, 6 (1974); Oxford University Press (Oxford), Figures 2.7, 10.4; Pergamon Press Ltd (Oxford), Figure 4.10; Table 5.7; Princeton University Press (New Jersey), Figures 9.1, 9.6, 9.7, 9.8; Royal Academy of Overseas Sciences (Brussels) and Geo-Eco-Trop, Figures 8.9, 8.26, 8.32, 8.33, 8.35, 8.38; Royal Scottish Geographical Society (Glasgow), Figures 8.4, 8.40; M. Fukuoka, South East Asian Geotechnical Society (Singapore), Figure 6.2; Government of Sierra Leone, Figure 4.12; Springer-Verlag, Figure 2.12; Tokyo Metropolitan University and H. Kadomura, Figures 7.8, 8.23; Table 7.5; UNESCO (Paris), Figures 4.8, 4.13; University of Chicago Press, Figure 9.5; Unwin Hyman, Figures 5.6, 5.14, 7.3, 7.4, 7.6, 7.7; Table 7.1; Van Nostrand Reinhold, Figure 10.30; Williams and Wilkins, Table 2.1.

INTRODUCTION

1

Geomorphology from a Tropical Perspective

Engineers, geochemists and soil scientists, as well as geomorphologists, are all concerned with the properties of tropical soils and saprolites, and the study, classification, treatment and management issues related to such materials have appeared distinctive to many writers (Young, 1976; Govett, 1987; Blight, 1990; Geological Society, 1990). Interest in lateritisation processes, for example, has led to a series of symposia (Trivandrum, 1981; Goudie & Pye, 1983; Melfi & Carvalho, 1983; Jacob, 1984; Sciences Géologiques, 1988), as well as to several monographs (Valeton, 1972; Goudie, 1973; McFarlane, 1976; Bardossy & Aleva, 1990; Valeton *et al.*, 1991). Other symposia on topics associated particularly with tropical conditions such as *tropical residual soils* (Wilson, 1983; Gidigasu, 1988; Geological Society, 1990), *dambos* (Thomas & Goudie, 1985), and *stone-lines* (Alexandre & Symoens, 1989) focus attention on issues almost totally neglected in the study of higher latitude landscapes. Yet doubts about the identity of a distinctive *geomorphology* of or in the tropics continue to persist (Stoddart, 1969; Douglas, 1978b; Ollier, 1983a; Gupta, 1993).

In view of the high level of interest in the status and history of the rainforests and savannas amongst ecologists and biogeographers (Prance, 1982; Whitmore & Prance, 1987; Furley, *et al.*, 1992), this uncertainty, which also breeds neglect, is puzzling. Tropical themes in hydrology and in soil studies have been specifically addressed in many special reports and monographs (e.g. Sanchez, 1976; Young, 1976; Greenland & Lal, 1977; Greenland, 1981; Lal & Russell, 1981; Balek, 1977; Keller,

1983). But there have been very few books on tropical geomorphology (Tricart, 1965; Thomas, 1974a; Faniran & Jeje, 1983; Goudie & Pye, 1983; Douglas & Spencer, 1985a; Wirthmann, 1987), and despite regular papers on arid, glacial and periglacial geomorphology in review journals, the humid tropics are seldom represented.

One possible reason for this is the lack of any specific process or set of processes that is considered unique to the non-arid tropics, and in an era of process-based geomorphology, this appears to rob the study of tropical landscapes of any urgent claim to separate consideration. This is hard to understand, when there has been so much interest in tropical forest ecosystems. After all, photosynthesis is no less universal a process than either hydrolysis or turbulent water flow, but this has not hindered the recognition of different ecosystem structures and rates of productivity. This study therefore seeks to elaborate those aspects of geomorphology in the humid and sub-humid tropics which appear central to an understanding of landscapes and some important resource issues in these areas.

Arid zone geomorphology has been the subject of separate studies (D. S. G. Thomas, 1989; Cooke *et al.*, 1993), and involves a wide range of topics which are not explored deeply in this study. However, no firm exclusion of arid conditions can be made, since palaeoenvironments in humid areas have been at least semi-arid as recently as during the last glacial maximum (Chapter 7), while many arid zones contain relict weathering profiles and landforms dating to more humid periods of the past.

SOME FEATURES OF TROPICAL CLIMATES

Air, soil and groundwater temperatures

Lowland areas within the tropics experience mean annual air temperatures $>20\,°C$ (WMO, 1983), and they range from $27-28\,°C$ down to $22-23\,°C$ in marginal tropical conditions, as at Rio de Janeiro. The coolest month remains $>20\,°C$ almost everywhere, though winter temperatures in the sub-tropics fall briefly to $15\,°C$, and the mean annual temperature may be $18\,°C$. The hottest month may exceed $30\,°C$ in the more continental, savanna locations.

Soil temperatures are characteristically high, at around $28\,°C$ most of the year below about $50\,cm$ under forest (De Hoore, 1964). At Ibadan, Nigeria a range from $25\,°C$ (August) to $30\,°C$ (March) was recorded under grass by Faniran and Jeje (1983). Surface soil temperatures can rise much higher, if unprotected by a vegetation canopy, and possibly reach $50\,°C$. There are many records of soil temperatures in the range $40-45\,°C$ (Monteith, 1979), but these relate to the top $1-5\,cm$ and are mainly recorded in cultivated soils. They do not apply to the environments of weathering at greater depths or under forest and woodland canopies. Many rock surfaces will become even hotter, exceeding $70\,°C$ during the day. However, the temperature of $\sim28\,°C$ at depth in the soil is probably also the normal temperature of the groundwater involved in chemical reactions.

Rainfall amount and seasonality

It is possible to divide the tropical climates according to rainfall amount and regime but this can be a somewhat arbitrary decision. WMO (1983), referring to Budyko (1963) and UNESCO (1978), defined the *humid* tropics as having no month with a mean temperature below $20\,°C$ and a mean annual precipitation of at least $1000\,mm$. The claim by Kinosita (1983) that Japan can be included within the humid tropics appears idiosyncratic, but illustrates well the need to extend attention to neighbouring areas with some similar climatic characteristics. Dubreuil (1985) offered a logical subdivision of the tropical climates of west Africa but did not consider wider global climates, such as

the monsoonal areas or extensions of tropical conditions into the wet east coast areas extending to some $30°$ latitude N and S. Categories (1)–(3) below are taken from Dubreuil (1985); (4) and (5) have been added to provide a more comprehensive scheme appropriate to this study:

(1) the arid and semi-arid, with annual rainfall of $100-750\,mm$, climax vegetation varying from herbaceous steppe to bush;
(2) transitional humid tropical and humid tropical, with annual rainfall above $750\,mm$ but not exceeding $1600-1700\,mm$; climax vegetation, bush to tree savanna;
(3) the wet equatorial climate (two rainy seasons) with annual rainfall above $1200\,mm$; climax vegetation, rainforest;
(4) the wet monsoonal climates with annual rainfall above $1600-1700\,mm$; climax vegetation, monsoon rainforest;
(5) transitional tropical to sub-tropical east coast climates, with annual rainfall above $1200\,mm$ (mean annual temperatures $>18\,°C$); climax vegetation, sub-tropical rainforest.

This study is concerned mainly with the humid tropical and transitional climates with rainfalls $>750\,mm$ year^{-1}. These regions embrace the Af and Am forest climates, Aw and As savanna climates and some warmer parts of the Caf humid sub-tropical climates of Köppen (1923), and were used by Faniran and Jeje (1983) for their study of *Humid Tropical Geomorphology*. Figure 1.1 provides a guide to the areas under consideration and is taken from WMO (1983).

The important topics of rainfall intensity and erosivity are discussed in relation to surface processes in Chapter 5, and it is rainfall *intensity* in the tropics which constitutes one of the most important parameters of the tropical environment. If the occurrence of precipitation intensities exceeding $50\,mm\,h^{-1}$ at least once every 2 years is mapped, it encompasses a very similar range of climates to those Köppen classes noted above (Reich, 1963). Moreover, many tropical climates are subject to large daily falls exceeding $400\,mm$ (Gupta, 1988), and also to prolonged heavy rains

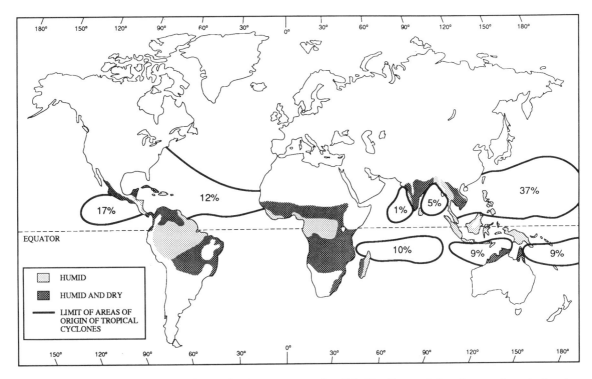

FIGURE 1.1   Tropical climates according to the Köppen system (modified by Trewartha) and areas of tropical cyclones showing the frequency of occurrence in per cent (after WMO, 1983; reproduced by permission of IAHS Press)

over periods typically lasting 3–6 days, during which falls exceeding 1500 mm are possible. A variety of tropical weather patterns are responsible for these events, including hurricanes, monsoon depressions and frontal systems (see Chapter 5). Areas between 5° and 20° latitude subject to tropical cyclones are indicated in Figure 1.1.

THE PROBLEMS OF CLIMATIC GEOMORPHOLOGY

Climatic regions for geomorphology

There have been many attempts to define *morphogenetic regions* on the basis of climatic controls over geomorphic activity and landforms (Büdel, 1948, 1970, 1982; Peltier, 1950, 1962; Corbel, 1957; Tricart & Cailleux, 1965; Birot, 1968; Wilson, 1973a; Figures 18.2 and 18.3 in Chorley *et al.* 1984), but this approach has not always been particularly enlightening. A useful summary of some of these

schemes has been offered by Doornkamp (1986). Ahnert (1987) has argued for a *magnitude frequency index* of extreme events to be developed for the definition of morphogenetic regions, but no widely applicable index of this kind is currently available. Chorley *et al.* (1984) recognised a *first order* group of morphogenetic regions, each experiencing a non-seasonal process regime, low average erosion rates, highly infrequent erosional activity, and a tendency for persistence during climate changes. These regions were considered to be the inner humid tropics, the arid zone and the glacial areas. A *second order* group of morphogenetic regions, subject to degrees of seasonality, in places having high erosion rates, some consistency of erosional events, and a tendency to enlarge, contract and shift location during climate changes, was divided into the warmer wet–dry climates and the cooler warm–cold climates.

The straightforward logic of this approach is attractive, and almost all schemes of classification make clear distinctions between the inner tropics and the outer tropics on the basis of increasing severity of the dry season. A problem with all of these schemes, however, is that they either spring from some a priori view of the effects of climate on intensity of different processes (Peltier, 1950), or from an unequal experience of the landscapes of particular parts of the tropics. Thus Büdel (1977, 1982) devoted just 15 pages to the inner humid tropics, and in his maps these are small core areas within the rainforests. In contrast, he used 65 pages to provide a full discussion of the peritropical zone which he extended to cover many semi-arid areas. Maps by Tricart and Cailleux (1965; also in Chorley *et al.*, 1984), provide a different view in which the humid tropics are more extensive and include some wet monsoonal areas, while the wet–dry tropics are far less extensive and exclude the semi-arid regions.

Both Bremer (1981) and Späth (1981) used 1650 mm year^{-1} as the rainfall limit for the perhumid tropics in their studies of Sri Lanka. In Nigeria, semi-deciduous rainforests persist into climates with rainfalls < 1500 mm year^{-1} and dry seasons of 3 months' duration (so long as moisture deficits are < 200 mm year^{-1}; Davies & Robinson, 1969), but in the more seasonal and monsoonal climate of Sierra Leone the transition from rainforest to savanna takes place at rainfalls of ~ 2000 mm year^{-1}. There is therefore much regional variation, not just in the boundaries of forest communities, but also in the character of the tropical climates within them. The seasonally wet, or 'wet–dry' tropics, excluding the semi-arid zone, were defined by Avenard and Michel (1985) as extending from 600–1300 mm annual rainfall in west Africa. Clearly these areas must be contiguous with the humid tropics *sensu stricto* and the boundary between them normally lies between 1300 and 1600 mm year^{-1} depending on the length and severity of dry season. However, in monsoonal areas with up to 5 dry season months the upper limit may reach 1800–2000 mm year^{-1}.

The most significant modification of the areas denoted in Figure 1.1 (WMO, 1983) is to extend the zone of broadly tropical conditions beyond the tropics of Cancer and Capricorn towards 30 °N or S latitude along the humid east coasts of Brazil, southern Africa, eastern Australia and China, and even perhaps into southeastern North America. These continental margins receive the bulk of their rainfall during the hot summer months and can support forest vegetation (Köppen Cf and Cw climates). Weathering and soil conditions in these areas show strong affinities with the tropics *sensu stricto* (Figures 1.4 and 1.5).

Cyclone activity (Figure 1.1) also rises towards and beyond the tropics in America, Asia and Australia. The impact of these tropical storms on flooding and mass movements is clearly great, and will be felt mainly in coastal regions subject to very high winds and coastal flooding, but very heavy rainfalls also occur in non-cyclonic tropical climates.

Global distribution of sediment yield

Another approach to the question of morphogenetic regionalisation has been based on regional differences in sediment yield (Fournier, 1960; Schumm, 1965, 1968; Wilson, 1973b; Walling, 1984), but this parameter is strongly influenced by non-climatic factors. Milliman and Syvitski (1992, p. 525) analysed sediment yields from 280 basins and found that 'sediment loads/yields are a log linear function of basin area and maximum elevation of the drainage basin'. Global maps of sediment yield therefore mainly reflect topography (smaller catchments occur in higher relief areas). These authors concluded that, 'it is probably the entire tectonic milieu of fractured and brecciated rocks, oversteepened slopes, seismic and volcanic activity, rather than simple elevation/relief, that promotes large sediment yields from active orogenic belts' (Milliman & Syvistki, 1992, p. 540). They noted that most very large rivers drain passive margins and also sequester large amounts of sediment in floodplains and deltas, while mountain rivers tend to deliver sediment direct to the ocean.

It is possible that this tectonic-relief factor operates at an order of magnitude above climatic–vegetation controls over sediment yield under natural conditions. However, present rates of erosion are affected by deforestation and farming, and may also be an order of magnitude higher than natural rates. This has been indicated for the Yellow River (Huanghe) in China (Milliman *et al.*, 1987), and although calibration is impossible for most rivers, it is a decisive factor in the interpretation of river loads. Douglas (1967a,b), for example, documented sediment yields from two catchments in the Cameron Highlands of West Malaysia, and was able to demonstrate a fivefold increase with a reduction in forest cover from nearly 100% to 64% of the catchment area (sediment yields $21.1 \, m^3 \, km^2 \, year^{-1}$ and $103.1 \, m^3 \, km^2 \, year^{-1}$ respectively). For these reasons, maps based on such measurements are omitted from this discussion. Those based in part on climatic parameters (Fournier, 1960; Strakhov, 1967) are equally likely to be misleading.

Morphoclimatic and climato-genetic maps

It must be concluded that available, process-based maps of global denudation are of little assistance in the discussion of climatic geomorphology. Equally serious problems arise with maps of morphogenetic regions based on climatic–morphologic data.

(1) We have really very little idea how to draw the boundaries on the basis of climatic means or extremes.
(2) Landform patterns are not adequately characterised and will in any case reflect lithologic, structural and tectonic influences as well as climate.
(3) Climate change in the medium term (10^3–10^4 years) has been profound across very large tracts of the tropics.
(4) Long-term secular changes of climate (10^6–10^7 years), accompanying continental drift and reflecting complex changes to the earth's relief and ocean/atmosphere circulation, have affected continents differentially according to their continental drift pathways.
(5) Many landforms and residual deposits persist in the landscape throughout these long-term changes and, as a consequence, there is a major element of inheritance amongst the landforms

and deposits on the earth's surface, and this factor renders attempts to characterise landforms solely in terms of present-day climates misleading.

This last statement does not, however, mean that the character of climate today and over decadial periods (10^1–10^2 years) is unimportant. This would be a travesty of the facts. Landform development is a continuum when viewed across long timespans, but episodic when viewed in terms of erosional events (Schumm, 1975; Douglas, 1980). The imprint of successive climates and of periods of rapid climatic transition will be an aggregate of landforming events, the magnitude and frequency of which vary as the climate changes. This imprint is witnessed by complex weathering profiles, mosaics of erosion scars, and sequences of slope deposits and alluvial sediments. Common patterns amongst these forms and deposits will exist in local areas, but, within larger regions, some elements of the pattern will change while others remain similar.

Although the temptation to provide maps of morphogenetic regions is resisted, the essence of many arguments can be expressed (simplistically) in terms of dominant landforms in Figure 1.2. Here the humid tropics (1) are characterised by thick saprolite, and the formation of multi-convex (*meias laranjas* or 'half orange') relief; the savanna or wet–dry tropics (2) by widespread duricrusts over variable depths of saprolite, cut by planar surfaces or *glacis* (slope pediments); and the *sahel* or semi-arid zone (3) by thin regoliths and extensive, often rocky pediments. But the categories 1, 2 and 3 could either be a *zonal* pattern or a *temporal sequence* ($1 \rightarrow 2 \rightarrow 3$), or more likely both. In other words, landscapes record each phase of development that is sufficiently prolonged to develop characteristic forms and deposits, and some parts of these features survive through each subsequent period of distinctive morphogenesis. Most landscapes are therefore palimpsests of features engraved on the earth's surface at different times (Ruxton, 1968a; Brunsden & Thornes, 1977, 1979; Brunsden, 1980; Starkel, 1987). This is also one of the tenets of the German approach to *climato-genetic geomorphology* (Büdel, 1980, 1982).

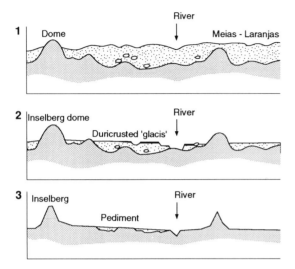

FIGURE 1.2 Outline for the evolution of tropical landforms. (1) Situation under dense forest with a clear separation of the 'double surfaces of levelling' of Büdel; formation of half orange (meias laranjas) relief in saprolite; (2) Situation under savanna with planation across saprolite and the formation of duricrusts; emergence of bedrock forms; rivers superimposed on the weathering front. (3) Situation in a semi-arid steppe environment (*sahel*) with the extension of pediments across fresh rock as well as saprolite. Situations 1, 2 and 3 may represent a sequence in time (see text) (from Demangeot, 1976 after Thomas, 1974a; reproduced by permission of Masson)

In many semi-arid parts of the world, notably in Australia, old weathered landscapes have progressed through the stages 1 → 2 → 3 (Figure 1.2) of stripping as the climate has become desiccated, and removal of the old saprolite has taken place. In the wet, monsoonal Freetown Peninsula of Sierra Leone, on the other hand, past dry climates are marked by bare rock surfaces, fans and glacis, while more recent humidity has led to duricrusting and renewed deep weathering. It is possible that within the inner tropics, but only perhaps in some very restricted areas, landscapes of type 1 have persisted for long-enough periods to develop a state of dynamic equilibrium in which the lowering of the weathered profiles and relief compartments takes place simultaneously. However, it is important to place such landscapes within a likely time frame for the development of their characteristic features which could have formed mainly during the

Quaternary (Chapter 3), but equally will inherit some parts of ancient saprolites. These remarks seem likely to apply to such areas as Cameroon, western Kalimantan, and southeastern Brazil, but also to many other humid tropical or near-tropical areas. It remains to be demonstrated whether the landforms of the so-called rainforest *refugia* (Haffer, 1969, 1987) demonstrate any greater continuity and geomorphic 'purity' than those of surrounding areas.

Bremer (1981, 1985b) took a different approach to climatic morphogenesis in Sri Lanka (which has striking climatic contrasts between dry and wet zones) and recognised that large-scale landforms were mainly developed before the Miocene under a perhumid climate (>1650 mm year^{-1}) that prevailed across the entire island. The medium-scale forms include intramontane and other basins and show diversity on the basis of regional climate and lithology, as do small-scale forms due to current processes.

In the present study, the landscapes of the semi-arid regions will not be considered in detail, but the impact of semi-arid conditions within the wet–dry tropics may have been very important during the later Tertiary and Quaternary, and, where possible, the effects of these incursions of aridity will be identified.

Pedogenetic maps and approaches

Another way to approach global environmental patterns is to consider the distribution of major soil clays. These reflect the overall rates of leaching and within the tropics divide between: (1) kaolinite + gibbsite (ferrallitic or allitic) in the wetter tropics, even where seasonality is considerable (e.g. northeast Queensland and southwest India); (2) kaolinite dominant, across a broad swathe of wetter savanna environments; and (3) smectite + kaolinite, in the drier savannas bordering the Sahelian climates at their arid margins. These transitions are expressed in the well known diagram first drawn by Strakhov (1967) and shown in Figure 1.3. Certain features of this model merit comment. First, the diagram can be misleading in its portrayal of the relative extent of forest (humid) and savanna (wet–dry)

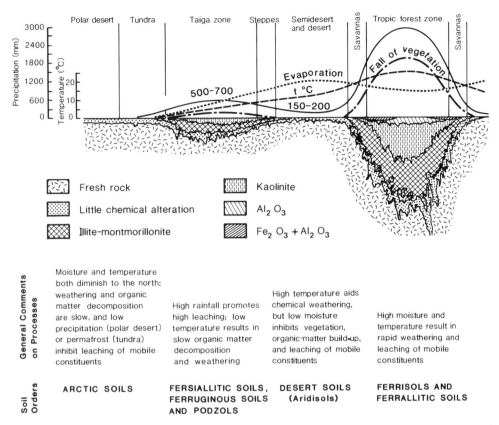

FIGURE 1.3   Diagram of the relative depth of weathering and weathering products as they relate to environmental factors in a transect from the equator to the north polar region (from Geological Society (1990) after Strakhov, 1967, modified by Birkeland, 1984; reproduced by permission of the Geological Society Publishing House)

areas. But it will be seen that, although the humid tropics are underlain throughout by kaolinitic soils, near-surface conditions change towards alumina-rich (allitic) residues in the inner tropics, with concentrations of sesquioxides of iron and aluminium occurring in the marginal areas and in the savannas.

World maps based on these criteria differ significantly, as comparisons of Figures 1.4 and 1.5 show (Pedro, 1968; WMO, 1983). Figure 1.4 groups both temperate and tropical *fersiallitic* soils together, but the latter tend to have much larger amounts of free iron (ferric acrisols; FAO, 1985) which can form laterite (plinthite) in favourable sites, while the temperate soils are more correctly termed *siallitic* (Duchaufour, 1982). The monsoonal climates of India have produced mainly fersiallitic soils (luvisols/ultisols but with some nitosols),

except for the vertisols of broad areas of the Deccan lavas. It is also evident that the high elevation and extensive Kalahari Sand cover of the central African plateaus removes this area from the normal grouping (many soils are arenosols). More significantly for this study, the *ferrallitic* soils (ferralsols; oxisols) are shown to extend well into the sub-tropics in the southeastern areas of Brazil and the USA and also in southern China, though many of the soils in these areas are nitosols and ferric acrisols rather than true ferralsols (FAO, 1985) or ultisols in the USDA scheme (Soil Survey Staff, 1975). This suggests good reasons to think that these regions, as far as ~ 30°N or S, will express other important features of humid tropical climates and geomorphology.

Tardy (1971) analysed natural waters draining both basic and granitic rocks, to predict the clay

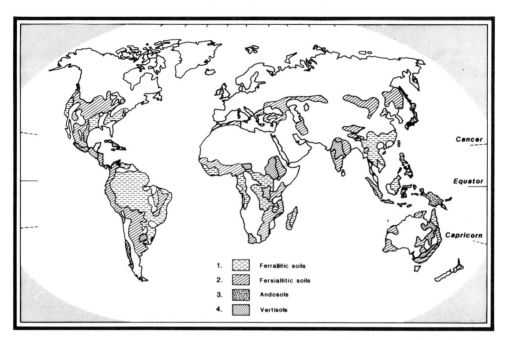

FIGURE 1.4   Simplified world distribution of the principal types of residual soils (based on the FAO World Soil Map). Note that ferrallitic soils extend into the sub-tropics along some continental east coasts. (Reproduced from Geological Society, 1990, by permission of the Geological Society Publishing House)

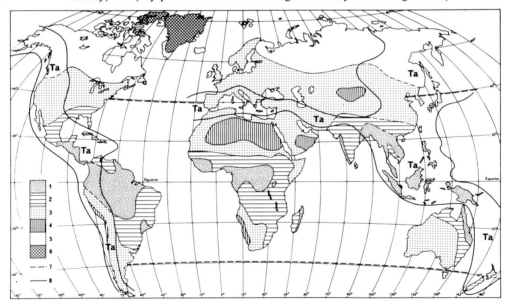

FIGURE 1.5   Distribution of the principal types of weathering on the earth's surface (from Thomas 1974a mainly after Pedro, 1968, with additions from Strakhov, 1967; reproduced by permission of The Macmillan Press Ltd). (1) Zone of allitisation (kaolin + gibbsite); (2) zone of monosiallitisation (kaolinite); (3) zone of bisaillitisation (smectite); (4) hyper arid zone without significant chemical weathering; (5) zone of podzolisation; (6) ice-covered areas; (7) approximate limit of red weathering crusts (after Strakhov); (8) approximate extent of tectonically active areas (TA) within which 'climatic' weathering types are modified

formation which would result from the molecular ratios present. His results also showed that fersiallitic (= bisiallitic) soils should result in almost all temperate areas and in the driest of the tropical areas. As rainfall increases in the tropics, kaolinitic (= monosiallitic) clays should be formed, and only in the wettest areas are the waters in equilibrium with gibbsite (= allitic).

Millot (1983) sought to recognise distinctive styles of landscape development under (1) kaolinitic mantles, (2) iron crust profiles, (3) smectite mantles, and (4) calcrete landforms. According to Millot (1983), as climate changes from humid to seasonally dry, the kaolinite mantle becomes destabilised with the mobilisation or iron, and the upper soil becomes sandy. The smectite landscapes were associated with pediments. In view of the importance of chemical weathering in tropical morphogenesis, these pedological considerations are central to many subsequent discussions.

EARLY IDEAS IN THE DEVELOPMENT OF TROPICAL GEOMORPHOLOGY

Many early geological pioneers were impressed by the extent and depths of weathered rock encountered in the humid tropics. Branner's (1896) account of the 'decomposition of rocks in Brazil' devotes 10 pages to the documentation of weathering depths, as great as 120 m (393 ft) and widely more than 30 m, throughout the wetter parts of the country. He quoted Charles Darwin (1876) and Louis Agassiz (1865) as being similarly impressed, and drew freely on previous observations by Pissis (1842). Included here are records going back to 1826 of early gold mines at Ouro Preto, where more than 50 m of 'soft rock' had been encountered. In Branner's accounts of weathering in tunnels and cuttings and in the gold mines, the reader is left in no doubt of the practical importance of this phenomenon, to which he added graphic descriptions of landslides in the saprolite. He saw clearly that heavy rainfalls (e.g. 140 mm in 2 h in 1859) triggered massive landsliding in the country around Rio de Janeiro and the Serra do Mar, and he made the important observation that despite their greater frequency, 'where the removal of earth at

the foot of the slopes has disturbed somewhat the natural equilibrium. They are by no means confined to such places, but occur also in the remote forests' (Branner, 1896, p. 266).

If Branner's (1896) account is distilled to its essence we can gain from it many of the major ideas relevant to modern geomorphology in the tropics:

(1) deep rock decay is widespread in a variety of rocks;
(2) the high rainfall is a dominant factor in rock weathering;
(3) the high temperature of the rainwater accelerates weathering;
(4) the agencies of weathering are dominated by organic acids, and carbonic and nitric acids;
(5) plant litter is spread through the upper saprolite by termites and other insects;
(6) the rock decay pre-dates recent dissection of the landscape;
(7) crystalline rocks may appear decayed beneath later sediments;
(8) the clay rich, kaolinised saprolite combined with high rainfall is responsible for widespread landsliding;
(9) landslides account for an important part of the total denudation, exposing fresh rock to further weathering.

In Nigeria, Falconer (1911) was no less perceptive about the role of weathering, and he attested the widespread mantle of saprolite and its duricrust cappings in West Africa. He paid more attention than Branner to the functional role of saprolite and its removal in the formation of the landscape in northern Nigeria. With regard to a deeply weathered profile near Kano, he observed that the hill, 'although deeply decomposed, still preserves in its lower parts detached boulders or cores of unweathered rock. If the subsequent erosion had continued until the weathered material had been entirely removed, the flattened hill would have been replaced by a typical kopje' (tor or boulder inselberg) (Falconer, 1911, p. 247). Falconer also understood the regional significance of the weathered mantle and the fact that it could be stripped and renewed over long time periods.

He stated (Falconer, 1911, p. 246) that,

> ... a plane surface of granite and gneiss subjected to long continued weathering at base level would be decomposed to unequal depths, mainly according to the composition and texture of the various rocks. When elevation and erosion ensued the weathered crust would be removed, and an irregular surface would be produced from which the more resistant rocks would project.... A repetition of these conditions of formation would give rise to the accentuation of earlier domes and kopjes and the formation of others of lower level.

Many other early writers reflected these ideas (Holmes & Wray, 1913; Cotton, 1917; Wayland, 1933; Willis, 1936), and in Germany there was similar interest in tropical landscapes and the formation of *inselberge* (Bornhardt, 1900; Passarge, 1904, 1923; Sapper, 1935; Jessen, 1936; Freise, 1938), and many of these ideas were summarised by Cotton (1942). A major contribution to the study of tropical weathering came from Harrison (1934), while studies of laterite, particularly in Australia, had made significant advances (Campbell, 1917; Woolnough, 1918, 1927, 1930). The foundations were also laid for the establishment of a climatic geomorphology.

These earlier studies became overtaken in the anglophone world by the forceful advocacy by King (1948, 1953, 1957, 1962) for pediplanation as the universal mode of slope and landscape evolution. King based his ideas on conditions of relief formation in Natal, South Africa, which is sub-tropical, in places semi-arid, and highly dissected. But he visited all of the Gondwana fragments and extended his ideas to virtually the entire globe, when he wrote his seminal work, *The Morphology of the Earth* (King, 1962). As a geologist who readily accepted the evidence for continental drift many years ahead of the new plate tectonic theory, King divided opinion sharply amongst his contemporaries. But, with the acceptance of plate tectonics in the 1970s, King's ideas appeared vindicated and his world-wide correlation of pediplains, though controversial, remains central to most debates about evolutionary geomorphology. It could be argued that the paucity of alternative ideas concerning the evolution of relief on the ancient Gondwana continents led many, who worked neither in the tropics nor in Gondwanaland, to grasp these concepts as a means to understand a vast and little known world. But King's work was more about the evolution of continental landsurfaces ('denudation chronology') than about the processes by which that evolution took place.

New ideas were emerging from the continent of Europe concerning these issues, and in France, Germany and the Netherlands an interest in the role of weathering in different climates was emerging. In Germany, Büdel (1957) and Wilhelmy (1958) were powerful voices, and Büdel's concept of 'double planation surfaces' (Chapter 9) has, belatedly, become as widely debated as King's version of pediplanation. This influential concept was brought to the anglophone community by the publications of Cotton (1961, 1962), who had previously alerted English readers to German research in the tropics during the 1930s (Cotton, 1942), and subsequently by many others, who, like the present author, found the arguments persuasive. Ollier (1959, 1960, 1969), Ruxton (1958), Ruxton and Berry (1957), Mabbutt (1961) and Trendall (1962), all made important contributions to debates about the significance of weathered mantles to the development of landforms in Africa, Hong Kong and Australia respectively. All these studies strongly influenced the present author, who found himself unable to interpret the landscapes of Nigeria in terms of the pediplanation concepts of King (Thomas, 1965a,b, 1966a).

In France, Birot (1958, 1968) and Tricart (1965), based many similar ideas on work undertaken in Brazil. On the other hand, the important findings of Rougerie (1955, 1960) which were based on detailed studies on the Ivory Coast have remained less well known. In Holland, Bakker (1960, 1967) was pioneering studies of granite weathering in different climates, and his publication with Levelt (Bakker & Levelt, 1964) on the nature of tropical denudation and the influence of near-tropical environments on the evolution of European relief remains a landmark in these fields. In Belgium, De Ploey (1964, 1965) pioneered work on processes and environmental change in tropical Africa.

Study of the Quaternary and its climatic oscillations in the tropics is considered in Chapter 7. But it can be observed here that the topic has been pursued more by non-geomorphologists, and the aims and objectives of multi-disciplinary Quaternary research have been directed in the main to understanding vegetation and climate change. In the glaciated lands of northern high latitudes, the modification of the landscape by ice led to a whole range of geomorphological studies, many of which go back to the 19th Century. Similarly, there has been much attention paid to deserts and desert marginal lands (hot and cold), and their processes and progression through the Quaternary changes in moisture status. But, where landforming events have appeared to retain some continuity under the influence of rain and water flow, enquiry into the geomorphic impact of Quaternary changes has not always been well focused. One of the major exceptions was the seminal work of Bigarella and Ab'Saber (1964), and Bigarella and Andrade, (1965), incorporated later in Bigarella and Becker (1975), on the Quaternary of southeastern Brazil. When a search is made for comparable studies from other areas of the tropics at this time, it is clear that examples are few (Brückner, 1955; De Ploey, 1965; Fölster, 1969). Of course a great deal has been written on the Quaternary since these papers were written, and the importance of these climate changes to the formation of landforms and surface deposits remains a matter to be explored further (Chapter 8).

EVOLUTIONARY GEOMORPHOLOGY

Geomorphologists everywhere face similar problems in the development of their science. Although classical geomorphology was based in part on the concept of the cycle of erosion (Davis, 1899a), with its promise of repeated uplift, dissection and planation, the landscape never returns to its previous state. Instead, as it evolves (Ollier, 1979) both the geology and the landforms become more complex unless finally obliterated by subduction. Even planation, that elusive geomorphological obsession, does not guarantee simple outcomes, and many cratonic surfaces are overlain by complex residual weathering profiles as

well as by a mosaic of surface forms and deposits of different ages. Such landsurfaces are truly *polygenetic* and may have formed as much by prolonged groundsurface lowering (Chapter 9) as by pedimentation.

Plate tectonic theory has revealed the progressive fragmentation of the former supercontinental landmass of Gondwana during the last 160 Ma, and the movement of many continental plates ever northwards. The histories of uplift and subsidence accompanying these changes have been too complex to allow simple reconstructions of past landscapes on the basis of elevation, and the progression of palaeoclimates consequent upon continental drift makes assumptions about long-term landform development under climatic conditions similar to those of today unrealistic. If Ollier (1988b) is correct in identifying a Precambrian erosion surface in the Kimberley Plateau of northwestern Australia, then we should not restrict our interest to the most recent phase of continental rupture and drifting which for many purposes implies the last 100 + Ma. In practice, however, it has not proved possible to explore the more ancient periods in this study. This task has recently been attempted by Ollier (1991).

The importance of cratonic terranes within tropical latitudes has greatly influenced the study of geomorphology in these environments and without doubt this has biased many results and conclusions arrived at. Many orogenic mountain areas in the tropics remain inaccessible even today. Equally, large tracts of the lowland rainforest areas in Amazonia and the Congo/Zaire basin remain poorly documented, and Baker (1983b, p. 460) has commented that, 'for fluvial geomorphology an irony of recent advances in the field is that the world's largest rivers remain poorly understood'.

Ease of access, good visibility, and long-standing historical (colonial) connections with Africa, India and Australasia, have ensured that the savanna regions of the world, along with the neighbouring arid lands, have received far more attention than have the truly humid tropics, and it is important to recognise that tropical geomorphology has been strongly influenced by studies in these seasonal savannas, and on the ancient cratonic areas of

Africa and Australia particularly. This bias was very evident in an earlier study (Thomas, 1974a) and an attempt is made here to redress this imbalance.

In recent decades, geomorphology has moved away from attempts to explain landscapes; away from its roots in the historical earth sciences, to embrace the precepts of hydrology and hydraulics. Baker and Twidale (1991) have reviewed the 'paradigms lost' in modern geomorphology, especially the Davisian trinity of structure, process and time. They questioned the problematic postulates of climatic geomorphology, and the conceptual and scalar limitations of experimental process studies. Their critique of a 'disenchanted' geomorphology led them to ask for the abandonment of ideologies, and for the investigation of anomalies; for discovery and creativity. Most of all they called for new hypotheses to challenge accepted theory.

A distinctive Australian tradition in geomorphology, strongly influenced by evolutionary principles, has done much to redress the imbalance in modern geomorphology. Australia nudges the tropics in the north yet penetrates the temperate latitudes in the south. It is predominantly an 'old' landmass and the stability of its landforms has received much recent comment which will be explored elsewhere. In Australia one is compelled to consider the longer timescales of landform evolution (Twidale, 1976; Ollier, 1979, 1983a; Gale, 1992). Although Northern Australia has moved towards and into the tropics since the Early Tertiary, it was warm and moist for most of this time, because this was the pattern of world climates. Its former, glacial climates are ancient (Permean) and largely of historical interest today.

Britain and Europe, on the other hand, have moved progressively *away* from the tropics and have been both humid and glacial. Some legacies of at least sub-tropical conditions remain in the European relief (Chapter 9) but they have been selectively removed and the debris redeposited as Tertiary strata (such as the London Clay). The pervasive glaciations of the Quaternary, while not everywhere determining the relief (Sugden, 1976), have nevertheless given temperate geomorphology a polar rather than a tropical perspective. Ollier

(1983a) has argued that, 'perhaps the most significant factor in the geomorphology of tropical regions is that they escaped Quaternary glaciation, and therefore have a very long geomorphic history'. That would be a simplification, but the absence of frost and ice, both today and in the last 160 Ma, has been a significant factor in landscape development. So too has the persistence of warmth since the late Mesozoic and the significant periods of humid climate, 'interrupted' by episodes of dryness that have repeatedly shifted the balance of the weathering and erosional systems.

In those parts of the tropics that have escaped recent orogenic earth movement (possibly >70% of the landmass area in tropical latitudes), the continuity of landform development has encouraged more extended time frames for the analysis of landscapes and deposits. It has also to be admitted that process studies and the development of theory based on these have been overwhelmingly concentrated in the middle and higher latitude countries. Land-based chronological data for the Quaternary are also concentrated in these areas. However, very significant advances have been made in all of these areas of study within the tropics during the last two decades, and it is perhaps timely to stand back from this body of work, to attempt both summaries of our understanding in particular areas and an evaluation of our existing knowledge concerning the geomorphic development of tropical landscapes. This will be done in the belief that such a review will be valuable not only to academic geomorphologists, but also to soil scientists, engineers, geologists and ecologists.

THEMES FOR A GEOMORPHOLOGY OF THE TROPICS

The argument for a particular study of tropical geomorphology is based largely on the *balance* between processes of weathering and accumulation of residual materials, and mechanisms of removal and deposition operating on these residual products of weathering in tropical environments. Many of these processes are chemical or more accurately biogeochemical and are commonly neglected in geomorphology. Because all processes operate in a time continuum, identifiable results in the landscape depend on the rates of landsurface change brought about by these processes. In many contexts these rates of change are slow, and their impacts on landforms and deposits must be inferred over periods much longer than modern measurements have been available. Inferences of this kind are hazardous and must depend either on unproven theory or on circumstantial (empirical) evidence that must be 'fitted' to the knowledge of process by a combination of extrapolation, theory and the use of ergodic reasoning (space–time substitution).

These issues are but sub-sets of wider problems of theory-building and prediction in the environmental sciences (Schumm, 1985), and it is perhaps pertinent to note that, after three decades or more of careful measurement in the landscape, scientists continue to encounter the twin problems identified by Ruxton (1968a), namely the multi-complexity of process and the inheritance of forms and deposits from past environments. Nevertheless what we *perceive* as happening (or having happened) in the landscape should address at least four aims:

(1) to understand those processes which it has become possible to measure, or which can be inferred from detailed study of earth materials (saprolites, soils, sediments, natural waters);
(2) to relate accurately dated sequences of earth materials to the operation of processes in the past, acknowledging the importance of magnitude and frequency amongst formative events;
(3) to understand the spatial relationships of forms and materials within the landscape and thereby identify the major factors leading to their formation;
(4) to infer, where possible, the sequential development and evolution of weathering profiles, depositional sequences and erosional landforms.

The reader must judge, however, how often the argument fails to identify the pitfalls inherent in seeking deterministic solutions to these problems in geomorphology.

In 1978, Douglas (1978a,b) called for geomorphologists in the tropics to, 'ascertain the problems of most importance to the people of tropical countries and endeavour to give them priority'. The present author fully endorses this dictum, and would argue that enquiry into the formational processes and histories of landscapes in the tropics merit thorough enquiry, for the insights which can be obtained into problems of land resources and environmental hazards of concern to the peoples of tropical areas. One example will help illustrate the approach to be taken in this book. Soil and land loss resulting from erosion and instability are problems of great magnitude and importance in tropical countries. Much effort has gone into establishing the parameters for solving the *universal soil loss equation* (USLE) of Wischmeier and Smith (1958), and erosion plot experiments have been repeated widely in the tropics (Chapter 5). Yet little attention is given in these studies to the substrate which forms the foundations of the soil and land resources. The presence in the tropical landscape of great (but highly variable) thicknesses of saprolite, residual soils of advanced alteration and chemical breakdown, and colluvial deposits of considerable depth and complexity, has produced a complex mosaic of earth materials and landforms. The formation of these materials and of the landforms developed on them is the focus for this book, and it is hoped that greater understanding of these and of the processes that formed them, will assist other scientists, engineers and land managers in their search for sustainable land use programmes.

PART I

PROCESSES AND PRODUCTS OF WEATHERING

2

The Alteration of Rocks under Tropical Conditions

INTRODUCTION

The importance of supergene or meteoric alteration, otherwise known as chemical weathering, of rocks under tropical conditions has long been known, and the remarkable statement on the 'Decomposition of rocks in Brazil' by Branner in 1896, together with the work of Falconer (1911) in Nigeria, stand today as major statements on the topic. Yet despite the early recognition of the significance of tropical weathering to geology and geomorphology, and the more recent findings of geochemists and pedologists working in this field, a full appreciation of the importance of weathering processes and products within the geomorphic system has not yet been achieved. Geomorphologists schooled in glaciated regions and in mountainous terrain, or working in semi-arid lands often find it difficult to conceive the role of a widespread mantle of saprolitic sandy clay, extending to depths of 100 m or more, in the formation of the relief or the control of present-day erosional processes. Yet the humid tropics and sub-tropics comprise vast areas of the earth and are home to some one-third of the world's population. In these areas such thick mantles are widely found, and any suggestion that they underlie only the ancient planation surfaces of continental interiors is misleading. Certainly such weathering mantles are thin or absent from recent orogenic relief forms, but they are not absent from other areas of steep relief, as along the faulted margins of Brazil or South Africa or in the Palaeozoic mountain chains of Australia, Malaysia or the Americas. In fact, subsequent argument will show that elevation and

relief formation are important factors driving the weathering systems, a fact well understood by karst geomorphologists.

Another common misconception is that most such weathering profiles are ancient and have been progressively thinned or removed by erosion since the late Mesozoic or early Cenozoic. This interpretation may correctly attach to the semi-arid interior of Australia (Mabbutt, 1961, 1965), and to parts of Africa, but it should not be extended to the wetter tropics, where weathering penetration continues unabated today.

The reasons for this situation are readily understood, when the origins of rock minerals are considered. The silicates which include all the common feldspars, and their associated ferro-magnesian minerals such as biotite and hornblende were formed at high temperatures and pressures, sufficient to ensure the exclusion of free water from the congealing rock melts. The energy of formation of these minerals is therefore very high, and they become unstable as they approach the surface and come into contact with water, air and dissolved organic acids at near atmospheric pressure. Such minerals undergo spontaneous reactions with these weathering agents and the reactions continue until new thermodynamic equilibria have been attained.

Rock partings (joints, microfissures) also appear spontaneously as rocks dilitate (expand) with the pressure release resulting from removal of superincumbent rocks and the formation of valleys. These partings allow the weathering agents access to the rock minerals, often to considerable depths, and are a major determinant of the rate at which rocks decay and disintegrate.

THE CLIMATIC FACTOR IN ROCK WEATHERING

Temperature control over weathering reactions

Climate is critical to rock weathering, because it forms the primary control over temperature and the availability of water and organic acids derived from the vegetation and soil. It has proved difficult to quantify these factors, but the temperature control over the rates of reactions is rather well known, and Curtis (1976) gives the relationship as the Arrhenius equation:

$$\text{Rate} = A \exp(-\Delta E_A / RT) \qquad (2.1)$$

where A is a rate constant, T is the absolute temperature, E_A the activation energy required to overcome the energy barrier and R is the gas constant. In practice this leads to a doubling of rate of reaction for every 10 °C rise in temperature.

Soil temperatures in the lowland tropics can be very high at or near the surface. De Hoore (1964) quoted a range of < 25–39 °C at a depth of 19 cm at a site in Yangambi (Zaire). Surface temperatures on dry soil and rocks can rise much higher (> 70 °C) but this is in the absence of water. At depths of > 50 cm a temperature of around 28 °C persists throughout much of the year which is some 20 °C higher than in cool temperate regions. The temperature factor alone therefore will accelerate chemical weathering reactions by a factor of four when compared to high latitude or high altitude stations.

Rainfall amount and rates of weathering

It is more difficult to quantify the effects of rainfall; the most important control that water inputs exercise over the reactions is to allow them to continue, by preventing equilibrium conditions from being established by a rise in the concentration of dissolved ions. This necessary disequilibrium in the weathering system also depends on the continual renewal of the water, which requires free drainage either vertically or laterally through the material. Humid tropical climates receive > 1200 mm year^{-1} and large areas have > 2000 mm year^{-1}, distributed throughout the year with only a short dry season, generally < 3 months. Once a mantle is established, the saprolite is kept moist throughout most of the year, and water moves through the material at frequent intervals.

Natural waters contain free H^+ ions and are therefore usually acid within a common pH range of 4–6. As Curtis (1976) and others have pointed out, the simple hydration of silicate minerals would produce negatively charged hydroxyl ions, leading to an alkaline reaction which is rare, except at the weathering front itself. Free hydrogen ions (protons) are produced by *proton donors* (Ugolini & Sletten, 1991) which are the various compounds capable of dissociation and proton donation to the soil system. The most important of these is carbonic acid which dissociates to form hydrogen and bicarbonate ions (see below). Humid tropical and sub-tropical forest environments supply large quantities of bicarbonate because of their high biological productivity, but it must also be recognised that partial oxidation of plant debris can lead to the formation of *organic acids* which play a key role in the mobility of metal cations entering the weathering system. Before exploring these reactions, tropical weathering should be viewed in a wider context.

Tropical weathering in perspective

Strakhov (1967) was clear that, when all factors are combined, weathering in the humid tropics could proceed at rates of 10–20 times those of cold climates. Many doubt these figures and they remain speculative. However, few would dispute that thin and stony regoliths typify the arid zone and the frost climates of high latitudes and altitudes. In cool climates where high rainfall occurs, as in western Scotland or Norway, there is evidence of strong podzolisation in permeable materials, but renewal of any saprolite removed by glaciation from crystalline rocks has hardly begun. If temperatures remain low when rainfall occurs then rates of weathering will be retarded, and this observation is confirmed by a comparison of sub-tropical landscapes with winter rainfall regimes (Mediterranean climates) with those (mainly east coast) areas with summer rainfall. Thus there is abundant deep weathering in eastern Australia, North and South America and in southern Africa to more than 35° S latitude.

In formerly glaciated areas, there may be a problem of getting the system started, because of the near total removal of a moisture-retaining regolth. However, there is abundant saprolite present to depths of 6–8 m in northeast Scotland (Hall, 1986, 1987), where glaciation was ineffective, and also in many warm temperate areas such as the Appalachian Mountains of the USA (Pavich, 1985; Velbel, 1985). Over most of the humid tropics, complete removal of the saprolite, due to past climatic crises, may have been localised, incidentally giving rise to prominent rock landforms such as the *bornhardt* domes (e.g. the famous Pão Açucar in Rio de Janeiro). This has allowed the saprolite blanket itself to influence the course of weathering by retaining moisture at the weathering front.

Thus a rate of weathering computed from thermodynamic principles or measured over a short period of time in the field is unlikely to be an accurate guide to the historic rates of rock decomposition which, even if temperature and rainfall are considered as constants (and there is good reason *not* to assume this), will change as the saprolite thickens and hydraulic conductivity alters. In other words, weathering profiles evolve over long time periods, and during these the rate of development will vary considerably. Nevertheless, all that we know about the warm and humid environments indicates favourable conditions for chemical weathering, and the advanced state of rock decay and deep profiles found in such areas, confirm impressions that rates of weathering must be higher, perhaps even an order of magnitude higher, than in cold climates and in arid environments.

THE ORIGINS OF DEEP SAPROLITES

Interest in contemporary processes has meant that many recent studies of weathering have concentrated on the weathering of tablets placed in the soil for short periods and on the weathering of exposed rock surfaces, using micro-erosion meters or historic dates for weathering rinds. But these approaches fail to tackle the major questions of weathering mantle formation. In the first place, this is because what is measured is the weight loss of (usually) limestone tablets, and this yields no information on the formation of clays. Secondly, the environment of weathering on an exposed boulder or rock face is quite different from that of beneath a soil/saprolite cover.

Within the weathering profile, gravitational transfer is confined, in the main, to groundwater movements and to the transport of fine particles (mainly clays and oxy-hydroxides). By contrast, weathering of exposed rock faces contributes to the formation of coarse sediments arising from spheroidal weathering and from rock falls that exploit rock partings, including those arising from pressure release (sheeting, exfoliation).

It remains difficult to describe the transition from the weathering of an exposed rock towards the development of a continuous or semi-continuous mantle of saprolite and soil. However, few rocks are entirely devoid of internal porosity or partings, and therefore what appears to be surface weathering can be transmitted quite rapidly into the rock fabric, before any visible physical disintegration occurs. Furthermore, very few weathering profiles on older rocks and landsurfaces can be traced to the time of first rock exposure to alteration, and some form of weathered rock may have existed on most present-day sites of deep weathering for many millions of years.

The complex nature of residual and transported mantles

The geomorphologist and pedologist are therefore confronted with landsurfaces that are commonly underlain by both a complex of different materials with separate origins *and* by materials which have been transformed over time by a succession of changing environments (i.e. they are polygenetic). Some emphasis is given to these ideas at this early stage in the argument, because most tropical landsurfaces which are deeply weathered are also mantled by transported materials of varied and often uncertain origins.

These materials include the debris resulting from landslides, but also sediments derived by sheetflow (Watson *et al.*, 1983), and by alluviation (Fairbridge & Finkl, 1984), as well as from previous aeolian deposition. The characteristics of these near-surface deposits will be explored in detail subsequently (Chapter 8). Here, it is important to recognise their widespread occurrence.

As a final speculation, it is possible to suggest that the initiation of weathering profiles may commonly have followed upon the deposition of transported materials derived from exposed rock faces during periods of landform development involving high relief, and forming talus slopes, or colluvial and alluvial deposits. The incorporation of transported resistate minerals such as vein quartz, quartzite, corundum, zircon, and other heavy minerals including diamonds, into duricrusts may also indicate such a sequence of events.

Partition of the weathering profile

Before embarking on a discussion of weathering processes under tropical conditions, a general view of the morphology of weathering mantles is appropriate, in order to provide a framework for later discussion. This is, inherently, a very difficult procedure, because of the great variety of rock types, fabrics and structures on the one hand and the variability imposed by relief development on the other. Some profiles are also very deep (>100 m) and this makes observation more difficult, and limited to relatively few occurrences. The description of the weathering profile usually depends upon, or corresponds with, the recognition of different *weathering grades*, based on qualitative criteria (Thomas, 1974a; Dearman *et al.*, 1978; Fookes *et al.*, 1971; Ollier, 1984; Gerrard, 1988; Geological Society, 1990). Other methods for describing this transition will be considered in Chapter 3.

The clearest partition is often seen in well jointed granites, and descriptions of profiles with rounded joint blocks or corestones have been used many times; Ruxton and Berry (1957) are often quoted in this context. Engineers in particular have adopted a sixfold classification of weathered materials that is now widely used (British Standards Institution, 1981), and can be applied in a variety of contexts (Geological Society, 1990; Lee & De Freitas, 1991). Slope stability analysis has also required type sections to be described for metamorphic and both clastic and carbonate sedimentary rocks (Deere & Patton, 1971). All descriptions are schematic versions of natural complexity, and boundaries between weathering zones can be difficult to define in the field. All recognise that fresh rock is usually transitional to a progressively more weathered material as the landsurface is approached. This transition may take place over many tens of metres or within one metre. In banded or bedded rocks, vertical profiles will commonly cross different lithologies at contrasting stages of alteration, while in jointed rocks it is not uncommon for boreholes to encounter 'fresh rock' which is in reality a large core boulder below which highly weathered rock may recur. Idealised or 'standard' profiles with and without corestones are illustrated in Figure 2.1.

The descriptive term 'saprolite', first used by Becker (1895), is commonly applied to the highly weathered Zones IV and V, while the transitional Zones II and III below are called simply 'weathered rock'. However, not everyone would respect this distinction. Zone VI within which the rock fabric is completely obliterated is usually regarded as 'residual soil', above which transported (allochthonous) material and the topsoil are found.

To describe the advance of weathering agents into the fresh rock Mabbutt (1961) introduced the term 'weathering front' which has become almost universally adopted in preference to the concept of a 'basal surface of weathering' as proposed by Ruxton and Berry (1957). However, when Büdel (1957) advanced his concept of 'double levels of planation' (*doppelten Einebnungsflächen*), he contrasted the upper 'wash surface' with the lower 'basal weathering surface' which was immune from abrasion but acted as a second level of planation by weathering agents. Two main objections to the term 'basal weathering surface' have been

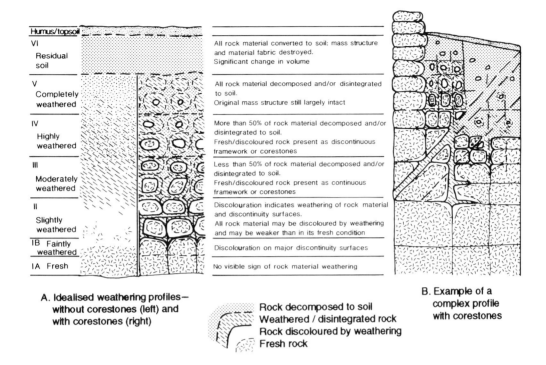

VI Residual soil	All rock material converted to soil: mass structure and material fabric destroyed. Significant change in volume	
V Completely weathered	All rock material decomposed and/or disintegrated to soil. Original mass structure still largely intact	
IV Highly weathered	More than 50% of rock material decomposed and/or disintegrated to soil. Fresh/discoloured rock present as discontinuous framework or corestones	
III Moderately weathered	Less than 50% of rock material decomposed and/or disintegrated to soil. Fresh/discoloured rock present as continuous framework or corestones	
II Slightly weathered	Discolouration indicates weathering of rock material and discontinuity surfaces. All rock material may be discoloured by weathering and may be weaker than in its fresh condition	
IB Faintly weathered	Discolouration on major discontinuity surfaces	
IA Fresh	No visible sign of rock material weathering	

A. Idealised weathering profiles— without corestones (left) and with corestones (right)

Rock decomposed to soil
Weathered / disintegrated rock
Rock discoloured by weathering
Fresh rock

B. Example of a complex profile with corestones

FIGURE 2.1 Weathering profiles and weathering grades of a rock mass as adopted by engineering geologists (after Geological Society, 1990. Reproduced by permission of the Geological Society Publishing House)

argued. First, weathering does not advance unidirectionally, but may attack divided rock masses from all sides. It is therefore misleading to use the term 'basal' in this context. Second, the transition to fresh rock may be gradual and no identifiable *surface* will then separate the unaltered rock from the weathered mantle.

There is no doubt that the term 'weathering front' is preferable, but on a spatial scale of 10^2–10^3 m the weathering front does appear as a basal surface and in many situations the transition from fresh rock to saprolite is accomplished across a remarkably narrow layer, perhaps 0.02–1 m thick (Figure 2.2, and Plates 2-I and 2-II). Elsewhere this transitional zone can be tens of metres thick, particularly over fissile or poorly consolidated rocks, usually sediments or low-grade metamorphic phyllites or schists.

It is also important to characterise the transition from saprolite to soil in the upper levels of

weathering profiles, for two different but equally important reasons.

(1) The transformation of rock to saprolite is essentially an *isovolumetric substitution* which can leave the rock fabric and structures at least partially visible, but the changes affecting the upper levels of the profile can involve a loss of volume of as much as 70–75%, accompanied by a collapse of original structures.
(2) The upper soil layer(s) are seldom truly autochthonous, but exhibit many signs of lateral transfer of material both physically and in solution.

Millot (1983), Pavich (1986, 1989), and Johnson (1993) have all stressed the importance of this upper level, where kaolinitic mantles can be destabilised with loss of iron and clay, to form sandy topsoils, with marked loss of volume. Aleva (1983) referred

FIGURE 2.2   Core boulders over a sharply defined weathering front (basal weathering surface) in granite, Jos Plateau, Nigeria

to *triple planation surfaces* below tropical interfluves. Beneath the ground surface, the base of the soil is seen to be a stone layer, generally termed a 'stone-line' but called a 'carpedolith' by Aleva (1983), marking the base of the pedogenetic horizon and the top of the structured saprolite. The third surface is at the weathering front. The origins and significance of stone-lines are discussed elsewhere (Chapter 8), but it can be noted here that they are considered by Aleva (1983), and also by McFarlane and Pollard (1989) as a function of the systematic development of saprolitic weathering profiles, and in effect a *denudation level* below the landsurface, since they imply loss of fine material from the saprolite by a variety of processes.

This notion, that denudation can occur at a level *below* the landsurface, is important, because it also focuses attention upon subsurface water flows which may be concentrated within the stone layer or by macropores and other structures in such a way

as to bring about substantial removal of fine sediment and the development of so-called *pseudo-karst* forms in crystalline rocks, clastic sediments and lateritic weathering covers. Where major mass movement has occurred, there may also be multiple weathering fronts, advancing within both the *in situ* and in the transported debris.

HYDROTHERMAL ALTERATION

Debate concerning a possible hydrothermal origin for deeply kaolinised granites, such as those found in southwest England (in the Cornubian batholith), remains inconclusive despite the hydrogen and oxygen isotope studies by Sheppard (1977). This study concluded that the kaolinite from the china clay deposits is of a low-temperature weathering origin, and Ollier (1983b) has reviewed a number of similar occurrences and concluded (p. 58) that, 'since weathering is common and hydrothermal alteration rare it seems reasonable to assume the former unless the latter can really be demonstrated.' Unfortunately the topic is not so easily discarded. Alderton and Rankin (1983) examined the fluid inclusion assemblages of the St Austell granite, and concluded that the low-temperature, low-salinity fluids present were in agreement with Sheppard's (1977) isotopic data, but that the temperature of formation could have been as high as 70 °C which is far higher than normal meteoric groundwater. A conclusion from this study was that, 'there is a continuum between cool groundwaters and very low temperature hydrothermal fluids of dominantly meteoric origin'. Conceptually, this is an important statement, because it emphasises the interaction between meteoric waters and late-stage hydrothermal fluids at relatively low temperatures. The deep penetration of surface water (most often ocean water) to levels where it becomes incorporated within hydrothermal mechanisms is necessary to explain many water:rock interactions (Taylor, 1977), and such processes should also drive home to the geomorphologist that meteoric groundwater is not confined to any arbitary upper zone of the lithosphere.

These results offer three possible explanations for problem profiles such as those containing the St Austell kaolinites:

(1) that there was a period of initial hydrothermal alteration or 'softening-up' followed by exploitation by percolating surface water, as favoured by Bristow (1977);
(2) that the alteration should be regarded as hydrothermal, but bordering on the supergene fluid regime (Alderton & Rankin, 1983); or
(3) that the alteration was achieved solely by meteoric waters and is the result of weathering processes (Sheppard, 1977).

Supporting evidence for supergene (weathering) alteration under warm climatic conditions is found elsewhere in southwestern England (Isaac, 1983; see also Chapter 9).

Diagnosing hydrothermal alteration

Given the nature of this debate, it is unlikely to be finally resolved, despite arguments to the contrary.

However, as Ollier (1983b) pointed out, late-stage hydrothermal alteration should be accompanied by the presence of volatiles such as Au, Sn, Ag, Cu, Zn, Pb cations and S, Cl, and F anions, and these are seldom identified in connection with deep weathering over granites. It is interesting that Ollier (1983b) was responding to a study of the arenaceous weathering mantle in the Bega Batholith, New South Wales, Australia, studied by Dixon and Young (1981), who advocated option (1) above (hydrothermal alteration followed by weathering) for the relatively little altered saprolite found in this area. The location is interesting because it forms one of many characteristic granite embayments in the Great Dividing Range of eastern Australia. Although Dixon and Young (1981) remarked on the absence of tors in this area, there are widespread corestones both on the surface and visible in road sections and some notable tors do occur (Figure 2.3). In this context it is possible to agree with Ollier (1983b) that no hydrothermal indicators appear to be present. A similar case was studied by Eggler *et al.* (1969) from the Laramie Range in Colorado and Wyoming, USA, where certain granites were subject to deep arenaceous alteration to 'gruss', while others formed bold relief (Cunningham, 1969). They attributed contrasts of this kind partly to differences in mineralogy and texture but also suggested an early (hydrothermal) phase of biotite oxidation that had weakened the fabric of granite which was subsequently deeply weathered.

A more systematic approach to this problem was taken by Meunier *et al.* (1983), who advocated very precise identification of clay minerals in order to distinguish hydrothermal environments. This entails the use of very small quantities of the material (a few grains only) for X-ray diffraction and electron microscopy. Meunier *et al.* concluded that, while supergene alteration is characterised by a kaolinite–montmorillonite product, hydrothermal alteration is characterised by a beidellite–vermiculite system in which the beidellite is potassic and the trioctohedral vermiculite is Mg rich. Fe-rich vermiculite, on the other hand, forms in weathering environments. The abundance of pure, white kaolinite (which can result from either alteration mechanism) from which Fe has been lost is also

FIGURE 2.3   Tor monoliths in the Bega Granite, New South Wales, Australia

considered a significant indicator of hydrothermal alteration. They noted that *argillan* vein and pore coatings are typical of weathered rocks.

This problem does not only affect granites, but is central to arguments about copper ore bodies and base metal sulphide *gossans*. Detailed studies of porphyry copper deposits at La Escondida, Chile, are revealing in many respects, but particularly for their record of early hydrothermal alteration followed by deep supergene weathering and Cu enrichment (Alpers & Brimhall, 1988, 1989), to depths as great as 600 m. The case is also important as a record of fundamental climatic change, from humid to the arid conditions that have persisted from the middle Miocene until the present day. Sulphide weathering was reviewed by Blain and Andrew (1977).

It is possible to conclude from this brief review that the extensive weathered mantle present across humid tropical and sub-tropical regions must result primarily from weathering by meteoric waters (supergene alteration). It transgresses rock formations, and is present equally in sediments and in crystalline rocks, where it is particularly well preserved over cratonised Precambrian and Archaean gneisses and similar rocks. This does not exclude hydrothermal alteration as a process affecting high-level intrusive rocks during the later phases of emplacement. This process may then mark an interaction between meteoric waters and hydrothermal brines at depths of 1–1.5 km from the surface. Deeply eroded plutons and high-grade metamorphic rocks are unlikely candidates for this explanation.

CHEMICAL WEATHERING PROCESSES

In this study, preliminary remarks about chemical weathering will be rather brief and readers are advised to consult standard texts (Keller, 1957; Garrels & Christ, 1965; Krauskopf, 1967; Loughnan, 1969; Carroll, 1970), and also the useful summaries by Curtis (1976), Birkeland (1984) and in Chorley *et al.* (1984). Initially, some basic weathering reactions will be defined, and the characteristic products of tropical weathering discussed. Reference will then be made to the major pathways of weathering under tropical and sub-tropical conditions, following mainly the work of Pedro (1966, 1968, 1984), Tardy (1971) and Duchaufour (1982). The development of weathering profiles will be discussed and specific reference will be made to bauxitisation in the next chapter (including the formation of lateritic bauxites), and questions concerning the rate of weathering processes will be considered.

Proton sources or donors

Although pure water will provide protons (H^+) which can react with rock minerals its *dissociation constant* is very low ($pK_a = 1 \times 10^{-14}$ at 25 °C) and this will limit the rates of reaction, despite the fact that water is almost ubiquitous in weathering environments. In practice, natural waters contain solutes and dissolved gases which may be described as *proton donors* (Ugolini & Sletten, 1991), the precise combinations of which determine the course of weathering reactions and pedogenesis. The most important of these proton donors is carbonic acid. Carbon dioxide is present everywhere as a component of both tropospheric and soil atmospheres and will dissolve to form carbonic acid which dissociates to form hydrogen and bicarbonate ions:

$$H_2O + CO_2 \rightarrow H_2CO_3 \rightarrow H^+ + HCO^{3-} \quad (2.2)$$

The dissociation of the carbonic acid permits hydrogen ions (protons) to substitute for metal cations in the silicate lattice, forming hydrated alumino silicates (clays). As this occurs, the equilibrium is displaced to the right, providing there

TABLE 2.1 pH of carbonated waters at different levels of P_{CO_2}

P_{CO_2}	pH
0.0003 (atmospheric)	5.65
0.01 (1%)	4.9
0.10 (10%)	4.4
1.00 (100%)	3.9

(After Ugolini & Sletten, 1991; reproduced by permission of Williams and Wilkins)

is sufficient water and soil drainage allows the removal of the ions in solution.

The concentration of carbonic acid depends on the partial pressure of carbon dioxide, and Ugolini and Sletten (1991) have calculated that at 25 °C this gives a dissociation constant (pK_{a1}):

$$H_2CO_3 \leftrightarrow H^+ + HCO_3^-, \qquad pK_{a1} = 6.35 \quad (25\ °C)$$
$$(2.3)$$

where the partial pressure of CO_2 (P_{CO_2}) = $10^{-3.5}$. This would give a theoretical pH for rainfall in equilibrium with an unpolluted atmosphere of 5.65.

However, in the soil atmosphere the partial pressure of CO_2 may be 60–100 times (perhaps as high as 400 times) greater than that of air (Johnson *et al.*, 1977; Holland, 1978). The expected pH values for different values of P_{CO_2} given by Ugolini and Sletten (1991) are shown in Table 2.1, indicating that moderate acidity can result from this source of protons alone. It is convenient at this stage therefore to characterise some of the common weathering reactions in terms of carbonic acid as the proton donor.

Hydrolysis reactions

Reactions with carbonic acid come under the heading of hydrolysis, in which metal cations are selectively replaced in the crystal lattice. In silicate minerals this is known as *incongruent dissolution*, by which the most mobile cations (Na^+, K^+) are lost first, then the divalent cations (Ca^{2+}, Mg^{2+} and Si^{2+}), followed by the polyvalent ions of Fe and Al. According to the progress of the weathering reactions, so the materials remaining in the solid phase will be variously depleted of these ions.

Before discussing the silicate minerals further, however, a brief summary of carbonate weathering relevant to limestone rocks will be given. In this case *congruent dissolution* takes place, which means that residual clays or other clastic components arise only from the presence of impurities in the rock and not from the weathering of the pure limestone itself.

Carbonate weathering

Limestone rocks are predominantly calcium carbonate ($CaCO_3$), and karstic landforms are usually limited to limestones with less than 20–30% clay or sand impurities. The dissolution rate of limestone is greatly enhanced by its purity, so that where $CaCO_3$ increases from 80% to 95%, dissolution rates rise by 3 to 5 times. Dolomite ($CaMg(CO_3)_2$) is also involved in karstic processes and landscapes but will not be discussed further here.

Karst geomorphology has always been pursued as a very specific line of enquiry which has led to an almost independent literature on the subject. This is unfortunate, because many themes in limestone dissolution and in the formation of karstic landforms have striking parallels with the behaviour of other rocks in the tropics. The chemical equilibria and reactions involved in limestone weathering, although deceptively simple in outline, are in detail rather complex. This complexity becomes particularly evident when the influence of climate through temperature controls is considered. Details of limestone weathering are given in Ford and Williams (1989); only the bare outlines will be offered here.

The solubility of $CaCO_3$ in pure water is given as 13 mg litre^{-1} at 16 °C and 15 mg litre^{-1} at 25 °C (Jennings, 1985), while Ford and Williams (1989) give 14 mg litre^{-1} for ionised water ($H^+ + OH^-$), and this is not much above comparable figures for quartz. Much higher solubility comes about in the presence of acids, especially carbonic acid (equation (2.3) above). Calcium carbonate dissociation is very weak, but in the presence of free protons (H^+), the reactions can be represented:

$$CaCO_3 \rightarrow Ca^{2+} + CO_3^{2-} \qquad (2.4)$$
$$\text{solid}$$
$$Ca^{2+} + CO_3^{2-} + H^+ \rightarrow Ca^{2+} + HCO^{3-} \quad (2.5)$$

So long as protons are available at the limestone surface this dissociation reaction will continue to the right, leading to progressive, acid dissolution of the rock. This in turn depends in large measure on the availability (or partial pressure) of CO_2 in the water (equation (2.3)). The reactions are all reversible, and with loss of CO_2 by so-called *degassing*, calcite will be deposited. This commonly occurs as phreatic water enters subterranean caverns, and where evaporation is enhanced in turbulent water, especially at waterfall sites, where calcite *tufas* are common.

The equilibrium partial pressure (P_{CO_2}) exhibits an *inverse* relationship with temperature, so that an increase in temperature from 0 °C to 20 °C will reduce P_{CO_2} by half, and at 35 °C it has fallen to one-third. An emphasis on this relationship has led some karst geomorphologists to hold the view that limestone dissolution will be at a maximum in *cold* wet climates (Corbel, 1959). However, this ignores a number of other very important controls on the system.

This relationship can be expressed as Henry's Law:

$$CO_2 \text{ (aqueous)} = C_{ab} \times P_{CO_2} \times 1.963 \quad (2.6)$$

where CO_2 is in g litre^{-1}, P_{CO_2} = partial pressure of CO_2, 1.963 is the weight of 1 litre of CO_2 at 1 atmosphere at 20 °C, and C_{ab} is the temperature dependent absorption coefficient. This declines from 1.713 at 0 °C to 0.878 at 20 °C and 0.665 at 30 °C.

As stated above, P_{CO_2} rises by two orders of magnitude within the soil, and studies have shown a rise from 0.03% in open air to 0.5–>2.0% in soil and cave air (quoted by Jennings, 1985). This is due to plant root respiration and to microbial processes (soil fauna and microfauna, bacteria, actinomycetes and fungi). The productivity of these organisms rises with temperature and the optima will normally be >20 °C. Figures for soil P_{CO_2} in tropical areas range from 0.2 to 11.0%, with a possible extreme value of 17.5% (Ford & Williams, 1989). Comparable figures for temperate areas are 0.1–3.5% (10% extreme value). It should be noted that the theoretical maximum value for P_{CO_2}, when all oxygen in the air has been replaced, is 21%.

Thus the aggressiveness of water involved in karst processes cannot be wholly predicted from atmospheric conditions, and this is especially true in wet tropical environments with abundant vegetation. Organic acids formed in peat bogs may also be effective in limestone dissolution, and sulphuric acid can also be present as a consequence of bacterial oxidation of sulphur (action of *Thiobacillus oxidans*). Carbon dioxide from the soil seeps into the cave air, since CO_2 is a heavy gas, and in wet areas a lot of vegetal matter is washed in. This means that cave air can have P_{CO_2} amounts 2–20 times greater than in a standard atmosphere. This 'booster' factor is not always adequately acknowledged in discussions of karst processes in the tropics.

The kinetics of equilibria and flow within the karst system are complex and will not be explored here in detail. As in other weathering systems, rock dissolution advances so long as equilibrium conditions are not established. This is a compromise between the rate of reaction and the rate of transport of the reactants. The reaction of acidified waters and carbonate rock *is* accelerated by temperature increases, providing that sufficient protons exist in the system. This means that the rate of limestone dissolution *can* increase with water temperature and flow velocity.

In addition, both velocity and thickness of flow must also be considered, particularly since turbulent flow abruptly increases limestone solution. These factors substantially alter simple attempts to correlate rates of karstification derived from temperature control over P_{CO_2}, and Corbel's (1959) contention that this rate would be 10 times higher in cold climates than in the tropics cannot be accepted. White (1984) used a theoretical approach to show that temperature control alone would increase limestone solution rates by only 30%, when temperature is reduced from 25 °C to 5 °C (Figure 2.4). On the other hand, correlation

with runoff is high and Presnitz (1974, quoted by Ford & Williams, 1989) obtained a figure of $r = 0.74$ for this relationship, and Ford and Williams (1989) considered that 50–77% of the total solution rate is accounted for by runoff.

Rossi (1976) considered the situation in the seasonal tropics of Madagascar, and concluded that dissolution values do not obey the principles worked out for temperate areas (heavily dependent on P_{CO_2} figures for atmosphere). The concentrations of Ca^{2+} in karst streams were considered less important than the quantity of runoff which is also largely concentrated into the wet season when biological productivity is at a maximum (therefore CO_2 is at its maximum).

Karst landforms will be discussed elsewhere, but there is little doubt that *surface* (as opposed to subterranean) karst landforms have their most spectacular expression in wet climates, most of them either warm or equatorial montane, as in Papua New Guinea. This observation emphasises the role of the forest cover and the high rate of CO_2 productivity in the forest soil, combined with a high input of rainfall, and enhanced by the modest temperatures at intermediate altitudes in the tropics.

Silicate weathering

Most common rock minerals are metal silicates; either feldspars or ferromagnesian minerals. Plagioclase feldspars (either as calcic anorthite, $CaAl_2Si_2O_8$ or as sodic albite $2NaAl_2Si_3O_8$) typify this group. The hydrolysis of albite can be represented:

$$2NaAlSi_3O_8 + 3H_2O + CO_2 \rightarrow$$
albite
$$Al_2Si_2O_5(OH)_4 + 4SiO_2 + 2Na^+ + 2HCO_3^- \quad (2.7)$$
kaolinite   silica ions in solution

In this case the highly mobile Na^+ ion is lost in solution along with some proportion of the silica which is not recombined to form clay (kaolinite). In a similar way anorthite hydrolyses:

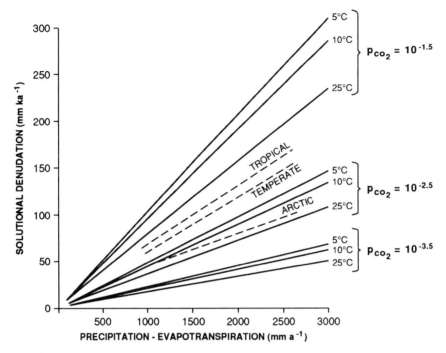

FIGURE 2.4 Theoretical relationships of solutional denudation of limestone to water surplus and CO_2 availability under coincident conditions from White (1984) with empirical relationships derived by Atkinson and Smith (1976) superimposed as dashed lines. (Reproduced from Ford & Williams, 1989 by permission of Chapman & Hall)

$$CaAl_2Si_2O_8 + 3H_2O + CO_2 \rightarrow$$
anorthite
$$Al_2Si_2O_5(OH)_4 + Ca^{2+} + 2HCO_3^- \quad (2.8)$$
kaolinite   ions in solution

The fate of the calcium in this equation is less certain, and may depend on the rate of leaching and the availability of other ions in the system (from the weathering of neighbouring minerals).

Similarly, the common mineral biotite may be hydrolysed:

$$K_2Mg_6Si_6Al_2O_{20}(OH)_4 + 14H^+ +$$
biotite
$$14HCO_3^- + H_2O \rightarrow Al_2Si_2O_5(OH)_4 +$$
kaolinite
$$4Si(OH)_4^\circ + 2K^+ + 6Mg_2 + 14HCO_3^- \quad (2.9)$$
silica and cations in solution

The silica not combined as kaolinite in these reactions goes into solution as silicic acid:

$$SiO_2 + H_2O \rightarrow 4Si(OH)_4^\circ \quad \text{(or } H_4SiO_4) \quad (2.10)$$
silica   silicic acid

Under weakly acid conditions with sufficient water and free drainage, more silica may be removed, permitting gibbsite to form from kaolinite (incongruent dissolution of Al and Si):

$$2Al_2Si_2O_5(OH)_4 + 105H_2O \rightarrow 42Al(OH)_3 +$$
kaolinite   gibbsite
$$42Si(OH)_4^\circ \quad (2.11)$$

or direct from plagioclase (albite):

$$2NaAl_2Si_3O_8 + 3H_2O + CO_2 \rightarrow Al(OH)_3 +$$
albite   gibbsite
$$3H_4SiO_4 + Na_+ + OH^- \quad (2.12)$$
silicic acid

The formation of gibbsite is commonly regarded as the end point of hydrolysis and the product of extreme weathering conditions, and although it can form direct from plagioclase feldspar, it is thought to form more commonly from the breakdown of kaolinite in profile. According to Kronberg *et al.* (1982), the formation of gibbsite is accompanied by a great increase in porosity due to the loss of molar volume, approaching 70%. Gibbsite is the principal mineral in bauxite which is discussed further in Chapter 4.

By contrast, in conditions where water is scarce, the hydrolysis reaction may be retarded and intermediate clay products will be formed, retaining a proportion of the metal cations liberated:

$$8NaAl_2Si_3O_8 + 6H^+ + 28H_2O \rightarrow$$
albite
$$3Na_{0.66}Al_{2.66}Si_{3.33}O_{10}(OH)_2 + 14H_4SiO_4 + 6Na^+$$
smectite $\qquad (2.13)$

Or possibly:

$$6KAlSi_3O_8 + 4H_2O + 4CO_2 \rightarrow$$
$$4K^+ + K_2Al_4(Si_6Al_2O_{20})(OH_4) + 4HCO_3^- + 12SiO_2$$
illite $\qquad (2.14)$

It will be seen that these clay minerals are more complex than kaolinite and their physical structure reflects their composition.

The presence of iron in the weathering reactions is also important and derives from the weathering of iron silicates such as fayalite and of highly reduced minerals such as pyrite (FeS_2). According to Curtis (1976) a simple oxidation and hydration reaction is possible:

$$2FeSiO_3 + (O) + 4H_2O \rightarrow Fe_2O_3 + 2Si(OH)_4° \quad (2.15)$$
iron $\quad$ oxygen water $\quad$ iron $\quad$ solution
silicate $\qquad\qquad\qquad$ oxide

However, in the presence of carbonic acid, Fe^{2+} is released into the weathering system by hydrolysis and may then be oxidised to Fe_2O_3. Following Krauskopf (1967):

$$Fe_2SiO_4 + 2H_2O + 2H_2CO_3 \rightarrow$$
fayalite
$$2Fe^{2+} + 2HCO^{3-} + H_4SiO_4 + 2OH^- \quad (2.16)$$

then:

$$2Fe^{2+} + 4HCO_3^- + \tfrac{1}{2}O_2 + 2H_2O \; Fe_2O_3 + 2H_2CO_3$$
$\qquad\qquad\qquad\qquad$ ferric $\qquad (2.17)$
$\qquad\qquad\qquad\qquad$ oxide

The formation of insoluble ferric iron in the course of weathering is the process fundamental to lateritisation. However, the behaviour of iron in weathering systems can be complex and is strongly affected by *redox* (reduction and oxidation) conditions as well as by pH. *Oxidation* involves the loss of an electron and thus imparts a negative charge, while *reduction* requires the gain of an electron and induces a positive charge:

$$Fe^{3+} + e^- \; \text{reduction} \rightarrow Fe^{2+} \qquad (2.18)$$
$\qquad\qquad\qquad$ ← oxidation
trivalent $\quad$ electron $\qquad\qquad$ divalent ion
ion

These conditions are usually measured as small voltages on the Eh scale (Figure 2.5). The presence of other proton donors may also be important (Ugolini & Sletten, 1991). Fe^{2+} may also enter into the lattice of clay minerals, particularly montmorillonite (smectite) (Figure 2.9).

Although the following sequence is not the product of a succession of aqueous phase reactions in natural systems, the degree of weathering (hydrolysis) can often be indicated by the nature of the clay formation:

alumino-silicate mineral →
$\quad$ →chlorite (Mg^{2+}, Fe^{2+} and Al^{3+} substitution
$\qquad\qquad$ for Si^{4+})
$\qquad$ → smectite (Na^+, Mg^{2+}, Fe^{2+} substitution for
$\qquad$ Al^{3+})
$\qquad\qquad$ → kaolinite (and halloysite)
$\qquad\qquad\qquad$ → gibbsite (and diaspore)

Chlorite is common in saprolites and involves Al^{3+} substitution for Si^{4+} in the early stages of weathering. It usually evolves to other forms of clay mineral. Smectites are a complex group of clay minerals in which Mg^{2+} and Fe^{2+} substitution for Al^{3+} in the octohedral sheet is characteristic. Since

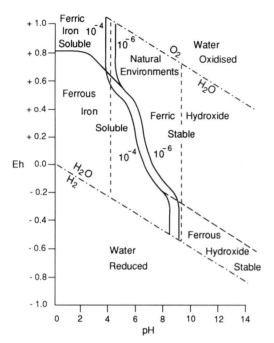

FIGURE 2.5 Solubility of iron as a function of Eh and pH. Labelled contour lines refer to activities of dissolved iron species of the indicated values, $T = 25\,°C$, total pressure = 1 atm. (Reproduced by permission from Douglas, 1976)

neither Mg^{2+} nor Fe^{2+} is present in the feldspar group of minerals these ions must come from the ferromagnesian minerals such as olivine and biotite which are also common minerals in silicate rocks. This also emphasises the fact that rock alteration is a function of the weathering of many minerals together, so that simple mono-minerallic representations of the process can be misleading.

Some indication of the different pathways taken by weathering reactions affecting specific rock minerals has been offered by Tardy et al. (1973):

K-feldspar → kaolinite → gibbsite

Biotite → (chlorite) → montmorillonite →
    vermiculite → Al-vermiculite → kaolinite

Plagioclase → (sericite) → vermiculite →
    montmorillonite → Al-montmorillonite →
    kaolinite → gibbsite

These transformations can also be indicated graphically as in Figure 2.6 which also shows the derivation of Fe-sesquioxides (goethite and haematite) from ferromagnesian minerals (hornblende).

One of the most common forms of smectite is montmorillonite in which Mg^{2+} is dominant. The chemical composition and physical make-up of smectites are complex and not easily represented in simple terms. Their structure allows absorption of water and they behave as *swelling clays*, the most common of which in tropical climates is montmorillonite (beidellite and nontronite are other forms of smectite).

Kaolinite is often regarded as the most common product of hydrolysis and it is certainly very widely found in free-draining soils. As can be seen from the formulae given above, the reaction progressively separates silica in aqueous solution from aluminium remaining in the solid phase (as kaolinite clay or as gibbsite). This is *incongruent dissolution* (in this case of alumino-silicate minerals) and is the basis for much of any discussion of saprolites and other weathering products. Close to the site of weathering of the primary mineral (at the weathering front), conditions may favour the initial formation of micaceous clays (chlorite, vermiculite) and smectite, but small quantities of gibbsite may also be formed here and kaolinite can progressively come to dominate the profile in free-draining sites with adequate water supply.

There are many problems with these simplified representations of weathering reactions, not least because so little is really understood about reactions taking place at the crystal surface under field conditions. According to Jenny (1980), unsatisfied bonds at the crystal surface attract the water dipoles, causing dissociation such that the hydroxyl ions can combine with silicon or aluminium while the hydrogen ions may combine with oxygen at the crystal surface or displace cations from the crystal lattice. This exchange of ions is disruptive to the rock fabric because of the high charge-to-size ratio of the hydrogen ion. Also the aluminium and silicon polyhedra become easily released from the mineral so that they can move into the soil solution (Birkeland, 1984).

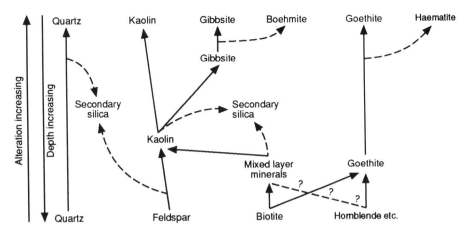

FIGURE 2.6   Mineral transformations during lateritic weathering of granite. Full lines represent major pathways; broken lines represent minor pathways (after Gilkes *et al.*, 1973; reproduced by permission of Blackwell Scientific Publications)

Solubility of compounds in weathering reactions

The products of weathering reactions do not all go into an aqueous phase or remain in solution. Some common bases such as Na^+ and K^+ remain in solution in the presence of OH^+, HCO_3^- and other anions at most concentrations found in soil solutions. Exceptions are found in enclosed depressions where evaporation rates are high, and more widely, in the semi-arid zone. The situation with Ca^{2+}, $Fe^{2+/3+}$, Al^{3+} and Si^{4+} is more complex and all of these ions are very sensitive to different properties of soil solutions, and particularly to the pH value. It can be seen from Figure 2.7 that, within the common range of soil pH, Fe_2O_3 and Al_2O_3 have very limited mobilities, while the solubility of SiO_2 (as quartz) remains low. Because Fe may be held in the reduced, ferrous form (Fe^{2+}) within rock minerals, the Eh of the soil solution is also important.

What is clear from these figures is that mobilisation of iron and aluminium within the common range of soil pH requires some additional mechanism which involves biochemical agents or proton donors. Highly alkaline environments (pH >8) only occur at the weathering front in moist climates, and it is only here (at close to abrasion pH) that Al^{3+} can go briefly into alkaline solution. On the other hand, low pH values (pH<4) are found in peat swamp environments, where a pH of

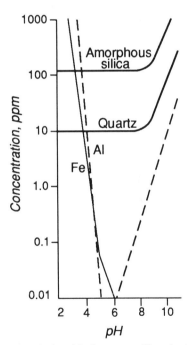

FIGURE 2.7   Relationship between pH and solubility of aluminium, iron, amorphous silica and quartz. Data for amorphous silica and quartz are from Krauskopf (1967), and data for iron and aluminium are from Black (1957). (Reproduced by permission of Oxford University Press from Birkeland, 1984)

3, sufficient to mobilise Fe^{3+}, is sometimes encountered. These relationships are best expressed on an Eh–pH diagram (Figure 2.5).

TABLE 2.2   Aluminium solubilities at different pH values

pH	Total Al as Al_2O_3 (mg litre^{-1})
4.9	0.0–2.0
4.5	0.0–12.0
4	1.5–23.0

(After Butty & Chapallaz, 1984; reproduced by permission of the American Institute for Mining, Metallurgical and Petroleum Engineers)

At the weathering front, iron is relatively immobile and tends to form complex gels containing Fe, Al and Si in various combinations. These are thought (Butty & Chapallaz, 1984) to evolve towards a crystalline phase of mainly aluminous goethite. However, the iron may be reduced and leached once bases have been removed in solution at some distance from the weathering front, leaving the residual gels to evolve into kaolinite or gibbsite. Aluminium ions appear as Al^{3+} below pH 4.5 and as $Al(OH)_4^-$ above pH 8, but the mobility of aluminium remains very limited in natural waters. Butty and Chapallaz (1984) quote from Bushinsky (1975) some measured solubilities of aluminium in Table 2.2.

Percolation experiments carried out near Manaus in Amazonia by Lucas *et al.* (1987) established the extraction of Al at 20–30 ppm over a 10 day period with a percolation rate of 10 mm day^{-1}.

Fe^{2+} solubility in natural groundwaters (in the absence of organic matter) commonly ranges from 0.01 to 2.8 ppm. In the presence of organic matter the figures for Si, Fe, and Al all change considerably as shown below, and it is becoming increasingly clear that weathering processes cannot be discussed solely in terms of inorganic geochemistry.

Biogeochemical processes

Although carbonic acid is used almost universally in texts on weathering to illustrate hydrolysis in weathering reactions, it is not the only proton donor present in the weathering systems. Ugolini and Sletten (1991) list the following: water, carbonic acid, biochemical compounds, nitric acid, sulphuric acid, and hydrolysis of polyvalent metals as the principal sources of protons. Of these perhaps biochemical compounds are the most important group, and also the least understood. Many of the

weathering products and profile features appear to owe their origins in part to the operation of biochemical compounds. In tropical environments this is particularly evident in all discussions of iron and aluminium mobility and the lack of precision in discussions of these topics is responsible for many major disagreements in understanding laterites and bauxites.

Tropical and sub-tropical forests are similar in many characteristics; both have a high total biomass, net primary production and total litter fall. In the case of the tropical rainforests values of 517 000, 34 200 and 27 500 kg ha^{-1} year^{-1}, respectively, have been given (Rodin & Basilevic, 1967), while sub-tropical forests return values of about four-fifths or 80% of these quantities. By contrast, although temperate forests may have biomass values of 50–60% of the tropical rainforest, their net productivity is much lower, less than one-third (26–33%).

Rapid decomposition of organic materials under rainforests is a central feature of their efficient cycling of nutrients, and it comes as no surprise to find that organic acids are widely present as a result of the leaching and breakdown of the large amounts of litter that otherwise would accumulate to great depths on the forest floor. In addition the metabolic products of living plants and microbes (e.g. fungi, algae, lichens and bacteria) also enter into the equation. Biochemical compounds found in soils include aliphatic acids, sugar acids, amino acids, aromatic acids and phenols. Most of these are unstable and therefore ephemeral and it is not certain how far they enter into mineral weathering as opposed to subsequent pedogenesis. However, from a long list of organic acids, *humic* and *fulvic* acids appear to be the main components of dissolved organic matter. The dissociation or acidity constant (pK_{a1}) for fulvic acid is not precisely known, but an average value of 4.5 is often accepted (Ugolini & Sletten, 1991). As these authors point out, this makes fulvic acid around 100 times stronger than carbonic acid. However, the functional importance of organic acids lies in their ability to form *chelate complexes* with polyvalent metals such as iron and aluminium, and also with silica, by utilising carboxylate and phenolate groups also present, to form ring structures incorporating the metal ion which is capable of forming more than one bond to the molecule of the complexing agent (Birkeland, 1984; Lehman, 1963; Huang & Keller, 1970). Such organo-metallic complexes are also deactivated in terms of the electrolytes present in the weathering environment.

Chelation

The most important effect of chelates is to allow congruent dissolution of polyvalent metals at common soil pH values. Within the normal range of soil pH values (pH 4–8), and in oxidising conditions (Eh usually positive), oxides of iron and aluminium appear bound to remain as insoluble residues of weathering, but in reality they have a limited mobility mediated by these chelating compounds. According to Ugolini and Sletten (1991), 'this may occur due to carboxyl and phenolic functional groups of fulvic acid being ligands for Al and Fe', and 'the solubility of the Al- and Fe-fulvates depends on the C/organically-complexed metal ratio; when a critical value is reached precipitation of the complexes may occur.' In addition, the greater strength of carboxylic acids as compared with silicic acid permits the replacement of adsorbed silica and an increase in silica mobility.

Because the breakdown of organic complexes is more rapid at higher temperatures, the mobility range of complexed metal cations is generally low. In other words most chelates are metastable. In the humid tropics and sub-tropics, therefore, there is a massive production of organic acids due to the large biomass available, but also a rapid breakdown of the fulvic (and other) acids produced, such that Fe and Al are often precipitated close to their weathering sites. Ong *et al.* (1970) gave some indication of the increased solubility of Al and Fe by laboratory experiments using peat extracts (Figure 2.8). Since peat swamps in the tropics appear to have a very important function in this respect (Sieffermann, 1988), this is very relevant to some later arguments. Equally important, however, is the notion that chelating agents may be in contact with fresh rock and able to promote much more rapid chemical change at the weathering front.

Schalascha *et al.* (1967) subjected pulverised granodiorite to reactions with several organic acids, including salycilic acid, at pH values well below those allowing Fe^{2+} to go into solution. A red Fe-salycilate was formed and they concluded that the organic acids acted by direct attack on the minerals. Skrynnikova (in Bushinsky, 1975) gave the following figures for enhanced solubility of iron in the presence of organic acids:

Organic matter, mg/litre^{-1}	5	5 – 10	35
Total Fe (expressed as Fe_2O_3), mg/litre^{-1}	0.6–1.4	4.1	12–27

But even these figures are low beside the 1700 mg/litre^{-1} Fe^{2+} recorded from swamp marginal soils in Sierra Leone by IITA (1975). It is clear from observations of seepages and where open water flow emerges from saturated pipe flow in river sediments that the precipitation of gels from metastable organic complexes is taking place, and this suggests that long distance transport of iron under saturated (hydromorphic) conditions is taking place.

McFarlane and Bowden (1992) have clearly established that Al levels measured in groundwater are strongly dependent on sample preparation and are much reduced by filtration and acidification, and this is consistent with the presence of Al in an organically bound form. Complexed forms of aluminium are known to reach concentrations of 2–13 ppm, in environments deficient in clay (thus avoiding fixation) and which are biologically inactive. Even higher levels, of 20–30 ppm, were obtained by Lucas *et al.* (1987) during percolation experiments in Amazonian podzols. Butty and Chapallaz (1984) concluded that, 'as a rule, the mobility of complexed forms of iron exceeds those of aluminium in a hydromorphic environment, a trend reversed in a non-hydromorphic environment.' Al-organic complexes usually biodegrade rapidly and aluminium tends to have a very short mobility range in tropical environments.

Gardner (1992) analysed saprolite samples from locations in Brazil and South Carolina, USA, to establish Al loss during weathering. By plotting volumetric concentrations of Al_2O_3 and SiO_2

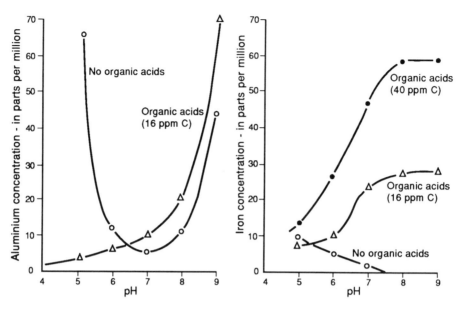

FIGURE 2.8   Solubility of (a) aluminium and (b) ferric iron (B) as functions of pH in the presence and absence of organic acids. (Reproduced by permission of the Macmillan Press Ltd from Thomas, 1974a; after Ong *et al.*, 1970)

against bulk density, molar losses of Al_2O_3 were established on the assumption that saprolite formation was isovolumetric. The molar loss ratio Si/Al gave figures ranging from 2.90 to 34.4, with an average of 8.34.

The problem posed in this study is to reconcile the loss of Al from the profile with very low concentrations of Al^{3+} in stream water. Some possible explanations were listed as:

(1) if the alteration had involved expansion then the calculations would be misleading, but this is thought unlikely;
(2) Al is precipitated from groundwater before reaching the river due to degassing of CO_2;
(3) organic complexes containing Al decompose before entry to the river;
(4) if the Si/Al ratio in river water is higher than in the saprolite then the hydrochemical conditions that produced the saprolite are relict.

The Rio Negro in Amazonia is cited as anomalous, having a Si/Al ratio an order of magnitude lower than other samples. This was attributed to the combination of: (1) catchment geology (crystalline);

(2) large organic matter content of water; (3) abundant CO_2 and intense leaching/complexing of Al; (4) with low concentrations of both Si and Al there is little precipitation of clay. However, the dilute chemistry suggested rapid groundwater movement and short contact time with weathering rocks; conditions that do not exist in this extensive catchment with low relief. Some of these points will be taken up later, but it is clear from the occurrence of laterites and bauxites in weathering mantles that Al and Fe *do* become separated during weathering and transportation in solution (Chapter 4).

Work by Bennett *et al.* (1988) also confirms the enhanced solubility of quartz in dilute aqueous solutions of organic acids at 25 °C. These authors observed concentrations of dissolved silica exceeding $1000\,\mu\text{mole kg}^{-1}$ in oil-contaminated groundwater in Minnesota, where theoretical solubility should have been only $75\,\mu\text{mole kg}^{-1}$. Quartz grains had also been strongly etched in the contaminated waters, but not elsewhere. This suggested an interaction between organic compounds and quartz at pH 7. Bennett *et al.* (1988) then conducted laboratory investigations which confirmed dissolution rates for quartz

8–10 times faster than in pure water when treated with a citrate solution (salicylic, oxalic and humic acids were also effective).

Although recognising the possibility that the organic complexes could derive silica from solution or from adsorbed ions, they favoured the direct interaction of organic acid anions with framework Si—O—Si bonds. Widespread field evidence for silica dissolution exists and such experiments go far towards an understanding of these phenomena, though they cannot specify the field conditions under which mobilisation takes place.

The main implications of chelating processes are listed by Butty and Chapallaz (1984):

—limitation of clay neoformation by deactivation of Al^{3+};
—decomposition of clays by protonation and replacement of adsorbed silica;
—promotion of Al and Fe mobility in the form of organo-metallic complexes.

It can readily be seen from this list that such processes may be very important as partial explanations for the occurrence of Al and Fe oxides and the destruction of clays in certain weathering environments.

WEATHERING PRODUCTS: THE FORMATION OF NEW PHASES

Ions liberated in weathering reactions or briefly bound within organo-metallic complexes may be lost to the system as solutes in soil and stream water. These are generally the more mobile constituents of weathering equations such as the cations Na^+, K^+, (possibly Ca^{2+}, Mg^{2+}), and anions including Cl^-, SO_4^{2-}, and NO_3^-. Others initially form amorphous products that may be difficult to characterise precisely, but usually alter to crystalline phases that can be described in terms of hydrous oxides (mainly of Al and Fe) or as specific clay minerals.

Clay minerals

Some common clay minerals have been characterised by chemical formulae (2.7)–(2.14)

TETRAHEDRAL LAYER

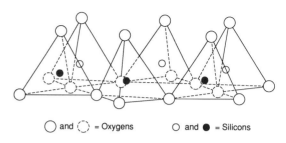

○ and ⦂ = Oxygens   ○ and ● = Silicons

OCTOHEDRAL LAYER

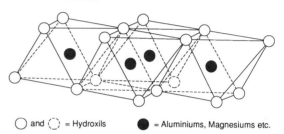

○ and ⦂ = Hydroxils   ● = Aluminiums, Magnesiums etc.

FIGURE 2.9 The atomic structure of tetrahedral and octohedral layers in clays after Grim. (Reproduced by permission of Methuen & Co. from Chorley *et al.*, 1984)

above. All of these may be described as phyllosilicates (or layer silicates) in which hydrated silicates, principally of aluminium, iron and magnesium, form layered or sheet structures which may consist either of silica tetrahedra, or octohedra centred on Al^{3+}, Fe^{3+}, or Mg^{2+} ions. The silicon tetrahedron is based on four O^{2-} ions surrounding one Si^{4+} ion, while in the octohedron six O^{2-} or OH^- ions are arranged around one Al^{3+}, Fe^{3+}, or Mg^{2+} ion. The linking of these sheets determines the character of the clay mineral and depends on the sharing of the O^{2-} ions in the tetrahedral sheets (Figure 2.9).

In a 1:1 layer clay a tetrahedral sheet links to one side of an octohedral sheet, and this characterises the kaolin/halloysite group. The 2:1 clays have a symmetrical arrangement of two tetrahedral sheets linked to a central octohedral sheet, and this characterises the illite and smectite groups. More complex arrangements arise from the linking of one octohedral sheet between two 2:1 layers forming a 2:1:1 layer phyllosilicate. Bonding between the sheets is achieved in different ways, for example between external O ions in one sheet with OH ions of another or by sharing a H ion between O ions of both sheets, forming a hydrogen bond. The latter usually comes about as a result of water molecules entering between the layers. Birkeland (1984) described this: 'OH ion (one sheet) to H ion (positive pole of water molecule) to OH ion (negative pole of water molecule) to H ion (adjacent sheet).'

Isomorphous substitution within the lattice is also possible if cations of similar size to the common Al^{3+} and Si^{4+} are available. Commonly Al^{3+} will substitute for Si^{4+} in the tetrahedral sheet, and Fe^{3+}, Fe^{2+}, and Mg^{2+} for Al^{3+} in the octohedral sheet. Where the substituting ion is of lower valency the layer is left with a negative charge and will attract other ions into the lattice.

In terms of the weathering reactions already demonstrated, the main clay minerals of interest are kaolinite and halloysite, and the smectite group. Illites are much less common in tropical saprolites. Kaolinite does not usually experience isomorphous substitution and is bonded by hydrogen ions, while hydrated halloysite is a 1:1 clay containing a water interlayer, also hydrogen bonded. Smectites are 2:1 clays which experience a variety of isomorphous substitutions in the octohedral layers and thus have a range of chemical compositions. The negative charge which results leads to a readiness to exchange cations and water layers, and this in turn contributes their well-known capacity to swell and shrink according to the amount of water present. The Mg/Ca smectite (montmorillonite) which is common in semi-arid areas, and in some poorly drained sites elsewhere in the tropics, is particularly associated with swelling properties. Vermiculite is a 2:1 clay in which substitution is in the tetrahedral layer which allows stronger bonding and expansion is small. Chlorite is a common 2:1:1 mineral in soils. It possesses positive and negative charges in adjacent sheets and is thus strongly bonded and immune to expansion. Its complex structure contains Fe^{2+}, Mg^{2+} and Al^{3+}, and this only occurs where hydrolysis has been arrested or is at an early stage. Chlorite is not a usual end product of weathering and clay formation in the tropics.

TABLE 2.3   Clay swelling in the presence of unlimited moisture

Clay mineral	%
Na montmorillonite (smectite)	1400–1600
Ca montmorillonite (smectite)	45–145
Illite	5–120
Kaolinite	5–60

Compiled from data in *California Div. Mines, Bull.* 169 (from Chorley *et al.*, 1984; reproduced by permission of Methuen & Co.)

The importance of clay swelling is indicated by the figures in Table 2.3 quoted by Chorley *et al.* (1984).

A simple representation of the succession of stages through which the weathering of a rock (basalt) may pass illustrates how the intermediate minerals will give way to more leached products with time, availability of water and adequate drainage. Thus Craig and Loughnan (1964) offered a weathering sequence for basalts in New South Wales (Figure 2.10).

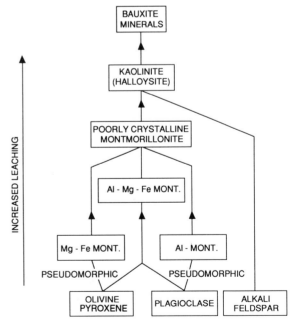

FIGURE 2.10 Diagrammatic representation of the weathering sequence in Tertiary basalts in New South Wales, Australia (after Craig & Loughnan, 1964; reproduced by permission of CSIRO Editorial Services)

GEOCHEMICAL PATHWAYS OF WEATHERING

Weathering mantles developed under tropical conditions cannot, however, be characterised solely by simplified chemical formulae. There are many reasons for this, including:

(1) Most parent rocks contain a variety of minerals that alter in different ways; Fe_2O_3 for example forms from the ferromagnesian group of minerals, while Al_2O_3 usually forms from the feldspar group.

(2) The weathering environment therefore contains ions and dissolved compounds from all the minerals present, and these may 'interfere' with the simplified reactions described above.

(3) The weathering products forming at a site will be determined by the stability relations arising from thermodynamic equilibria which are strongly influenced by the soil solution and its proton donors (which may be biochemical in origin or even microorganisms).

(4) The saprolite also contains resistant minerals such as quartz that may experience leaching but remain substantially unaltered in the profile (other *resistates* are mainly heavy minerals such as corundum, tourmaline and zircon, and include some economic minerals such as cassiterite, diamond and gold);

(5) The saprolite is not uniform with depth but develops distinct profile characteristics, reflecting the changing chemical equilibria at different depths from the ground surface.

(6) The saprolite will also vary spatially, due to changes in parent rock mineralogy and lithology, elevation and site conditions.

(7) Saprolites of more than 1–2 m depth will have experienced changes in the controlling factors through time, particularly as a result of varying rainfall/evaporation relationships (climate change) and as a consequence of changing soil hydrological conditions (often a fall of water-table resulting from dissection).

Even when these factors are accounted for, wider regional variations in saprolite characteristics are not wholly accounted for. These arise mainly from the patterning of regional climates, but also reflect major structural and geomorphological features, such as the large basins and swells found on the Gondwana continents, or the contrasts between orogenic and passive areas of the crust. Equally, there can be major differences between areas underlain by crystalline (silicate) rocks and those underlain by sediments.

Plasmogenic and non-plasmogenic alteration

It is helpful to consider the major geochemical pathways of weathering as described by French pedologists such as Pedro (1966, 1968, 1983, 1984), Tardy (1971), and Duchaufour (1982). Pedro (1984) distinguished *plasmogenic* from *non-plasmogenic* alteration. In the former the neoformation of clays is characteristic, while in the latter the destruction of the primary minerals is so thorough as to form only a residue, mainly of quartz sand. Such conditions depend on extremes of pH (or on chelation) that greatly increase the solubility of the less mobile ions such as Fe^{3+} and Al^{3+}. These may be highly acid (pH < 4) which leads to *podzolisation*, or highly alkaline (pH > 8.5) which can lead to sodic soils (*solods*). In addition, soil environments subject to extremes of seasonal inundation and desiccation can lead to the breakdown of clays by the process of *ferrolysis* (Brinkman, 1970, 1979).

Plasmogenic environments and clay formation

Most weathering environments are plasmogenic and Pedro (1984) distinguished between *altération massive* within humid tropical climates, where intense weathering reduces the bulk of the rock to clays and hydrated iron oxides which are neoformed from an aqueous phase, and *altération ménagée* in less humid or colder climates, where the clay minerals inherit the cations present in the parent rock by a process of transformation in the solid phase. This leads to the occurrence of 2:1 lattice clays (especially smectites). Fe and Al remain *in situ*, but calcium carbonate ($CaCO_3$) may be precipitated in the profile. These categories may be subdivided (Tardy, 1971) (see also Figure 1.5):

(1) *Allitisation* occurs when all basic cations and most of the SiO$_2$ are removed in solution, leaving Al and Fe present as hydroxides (gibbsite and goethite) (SiO$_2$:Al$_2$O$_3$ < 2).

(2) *Monosiallitisation* results from the partial removal of silica and the neoformation of kaolinite (SiO$_2$:Al$_2$O$_3$ = 2).

(3) *Bisiallitisation* occurs when the bulk of the SiO$_2$ is retained the clay (SiO$_2$:Al$_2$O$_3$ > 2), leading to 2:1 layer clays and retention of some cations released in weathering.

Some special conditions prevail with the weathering of volcanic deposits, leading to amorphous material and allophanes which have mineralogical characteristics between (1) and (2) above.

Other ways exist to describe these materials and the processes of transformation. For example, Duchaufour (1982) more specifically recognised the role of iron in tropical weathering by referring to the following:

(1) *Fersiallitic soils* of sub-tropical and some temperate regions (mean temperature 13–20 °C, rainfall 500–1000 mm) correspond with the process of bisiallitisation. They exhibit mainly smectite clays and with some lessivage and accumulation of SiO$_2$ and bases in the lower profile. Fe will be present as hydrated iron oxides, or held within the clay lattice.

(2) *Ferruginous soils* occur in hotter climates with higher rainfalls, and are dominated by kaolinite clays. Gibbsite is usually absent and 2:1 clays subsidiary; some weatherable minerals such as orthoclase and muscovite remain little altered.

(3) *Ferrallitic soils* characterise the humid tropics (mean temperature > 25 °C, rainfall > 1500 mm) and show advanced hydrolysis of all silicate minerals. Leaching of SiO$_2$ provides excess Al^{3+} in the profile and this combines to form gibbsite which accompanies kaolinite within the saprolite. Iron may be mobilised by chelation and is usually present as goethite which may become concentrated as a lateritic horizon within the profile.

Segalen (1971) has pointed out that the segregation of iron oxides, often in amorphous form, must be an inevitable result of the widespread formation of kaolinitic clays. The stability of kaolinite can be attributed to the cycling of Si within the forest ecosystem, thus preventing the leaching of SiO$_2$ from the topsoil, gibbsite being produced at greater depths (Lucas *et al.*, 1993).

While some authors have proposed rainfall thresholds for the appearance of characteristic clay mineral assemblages, these are often misleading. First, because the distribution of rainfall can be as important as the annual mean. In sub-tropical climates there is a clear distinction between winter rain and summer rain environments, and only the latter show affinity with the tropical environments, where temperature is always high enough to promote weathering when moisture is present. The duration and severity of drought seasons are also important to the immobilisation of bases in the soil. Second, because of the operation of other factors of soil formation, especially those of topography and time which are largely independent of climate and vegetation.

One deficiency of most schemes is the lumping together of all climates within the humid tropics, when there are major contrasts of environment between marginal forest areas receiving ~ 1400 mm year^{-1} of rainfall annually and those monsoonal, equatorial or orographic areas experiencing 2000–5000 + mm year^{-1}. It is in the latter that gibbsite becomes most abundant. West and Dumbleton (1970) for example found a steep rise in gibbsite with rainfall increases from 2000 mm year^{-1} to 3500 mm year^{-1} in Malaysia. The percentage of total clay, measured as particles < 2 μm, also rises with rainfall and with temperature. This was demonstrated by Sherman (1952) from studies of basalt in Hawaii which showed clearly how clay content and clay mineral dominance vary together across the rainfall range, the former rising to nearly 60% in soils with ~ 50% hydrated Fe$_2$O$_3$ and Al$_2$O$_3$ (Figure 2.11). Jenny (1941, 1980) was also able to demonstrate the effects of temperature in increasing the clay content of soils from basic rocks, from around 15% at 10 °C to 50% at 16 °C (Figure 2.12).

A further complication of these relationships is the difference between the chemical composition of water circulating in the saprolite and the

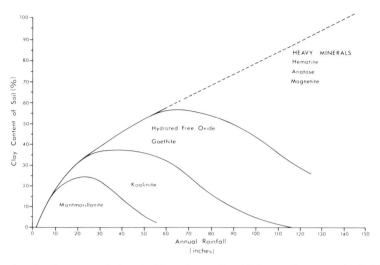

FIGURE 2.11   Progressive development of clays in Hawaiian soils with increasing annual rainfall in a seasonal wet–dry climate. (Reproduced by permission of the Institute of Mining and Metallurgy from Sherman, 1952)

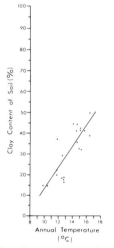

FIGURE 2.12   The clay content of soils derived from basic rocks in relation to mean annual temperature (after Jenny, 1980; reproduced by permission of Springer-Verlag)

composition of the soil clays. Thus Tardy (1971, p. 253) has commented that, 'the chemical compositions of water are sensitive to seasonal climatic variations. They reflect the evolution of weathering at the moment of sampling. In contrast, the different neoformed minerals, usually stable enough, integrate the seasonal variations and resist the succession of climates which they undergo.' This rather important observation emphasises the

temporal factor in saprolite formation, and the fact that there may be a condition of disequilibrium between the soil solutions and the materials within which they circulate.

Degrees of alteration in the weathering products

The method of total chemical analysis in which results are expressed as molar ratios is also widely used to describe the weathering products in the solid phase (for example, see Birkeland, 1984). According to Grant (1969), *abrasion pH* offers a simple expression of these relationships, as a function of the ratio between the mobile ions remaining and the clay minerals formed in the saprolite:

$$\text{Abrasion pH} = f\,\frac{\text{Na} + \text{K} + \text{Ca} + \text{Mg}}{\text{Clay minerals}} \quad (2.19)$$

Abrasion pH will fall with the leaching of bases from the original rock. Thus Hendricks and Whittig (1968) gave the data shown in Table 2.4 for the weathering of andesite in coastal California, including the SiO_2/Al_2O_2 ratio. Such methods for determining *degrees* of weathering are important signposts towards the physical characteristics and properties of saprolites which directly affect denudation processes, and other indices have been developed by engineers.

TABLE 2.4   Weathering of andesite in coastal California

	Hypersthene andesite	Saprolite samples			
		least weathered	$\rightarrow$		most weathered
Molar $SiO_2 : Al_2O_3$	5.79	3.47	2.40	2.09	2.13
Abrasion pH	8.9	5.5	5.1	4.9	

(After Hendricks & Whittig, 1968; reproduced by permission of Blackwell Scientific Publications)

One of the problems faced by these methods is to express alteration in relation to a constant, unchanging proportion of a single rock mineral. As research into weathering has progressed, each possible candidate for this constant has been criticised as subject to change and loss from the rock itself. However, the *silica-sesquioxide ratio* (K_r) calculated from major element analyses (Geological Society, 1990):

$$K_r = \frac{\%SiO_2/60}{(\%Al_2O_3/102) + (\%Fe_2O_3/160)} \quad (2.20)$$

is widely used to express relative differences between materials.

One alternative is to compare weathered residues with the fresh rock. This obviously depends on having access to the fresh rock which is seldom possible in the tropics unless deep excavation has taken place. The *Decomposition Index* (X_d) of Lumb (1962) is calculated on this basis:

$$X_d = \frac{(Nq - Nq_o)}{(1 - Nq_o)} \quad (2.21)$$

where Nq is the weight ratio of quartz to quartz + feldspar in the soil, and Nq_o is the same ratio in the fresh rock. This gives values from 0.1 to 1.0. However, this index is subject to the objection that there are many states of partial alteration of feldspar, and iron oxide cementation of altered feldspar pseudomorphs may affect the results (Irfan, 1988).

More recently, Sueoka (1988) has proposed a new *Chemical Weathering Index* (CWI), based on the molecular weight percentages of sesquioxides plus H_2O (loss on ignition) as a proportion of all chemical components (alkalis, alkali earths, silica):

$$CWI = \left(\frac{(Al_2O_3 + Fe_2O_3 + TiO_2 + H_2O(\pm))}{(\text{All chemical components})} \right) mol \times 100(\%) \quad (2.22)$$

According to Sueoka (1988), this formula produces figures that correspond closely with physical properties of weathered granite. Thus for fresh granite, CWI = 13–15%; for weathered granite, CWI = 15–20%; for residual soil, CWI = 20–60%; and for laterite, CWI > 60%. The index can also be matched to the conventional sixfold division of the weathering profile, as shown in Table 2.5.

Another approach was adopted by Irfan and Dearman (1978), who proposed a *Micropetrographic Index* (Ip) to represent the proportion of sound constituents to unsound materials in the rock:

$$Ip = \frac{\% \text{ (quartz + unaltered feldspars + unaltered biotite)}}{\% \text{ (altered minerals + voids + microcracks)}} \quad (2.23)$$

When applied to Hong Kong granites this (Ip) index gave values ranging from 33.4 (fresh granite) to 0.76–0.39 (decomposed granitic soil) (Irfan, 1988).

TABLE 2.5   Weathering profiles and the chemical weathering index (CWI)

CWI (%)	Profile zone	Extent of weathering
60–90		Duricrust
40–60	VI	Residual soil
20–40	V	Completely weathered
	IV	Highly weathered
15–20	III	Moderately weathered
	II	Slightly weathered
13–15	I	Fresh rock

(After Sueoka, 1988; reproduced by permission of Balkema)

Non-plasmogenic environments and podzolisation

Mention has been made above of certain processes that mobilise the ions of Al, Fe and Si. These ions normally enter into the processes of clay formation, but it is well known that silica has a significant solubility within the normal range of temperature and pH in natural waters (100–200 mg litre^{-1} at 25 °C). The behaviour of iron and aluminium is more complex, however, and will be discussed first in terms of podzolic and hydromorphic environments, and subsequently in terms of laterite and bauxite formation as products of (mainly) ferrallitic/allitic weathering systems.

Podzolisation may be a process which appears to be remote from the ferrallitic weathering associated with tropical environments, but although the latter is usually held to be the process responsible for lateritisation (Pedro, 1983, 1984), including the formation of *lateritic bauxites*, this often rests on the assumption that aluminium is residual and does not readily mobilise in the profile. But there is evidence of aluminium transfer (absolute accumulation) within bauxite profiles, and this suggests parallels with podzolisation. There is ample evidence for the operation of podzolisation processes in humid tropical environments, and *hydromorphic tropical podzols* are formed, where the parent materials are sufficiently permeable and water-tables high. These conditions are typified by the coastal and riverine terrace sands of Borneo, the Congo/Zaire basin and Amazonia for example (Andriese, 1968, 1970; Duchaufour, 1982; Butty & Chapalaz, 1984; Schwartz, 1988) but hydromorphic environments are also found on the flat summits of some crystalline massifs. In western Kalimanatan, where podzolised sandy terraces are widespread, Buurman and Subagjo (1978) described soils over adjacent granodiorites as intermediate between latosols and podzols. It has been suggested (Bravard & Righi, 1990) that losses of clay particles from the A1 horizons by *lessivage* and selective erosion would proceed *before* podzolisation could take over as the dominant soil forming process. In the oxisols, fulvic acids and polysaccharides are abundant but adsorbed on to clay particles; however, following loss of clay, biodegradation and migration of chelates becomes possible, and this was demonstrated by the presence of Al and Fe in organic complexes. Evidence from other equatorial environments suggests that clayey kaolinitic saprolites form over most parent materials and that further hydrolysis of kaolinite would be dependent on enhanced water movement following loss of clays from the profile. However, this may not be due so much to eluviation, but more often to lateral water flow in a hydromorphic environment.

Support for these observations also comes from the analysis of river waters, and Stallard and Edmond (1987, p. 8300), on the basis of work in Amazonia, have concluded that, 'where weathering is sufficiently rapid, cation and silica concentrations are high, kaolinite forms, and quartz is stable. Where weathering rates are lower, perhaps because of thicker soils . . . silica concentrations decrease, and quartz dissolution can start contributing additional silica.' They continued, 'eventually, with sufficiently low concentrations of silica, kaolinite is no longer stable, and any rock that is exposed to weathering can presumably have all its silica dissolved, leaving either gibbsite, or nothing.' In the lowland rivers of Amazonia, the proportions of Fe and Al to other elements are close to those in the crystalline rocks of the subjacent shield, and this suggested to Stallard and Edmond (1983, 1987), 'that the bedrock is actually dissolving completely'. This circumstantial evidence for chemical denudation is very difficult to quantify, but remains a pivotal observation in the discussion of landscape dynamics of the humid tropics. This might throw light on the problems posed by Gardner (1992) (see above, p. 36), because present-day hydrochemical conditions are a product of a long history of leaching and evolution of weathering profiles, and it is doubtful whether the river chemistry can be held to reflect conditions for isovolumetric saprolite formation.

The role of hydromorphic conditions

The development of hydromorphic conditions of weathering may provide the vital link between saprolite formation and the later stages of chemical

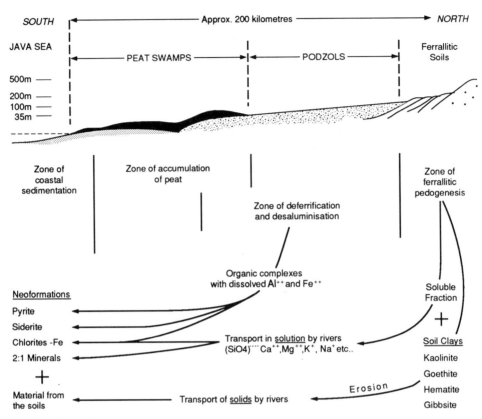

FIGURE 2.13   Transfers of material in the peat landscapes of Kalimantan (after Sieffermann, 1988; reproduced by permission of Armand Colin)

evolution. Hydromorphic weathering processes are also probably necessary to explain podzolisation in the central parts of many tropical plateaus (Schnütgen & Bremer, 1985; Brabant, 1987; Lucas *et al.*, 1988, 1989), and in coastal plain, sedimentary environments of low relief and high water-table, such as is found within the Pleistocene 'white sands' of Western Kalimantan (Thorp *et al.*, 1990). In these environments, reduction of iron and its migration as Fe^{2+} would increase the proportion of highly dispersible, iron-free kaolinite that could migrate over long distances, leaving a sandy residual material (Chauvel, 1977). Such long-distance transfers of solutes and dispersible clays are also called for by Sieffermann (1988) in a study of equatorial peat-swamp systems. Using Kalimantan to exemplify a model, Sieffermann (1988), noted the transition from ferrallitic soils inland, to giant podzols (white sands) that pass beneath the peats

and recent river sediments near the coast (Figure 2.13), and sought a link between the *deferrification* of the ferrallitic residues and the occurrence of the peat-swamps. In the course of a study of the Pleistocene terraces (white sands) in western Kalimanatan, Thorp *et al.* (1990), observed a major change in character between the white sands (podzols) of the Pleistocene terraces which, today, have slopes of $4\,m\,km^{-1}$ and rise inland from (below) sea level to $+25m$ asl, and ferruginised terrace remnants continued inland and at higher elevations. The white sands can be regarded as products of hydromorphic podzolisation. According to Sieffermann (1988), similar white sand formations surround the peat-swamps of the Amazon and the Orinoco, and he speculated that the reversal of drainage consequent upon the uplift of the Andean Cordillera in the early Tertiary could have led to the formation of a vast lake bordered by swamps. Into such an environment the red clays and sands from erosion of the headwaters would have accumulated and been leached of iron. Even without this unproven hypothesis, it seems likely that epeirogenic movements and sea level changes will have altered the hydrological regimes within the porous Tertiary and Pleistocene sediments, at least during the Neogene, so that white sand terraces of today may have undergone their formative pedogenesis at lower relative elevations in the past.

Ferrolysis

The processes involved in hydromorphic conditions can be complex, depending largely on the water regime. Where lateral or vertical leaching is dominant throughout the year, lessivage and podzolisation occur, but where seasonal inundation alternates with strong desiccation a process of *ferrolysis* has been described (Brinkman, 1970, 1979). According to Butty and Chapallaz (1984), ferrolysis can be considered a particular form of *acidolysis*, which leads to the congruent dissolution of kaolinite:

$$Al_2Si_2O_5(OH)_4 + 6H^+ \rightarrow 2Al^{3+} + 2H_4SiO_4 + HOH$$

kaolinite $\qquad\qquad\qquad\qquad$ (2.24)

Conditions necessary for this to take place are, acidity maintained at pH < 4 and with log $[H_4SiO_4] < -4.6$. This equation should be compared with that for *in*congruent dissolution of kaolinite which takes place under weakly acid conditions leading directly to the formation of gibbsite. Under these conditions iron is reduced to Fe^{2+} and also becomes mobilised. Swamp environments are among the most likely sites for acidolysis to occur and they may therefore be critical to the vertical and lateral transport of the less mobile ions of Si^{4+}, Al^{3+} and Fe^{2+}.

If the pH > 4 the Al^{3+} will recombine with soluble silica to form amorphous clay products. This can arise with desiccation and the formation of siliceous, ferruginous and aluminous gels.

However, Brinkman (1979) argued that podzolisation as a process should be restricted to conditions where cheluviation takes place, and

referred to surface water gleys, within which the eluvial horizon contains aluminium interlayers in the clay fraction and these also contain trapped ferrous iron. A full description of the process of ferrolysis will not be attempted here, but it is a process of clay decomposition by alternate reduction and oxidation of iron. Under saturated conditions the iron is reduced to Fe^{2+} which displaces exchangeable cations, all these being partially leached by the moving groundwater. H^+ ions are consumed during this process and the pH rises from around 4 to 6–7. As this occurs, some of the Al^{3+} ions are fixed as interlayers in swelling clay minerals and Fe^{2+} ions are also trapped and become oxidised to the immobile ferric Fe^{3+} as the profile dries out. Mobilised silica may also be re-precipitated in amorphous form, while quartz grains are likely to become deeply etched (Eswaran & Stoops, 1979).

In summary, the breakdown of clays can occur in different ways:

(1) near neutral *hydrolysis* → incongruent dissolution with the formation of gibbsite beneath well drained sites;
(2) *acidolysis* → congruent dissolution promoted by chelation, but dependent on pH < 4 and is associated with swamps;
(3) *ferrolysis* → alternate congruent dissolution and the formation of interlayered clays, in seasonally inundated soils.

Case studies of podzolisation from Amazonia and Guyana

In the headwater regions of the Rio Negro, Schnütgen and Bremer (1985), building on the work of Klinge (1965), have studied the sandy substratum supporting patches of savanna vegetation within the rainforest and have been able to demonstrate that the quartz sands are the residue of granite weathering, and that they have developed due to rising water-tables, where the relief has been reduced to minimal values.

Further studies by French pedologists on the transformation of ferrallitic soils to podzols in South America have provided valuable models to describe relationships between pedological processes and the distribution of ferrallitic (clayey) and podzolic (sandy) mantles (Chauvel, 1977; Boulet *et al.*, 1984; Chauvel *et al.*, 1987; Dubroeucq & Blancaneaux, 1987; Lucas *et al.*, 1987, 1988). Moreover these models have very important implications for the study of geomorphology in humid tropical climates.

Three different cases were described by Lucas *et al.* (1987): the first from the North-Manaus area of Amazonia (3 °S), on the Barreiras sedimentary formation; the second on the crystalline Precambrian shield in Guyane Française (5 °N), and the third on the Guyanese coastal plain Quaternary sediments. In the latter two areas, a ferrallitic mantle is described as intact around the margins of low plateaus giving way to podzolic soils in interior depressions. The authors described ferrallitic clays (40–50% < 2 μm) containing *in situ* lithorelicts around the margins and bounding upper slopes of the plateaus, but with soils becoming very sandy (podzolic) in the plateau centres. A sandy, ferrallitic (and permeable) horizon is found below the clays and laterally between them and the podzols.

The situation was interpreted as indicating a lowering of the weathered landsurface, by replacing the saprolite at depth, and by transformation of the ferrallitic soil to white sand, first in the lower (hydromorphic) horizon, gradually intersecting the ground surface and spreading laterally as the landscape is lowered (Figure 2.14).

Over the coastal sediments, on the other hand, the ferrallitic mantle appears to have been subjected to strong leaching and loss of clays (lessivage) from the surface, gradually developing hydromorphic conditions at depth and in the centre of the landform. In this way the ferrallitic mantle has been transformed to a podzolic white sand *without* significant lowering of the ground surface (Figure 2.15). This difference between the two, is considered to result from the great depth of dissection and thickness of mantle over the crystalline shield.

The example of podzolic white sands from the Barreiras sediments has occasioned more problems. It was suggested that the transition from a ferrallitic to a podzolic mantle arose from a transformation starting within lower members of the toposequence

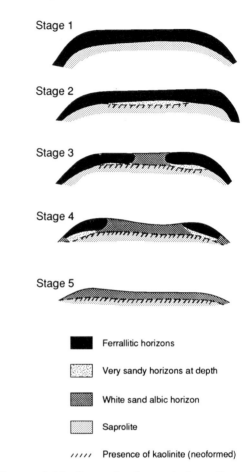

FIGURE 2.14 Stages in the evolution of weathered mantles on the crystalline shield of French Guyana (after Lucas *et al.*, 1987)

and extending laterally upslope (Lucas *et al.*, 1987), but this ignores (Lucas *et al.*, 1988) the possible influence of lithological contrasts within the Barreiras sediments. Bravard and Righi (1990) also analysed a toposequence of soils on the same parent materials, where they found ferrallitic weathering (*oxisols*) below the upper slopes, graduating via *ultisols* to Giant Podzols (*spodosols*) on the lowest member of the sequence. Here, the footslope soils were described as very sandy and there was evidence of accumulation of quartz. The oxisols showed a decline in SiO_2 with an increase in Fe_2O_3 downslope, while Al_2O_3 (as gibbsite) also increased from 2 to 25% downslope. Evidence for chelation

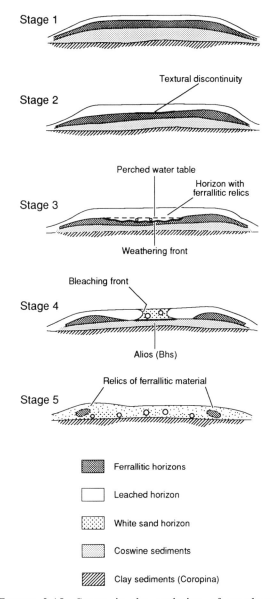

Stage 1

Stage 2

Textural discontinuity

Stage 3

Perched water table

Horizon with
ferrallitic relics

Weathering front

Stage 4

Bleaching front

Alios (Bhs)

Stage 5

Relics of ferrallitic material

Ferrallitic horizons

Leached horizon

White sand horizon

Coswine sediments

Clay sediments (Coropina)

FIGURE 2.15 Stages in the evolution of weathered mantles on the coastal sediments of French Guyana (after Lucas *et al.*, 1987)

of Fe^{2+} was obtained from resin experiments. Sand increased from 10–20% in the oxisols to 90% in the spodosols, but no adequate explanation for this increase was found. The possible explanations included: (1) coarser sediments outcropping at the base of the slope; (2) colluviation of sand from upslope; (3) direct transformation of parent material; or (4) progressive transformation of the oxisols to spodosols. Bravard and Righi (1990) demonstrated the similarity of the particle-size curves for quartz in both the oxisols (10–20% sand) and the spodosols (90% sand), suggesting a common inheritance and/or similar conditions of transformation, involving advanced etching, pitting and fragmentation of the quartz grains.

Although different explanations are available, Lucas *et al.* (1987, 1988) remained convinced of the principle of transformation of ferrallitic residues to podzols under hydromorphic conditions within the profile at depth or at lower levels on a toposequence. The models involve the lowering of the groundsurface by collapse of the soil structure and liberation of aluminium.

These processes are of central importance to the *chemical denudation* of the continents, because they are capable of reducing the bulk volume of other weathering products, by removal of the least mobile ions (Si^4, Al^{3+} and Fe^{2+}) which in all other circumstances recombine immediately within the weathering environments where they are produced, to form clays and hydrated oxides. Although chelating compounds promote the movement of these ions (mainly) down-profile or down-slope, our present knowledge suggests that large scale and long distance transfers will occur under rather special conditions that are found especially around the coastal fringes of continents and within major drainage basins of low relief. However, localised transfers of mobilised elements from saprolite breakdown can occur in many other situations, including the *dambo valleys* (see Chapter 8), and similar seasonally flooded depressions on elevated plateaus. In these cases highly seasonal rainfall regimes and low rainfall totals combined with high evapotranspiration can lead to precipitation of ions on lower slopes and in basins, where even sodic soils can be found over certain lithologies.

RATES OF CHEMICAL WEATHERING

In geomorphology, the rates of processes involved in the formation of the landsurface must always be of central importance, and it might be thought that this was a topic capable of precise numerical expression. Unfortunately this is seldom the case, unless we are content with loss of material per unit area/volume over experimental time periods (often 1–5 years). Such losses are most frequently and easily measured as erosional transfers of sediment and solutes in rivers, across slopes and, occasionally, in groundwaters. In the context of chemical weathering, the solute load of groundwaters is particularly important, and analysis of stream base flow is widely used as a surrogate measure for this quantity.

However, such measures tell only a part of the story. Most weathering reactions involve the neoformation of hydrated minerals, clays and oxyhydroxides, and within the saprolite this is usually considered to be an isovolumetric transformation. In this context, therefore, there should be no net lowering of the upper surface of the saprolite. Further weathering takes place in the residual soil (Zone VI) and may bring about a major change of volume, estimated on theroretical calculations at 70–75% (Kronberg *et al.*, 1982; Pavich, 1989). This type of calculation, however, ignores the lateral transfers of ions such as Fe^{2+} in solution and their subsequent precipitation as hydrated oxides. The loss of volume during pedogenesis, the export of solutes as well as sediments derived from weathering of exposed rock surfaces (Cleaves *et al.*, 1970), and the lateral transfer of ions in solution, all add up to a major chemical component of denudation which must also be addressed.

Measures of weathering penetration and rock transformation

It is, nevertheless, legitimate to consider weathering rates in terms of *either* the rate of penetration of weathering into the rock mass (rate of saprolite production) *or* the rate of transformation of one mineral phase to another, more in equilibrium with the groundwater environment (rate of alteration). These two approaches produce results which are not strictly comparable, a fact sometimes omitted from accounts of this topic. A further complication concerns the extrapolation from present-day rates of weathering to longer time periods, often expressed in $m\,Ma^{-1}$ ($mm\,ka^{-1}$). The validity of such extrapolations is highly questionable.

The most important method for calculation of present-day rates of weathering is from geochemical mass balance (input–output) studies, using small watersheds. These have come particularly from the Appalachian piedmont area and were pioneered by Cleaves *et al.* (1970, 1974). More recently, several scientists have pursued different aspects of this question (Pavich, 1985, 1989; Velbel, 1985, 1986; Dethier, 1986; Cleaves, 1989) in the context of humid temperate forested areas of the USA. No such studies are known to the author from humid tropical regions.

Different results are obtained from the measurement of weight loss from buried rock tablets (usually limestone) (Trudgill, 1985), and from the measurement of weathering rinds on exposed rock surfaces. Neither of these methods adequately simulates the formation of saprolite, although they throw valuable light on other questions.

Methods which give rates of weathering over longer time periods depend on the dating of geological materials, such as volcanic tephra and glass, or capping lava flows; the stratigraphic or other age relationships of planation surfaces below which saprolite has formed, or on the isotopic analysis of the saprolite itself. This last technique has not so far provided rates of weathering, but has given evidence regarding the *age* of weathering profiles. In appropriate circumstances remanent palaeomagnetism can be related to geological age.

Data given in Table 2.6 (A–C) are mainly confined to (A) rates of saprolite production, and (B) rates of mineral alteration, though some other calculations are included for comparison. In the former, the data mostly derive from the southern Appalachian region, where the forest climate receives $> 1000\,mm\,year^{-1}$, and has summer temperatures between 30 and 40 °C, but cold winters and average annual temperatures of ~ 10 °C. Latitude is 36–39 °N.

One salient observation emerges from the data in Table 2.6 (if the exceptional values for volcanic materials are omitted) and that is that whichever

TABLE 2.6   Rates of weathering and surface lowering in silicate rocks

(A)   *Rates of saprolite production and surface lowering*[a]

Experimental area	Author(s)	Method used	Rate
Smoky Mountains, USA	Velbel (1985)	Mass balance	38 m Ma^{-1}
S. Blue Ridge	Velbel (1986)	Mass balance	37 m Ma^{-1}
Pacific NW USA	Dethier (1986)	Mass balance	33 m Ma^{-1}
Masanutten Mtn, Virginia, USA	Afifi & Bricker (1983)	Mass balance	2–10 m Ma^{-1}
S. Piedmont, USA	Pavich (1986)	Mass balance	4 m Ma^{-1}
	Pavich (1989)	Residence time	20 m Ma^{-1}
Baltimore piedmont	Cleaves *et al.* (1970)	Mass balance	4–8 m Ma^{-1}
	Cleaves (1989)	Equilibrium model	25–48 m Ma^{-1}

[a]All the above refer to crystalline, igneous and metamorphic rocks, though Dethier (1986) included varied lithologies.

(B)   *Estimated rates of alteration*

Area	Authors	Comment on calculation	Rate
Ivory Coast	Leneuf (1959)	Ferrallitic alteration of granite (leaching of Si, Ca, Mg, Ka, K)	5–50 m Ma^{-1}
	Leneuf & Aubert (1960)		13–45.5 m Ma^{-1}
	Boulangé (1983)		14 m Ma^{-1}
Chad	Gac (1979)		13.5 m Ma^{-1}
New Caledonia	Trescases (1973)	Ultramafic rocks	29–47 m Ma^{-1}
Uganda	Trendall (1962)	180 m surface lowering and lateritic weathering	5 m Ma^{-1}
	Fritz & Tardy (1973)	gibbsitic alteration	3 m Ma^{-1}
Papua New Guinea	Ruxton (1968b)	Profile depth in 650 ka^{-1} volcanic ash	58 m Ma^{-1}

(C)   *Other estimates and calculations*

Area	Authors	Comment on calculation	Rate
New Guinea	Ruxton (1968b)	Alteration of volcanic glass to clay	8–27 ka^{-1}
New Zealand	Lowe (1986)	Alteration of tephra to >85% clay	>500 ka^{-1}
Papua New Guinea	Haantjens & Bleeker (1970)	Alteration of volcanic rocks	
		—skeletal	~5 ka^{-1}
		—smectite	5–20 ka^{-1}
		—kandite	> 20 ka^{-1}
Brazil	Kronberg *et al.* (1979)	(1) Rate from 100 ppm river solute load	50 m Ma^{-1}
	Kronberg *et al.* (1982)	(2) Bauxite formation	~2 m Ma^{-1}
Uganda	Trendall (1962)	Ferricrete formation	0.3 m Ma^{-1}
Senegal	Nahon (1977)	Nodular ferricrete	~2–3 m Ma^{-1}
		—100 m profile	10^7 years

TABLE 2.7   Means and extremes of weathering rates from Table 2.6 (A) and (B)

	Minimum	Mean	(Median)	Maximum
Table 2.6(A)	2	20.8	(20)	48
Table 2.6(B)	2	22.5	(13.75)	50

method is used (from Tables 2.6 (A) and (B) the figures obtained have a similar range and similar mean values (m Ma^{-1}), as shown in Table 2.7. When it is considered that the figures in Table 2.6(B) are for different measures of weathering as compared with Table 2.6(A), this is surprising. If comparisons here are not spurious, then maximum figures for the long-term weathering of crystalline rocks appear to lie in the range 20–50 m Ma^{-1}. This is sufficient for most observed profiles (where other age criteria do not apply) to have been formed during the Quaternary. On the other hand, where lateritic or gibbsitic residues are involved, there is a small group of estimates between 2 and 9 m Ma^{-1}, and since most laterite profiles attain thicknesses of >20 m, their periods of formation may need to be >10 Ma, and if the lower estimate is used for profiles of ~50 m, then the period of formation would be 25 Ma. Kronberg *et al.* (1982) agree with this estimate but recognise that the lateritic bauxites on the Brazilian plateaus may have been evolving since the opening of the South Atlantic towards the end of the Cretaceous.

Such estimates, however, need to be treated with extreme caution. One reason for this is that, in the case of Table 2.6 (A), all the figures are for present-day rates of saprolite formation, although they have been compared with geological constraints over long-term rates. The chemistry of ground and river waters today will reflect, not only the present state of evolution of relief and weathering profiles, but also the prevailing bioclimate. Past conditions will have been very different both in terms of rainfall inputs and also hydraulic gradients which decline with relief values.

Possible rates of weathering in the humid tropics

What these figures do not indicate are possible rates of weathering for perhumid lowland tropical rainforest regions, where soil temperatures are two to three times those of the temperate areas, and rainfall amounts are two to five times greater. If we make a crude assumption that increased base flow will accelerate the weathering rate by a factor of two, and that this might be increased to three by the temperature increase, then 60–150 m Ma^{-1}

of saprolite might be formed. The higher figure of 150 m appears high when compared with published accounts and evidence, particularly since figures so far obtained from tropical regions do not exceed 60 m Ma^{-1} (Thomas, 1994).

When it is realised that geological, tectonic, topographic and other factors all complicate climatic comparisons, the tenuous hold which we presently have on rates of weathering and saprolite formation will be evident. It is tempting to accept a higher rate of weathering from some wet tropical areas such as southeastern Brazil, where 50–80 m of weathering is not uncommon at many different levels in the landscape, from the low convex hills of the coastal zone, to the intermediate and higher plateau levels inland. According to Ab'Saber (1991) and Lichte (1989), aridity during the last glacial maximum was sufficient to remove much of the saprolite, during the formation of transported mantles and stone-lines (see Chapter 8). Such a claim could not be applied to the formation of very deep profiles, since they would require rates of decomposition two orders of magnitude higher than present calculations suggest are reasonable. It is, however, interesting to reflect that a rate of even 1 mm year^{-1}, which does not (intuitively) appear impossible, would imply the formation of 1000 m of saprolite during 1 Ma, but would only form 10 m during the Holocene (10 ka).

The possibility of such very high rates of weathering appears limited, however, by two sets of field evidence: (1) widespread mantle thicknesses of more than 100 m are rare; (2) long-term erosion and sedimentation rates derived from off-shore sedimentary volumes, do not support correspondingly high rates of removal of saprolite which could account for the surviving profiles.

In subsequent discussions, the possibility of rates exceeding a figure of ~60 m Ma^{-1} for humid tropical regions will be kept in mind, but, so far, we have no clear evidence in support of this idea. One factor that should be noted, however, is that during the last 0.5 Ma, only a part of this period was humid tropical in many areas, and periodic (glacial) desiccation might require a reduction of the long-term weathering rate (thus the 1 mm year^{-1} might be reduced to 0.5 mm year^{-1} or 500 m Ma^{-1}

which is still very high). Where thick bauxite deposits or lateritic ferricretes are associated with deep and complex weathering profiles then a time scale of 10–25 Ma should still be considered.

Where thick ferrallitic mantles occur widely, as in parts of central Amazonia, river water is found to contain very few solutes (see Chapter 5). Weathering rates in these circumstances must be very slow. On the other hand, the high solute load of the Andean tributaries to the Amazon indicates rapid weathering (Stallard & Edmond, 1983, 1987). It seems likely therefore that reaction rates become slower as rocks are progressively altered, and as relief declines and mantle thickness increases. However, when plains mantled with ferrallitic weathering products are uplifted, hydraulic conditions are changed and weathering attack at depth will become renewed. Such sequences probably help account for the complex weathering profiles that underlie many old landsurfaces in the tropics.

3

Weathering Profiles and Saprolites

THE FORMATION OF THE WEATHERING PROFILE

The physical properties and profile characteristics of saprolites depend critically on the type and degree of alteration by chemical weathering processes. Yet no agreed system of equivalence between chemical and physical properties has so far been formulated. Some attempts have been made to relate degrees of weathering, as indexed by chemical properties (Chapter 2), to engineering characteristics, but this field is not well developed or documented (see Geological Society, 1990). In its simplest expression the weathering profile will exhibit a progressive increase in the degree of alteration from the weathering front towards the ground surface. Unfortunately, however, this situation can be complicated by a variety of conditions:

(1) in banded or stratified rocks more susceptible lithologies may occur at any depth below the surface and become preferentially altered;
(2) the most intense or aggressive weathering may take place at some depth, controlled by the groundwater regime;
(3) most deep profiles have been affected by uplift or by depression, both of which affect the water-table levels and groundwater regimes such that the locus of intense weathering may have shifted within the profile over time;
(4) climate and vegetation changes during the progress of weathering may superimpose the effects of one regime on those of the previous one;

(5) changes in the topographic characteristics of the ground surface, due to periods of dissection or planation for example, will alter the *catenary position* of the site and therefore its near-surface weathering environment;
(6) the catenary position will also strongly influence the *lateral* inputs and outputs of solutes and the accumulation of neoformed clays and other compounds.

The effects of surface translocation and disturbance of the upper horizons of the profile are not included here. These effects may be very important, but they do not relate to the *in situ* saprolite mantle.

If the foregoing qualifications are accepted, then a simple progression of weathering characteristics up-profile from the weathering front will only occur where geology is more or less uniform and the site history has not led to any reversals of weathering trends. In such cases the expectation might be that the degree of weathering up-profile will mirror the phases recognised by Duchaufour (1982) and Pedro (1983, 1984). In other words, the vertical profile will contain at different depths the characteristics of fersiallitic, ferruginous and ferrallitic soil materials. This is shown in Figure 3.1(A) (which should be read alongside Figures 1.3 and 2.1), but it will be noticed that all these categories together only describe the completely weathered rock (Zone V), and that four engineering weathering grades occur at greater depths. None of these is adequately described by the pedological literature, mainly

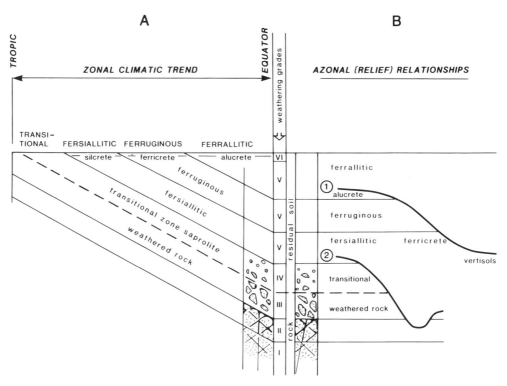

FIGURE 3.1 Schematic diagram to relate climatic weathering stage (adapted from Strakhov, 1967 and Figure 1.3) to the vertical profile and weathering grade (A). In (B) two hypothetical slopes cut into a pre-weathered profile are illustrated to show how different weathering grades may outcrop across the slopes (drawn by the author for Geological Society, 1990. Reproduced by permission of the Geological Society Publishing House)

because they generally lie at depths well below the *solum* as analysed by soil scientists.

Conditions of weathering at depth

At these greater depths important changes in the environment of weathering occur:

(1) the saprolite profile contains an increasing proportion of unweathered rock fragments;
(2) in many profiles the deeper levels will lie below the water-table for most if not all of the year, inducing saturated conditions;
(3) the deeper levels may lie beyond the reach of metastable biochemical compounds that are active in the zone of residual soil;
(4) water movement and the flushing of ions in solution may be slower, due to reduced

permeability or confinement of local aquifers in the saprolite. This will cause a rise in pH.

Not all of these conditions necessarily prevail at depth, and this is especially true in areas of accentuated relief. However, if they do then several consequences follow. First, the pH may rise because of the lack of proton donors and the concentration of bases near the weathering front and within slow moving solutions. These conditions can mobilise Al^{3+} and Fe^{2+} (at $pH > 8.5$) at the site of weathering, but equilibrium concentrations will quickly develop and the stability fields will favour the formation of sericite from plagioclase and chlorite from ferromagnesian minerals, possibly in company with smectites or illite. Reactions will continue only if dissolved ions can be evacuated from the lower profile. The existence of very deep weathering profiles, in excess of 100 m thick

(Plate 3-I), requires some mechanism for the continuation of weathering reactions to considerable depths.

If water can move laterally or vertically to progressively deeper levels then there is no serious problem, and despite the confining pressure of rocks at such depths, some water flow undoubtedly continues. But it takes place along rather few open fractures or within shatter zones, and it must respond to the hydraulic gradient which in turn depends on some external relief control, such as deep valley dissection or proximity to an escarpment zone.

The process of ionic diffusion

In the absence of such features, the renewal of water at depth is at best very slow. However, a strong gradient of concentration is set up within the groundwater as a result of the situation described here and this may permit an upward diffusion of ions in solution. This *ionic diffusion* was proposed by Lelong (1966) and reported in an earlier study (Thomas, 1974a). Since then it has also been studied by Mann and Ollier (1985) and strongly advocated as an important mechanism in weathering by Ollier (1988a).

As proposed by Lelong (1966), ionic diffusion arises as a consequence of the gradient of concentration of dissolved ions in the groundwater, and was linked to his observations of groundwater behaviour in four experimental wells at Parakou, Benin (formerly Dahomey), where rainfall is 1200 mm year^{-1}, concentrated between May and October. In a saturated regolith, the dissolved ions close to the weathering front were thought to rise upwards in the profile by diffusion, responding to the lesser concentrations within the weathering solutions at the water-table itself, where lateral flow and renewal of groundwater takes place frequently. Mann and Ollier (1985) have claimed that reduced Fe^{2+} may be transferred upward in the profile by this process, with oxidation and precipitation as Fe^{3+} within hydrated oxides occurring at the water-table. According to these authors this process can form ferricrete at a rate of 1 m year^{-4}. However, this remains a controversial mechanism

for ferricrete formation, and the range of depths over which ionic diffusion transport can occur remains unknown.

Down-profile changes in saprolite mantles

It is obvious that saprolites will contain increasing amounts of unweathered material down-profile. The manner in which this is presented to the observer depends very much on the lithology of the parent rock, and probably also on the weathering history of the site. It is the author's view that, where rapid weathering is encouraged by a strong hydraulic gradient, as in hilly areas with high rainfall, thick zones of partially altered saprolite will commonly occur within weathered compartments, except where slopes are very steep, generally >25 °. These often form the multi-convex *demi orange* (French) or *meias laranjas* (Portuguese) hills, typical of such areas (Figure 3.2). In this situation, the following are examples of the transitional zone properties (Zones 3 and 4, Figure 2.1).

(1) Massive granite or migmatite may exhibit a sharp weathering front, appearing as a basal surface with transition to highly weathered (Zone 4) characteristics over a distance of 1–3 m (Plates 2-I and 2-II).

(2) Jointed granite and granodiorite commonly present an increasing proportion of corestones in the profile with depth, and the weathering front is abrupt around each boulder as well as along an irregular basal surface at depth (Figure 2.1; Plates 3-II to 3-IV).

(3) Fissured granites (with dense microcrack systems) and other fissile rocks will show a gradual transition towards partially altered 'gruss', with individual feldspar crystals separated as a gritty sand forming a matrix with a low (2–10%) clay content.

(4) Banded rocks such as schists may display a highly irregular profile, as decomposition follows bands of dark minerals to much greater depths than in micaceous or quartzitic layers. A sharp basal surface is seldom present.

(5) Stratified sedimentary rocks can display complex features depending on the lithological

FIGURE 3.2 Multi-convex, half orange (*demi orange*; *meias laranjas*) relief on granite about 1500 m in the forest/grassland zone of West Cameroon. Some outcropping tors or corestone groups can be seen

sequence and structural arrangement; most clastic sediments are predominantly quartz which is slow to weather and the profile may reflect the dissolution of cementing materials including $Fe_2O_3nH_2O$; $CaCO_3$; and amorphous or cryptocrystalline SiO_2 which can go into solution more readily than quartz.

Limestone weathering is not explored here, because it is not generally associated with a saprolite profile; the rock being almost completely dissolved by the weathering process and leaving only a thin layer of residual clay which is commonly washed down into enlarged joint cavities (Figure 3.3).

Fissured rocks and microcrack systems

Where the rock minerals are highly fissured or contain many fluid inclusions, and also in coarse granitoid rocks with abundant plagioclase and mica, weathering penetration may take place throughout the rock mass, leading to forms of early granular disintegration. Analyses of such 'gruss' materials show that clay formation has hardly begun, but hydration and expansion of feldspar crystals together with early weathering of biotite have disrupted the rock fabric. In this situation no marked basal surface of weathering appears. It is also these rocks which commonly weather to greater depths. Wherever shatter zones with concentrated faulting and jointing occur, a tendency for greater weathering depths and an indistinct lower boundary will also be evident.

In granites, microfracturing is consequent upon the opening of grain boundaries, and the development of intergranular porosity, partly due to the formation of large solution features in the feldspars, has been widely reported (Baynes &

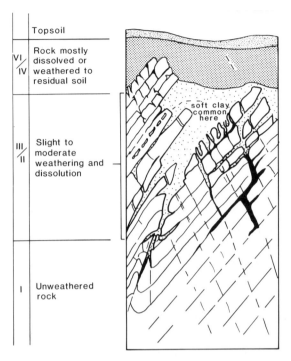

	Topsoil
VI / IV	Rock mostly dissolved or weathered to residual soil
III / II	Slight to moderate weathering and dissolution
I	Unweathered rock

soft clay common here

FIGURE 3.3 Typical weathering profile on carbonate rocks. The occurrence of a saprolite zone (V) is restricted to impure sandy or silty rocks, while Zone IV characteristics occur mainly in chalky limestones. Soft clays often occur as cavity fillings and are not strictly *in situ*. (Reproduced from Geological Society, 1990, by permission of the Geological Society Publishing House; adapted from Deere & Patton, 1971)

Dearman, 1978; Collins, 1985). According to Vaughan and Kwan (1984) weathering is effectively a weakening process leading to the formation of a large number of holes in the parent rock. These holes are therefore due in part to the removal of ions in solution, leaving cavities in the original minerals.

Microcrack (microfissure, microfracture) intensity is sometimes expressed as an index, such as the *microfracture index* employed by Irfan and Dearman (1978) which uses the number of microcracks seen in a 10 mm traverse of a thin section. It has been claimed (Ehlen & Zen, 1986) that microcracks are more frequent in igneous rocks with important admixtures of mafic minerals, particularly where these form isolated clumps. This is thought (Gerrard, 1988) to result from the different thermal expansions and compressibilities of pyroxenes and plagioclase feldspars—the former shrink more on cooling and expand less on decompression. Such differences set up stresses within the the rock fabric when the rocks are cooled and also, later, when unloading of country rock takes place. When stress is released along grain boundaries and other small-scale discontinuities, then microcrack systems spread through the rock mass, rendering it susceptible to water entry and to chemical weathering.

The nature of the weathering front

Studies of alteration at the weathering front have been few, and in some situations this feature is not readily recognised. Yet, in others, it occurs as a sharp discontinuity or 'basal surface', which is functionally important in marking a major change in erosional mobility (Figure 3.4). The control here is probably lithological; rocks with minimal porosity tending to decompose thoroughly across a narrow band of saprolite formation. Nossin (1967) found that thorough decomposition of granite around corestones on Singapore Island was accomplished within 50 cm of the fresh rock. However, the detachment of the corestone during weathering promotes the leaching of ions formed in solution, allowing the weathering reactions to proceed along the pathways established above. Similar sharp transitions are often found where massive igneous rocks undergo sheeting due to pressure release (*dilatation*), which of course is also responsible in part for creating joint blocks from which corestones are formed. These cases relate to rocks within which microfissures or other weaknesses in the fabric are largely absent, leaving the soil solutions to penetrate only the larger open joints which may include partings due to sheeting (Plate 3-V).

It is tempting also to suggest that a sharp weathering front will typify weathering environments with free drainage which promotes the advance of weathering reactions. This often applies to up-slope positions, away from depressions and valley floors. But, unfortunately, there is little good evidence to prove or disprove this

FIGURE 3.4 Sharp interface at a weathering front over gneiss in a southern Nigerian forest zone. The almost flat rock surface on the left is part of a natural exposure emerging as the rotten rock is removed slowly by erosion

view, partly because most excavations occur beneath hillsides (road traces do not need to excavate valley floors), and because drilling logs from other sites are often imprecise in this respect.

Some specific observations may be helpful here. Sharp weathering fronts, associated with sheeting, are well exposed in deep-seated intrusive granites (usually Proterozoic or Archaean) and in some granitised gneiss. In these cases they are associated also with large domal outcrops or *bornhardts*. These are found around Rio de Janeiro (migmatised gneiss), Idanre in Nigeria (Archaean granite), the Gbenge Hills in Sierra Leone (porphyroblastic Archaean granite), and many places in India. Similar features are found in the Freetown Peninsula basic igneous complex which contains gabbro. Sharp weathering fronts are also associated with some low relief areas, where there is relatively shallow weathering and low, convex outcrops occur.

This is seen clearly in the granites above the Darling Scarp, on the weathered landsurface inland from Perth, Australia (Plate 2-II), and in many parts of southern Nigeria over the Precambrian Basement gneiss and migmatite. Such areas will be described later as *stripped etchplains* (Mabbutt, 1961).

THE PHYSICAL PROPERTIES OF SAPROLITES

The weathering of crystalline rocks to form kaolinitic residues leads to a breakdown of the rock fabric and to a loss of cohesion. The physical properties of the saprolite and residual soil will commonly be in stark contrast to those of the original rock. Some of these properties which are important to geomorphological processes and to land stability are considered here.

Clay mineralogy

Clay mineralogy is important, because the amount and type of clay minerals present will greatly affect the mechanical properties and behaviour of the saprolite. Most clay minerals are platey and have a low coefficient of inter-particle friction. They can orientate under shearing, resulting in polished shear surfaces and a low residual frictional strength (Lupini *et al.*, 1981). If smectite clays are present then expansion and shrinkage with the formation of deep desiccation cracks, take place with varying moisture and stress conditions.

Particle-size composition

Particle size (*grading*) is a property most widely used to describe sediments, but it is also used on saprolites. However, in materials with a high clay-sesquioxide content, the pretreatment of the sample can greatly influence the experimental results, especially in respect of the clay-size percentage ($<2\,\mu m$) and this is due to the presence of micro-aggregates (see below) which may not be completely broken down. This makes comparisons between different accounts difficult and sometimes misleading.

The rise in clay *mineral* content with the advance of weathering has already been documented in Chapter 2, and if translated into clay-sized particles this would produce values of $>60\%$ in mafic rocks. But in zones of active weathering and erosion, the clay-sized fraction may be $<10\%$ in granites, within 5 m of the ground surface (Lumb, 1975), so that the saprolite is predominantly silt and sand and its mechanical properties are controlled by the dominant particle size which may be coarse sand in many granites, and silt in volcanic rocks. In general, regolith with $<15\%$ clay behaves as a granular material, but when the proportion rises above 40% its properties are dominated by the clay (Geological Society, 1990).

Voids ratio

The voids (or void) ratio is seen as a key property in engineering studies and is equally important to

TABLE 3.1   Typical void ratios of Brazilian residual soils

Parent rock	Unit weight of grains (g cm^3)	Voids ratio
Gneiss	2.60–2.80	0.30–1.10
Quartzite	2.65–2.75	0.50–0.90
Schist	2.70–2.90	0.60–1.20
Phyllite and slate	2.75–2.90	0.90–1.30
Basalt	2.80–3.20	1.20–2.15

From Geological Society (1990), after Sandroni (1985), reproduced by permission of the Geological Society Publishing House.

the operation of natural erosion processes. Although it may vary in saprolites derived from different rocks, it is more a function of the weathering stage. Sandroni (1985) has given typical void ratios for residual soils in Brazil (Table 3.1).

Although Lumb (1962, 1975) has shown an *increase* of void ratio towards the ground surface through weathering profiles in granite and in volcanic rocks from Hong Kong (Figure 3.5), Prusza *et al.* (reported in Geological Society, 1990) found that the void ratio *decreased* markedly towards the surface at the Guri Dam in Venezuela. Such variations with depth were also discussed by Radwan (1988). In fact Lumb's (1962) own results show a very wide scatter when plotted against the degree of decomposition (X_d, formula (2.22)), and when X_d approaches 0.8–1.0 values decline sharply (Figure 3.6). Clearly this property is influenced by factors other than grading, including the formation of aggregates and of Fe segregations leading to nodular ferricrete (Chapter 4).

Structure and inter-particle bonding

Structure and inter-particle bonding are linked: the former refers to the presence of aggregates and pores dependent on properties other than particle size and void ratios, and commonly includes large macropores (pipes, termite passages); the latter encompasses charge bonding and the precipitation of cementing compounds during weathering, an extreme case being the Fe_2O_3 (as goethite or haematite) in duricrusts. These physical properties confer a measurable *yield stress* to otherwise non-cohesive soil materials, and this may be destroyed or greatly reduced by any remoulding of the

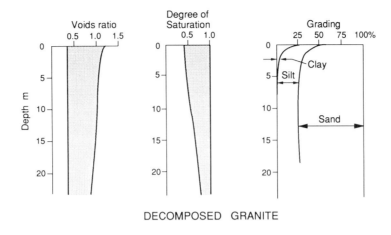

DECOMPOSED GRANITE

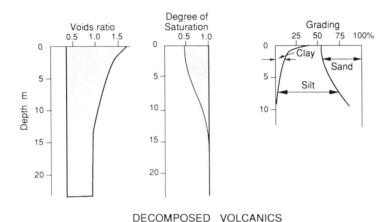

DECOMPOSED VOLCANICS

FIGURE 3.5   Bulk properties of residual soils in Hong Kong (after Lumb, 1975; reproduced by permission of the Geological Society Publishing House)

material. Thus most saprolites exhibit a *peak shear strength*, which implies that there is a discontinuity in the stress–strain behaviour, such that the material will become 'destructured' once this stress is exceeded, and the porosity will revert to that appropriate to the unbonded particle-size distribution. Measured yield stress is test dependent, but Vaughan (in Geological Society, 1990) listed a number of cases, that indicate maximum figures ranging from 150 to 1400 kPa, and with a mean of 424 kPa.

An interesting example of this effect has come from a study of weathering in overconsolidated mudrocks in Barbados (Fan *et al.* 1994). Weathering of the scaly-clay fabric with its preferred orientation initially leads to the formation of a more open fabric having increased pore volume and greater shear strength. But with increased intensity of weathering the material behaves as if remoulded and shear strength is reduced. The authors considered that the effect of weathering on steep slopes in Barbados was to increase their stability, and this observation may be of much wider significance.

Clay aggregates and permeability

The presence of amorphous Al and Fe sesquioxides in Zones IV and V of tropical weathering profiles contributes to the aggregation of clay particles. The

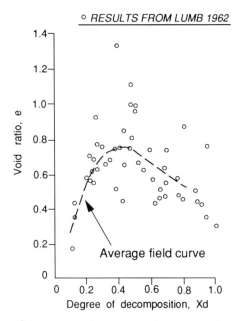

FIGURE 3.6   Observed void ratios in weathered granite as related to the degree of decomposition (*Xd*, see text) (adapted from Vaughan & Kwan, 1984; reproduced by permission of the Geological Society Publishing House)

negatively charged kaolinite attracts the positively charged sesquioxides which form a coating to the clay mineral, sometimes cementing grains together to form aggregates (Geological Society, 1990). This coating also reduces the ability of the clay minerals to absorb water, while at the same time the aggregates promote high permeability.

Microfabric analysis

It is recognised that the *microfabric* of a saprolite or residual soil will reflect the duration and intensity of weathering at that site (there is commonly a high variability of weathering microenvironments within a small area or volume of rock). In particular, it will reflect feldspar decomposition, clay formation, and degree of eluviation, but the influence of iron oxides as binding substances in cementing clay particles into aggregates is also extremely important in tropical soils. In the near-surface soil, aggregates are also stabilised by organic matter.

Fabric organisation involves the arrangement of clays and unaltered mineral grains (mostly quartz). Collins (1985) recognised three levels of organisation:

I   The Elementary Level of particle arrangements
II The Assemblage Level of particle aggregates and matrices
III The Composite Level of heterogenous and anisotropic 'pedal' microfabrics

These properties are illustrated in Figures 3.7 and 3.8. A pedologist's approach to this question is illustrated by the work of Oades and Waters (1991) who recognised macro-aggregates ($>250\,\mu$m), micro-aggregates (between 20 and $250\,\mu$m), and particles ($<20\,\mu$m). These macro-aggregates were maintained by living roots and fungal hyphae, often clustered around plant debris and stabilised by microorganisms, but these observations relate to soil rather than to saprolite.

The arrangement of clays into *random*, *parallel* and *clustered* fabrics is particularly important in determining the engineering properties of saprolites, while subdivisions of these categories such as tightly packed, 'open card house' and 'porous networks' may determine voids ratios (and size of pores) and the susceptibility of the saprolite to collapse under loading. The tendency for clays and grains to form aggregates at a higher level of organisation is equally important. Collins (1985, p. 87) has observed that 'porous aggregate textures seem to be associated with the presence of sesquioxides which coat clay internal surfaces, binding them together'. He noted that rhyolite lavas near Curitiba in Brazil had weathered to an open spongy texture 18 m above the fresh rock, reflecting *in situ* mineral decomposition and the aggregation of clay-sized particles. These aggregates act as grains with rough surfaces, such that the weathered lavas exhibit high voids ratios and high liquid limits despite their small particle size. These features impart low compressibility, high permeability and high angles of shearing resistance to materials high in clay percentage (Collins, 1985). On the other hand, the presence of aggregates such as the tubular forms of halloysite may be responsible for the high *sensitivity* of many tropical soils to drying and to manipulation during construction or excavation.

The *role of iron oxides* in binding the clay particles is widely acknowledged, and Rao *et al.* (1988) have demonstrated this effect from studies

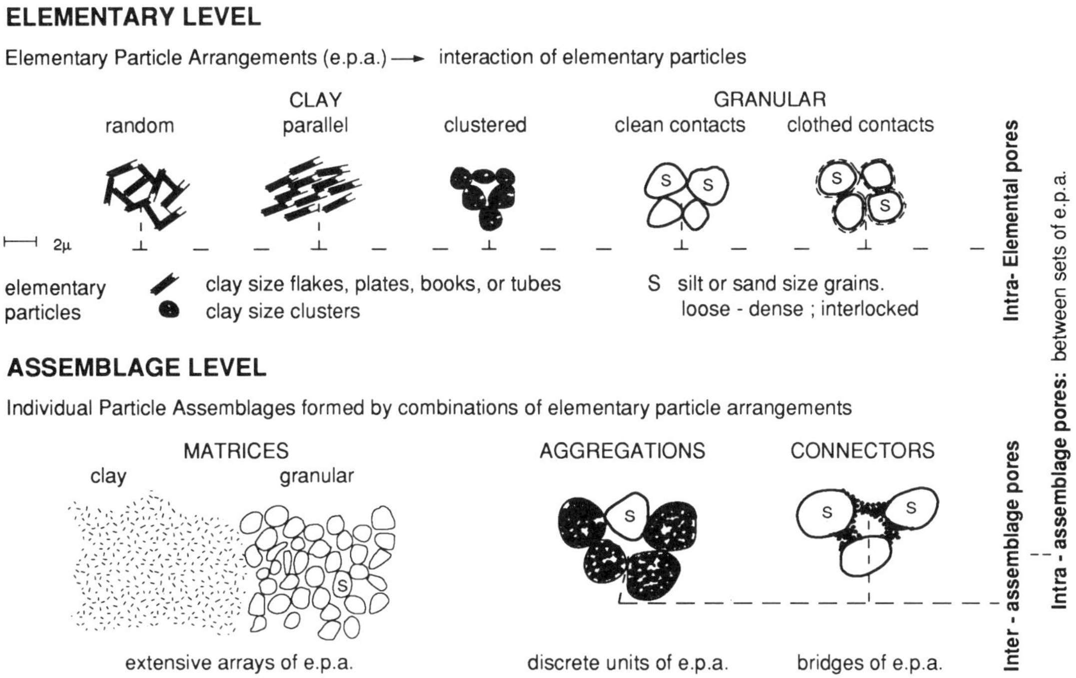

ELEMENTARY LEVEL

Elementary Particle Arrangements (e.p.a.) ⟶ interaction of elementary particles

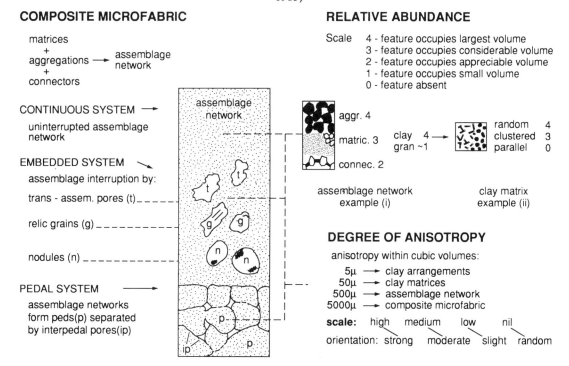

FIGURE 3.7   Common particle arrangements and aggregate structures found in tropical residual soils (after Collins, 1985)

FIGURE 3.8   Composite diagram illustrating tropical residual soil microfabrics (after Collins, 1985)

TABLE 3.2   Index properties of clays before and after removal of iron oxides

Soil	SGª	Grain size %			Liquid limit	Plastic limit	Plasticity index	Shrinkage limit
		Sand	Silt	Clay				
K	2.57	14.0	19.5	66.5	80	35.3	44.7	21.9
K'	2.50	10.0	11.5	78.5	66	31.3	34.7	16.7
M	2.79	35.0	17.0	48.0	132	41.0	91.0	11.8
M'	2.71	23.0	16.0	61.0	145	52.7	92.3	9.0

ªK is untreated kaolinitic red soil and K' is the same soil after removal of iron oxide; M is untreated montmorillonitic soil and M' is the same after iron oxide removal (after Rao *et al.*, 1988; reproduced by permission of Balkema).

in Karnataka State, India. Samples of both kaolinitic and montmorillonitic (smectite) soils were tested before and after treatment with dithionite. The results are given in Table 3.2.

It can be seen that removal of the Fe_2O_3 leads to an increase in measured clay percentage in both samples. In the case of the kaolinitic material, liquid and plastic limits are reduced as the void spaces become smaller, and the shrinkage limit also falls due to the loss of aggregates and less random particle orientation. On the other hand, the smectite residue, while also increasing in clay percentage, shows a rise in plastic and liquid limits. This is probably due to the more consistent orientation and exposure of the double clay layer, leading to inter-particle repulsion as a consequence of the electrical forces present in these clays, and to the enhanced adsorption of water. Thus in kaolinitic soils loss of the iron lowers the voids ratio and compressibility, while in montmorillonitic (smectite) soils both voids ratio and compressibility are increased.

Heterogeneity in weathered materials is also widely observed, and this is not tested by indices of weathering. Relict discontinuities are inherited from the parent rock and may persist in highly weathered saprolites, even where they are not visible to the naked eye. Also, both macro and micro stuctures in weathered materials vary widely, influencing the soil matrix strength (Irfan, 1988). The real problem identified here is that conventional chemical and mechanical tests can fail to detect many essential elements of residual weathered materials.

Permeability

The permeability (and other hydraulic properties) of saprolites may turn as much on the presence of macropores and cracking as on the grading properties, and can be high in tropical saprolites such that saturated conditions and surface runoff are rare. Materials with the grading characteristics of a clay may have measured permeabilities of 10^{-4}–10^{-5} m s^{-1}. The curve of permeability with depth (Figure 3.9) may show high values in Zone VI (soil A and B horizons), low values in the upper part of the saprolite where high clay content is found (Zone V), gradually increasing through Zone IV, possibly reaching a maximum near the Zone III/II boundary, where it is enhanced by the weathering of fissures and the coarseness of the little-altered regolith (Jones, 1985; Acworth, 1987).

Collapsible and expansive tropical soils

It follows from this discussion of multi-level pore systems (Collins, 1985) that random or 'cardhouse' arrangements of clays, and the existence of 'bridge structures' of Fe_2O_3 or other cementing material between quartz grains (Figures 3.7 and 3.8), must be subject to realignment or collapse under 'critical' loads. This will commonly occur when the material becomes softened by wetting under load. The importance of Fe-sesquioxides and colloidal clays as weak cements has also been emphasised in studies from Brazil (Vargas, 1990), where collapsible soils occur in porous ferrallitic layers, commonly

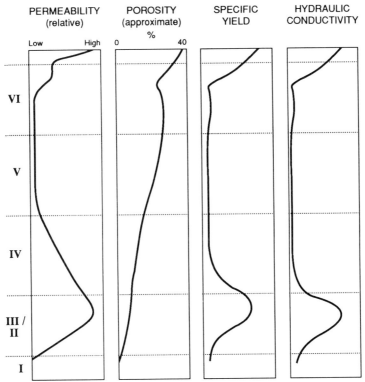

FIGURE 3.9 Typical curves for saprolite properties and hydraulic behaviour, based on experience with Basement Complex rocks in Africa. Weathering grades as used in Figure 2.1 are shown I–VI; thickness may be highly variable (Zones II + III are commonly 1–20 m; Zones IV and V may range from 1 to 30 m; Zone VI is usually less than 15 m). Hydraulic conductivities may range from 10^{-2} to $10 \, \text{m day}^{-1}$ (derived from Jones, 1985, and Acworth, 1987; reproduced by permission of the Geological Society Publishing House)

developed on weathered basalts, but are also observed in colluvium and by the pedogenic transformation of Tertiary clay beds.

Considerable settlement of such soils was detected where clays and oxides can be dissolved or eluviated from between sand grains. It is therefore the pedogenic processes of leaching and eluviation (*lessivage*) that can lead eventually to collapse, and these processes are most active in the upper profile, dominated by kaolinite and gibbsite. Below this level, clay *illuviation* has frequently stiffened the soil materials, while within the saprolite leaching has been less severe and voids ratios are lower.

A *collapse potential formula* has been based on the change in voids ratio on saturation:

$$\frac{\Delta e}{1 + e_0} \times 100\% \qquad (3.1)$$

where Δe is the change in voids ratio on saturation and e_0 is the initial voids ratio of the sample (see Geological Society, 1990).

In soils rich in smectite (montmorillonite) swelling and shrinkage are generally much greater problems than settlement due to microfabric collapse. The swelling characteristics may be predicted from the Atterberg Limits; swelling increasing with increasing liquid limit and plasticity index (Geological Society, 1990). The shrinkage limit, on the other hand, indicates the moisture content at which no further shrinkage takes place.

In addition to the existence of collapsible microstructures in residual soils, the widespread presence of colluvium susceptible to these processes can be a problem. The largely structureless and homogenous, fine clayey sand so widely encountered as colluvium (commonly 5–15 m thick) may have a very high *compressibility ratio*. This was found to lie between 72% and 90% in the colluvial soils covering as much as 50% of the land area in Sao Paulo State (Ferreira & Monteiro, 1985). These materials were mainly derived from sandstones and are associated with high voids ratios and porosities; they comprise 20–25% clay, 10–20% silt and 60% sand, and their Fe_2O_3 content is around 10%.

The influence of initial voids ratio on magnitude of collapse was also demonstrated by Ferreira and Monteiro (1985), who found that remoulding of the soils contributed to greater collapsibility. A typical undisturbed sample with a voids ratio of 0.9 was found to collapse by 5–6%. In Java, Heath and Saroso (1988) recorded many landslide failures in colluvium derived from volcanic breccia and ash deposits which had both a high (60%) smectite content and a porous, collapsible structure (voids ratios 0.8–1.2).

The leaching of Fe_2O_3 cements will reduce the shear strength of the soil mantle, while the collapse of the microfabric may reduce the permeability of the material. Such changes may promote instability on slopes, but will also lead, on level surfaces, to settlement of several per cent (5–6% has been measured). Such settlement may be uneven and have potentially serious results for the stability of building structures.

Dispersive and erodible soils

Dispersion refers to the deflocculation of clay in the presence of relatively pure water. Soils with a high exchangeable sodium percentage (ESP) are usually dispersive and liable to rapid erosion on saturation (Stocking, 1981a,b). This is also reflected in a strong relationship between ESP and liquid limit. The reason for this is the high activity of the Na^+ ion in the presence of ionised water. Dispersive soils are commonly associated with smectites and non-crystalline allophanes, but soils containing high amounts of halloysite (though not kaolinite) are also susceptible. Fundamentally, conditions for the formation and retention of Fe-cemented aggregates and for the dispersion of clays with high ESP values lie at opposite ends of a spectrum of weathering environments. Dispersive clays are associated with

the formation and collapse of soil pipes (Chapter 5) and with gully erosion (discussed below).

Low-strength discontinuities

The properties of clay-rich saprolites are modified by the presence of low-strength discontinuities. These derive from structures in the parent rocks, such as joints, faults, foliation and mineral banding, and may be coated with low-friction iron and manganese compounds. These in turn can reduce the angle of shearing resistance from perhaps 22–30° for kaolinitic soils to 15–20° in the unsheared state, but after shearing this can drop to 10° if slickensiding (polishing of the sheared surfaces) has occurred. The most important of these discontinuities in terms of land stability are:

(1) planes of schistosity in schist-derived soils (De Freis reported by Deere & Patton, 1971);
(2) joints parallel or sub-parallel with steep slopes and also relict joints within the saprolite (Koo, 1982a,b);
(3) the weathering front itself, along which there may be rapid changes in the properties of the materials parallel with the slope;
(4) boundaries between soil horizons which possess different texture, structure and permeability, including buried soil B horizons; and also
(5) the presence of old shear planes from previous landslides (Nilsen & Turner, 1975).

Joint planes (2) may be coated with thin films of manganese or iron oxide, and more open partings often contain clay fillings. This is particularly common in deep-seated intrusive rocks that have been subject to unloading and dilatation. Outer sheets of parted rock from 0.3 to 2 m thick may become separated from the main rock mass, facilitating the movement of water clay and other particles. Such sheets also become subdivided into blocks that weather to form a regolith cover on slopes that may exceed 40°.

Relict discontinuities may also act as preferential seepage paths, and according to Irfan and Woods (1988) this can lead to the development of high pore water pressures, where inflow exceeds

65

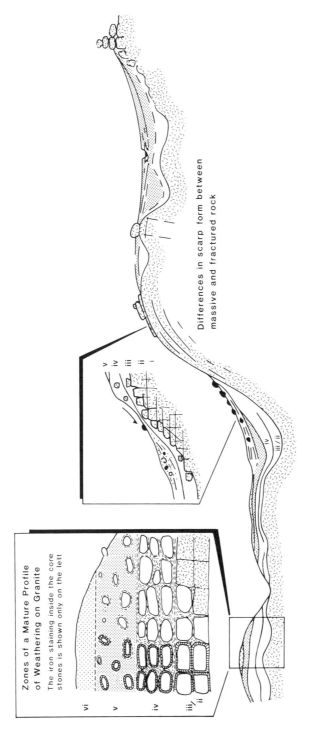

Zones of a Mature Profile
of Weathering on Granite

The iron staining inside the core
stones is shown only on the left

Differences in scarp form between
massive and fractured rock

FIGURE 3.10   Weathering zones in granitoid rocks across a stepped (two-storey) landscape (Geological Society, 1990, adapted from Ruxton & Berry, 1961, and field sources; reproduced by permission of the Geological Society Publishing House)

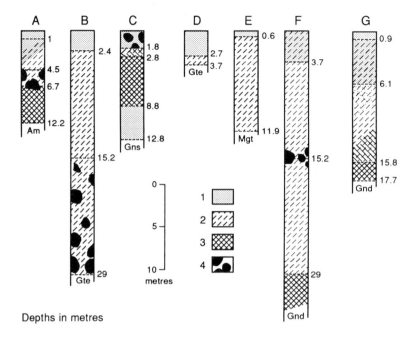

Depths in metres

Some characteristic weathering profiles over crystalline rocks in Nigeria

1. Sandy or clayey material without surviving rock structures.

2. Decomposed rock appearing as sand and clay but with visible structures largely intact.

3. Partially weathered rock : fractured and stained.

4. Unweathered or little altered core-stones.

Unaltered, massive rock is left uncoloured, but is identified :

Am ---- amphibolite

Gte ---- granite

Gns ---- granite-gneiss

Mgt ---- micro-granite

Gnd ---- grano-diorite

FIGURE 3.11   Examples of different weathering profiles recorded on the Basement Complex rocks of Nigeria in the vicinity of the Niger Dam (from Thomas, 1967; data compiled from Balfour Beatty & Co. Ltd, 1961)

transmissivity. These features play a central role in the generation of landslides.

WEATHERING DEPTHS AND PROFILE CHARACTERISTICS

The foregoing sections have explored the processes and progress of chemical weathering into rocks, and given some indication of the physical properties of these weathered materials. In the course of this discussion 'typical' weathering profiles have been illustrated, mainly in connection with the transition to fresh rock. But little attention has been given to the upper zones of profiles where *duricrusts* may occur. In addition, the patterns and distribution of weathering, and the evolution of weathering profiles through geological time require comment.

Characteristic weathering profiles

Ruxton and Berry (1957) proposed a fourfold zonation of the granite weathering profiles of Hong Kong, based on the proportions of corestones in a well-jointed granite (Figure 3.10), and this has become the basis of the scale of six weathering grades commonly used by engineering geologists (Geological Society, 1990; Figure 2.1). Roe (1951) observed that granites in Malaysia weathered, '(i) to form core boulders, (ii) to form clayey soil and soft rotted rock which is separated by a well defined line of contact from hard fresh granite, (iii) to give a gradual transition from clayey soil to fresh rock'. These categories are not limited to granites, but apply to a wide variety of, mainly crystalline, rock types. Some variations of this kind are shown in Figure 3.11, compiled from the report (1961) on the foundation conditions for the Niger Dams.

Deere and Patton (1971) illustrated a wide range of profile types encountered on different rocks, but there has also been an unfortunate concentration elsewhere in the literature on examples from granitoid rocks and on the so-called 'laterite profile', and this study cannot altogether shed that bias.

If the weathering profile is viewed against the classification of residual soils used in Chapters 1 and 2 (after Duchaufour, 1982; Pedro, 1968, 1984)

then the weathered rock (Grade III) and transitional zone saprolite (IV) are seen (ideally) to grade upwards into fersiallitic, ferruginous and ferrallitic/ allitic residues capped by either duricrust (alucrete, ferricrete, silcrete) or a soil B horizon (Figure 3.1(A)). This figure also shows that, where deep pre-weathered material has been formed, dissection of the weathered landscape will produce outcrops of saprolite of different weathering grade across the hillslope (Figure 3.1(B)). This has implications for modern soils, since these are developing on different zones of the weathering profile which have become the parent materials.

Duricrusted profiles

The sesquioxides (or oxyhydroxides) of iron and aluminium often become concentrated as crusts or '-cretes' on or within the weathering profiles of humid tropical or formerly humid tropical regions, while silica and calcium carbonate play a similar role in highly seasonal climates. The duricrust family arranged according to the mobilities of the dominant cations is:

$$\text{HUMID} \quad \rightarrow \quad \rightarrow \quad \rightarrow \quad \rightarrow \quad \rightarrow \quad \text{ARID}$$

alucrete or 'bauxite' (Al_2O_3)
   →ferricrete or 'laterite' (Fe_2O_3)
      →silcrete or 'porcellanite' (SiO_2)
         →calcrete or 'caliche' ($CaCO_3$)
           →gypcrete ($CaSO_4 \cdot 2H_2O$)

The formation of these materials is discussed in Chapter 4, but their appearance within weathering profiles should be noted here. A so-called 'standard' laterite profile has been described (Figure 3.12(A)) but this is a misleading concept, because of the wide variation amongst actual occurrences (Figure 3.12(B)). However, Bardossy and Aleva (1990) commented that, while horizon thicknesses and chemical compositions will vary from place to place, a 'typical' or 'average' profile may be described, in terms of the *succession* of horizons. This succession is described for a 'bauxite' profile in Table 3.3.

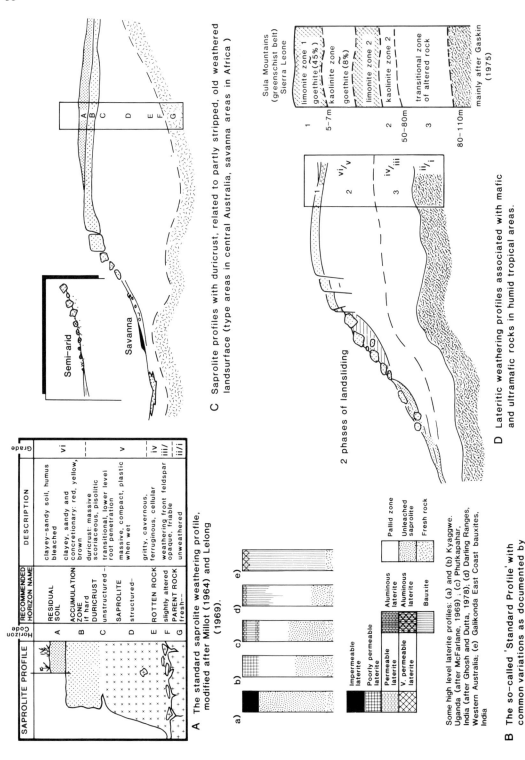

FIGURE 3.12 A range of duricrusted profiles from different tropical environments. (A) The so-called 'standard' laterite profile (see text); (B) common variations from the 'standard' laterite profile; (C) landscape associations in drier, savanna climates; (D) landscape associations in humid tropical areas. (Reproduced by permission of the Geological Society Publishing House from Geological Society, 1990)

TABLE 3.3   The bauxite profile

Horizon	Depth range (m)	Most frequent thickness (m)
Soil	0–2	0.3–1.0
Duricrust	0–4	0.5–2.0
Bauxite horizon	1–54	4.0–8.0
Saprolite horizon	0–100	10.0–30.0

After Bardossy and Aleva (1990). Reproduced by permission of Elsevier Science Publishers BV.

TABLE 3.4   The laterite profile

Horizon	Depth range (m)	Most frequent thickness (m)
Soil	0–4	0–2
Duricrust	1–25	1–10
Mottled clay	1–10	1–5
Pallid zone	0–60	5–25
Saprolite	1–60	10–30

Bardossy and Aleva (1990) described the surface soil horizon as 'residual', but in many cases this contains significant allochthonous material, as may also the duricrust which underlies it. The duricrust, in this context, constitutes a hard, mainly crystalline (probably aluminous) ferricrete. This, typically, is underlain by the bauxite horizon which may be highly variable internally (Plate 3-VI). However, it tends to be softer and paler in colour than the duricrust, and is commonly white, pink or ochreous in colour. The great range of thickness reflects the position and age of deposits. High-level bauxites, often regarded as older, tend to have much deeper profiles, as in the Eastern Ghats of India.

A transition from concretionary ferricrete/ bauxite to a bauxitised rock retaining its textures and structures in an isomorphous transformation is not uncommon, and similar examples of alteration to ferricrete (*primary laterite*) may occur over iron-rich rocks, such as the gabbro–norites of Freetown in Sierra Leone (Plate 3-VII).

With respect to duricrusted laterite (or ferricrete) profiles, many early authors have referred to a *mottled clay horizon* occurring beneath the duricrust and to a leached (and bleached) *pallid zone* below this. Nahon (1986), for example, described an idealized ferricrete profile containing a mottled clay layer (discussed in Chapter 4; see Figure 4.7), from the semi-arid zone of West Africa (Senegal). Millot (1970) provided an 'ideal' laterite profile of this kind which has been noted in many locations, particularly within seasonally arid climates in West Africa and Australia. This profile is illustrated in Table 3.4 and Figure 3.12(A). The features of this zonation to which attention should be drawn are the mottled clay horizon which

suggests delicate near water-table relationships in a climate with seasonal drying out of the profile at this level, and the presence or absence of the pallid zone. Generally, if the pallid zone is well developed, the transitional (saprolite) zone is thin and vice versa. The occurrence of a pallid zone beneath a mottled clay horizon is likely to indicate a change of formative condition, from a low level, high water-table environment, to a high-level, low water-table position, following elevation and incision of the landsurface (McFarlane, 1983a,b). That this has been a common sequence is seen in the widely observed occurrence of duricrusts as plateau cappings (Figure 3.13). The formative processes leading to Al- and Fe-duricrusts are discussed in Chapter 4.

DISTRIBUTION AND PATTERNS OF WEATHERING

Controlling factors in weathering penetration

Variations in depths of weathering and spatial patterns of weathering types or phases must be examined at different scales and in relation to a number of different factors:

(1) weathering depths show strong relationships with regional climate and particularly with rainfall amount;
(2) weathering depth is affected by rock mineralogy and texture;
(3) weathering is accentuated and accelerated along shatter belts or other zones of intense jointing and/or faulting;
(4) deep weathering is characteristic of cratonised

FIGURE 3.13   Duricrusted residual hill with laterite profile over basic metamorphic rock, Kano, northern Nigeria. The duricrust (ferricrete) capping is 3–5 m thick and the complete profile is ~40 m. Corestones of unweathered rock can be seen in the foreground

landsurfaces but restricted in mobile orogenic terranes;

(5) weathering depth may be partly controlled by relief and promoted by dissection;

(6) weathering depth can sometimes be related to the age of landsurface, since time is required to develop deep profiles.

Often the factors that promote deep alteration also lead to advanced weathering, but this is not invariable. In particular, geomorphological controls may operate in an opposed manner. Thus, beneath old (pre-Neogene) landsurfaces, the advance of the weathering front at depth may be slow, because water movement is slow and the concentration of ions in solution is high. However, higher in the profiles leaching is able to continue and may have

produced advanced stages of decomposition in the rocks. On the other hand, beneath 'young' landsurfaces undergoing dissection and compartmented into hills and valleys, water flow is accelerated and the advance of the weathering front promoted. However, no part of the landscape is geologically old, and much of the saprolite may be in a state of partial alteration and liable to mass transfer as landslides.

The role of regional climate

The role of regional climate is implied by Figure 3.1. This draws heavily on the work of Strakhov (1967) and suggests that depth and phase of weathering vary together, ferrallitic weathering being associated with deep profiles; fersiallitic weathering with less deep profiles, and partially altered rock typifying the arid, continental and west coast tropics. Like all constructs of this kind, it should be used with caution.

One reason for this caution is the role of climate change over long geological time periods involving plate tectonic motion. Changes in latitude, world relief and climate patterns have had profound effects on the weathered mantles of the landmasses as we see them today. Some discussion of this topic will be offered subsequently (Chapter 11); here the issue is whether the weathering profiles in many areas reflect past rather than present conditions of formation. In fact, of course, most profiles of 30–50 m depth probably have formative histories of 10^6 years, perhaps 10^7 years, during which time important changes in climate and elevation will have taken place. Many old and duricrusted landsurfaces have been dated back 10^8 years, and over that period environmental conditions may have changed many times. Most weathering profiles are therefore palimpsests with new imprints obscuring older features and transforming them.

The profiles seen in the field are often relict, that is to say they are inherited from past morphogenetic systems, and the weathering products are often not in equilibrium with the prevailing climatic environment. Yet in many cases these materials are more or less stable and have little capacity for change. This is because the mobile ions that might once again combine as new clay minerals (for example) have been leached away and cannot be regained. Even the surface soils (Zone VI), under the influence of present-day climate and vegetation, remain strongly influenced by this inheritance (Ollier, 1959; Van Wambeke, 1962). Thus, deep weathering profiles with leached pallid zones can be seen in the semi-arid zone of northern Nigeria, exemplified in the residual hills in Kano city (Figure 3.13), and in the arid central heart of Australia (Mabbutt, 1965). By contrast, old duricrusts can be found broken and undergoing dissolution in rainforest environments such as central Sierra Leone (Figure 3.14) and the Haut-Mbomou of central Africa (Beauvais & Tardy, 1991).

Attempts to set climatic limits for the formation of such materials as bauxite, pallid zones, or ferricrete duricrusts have nearly always been met with serious criticism. However, the appearance of duricrusts containing progressively more mobile cations as sesquioxides ($Al_2O_3 \rightarrow Fe_2O_3 \rightarrow SiO_2$) usually corresponds with a rainfall/humidity gradient from wet to drier climates, with $CaCO_3$ and ($CaSO_4.2H_2O$) being found in still more arid conditions.

The roles of regional geology and geomorphology

The roles of regional geology and geomorphology in controlling deep weathering take place at scales covering distances of at least 10^2 km and covering areas of 10^3–10^4 km^2, and can only be described in very general terms. The recognition of different geological *terranes*, resulting from the cumulative effects of tectonic evolution is directly relevant to many geomorphological questions, including the nature of the weathered mantle.

As a general statement it is possible to claim that *cratonised terranes* have tended to be emergent over long periods of geological time, and are therefore exposed to chemical decay for equally long periods. Such areas are not confined to the Proterozoic and Archaean continental nuclei, but include the roots of Palaeozoic mountain chains and also some early Mesozoic terranes such as Sundaland, though this craton (Barber, 1985), which encompasses parts of Malaysia, Sumatra and Borneo, has been partly submerged. Many such terranes border the Atlantic and Indian Oceans and function as trailing edges of divergent continental plates. The post-Triassic opening of these new oceans, disrupted a vast continental landmass and produced high, warped and faulted coastal zones along the margins of Brazil, Angola, southern Africa, western India and Western Australia. According to Partridge and Maud (1987), Africa may have stood close to 2000 m asl at this time, while Wise (1972, 1974) has commented on an increasing continental *freeboard* through Phanerozoic time. The occurrence of thermally driven domal uplifts within the old Gondwana continental fragments (Le Bas, 1971; Dingle, 1982; Summerfield, 1985) is also relevant. These uplifts have been driven by mantle convection

FIGURE 3.14 Dissolution and disintegration of ferricrete on an old *glacis* surface near the Kasewe Hill (in the distance), Sierra Leone. A solution hollow in the ferricrete is seen in the foreground. Under natural conditions this would be forested (Rainfall ~2300 mm year^{-1})

plumes beneath the continental plates, and are associated with great outpourings of basaltic lavas in the Karroo (Triassic) and Deccan (late-Cretaceous). They also led to dissection, falling water-tables and to widespread stripping of saprolite (Thomas, 1989a,b).

All of these ideas converge to suggest that the continents may be divided into large structural provinces or geological terranes, that have experienced similar tectonic and geomorphic evolution. According to many authors (Penck, 1924; King, 1962, 1976; Meyerhoff, 1975; Ollier, 1985; Partridge & Maud, 1987) a broad synchroneity of events leading to up-doming or rifting of the continents has led to both regional and perhaps world-wide correlative planation, the advancing pediplains effectively decapitating the

structural domes and then facetting their flanks. However, in reality, these formative events were diachronous, governed by the differential movement of the continental plates.

This close correspondence between tectonics and evolutionary geomorphology is important for weathering studies, and some would argue that successive levels in the landscape carry diagnostic saprolitic residues and duricrust cappings (Grandin, 1976; Bardossy & Aleva, 1990) (Figure 4.3). However, reality is more complex; many local 'planations' are related to particular drainage basins or to block-faulted units, and the complex pattern of rock types within these large provinces has led to much differential denudation (Thomas, 1980, 1989a,b), and to differences of relief and weathering mantle.

It is often claimed that progression inland from coastal areas brings the observer to higher and older landsurfaces via a series of steps as in South America (Bigarella & Mousinho, 1966; Aleva, 1984; Briceño & Schubert, 1990), but such reconstructions have not gone unchallenged (Kroonenberg & Melitz, 1983). Certainly, it is true that in central Africa and Brazil there are large areas of ancient plateau landsurface beneath which deep and highly evolved profiles have been developed, and ancient weathered landsurfaces in Australia have also been clearly delineated and described (Mabbutt, 1966). But the old weathered landsurface of Sundaland appears to be warped down beneath the Sunda Sea (Batchelor, 1979a; Thorp *et al.*, 1990), and deeply weathered compartments containing bauxite profiles occur close to sea level (~ 25 m asl; Rodenburg, 1984). In Sierra Leone, on the flanks of the Leo Uplift in West Africa the situation is uncertain; some regional and local planations have been described (Dixey, 1922; Hall, 1969), but it seems possible that an old weathered surface is warped below the coastal plain sediments (Thomas, 1980; Thomas & Summerfield, 1987). In eastern Brazil, profound deep weathering occurs at all levels, from the interior plateaus to the *meias laranjas* landscapes of the coastal and intermediate areas. A major conclusion from the study of South African landscapes by Partridge and Maud (1987) was that the peripheral escarpment (Drakensberg in South Africa) came into being at a late stage (probably Pliocene) of landscape evolution and that it now separates landscape of similar age but very different elevation.

In summary, although profile differentiation (Thomas, 1989b) may be favoured by long, undisturbed time periods of evolution beneath ancient plateau surfaces, weathering *depths* are affected by other factors as well. Water circulation and the advance of the weathering front are possibly both retarded by the conditions at depth beneath extensive plateaus, but are accelerated where there is dissection and fissuration of the rocks, as along faulted margins of the continents. However, the most advanced leaching of rocks (to form tropical podzols) is associated with prolonged, low rates of weathering and cation removal via rivers (Stallard & Edmond, 1987) draining areas of low relief.

The importance of geological outcrop patterns

Geological outcrop patterns generally operate at more local scales, covering distances of 10^1 km and areas up to 10^2 km^2, though there are exceptions on some sedimentary platform areas. Weathering depths are greater where rocks are weakened by microcracks and other fractures, and especially along shear zones. But the mineralogy of rocks fundamentally affects their weatherability. Contrasts exist between members of the granitoid suite of rocks; those with predominant orthoclase and muscovite resist weathering, while those with plagioclase and biotite are more susceptible. Many deep weathering profiles in granitoids relate to granodiorite or even diorite rocks containing pyroxenes and hornblende. Brook (1978) was able to demonstrate the effects of varying mineralogy in a study of granite weathering from the Pietersburg Plateau, South Africa, and concluded that deep weathering preferentially affected the foliated biotite granodiorites, and that inselberg outcrops were mainly found amongst the porphyritic ademellite granites that had been subject to potash metasomatism. Trend surface analysis of weathering depths over distances of 10^2 km also demonstrated a regional basin and ridge pattern in the weathering depths, influenced by the occurrence of the ademellite plutons (Figure 3.15). This study is also remarkable for its demonstration of the widespread occurrence of weathering depths exceeding 30 m and reaching 60 m in an area of subtropical dry climate, where King (1972) had applied theories of pediplanation in unaltered rock to explain the inselberg forms.

Banded rocks with prominent schistosity weather deeply along the susceptible, darker, mineral bands, leading to a disintegration of the rock fabric. This is particularly effective where the foliation is near vertical. Some metavolcanics such as amphibolitic rocks are prone to rapid deep weathering, but their high Fe^{2+} content often leads to duricrust formation at an early stage, protecting the underlying saprolite from erosion. This is an

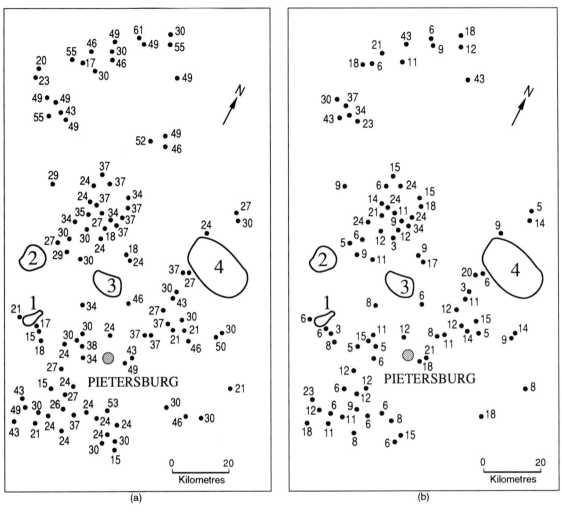

FIGURE 3.15   (a) Maximum and (b) minimum depths (m) of weathered rock in the Pietersburg region, South Africa. Areas 1–4 are biotite ademellite plutons (adapted from Brook, 1978; reproduced by permission of Gebrüder Borntraeger)

important factor in preserving saprolite profiles, and one reason for shallow profiles over many felsic rocks, low in iron which seldom form thick or extensive ferricrete layers, except in convergent zones of local concentration of groundwater carrying Fe^{2+}.

Clastic sedimentary rocks are usually more than 80% quartz and weathering mainly attacks the cementing substances binding the grains. These may be hydrated Fe_2O_3, amorphous SiO_2 or possibly carbonates, but all are susceptible to solution, and some sandstones may lose cohesion to great

depths. In other instances striking pseudo-karst or silicate karst forms have been created (Young, 1987a,b; Busche & Erbe, 1987). These features are discussed in Chapter 9.

The influence of structure

Structure is an important factor influencing weathering depths *within* major geological provinces and rock outcrops. The roles of faulting, jointing and microcracks have been discussed above, but it is often possible to relate the *pattern*

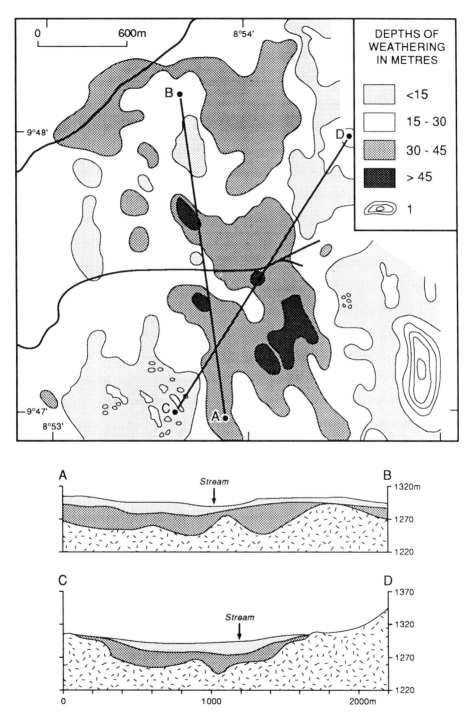

FIGURE 3.16   Deep weathering patterns in fine-grained, biotite granite, near Jos, northern Nigeria. 1 = Outcrops of unweathered granite showing contours at 16 m intervals. On the cross-sections (A–B, C–D) saprolite is ruled; alluvial–colluvial fill is stippled (after Thomas, 1966a; reproduced by permission of the Institute of British Geographers)

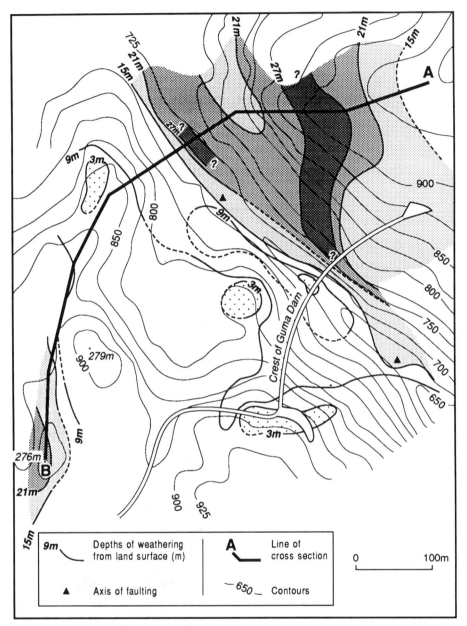

FIGURE 3.17   Uneven weathering depths in steep terrain over gabbro at the Guma Dam site in the Freetown Peninsula, Sierra Leone. The valley and the zone of deeper weathering are thought to follow a major fracture zone

of weathering specifically to this factor. The author was able to demonstrate this from drilling records in a biotite granite on the Jos Plateau (Thomas, 1966a, 1974a, and Figure 3.16), which were then related to lineations from neighbouring outcrops.

These lineations mark fracture directions and they show a remarkable congruence with the patterns of weathering. Intersections of the major joint sets appear to be marked by troughs or basins of weathering; intervening areas marking domal rises

TABLE 3.5   Depths of the weathered rock mantle at the Santa Rita and Puerto Velo sites, east of Medellin, Colombia

	Santa Rita site		Puerto Velo site		
Bedrock	Quartz diorite and hornblende diorite		Quartz diorite		
Mean elevation	1890 m		750 m		
Local relief	125 m		235 m		
Mean annual temperature	17 °C		27 °C		
Mean annual rainfall (estimated)	4950 mm		2400 mm		
Topographic location	Mean depth (m)	Number of borings	Mean depth (m)	Number of borings	Standard deviation (m)
Hilltop and ridge	46	28	45	9	13.5
Upper half of slope	36	27	43	1	8.0
Lower half of slope	26	47	31	2	8.0
Valley bottom	17	27	10	1	7.5
Mean	31	(129)	40	(13)	14.0

After Feininger (1971).

in the weathering front which may break the landsurface to form actual outcrops. The influence of a local fault zone on depths of weathering at the Guma Dam site in the Freetown Peninsula of Sierra Leone is shown in Figure 3.17, which relates to an area of high local relief with deep valleys excavated between high, rocky hills of gabbro–norite composition in a very wet environment (Ledger, 1975).

The relief factor

Relief or topography is a factor that undoubtedly influences weathering processes and some tentative relationships can be advanced. However, one salient feature of the weathered mantle is its variable depth and *lack* of surface expression. Put another way, there is not always any close correspondence between the undulations of the weathering front and those of the landsurface. Brook (1978) found that in 80 geological reports of well surveys, the weathering front had an average relief of 22 m across a 150 m traverse, and ranged from 3 to 55 m in individual profiles (maxima varied between 15 and 61 m, and minima from 3 to 43 m; Figure 3.15). In Nigeria, the author was able to plot a number of traverses from dam site surveys and these show clearly that the ground surface and the basal surface of weathering or weathering front undulate

independently in many traverses, but there is evidence for deeper weathering to occur beneath interfluves, even where these are not protected by duricrusts (Figure 3.18).

Feininger (1971) was able to corroborate this finding from studies in Colombia (Table 3.5), and Faniran (1974) obtained similar results from further studies in Nigeria. This tendency for deeply weathered compartments to occur between stream valleys is attested by countless road cuttings and excavations in the humid tropics and sub-tropics (Plates 3-VIII and 3-IX). It is also demonstrated by studies from the warm temperate areas of the US Piedmont (Costa & Cleaves, 1984; Pavich, 1986, 1989) (Figures 3.19 and 3.20). A question remains to be answered here, however, and that is whether the pattern observed is a consequence of dissection into a deeply weathered landsurface of formerly low relief (as suggested in Figure 3.20), or is a consequence of more rapid weathering penetration beneath interfluves within an undulating topography. The two possibilities are not exclusive, but distinguishing between them can be difficult.

A further observation is the occurrence of deep troughs of weathering below the levels of local river channels. In some cases the weathering extends beneath the channel itself, but this is usually confined to low order, seasonal watercourses. More

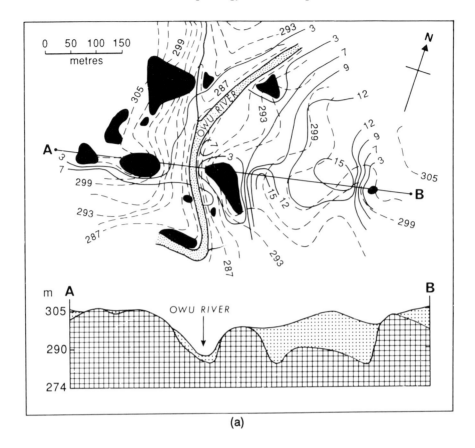

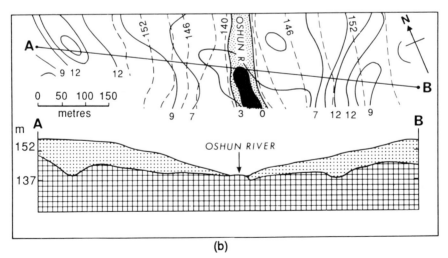

FIGURE 3.18 Contrasting patterns of weathering in two cross-valley profiles on Precambrian gneiss in southern Nigeria. (a) Uneven outcrop and weathering patterns indicate a small basin without surface expression. (b) Progressive thickening of saprolite zone with distance from river channel (after Thomas, 1966a; reproduced by permission of the Institute of British Geographers)

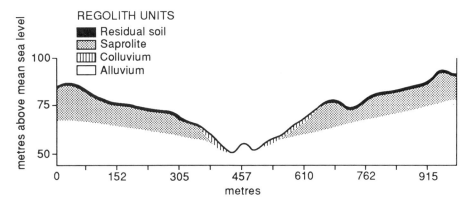

FIGURE 3.19 Cross-section showing saprolite thickness in relation to valley profile in the Virginia Piedmont, USA. The interface between saprolite and rock is shown to be flatter than the valley slopes; the steep slopes have significant colluvial cover, and the stream channel has incised bedrock (after Pavich, 1986; reproduced by permission of Academic Press)

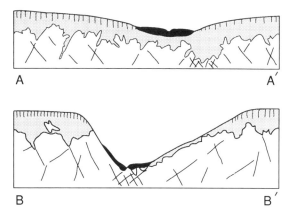

FIGURE 3.20 Two contrasting cross-valley profiles indicating saprolite thickness in the Maryland Piedmont, USA (compare with Figures 3.18 and 3.21 from Nigeria) (after Costa & Cleaves, 1984; reproduced by permission)

often the weathering extends below the channel level at some distance away from the river (Figure 3.21). There are not many published records of this phenomenon, but anecdotal evidence from engineers suggests that it is common. In parts of the Amazon, weathering is known to extend more than 10 m below the stream outcrops (Pedro, 1991). Clearly, groundwater movement must be possible in these situations and it will be continually renewed from the river bank and floodplain seepages. This water will acquire acidity from the saprolite environment and may become more 'aggressive' than the river water itself. Evidence for deep weathering beneath floodplain sediments is generally lacking, but the presence of moving groundwater in a non-abrasive environment beneath the floodplain sediments should promote rock decay. However, since most floodplain sedimentary units are thought to be less than 40 ka old, the depth of penetration may be too small to be noted. As a mechanism for preparing valleys for further deepening by tropical rivers, this process may nonetheless be important.

Ollier (1976) has suggested some catenary patterns, relating relief to ferricrete, saprolite and rock outcrops (Figure 3.22). Most of these indicate dissection into an older, pre-weathered landscape. In more dissected areas with rocky relief, weathering penetration will be very uneven and mainly confined to shatter zones and other lines of weakness (Figure 3.13). In these circumstances also boulder accumulations may form talus slopes, and there is good evidence for deep weathering to occur beneath these, where it can attack the base of the rock slope and possibly hasten its collapse or retreat. It must also be recognised that shallow weathering profiles exist widely in the tropics, beneath resistant and massive rocks, and that rock-cored compartments can be found underneath mature rainforests (Figure 3.23).

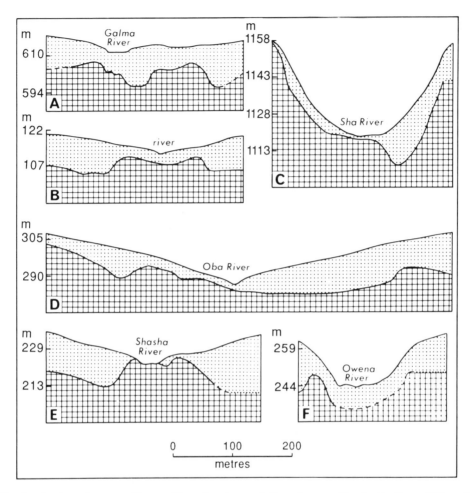

FIGURE 3.21   A selection of cross-valley profiles, showing variations in saprolite depths from different locations in Nigeria. Superimposition of river channels on the basal weathering front can be implied from these profiles (after Thomas, 1966b; reproduced by permission of the Institute of British Geographers)

Optimal conditions for deep weathering

From these considerations it is possible to offer a summary list of optimal conditions favouring deep weathering. However, many factors interact to determine the environment at a specific site and limit our ability to predict weathering depths from surface appearances. The final answer must come from sub-surface geophysical or drilling data. Nevertheless, some of the major factors are:

(1) wet equatorial or monsoon climates with rainfall $>1500\,\text{mm year}^{-1}$, together with rain-forest vegetation cover;

(2) cratonic terranes in continental interior or passive marginal plate positions;

(3) domal and plateau uplifts and warped or faulted continental margins with strong tensional fracture patterns;

(4) shear zones and intersections of dense fracture sets;

(5) hydrothermally altered or weakened rocks;

(6) fissile metamorphic rocks or igneous rocks with dense microcrack systems;

(7) old (pre-Neogene) landsurfaces at moderate to high altitude, even in sub-optimal seasonal climates, where long-term denudation rates have been low;

Rock at high levels   Rock at low levels   No rock at surface

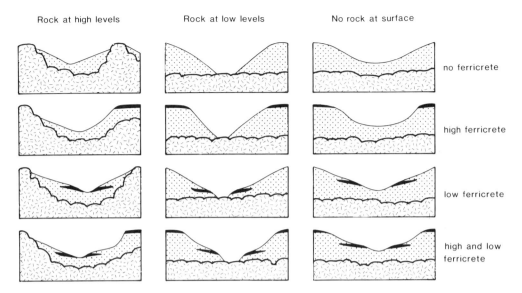

no ferricrete

high ferricrete

low ferricrete

high and low ferricrete

FIGURE 3.22   Schematic catenary relationships between duricrusts, saprolite thickness and form of the weathering front in areas of deep pre-weathering (after Ollier, 1976; reproduced by permission)

(8) proximity to structural depressions and escarpment zones promoting strong groundwater gradient;

(9) free-draining sites beneath interfluves and hillslopes < 20°, particularly where protected by duricrust cappings;

(10) predominantly warm and humid conditions during substantial periods (10^6 years) of palaeoclimatic history over 10^8 years.

Examples of weathered landscapes

Some specific examples can help illustrate the operation of these factors (those operating in each case are listed according to the numbers above, (1–10)):

The rifted margin of southeast Brazil

The Atlantic coastal margin of the Brazilian plateau, between 20°S and nearly 30°S, experiences rainfalls between 1800–3500 mm year^{-1}. The relief is tiered inland from the coast with strong faulting of the migmatite basement that contains large areas of biotite gneiss. The Serra do Mar in places rises to > 2000 m asl within 50 km of the coastline. Deep decay is ubiquitous except along the steep scarp fronts (Branner, 1896) and depths of 50–120 m are well documented (factors 1,2,3,4,6,9). The weathering is associated with the widespread development of multi-convex, compartmented relief (*meias laranjas*) which occurs at all levels from the coastal plain under 50 m asl to the high plateaus over 2000 m asl. It can be argued that this represents an old weathered landsurface, fractured and faulted along the passive plate margin, but many deep sections reveal immature sandy weathering and an explanation coupling the wet climate, fractured basement and stepped relief with steep hydraulic gradients, demands fresh enquiry (Plate 3-VIII).

Although faulted, the stepped topography is reminiscent of the account of stepped topography in the Sierra Nevada given by Wahrhaftig (1965). These optimum conditions are not met in southeast Africa or Australia, but despite aridity in the former and lower temperatures in the latter, some deep weathering is widely found in both areas.

The interior plateau of central Africa

The interior plateau of central Africa includes the Zaire copper belt, where rainfall is

FIGURE 3.23   Rock-cored granite inselberg in Johore, West Malaysia. The forest maintains a thin regolith over little fractured rock

~1200 mm year^{-1} at an altitude of 1200–1500 m. The plateau surface may date to the late Mesozoic or Palaeogene, and is deeply dissected. The metamorphic rocks are fissile and weathered to depths exceeding 150 m, with some alteration to more than 600 m. Deep ferrallitic weathering profiles with thick pallid zones and extensive duricrusts suggest very old weathering penetration beneath the plateau surface, but renewed weathering penetration following uplift and dissection is probably superimposed on the ancient system, and responsible for the very great depths of decay (Plate 3-I). Very deep weathering profiles are also found over much of Uganda which has a similar geomorphic position (possibly containing remnants of the 'African' planation surface). The greenschist belts in Sierra Leone, including the banded ironstones in the Sula Mountains, show similar depths of decomposition below duricrust cappings (Gaskin, 1975; Thomas, 1989b) (Figure 3.12(D)) (factors 1,5,6,7,8,9,10).

The porphyry copper deposit at La Escondida, Chile

The porphyry copper deposit at La Escondida in the northern Atacama desert of Chile (Alpers & Brimhall, 1988, 1989) shows kaolinisation to 600 m in Cenozoic rocks in an active tectonic setting and a desert climate. This deposit indicates the importance of major climatic change from humid to arid conditions in the mid Miocene, and the roles of both hydrothermal activity and strong uplift in promoting the penetration of supergene weathering. Hydrothermal alteration was estimated to have taken place between 33.7 and 31.0 Ma based on K–Ar dating of hydrothermal biotite and sericite, while the supergene weathering was active from 18.0 to 14.7 Ma based on K–Ar dating of supergene alunite (Alpers & Brimhall, 1988, 1989) (factors 5,6,9,10). Desiccation of the climate after 14.7 Ma appears to have halted further decomposition.

The Mesozoic andesitic province of northwest Kalimantan, Indonesia

The Mesozoic andesitic province of northwest Kalimantan in Indonesia has been upwarped inland but submerged between Borneo and Sumatra. A deeply weathered landsurface appears to follow this warping, being buried by Neogene and Quaternary sediments in the Sunda Sea (Batchelor, 1979a). Widespread deep weathering is present in Kalimantan under an equatorial climate (>3000 mm year^{-1}), and has been described as the 'Old Sundaland regolith' (Batchelor, 1979a). But this weathering is likely to be Neogene, since there are Palaeogene sediments also outcropping in the area. Low convex hills with bauxite profiles are found in the coastal zone (Rodenburg, 1984), and these *allitic* weathering horizons appear to conform to the present topography (Figure 4.11) and may be products of Quaternary as well as Tertiary weathering. Some very large, displaced blocks of ferricrete were seen at low elevation in this area, and multi-convex relief forms predominate in the more elevated, dissected interior, where there are also intrusive rocks of Miocene age (16 Ma). All of the evidence from this area suggests that any old weathered surface which may have existed has been strongly modified by continuous weathering under a perhumid climate following both Tertiary and Quaternary uplift (factors 1,7,8,9,10).

The metamorphic piedmont of Southern Appalachia, USA

The Southern Appalachian piedmont province in North America illustrates the role of deep weathering in metasedimentary and metavolcanic rocks in a warm temperate to sub-tropical climate (rainfall 800–1500 mm year^{-1}). Epeirogenesis from isostatic rebound has taken place in response to denudation during the Cenozoic, permitting weathering penetration to keep pace with erosion. Saprolite is widely found, with a mean depth of 18 m and a maximum of 90 m (Pavich, 1986), and this area has been subject to detailed geochemical and geomorphic research which is touched on elsewhere (p. 48, p. 307) (factors 2,3,6,7,8).

Conclusion

These examples illustrate some common problems of interpretation including, the evaluation of palaeoclimates in the formation and modification of the profiles, changes to the rate of denudation which can lead to truncation of profiles or to the accumulation of transported materials, and our present uncertainties regarding the long-term rates of weathering penetration.

THE EVOLUTION OF WEATHERING PROFILES

The recognition that not only the prevailing climate and the local rocks influence saprolite formation, but that climatic, tectonic and geomorphic evolution through perhaps 10^6 years can be equally important, makes it clear that weathering profiles do not simply develop in equilibrium with a single set of prevailing environmental conditions. Equally, the profiles often have no definitive beginning or ending of evolution, though the Escondida copper porphyry example comes close to this; beginning with the 31 Ma intrusive activity and ending with the 15 Ma desiccation of climate.

The geochemical evolution of saprolites has been seen as a progression towards greater leaching and bauxitisation (or podzolisation), the most evolved materials normally appearing in the upper part of the profile (Zone VI, Figure 3.1). But it will be pointed out (Chapter 4) that bauxitisation (allitisation) may take place during the progressive deepening of profiles, below a ferricrete duricrust capping. The end product of one pedogenesis was therefore laterite (or ferricrete); that of the succeeding period was bauxite—a view recently expressed by Valeton *et al.* (1991). This evolution implies elevation and dissection, while depression and inundation may lead to *ferrolysis* (Brinkman, 1970, 1979) under seasonal waterlogging or to *deferrification* (Sieffermann, 1988) in organic peat swamps. These processes equally imply further evolution of ferrallitic profiles (Lucas *et al.*, 1987, 1988).

Weathered mantles cannot be regarded as static profiles within which horizonation has taken place, but must be thought of as slowly descending columns, accumulating less mobile elements and some transported materials in the upper horizon, as the weathering front *and* the ground surface are gradually lowered by the denudation system, with loss of both solutes and fine particles from the surface and near-surface soil. It is now claimed that loss of volume due to this *lessivage* can amount to 70% (Kronberg *et al.*, 1987) or even 75% (Pavich, 1986) of the soil column (Zone VI). The field and experimental studies that have established the enhanced mobility of Al^{3+}, Fe^{2+}, and SiO_2 in the presence of organic acids also provide good reason for considering the lowering of the landsurface, even without physical erosion of particulate material, a reality. This process is also central to McFarlane's (1983a) view of laterite genesis (Chapter 4), and to the accumulation of nickeliferous laterites (Esson & Surcan dos Santos, 1978) (see p. 115 and Chapter 11). The fate of the saprolite mantle, therefore, will depend on the delicate balance between rates of weathering and of erosion.

The behaviour of weathering profiles through time

A summary of the ways in which weathering profiles may evolve is listed below (modified from Thomas, 1989b).

(1) *Profile lowering* will occur if weathering penetration and surface denudation are delicately balanced (when averaged over a substantial time period). This can permit the descent of the column without substantial modification, in a state of dynamic equilibrium between the rates of uplift and

denudation (Hack, 1960, 1979; Velbel, 1985; Dohrenwend *et al.*, 1987), but if Fe^{3+} is precipitated as goethite segregations which evolve towards crystalline haematite nodules, then these will accumulate in the soil B horizon, while resistant clasts of vein quartz (plus quartzite, tourmaline, corundum, etc.) may also accumulate as a stone-line. This twin process is probably self-limiting as the profile becomes effectively 'sealed' by a developing duricrust.

(2) *Profile deepening* will depend on a minimal rate of surface lowering, during continued weathering penetration. This may possibly occur beneath extensive plateaus lacking significant slope or drainage systems, but more often it will result from the protection of a duricrust capping which prevents the further removal of saprolite in situations where active dissection of an old landsurface is taking place. These are conditions for pallid zone or bauxite zone formation, providing that free drainage is maintained by uplift and dissection. The high plateaus of central Africa contain such profiles which are revealed in the deep copper ore workings in Shaba, Zaire (Plate 3-I). The deep profiles below the duricrusted Sula Mountains in Sierra Leone, first described by Gaskin (1975, and Figure 3.12(D)) occur in a different situation, on the dissected flanks of a crustal upwarp (Leo Uplift) on the Atlantic margin of West Africa.

(3) *Profile collapse* appears to occur where ferrallitic residues become subject to hydromorphic conditions, and widespread leaching of Al^{3+}, Fe^{2+}, and SiO_2 takes place, leading to podzolised white sands (Figures 2.13 and 2.14). This involves both podzolisation (Chauvel, 1977) and ferrolysis (Brinkman, 1970, 1979). Theoretically the entire rock could be dissolved (Chapter 2).

(4) *Profile thinning* is a process that results from ground surface lowering, when weathering-resistant rock compartments are encountered at depth (Twidale & Bourne, 1975). The rate of erosion may be slow, but the column, in effect, becomes compressed from below. In this way a highly evolved upper horizon (probably a duricrust) may be seen to rest on almost fresh rock.

(5) *Profile truncation* is distinct from (4) and arises from an acceleration of surface erosion,

which removes the more highly evolved topsoil and upper saprolite. This usually arises from climatic crises leading to opening of the vegetation canopy and increased runoff, but may also be due to the progress of dissection. This process can result in fersiallitic materials being exposed where the present climatic parameters suggest development of a ferrallitic mantle. According to Fairbridge and Finkl (1980) prolonged periods of stability (10^7–10^8 years) and weathering have been interspersed by periods of rapid erosion (10^4–10^5 years) within what they called the *cratonic regime* (see Chapter 9).

(6) *Profile burial* can occur if sedimentation is not accompanied by erosion of the pre-weathered saprolite. Circumstances favouring this situation include tectonic subsidence, though some trimming or *truncation* of the profile is likely to occur. This has taken place in Sundaland, Indonesia.

(7) *Profile renewal* probably continues alongside most other processes. However, soil systems energised by the dissection of old landsurfaces, under the protective cover of rainforest may display well organised profiles in sympathy with the prevailing relief, but developed in an ancient saprolite. This is the 'two cycle' theory of tropical pedology described by Ollier (1959).

Initially, *profile differentiation* was added to this list, but the differentiation of horizons, though related, is a separate topic and a question of *pedogenesis*. However, rapid denudation, leading to thin regoliths clearly opposes horizonation in the soil. The dynamics of soil profiles has been a central topic in pedology since the work of Nikiforoff (1949, 1959). Recently Johnson and Watson-Stegner, 1987) have defined *progressive pedogenesis* which leads to organised profiles (horizonation), in contrast to *regressive pedogenesis* which promotes simplified profiles (haploidisation). The factors at play in these two scenarios of soil development overlap substantially with those leading to changes in the entire weathering profile, but the time scales of concern may be different by one or even two orders of magnitude. When rates of erosion or accretion converge with rates of energy and mass fluxes in the soil, then horizonation becomes

opposed by the rates of surface transfer (cases (4) and (5) above). In a system at dynamic equilibrium, rates of denudation (morphogenesis) and rates of pedogenesis will be delicately balanced as envisaged in case (1) above. But, as has been shown, profiles also *evolve* through time, with accumulation of resistant clasts and oxides (mainly of Fe^{3+}). These may eventually prevent significant surface lowering, but encourage further deepening and horizonation of profiles (case (2)).

The accumulation of stone-lines and ferricretes is an example of a *feedback mechanism* at work, preventing surface erosion from continuing uninterrupted across the landscape, unless the energy of surface transport can remove the larger and denser (Fe_2O_3) clasts. When surface erosion is accelerated, as when the vegetation canopy is opened up by desiccation and fire, truncation may occur for a period, to be followed by a *renewal* of the profile, when the canopy is restored. This took place over periods of 10^3–10^4 years during the Quaternary glacial–interglacial (or interstadial) cycles. These ideas show how closely geomorphology and pedology are linked, yet concepts developed to serve the needs of each field of study cannot be interchanged without further discussion which is beyond the scope of this study.

GROUNDWATER AQUIFERS IN WEATHERED ROCKS

The availability of shallow groundwater has always been a major constraint over occupation and land use in sub-humid areas of Africa and other continents, where streamflow in all but the largest rivers dwindles to zero after 3–4 months of dry season conditions (Ledger, 1969; Buckley & Zeil, 1984). In Africa alone, it has been calculated that 50 million people occupy basement complex rock areas (Clark, 1985), and still greater numbers are involved in India. The importance of shallow, weathered-zone aquifers to these people is inestimable, and some thought has been given to the nature of these aquifers and to the best methods to search for them (Chilton & Smith-Carrington, 1984; Faillace, 1973).

Some mean figures for weathering depth have been published for parts of Africa. Thus Jones (1985) quoted figures of 50 m for Tanzania, and 35 m for Zambia, while Clark (1985) suggested a range of 30–90 m for the Karamoja area of Uganda, 40–70 m for the Ivory Coast and 50 m for Nigeria, and McFarlane *et al.* (1992) found a mean of around 30 m for the plateau areas of Malawi. Two, apparently contradictory, observations have been made regarding these figures; first that across the central plateau areas of Africa the weathered zone is remarkably uniform in thickness, and, second, that profile depths are subject to abrupt variations. Regional uniformity is interpreted by most observers as a consequence of prolonged weathering over long time scales (10^7 years) beneath well-preserved planation remnants. However, this supposed uniformity masks much variation in local depths of weathering. This variation was discussed above.

A study from the Jos Plateau of northern Nigeria (Thomas, 1966a, 1974a) also revealed the existence of enclosed basins and troughs in the weathering depths (Figure 3.16). This accords with observations made by Enslin (1961, p. 379) in South Africa, that the weathered rock, 'tends to vary in depth and lateral extent, and to form a large number of isolated, basin-shaped or trough-like groundwater compartments. These areas show practically no surface indications, but can be differentiated by geophysical methods.'

The nature of basement aquifers

The factors affecting groundwater flow through these weathered materials include (1) availability of recharge groundwater; (2) the permeability of the weathered material; and (3) hydraulic gradient between recharge and discharge areas mainly as a function of surface morphology (slope). The hydraulic conductivity of unweathered crystalline rocks is very low, but this increases rapidly as joints become widened and blocks are separated by weathering products. Permeability, conductivity and specific yield all peak close to the weathering front (Figure 3.9), across a zone of possibly 5–15 m thick, which separates the fresh rock from the clay-rich saprolite above. This is Zone II of the 'standard' six-zone weathering profile (Figure 2.1).

Transmissivity within this partially altered and fractured zone has been given as 10 times that for the weathered zone above (Rushton & Weller, 1985). However, if water is withdrawn from the lower zone, it can be recharged by two separate routes; first, by lateral flow into the site from surrounding areas (dependent on hydraulic gradient); second, by slow seepage from the overlying saprolite. The weathered mantle must therefore possess sufficient extent, thickness and hydraulic conductivity to ensure enough groundwater storage and the possibility of useful sustained yield (Grillot *et al.*, 1989). Many of the thicker profiles form almost-isolated, groundwater basins, and dewatering may only occur over a radius of 10–100 m (Clark, 1985). However, in many cases, if wells are pumped dry, they will recover over time, indicating that complete isolation is not in fact frequent, an observation which accords with the records that show continuity in the weathered layer over many basement landscapes.

Most wells have a low yield. Clark (1985) states that this is likely to be $< 500 \text{ m}^3 \text{ day}^{-1}$ (~ 35 litres min^{-1}), but many records suggest values can range upwards to nearly $1000 \text{ m}^3 \text{ day}^{-1}$ (McFarlane *et al.*, 1990). Against this limitation should be placed the possibility of close spacing of low yield wells, each tapping into restricted groundwater basins. Clark (1985) even suggested that wells could be a mere 100–200 m apart where small fissure systems occur.

Surface geomorphological indications

Opinions differ concerning the possibility of detecting groundwater potential from surface morphology. Beneath the high plateaus of the Gondwana continents, the extent and uniformity of the weathered mantle has led some people to suggest that recognition of these remnant planation surfaces is sufficient to offer good prospects of success. According to Clark (1985) such criteria have greater accuracy (70–80%) over metamorphic rocks as compared to granites (50–60%), and this may be due to the highly uneven, fracture controlled, weathering patterns in the latter.

However, in more dissected areas, the existence of discrete topographic basins and troughs in granitoid (and other plutonic) rocks, allows prediction at a different scale. In these rocks, groundwater reserves are likely to occur in colluvium and saprolite found in negative relief areas between inselberg domes or other outcrops. In fact, both the marginal area around large outcrops and the valley floors may be good targets; the former receiving frequent runoff from the bare rock during rainfall, and seepage from deep-seated rock fissures for some period afterwards, while the latter may provide longer-term storage throughout the dry season. Aquifers found within larger basins can yield water for small rural communities during runs of drought years. According to McFarlane *et al.* (1991), an inverse relationship between total regolith depth, saturated depth and yield on the one hand, and relative relief (< 80 m) on the other, can be demonstrated from borehole information in Malawi.

Opinion differs concerning the presence of weathering and groundwater aquifers below valley floors. According to Clark (1985, in discussion, p. 45), 'in those valley bottoms where structural control can be shown by photo-interpretation, weathering is likely to be much deeper than on either side of the valley'. This relationship is found at the Guma damsite in the Freetown Peninsula, Sierra Leone (Figure 3.17), and was predicted for zones of deep alteration in the Abel Tasman National Park in New Zealand (Thomas, 1974b). However, in many dissected areas, rivers, even where guided by structure, have bedrock channels. The wider question of bedrock weathering beneath floodplains and their potential for groundwater abstraction remains an area for further research.

The prediction of weathering patterns from surface morphology is notoriously difficult at a detailed local level and geophysical methods of determination have proved cost effective. These include shallow seismic and electrical resistivity techniques; the former being favoured by many contractors. None the less, a summary of positive indicators might include the following:

(1) occurrence of planation surface remnants forming low relief plateaus at varying altitudes;
(2) existence of weathered zones along structural alignments between outcropping rocks;
(3) valley floors apparently developed along structural trends;
(4) basinal areas in granitoid rocks;
(5) shallow valleys such as 'dambos' and similar depressions.

The search for weathered basement aquifers is not new (Bannerman, 1973), but it may become more urgent as rural water supplies become more precarious. However, small aquifers need frequent recharge and all depend on the promotion of infiltration by protection of weathered watersheds and receiving sites for rapid runoff around inselbergs.

4

Laterites, Bauxites and Duricrusts

INTRODUCTION:
THE SCOPE OF THE TOPIC

Lateritic weathering and duricrusts

Most ferrallitic soils contain considerable quantities of kaolinite, but this mineral is subject to further breakdown, or removal by *lessivage*. Horizons containing a high percentage of sesquioxides, combined with a low clay content tend to form compact, resistant *duricrusts*, when exposed to repeated wet–dry cycles that encourage oxidation of iron compounds and the formation of crystalline goethite and haematite.

The common ground occupied by ferrallitic weathering products can be illustrated by the use of ternary diagrams (Figures 4.1 and 4.2), to show the compositional variation of chemical residua in terms of Al_2O_3, Fe_2O_3 and kaolinite (Bardossy & Aleva, 1990), or of Al_2O_3, Fe_2O_3, and SiO_2 (Dury, 1969; Schellmann, 1981). The difference between these approaches is significant; the former indicates a genetic relationship between kaolinitic weathering and lateritisation, while the latter emphasises relationships between members of the duricrust family. This also emphasises the wide compositional range and continuous variation amongst deposits

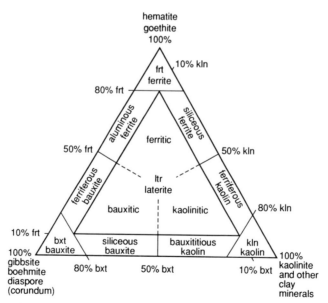

FIGURE 4.1 Classification of bauxite, ferrite, kaolin and transitional weathered rock types (after Bardossy & Aleva, 1990; reproduced by permission of Elsevier Science Publishers BV)

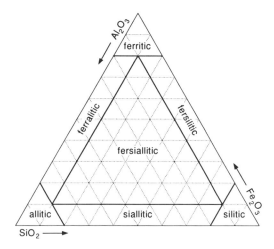

FIGURE 4.2 Ternary diagram suggesting boundaries among seven types of duricrust in the fersiallitic range (after Dury, 1969; reproduced by permission of the Macmillan Press Ltd)

which makes strict definition or categorisation a more or less arbitrary matter.

In order to distinguish the hardened duricrusts from the (lateritic) products of ferrallitic pedogenesis, terms such as laterite and bauxite should not be directly applied to them; however, there is a widespread and common usage in this sense. Ambiguity is avoided if duricrusts can be described as '-cretes' (there have even been objections to this) according to the dominant cation present (Chapter 3, p. 67). To the major group of duricrusts defined by their content of Al_2O_3, Fe_2O_3, and SiO_2 should be added those rich in $CaCO_3$, $CaSO_4.2H_2O$, and also saline crusts (see Goudie & Pye, 1983). Although some $CaCO_3$ accumulation is not uncommon in savanna soils, calcretes of importance to the understanding of landform development are rare outside of the semi-arid zone, and their investigation does not belong with the discussion of silicate weathering, even though SiO_2 impregnation of calcretes is found (Arakel *et al.*, 1989).

The limited mobility of Al^{3+} in most weathering systems implies that the majority of bauxite crusts (alucretes) will form as part of a free draining profile from which more mobile cations have been evacuated. On the other hand, because Fe^{2+} has sufficient mobility in groundwater to move laterally to form ferruginous (ferritic) crusts in a variety of materials and locations, ferricrete may not always be formed as part of the laterite profile. Silcrete formation reflects the still greater mobility of SiO_2 in weathering systems and is usually regarded as distinct from the process of lateritisation. Silcrete will therefore receive a separate discussion here.

Laterites and ferricrete

According to Schellmann (1981, p. 6):

Laterites are products of intense subaerial rock weathering whose Fe and/or Al content is higher and Si content is lower than in merely kaolinised parent rocks. They consist predominantly of mineral assemblages of goethite, hematite, aluminium hydroxides, kaolinite minerals and quartz.

This definition can be modified (Schellmann, 1983), and alternatives abound (McFarlane, 1983a), but it will be sufficient to guide discussion. The study of lateritisation processes thus develops from an understanding of ferrallitic weathering and pedogenesis, but podzolisation processes and the occurrence of hydromorphic conditions may also be involved in kaolinite dissolution and deferrification, and some would argue that Al enrichment of sesquioxide residues can occur in these environments (Valeton, 1983; Valeton *et al.*, 1991). Laterites (and lateritic or residual bauxites) can also form directly from Fe- or Al-rich parent rocks, without evolving through a (ferrallitic) soil phase. In these cases the laterite or bauxite replaces the bedrock and inherits bedrock textures, without the formation of intervening horizons (Plate 3-VII). These deposits are therefore difficult to incorporate into soil classifications. This difficulty becomes even greater when the ferrallitic material (laterite) evolves to form a duricrust. In this case the following arguments apply.

(1) Materials commonly distinguished as laterite may have compositions similar to certain tropical soils, but they have a greater degree of crystallinity and continuity of the crystalline phase (Young, 1976).

(2) Thick laterite crusts (ferricrete, duricrust) appear more as rock formations (perhaps >30 m thick) than as soil horizons (Senior & Mabbutt, 1979; Young, 1976); such duricrusts may be of considerable geological age.

(3) Laterites are not always associated with deep weathering or soil profiles; they can form by direct alteration of parent rock (Harrison, 1934; Bleackley, 1964; Boulangé & Millot, 1988), and often preserve its structures.

(4) *Ferricrete* may also develop in transported materials (Ollier & Galloway, 1990), and it is common in terrace sands and gravels and within piedmont sediments; *bauxite* can also be closely associated with sedimentary environments (Valeton, 1983).

(5) While some laterite is *pedogenetic* in origin, a category of *groundwater laterite* can be recognised (McFarlane, 1976, 1983a) which forms within the zone of water-table fluctuation.

It is not surprising therefore to find that Schellmann (1981) has asserted that, 'laterite is a petrographic term and lateritic weathering a geological phenomenon'. Bardossy and Aleva (1990) also considered laterites to be *residual rocks*, and Senior and Mabbutt (1979) actually proposed a method of mapping deeply weathered profiles (most of which had duricrusts) on stratigraphic principles. Separation of lateritisation from studies of pedogenesis is nonetheless unrealistic and untenable. Soil and weathering systems share common processes and overlap, except where spatially separated in very deep profiles. There is also increasing evidence for the importance not only of organic acids and the formation of organo-metallic complexes in the mobilisation and re-precipitation of the major elements involved in laterite genesis (Al, Fe, Si), but also for the action of bacteria in mobilising many ions (including Fe^{2+} and Al^{3+}) under pH and Eh conditions that would normally render them stable (McFarlane, 1976, 1983a, 1987a; Butty & Chapallaz, 1984).

Nevertheless, because many (if not most) lateritic crusts have developed over 10^6–10^7 years, and may have experienced phases of surface erosion and dismantling (Nahon, 1986), followed by renewal of a soil cover, they are often seen as the *parent materials* for contemporary soils. For these reasons laterites have often been studied separately from the soils with which they may be associated. The study of *lateritisation processes* therefore draws the concepts of soil science, but also depends critically on the methods of geochemistry and biogeo-chemistry that are common to both pedology and geology (Nahon, 1986; Tardy & Nahon, 1985). However, the under-representation of microbiology in laterite research may have biased many conclusions (McFarlane, 1987a).

Lateritisation processes and their products should occupy a central place in the study of tropical geomorphology. Too often, the subject has been seen as a specialised topic, though McFarlane (1971, 1976, 1983a), and Goudie (1973) have provided a geomorphological context for much recent work. This centrality derives only in part from the resistance of duricrusts to erosion, and the protection which they afford to deep weathering profiles and to weak rocks. Such deposits are equally significant in their formational processes and history, since they influence and record the history of the landsurface itself (McFarlane, 1971, 1976, 1983a; Thomas, 1989a,b; Bourman, 1993).

Laterite as defined above may, initially, be a relatively soft material which can be dug with a spade, but such lateritic horizons may become *indurated*, to form hardened, and partly crystalline *duricrusts* (French: *cuirasses*). The process of induration is complex and involves dehydration and an increase in the crystallinity of the sesquioxides present (especially of goethite and haematite). Further loss or breakdown of kaolinite clay is also common. It is probable that this process requires a change to the weathering environment as a result of stream incision and/or a less humid climate, both of which lead to changed groundwater regimes and usually to a fall in the level of regional water-tables, and leading first to frequent wet–dry cycles and then progressively to the desiccation of the deposit.

According to Nahon (1986), 'ferricretes (hard iron crusts), attaining several meters in thickness and consisting of several facies, are extremely widespread in tropical countries. They are pedogenic in origin and result from the lateritic weathering of a variety of sedimentary, meta-morphic and igneous rocks.' It should not, however, be thought that this implies that vast areas of land in the tropics are blanketed by ferricrete. Individual occurrences commonly cover quite small areas. This statement also ignores the fact that some ferricretes were probably never laterites *sensu stricto*, but were formed by the precipitation of Fe^{3+} from laterally moving groundwater, within the voids of sandy sediments or other host materials. According to Ollier and Galloway (1990), most ferricrete duricrusts lie *unconformably* over saprolite and are often comprised of Fe-cemented alluvium. These authors specifically divorce the formation of the ferricrete horizon from the evolution of the underlying mottled and pallid zones, usually considered to be genetically linked to the hardened laterite as part of a complete laterite profile (Nahon, 1986). Although this view is not widely held as a general proposition, many instances certainly exist where alluvial and colluvial sediments have become ferruginised (but not 'lateritised', since the Fe_2O_3 is precipitated directly between the sedimentary grains) to form ferricreted benches and glacis, and these can also occur as hilltop cappings following relief inversion. The distinction between laterite and ferricrete may therefore be more than a question of hardening and involves the complex formational histories that can differ between sites.

Debates about duricrusted laterites have often centred on the sources and processes of iron enrichment that lead to the formation of ferricrete. This problem subsumes questions about iron mobility, vertical and lateral movements of groundwater containing reduced ferrous, or complexed organo-metallic iron, and the landscape or relief control over such movements. The reader is directed here to some of the major pioneer works in this field (Campbell, 1917; Prescott & Pendleton, 1952; Alexander & Cady, 1962; Sivarajasingham *et al.*, 1962; Maignien, 1966) as well as to more recent discussions (Goudie, 1973; McFarlane, 1976, 1983a; Schellmann, 1981; Valeton, 1983; Valeton *et al.*, 1991; Butty & Chapallaz, 1984; Nahon, 1986; Bardossy & Aleva, 1990). Kronberg and Melfi (1987) have drawn attention to the wider setting or *context* of laterite formation in their discussion of

'the geochemical evolution of lateritic terranes', a topic also addressed by McFarlane (1976, 1983a). Similarly, a broader view is also able to bring together the study of laterites and bauxites into a coherent framework (see also McFarlane, 1983a,b; Bardossy & Aleva, 1990).

Notwithstanding dissenting voices from Australia (Paton & Williams, 1972; Ollier & Galloway, 1990; Taylor *et al.*, 1992), lateritisation phenomena are generally regarded as characteristic of tropical weathering environments. The view that laterites may develop in temperate environments given sufficient time is not widely held, but serious questions have been raised by research into the bauxitic profiles developed on early Tertiary basalts in New South Wales (Taylor *et al.*, 1992). The age and relationships of these profiles indicate development under moist temperate conditions at 1000 m asl in a palaeolatitude of 57.5 °S. However, palaeoclimates of the early Tertiary may have been tropical in character to latitude 45 °S and therefore warmer than today, especially during the summer months, at higher latitudes. Comparable profiles from Antrim, Northern Ireland were also formerly considered 'tropical', but a reassessment by Smith and McAlister (1987) proposed a warm, seasonal sub-tropical climate. In neither case are the weathering products truly ferrallitic, while the climates of the early Tertiary may have had seasonal regimes with low and high temperatures (leading also to ambiguous pollen spectra) (Crowley & North, 1991). However, bauxites are also found on alkaline rocks at over 1500 m in southeastern Brazil and it would be wrong to indicate that gibbsitic residues cannot form over suitable rocks in non-tropical climates.

Assessment of many of these problems will be influenced by the spatial and time scales used for enquiry. Viewed over long time periods (10^6–10^8 years) and for extensive terrains, lateritic residues appear as weathering products, associated with deep, kaolinised profiles (Nahon, 1986; Kronberg & Melfi, 1987). During such long-term evolution, a succession of palaeoclimates will have been experienced, and landsurface stability is unlikely to have been maintained everywhere. In some cases, ferrallitic residues may have become

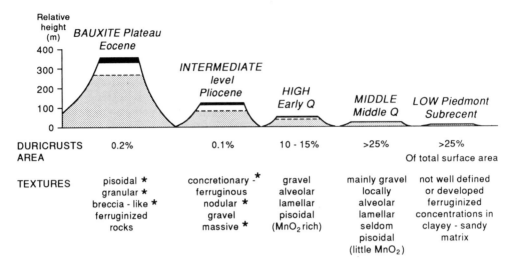

FIGURE 4.3   Schematic diagram through the Blafo-Guéto area on the Ivory Coast, indicating variations of surface area, and texture of duricrust/bauxite horizons at successive levels (after Grandin, 1976; reproduced by permission of Elsevier Science Publishers BV)

destabilised (chemically) leading to loss of iron and silica and the formation of allitic (bauxite) materials. Some physical disintegration, surface erosion and re-cementation by Fe_2O_3 can also occur, so that the upper lateritic layer becomes a complex structure incorporating allochthonous material, but is not necessarily detached genetically from the strictly *in situ* saprolite beneath.

Differences in the composition and profiles of lateritic duricrusts have been held to vary with the age of the deposit, particularly following a study of duricrusts on the Ivory Coast by Grandin (1976) which indicated the association of bauxites with the older (Eocene?) landsurfaces (Figure 4.3). Such observations are not refuted, but detailed studies have shown that the composition of lateritic weathering products varies greatly within a single deposit (Thomas, 1980; Figure 4.4) and also with bedrock type (Figure 4.5), and catenary position (Taylor & Ruxton, 1987). The dangers of premature generalisation or assertion, based on a few samples, are particularly well illustrated by the laterite family of deposits.

FERRALLITIC WEATHERING AND THE LATERITISATION PROCESS

In Chapter 2 the progress of chemical weathering has been discussed, and following the 'French' school of Pedro (1968, 1984) and Duchaufour (1982). In this scheme, *allitisation* (Pedro, 1984) is viewed as the final stage of plasmogenic alteration, forming a residue rich in gibbsite and ferric oxyhydroxides, but depleted in silica (and therefore kaolinite). Where kaolinite is dominant, along with ferric oxyhydroxides, gibbsite is present only in minor quantities.

In other words, in desilicified materials, variable proportions of Al_2O_3 and Fe_2O_3 combine to form aluminous laterites or actual bauxites. But if SiO_2 is retained in the saprolite, it is likely to remain in combination with Al^{3+} as kaolinite clay, in which case, the Fe_2O_3 may be dispersed throughout the material and weakly bonded to the kaolinite platelets. A ferricrete low in Al_2O_3 may form if the water regime favours movement of Fe^{2+} in the profile, while retaining most of the kaolinite.

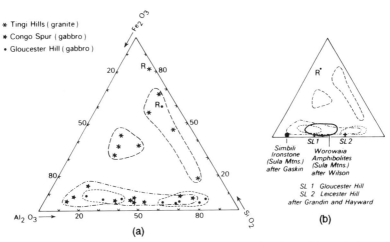

FIGURE 4.4 Composition of some duricrusts from Sierra Leone, showing wide variation within small areas. (a) Author's samples from Tingi Hills (granite) and Freetown Peninsula (gabbro). (b) Published data for Sula Mountains (amphibolitic schists) (after Wilson & Marmo, 1958) and Freetown Peninsula (after Grandin & Howard, 1975). (Reproduced by permission from Thomas, 1980)

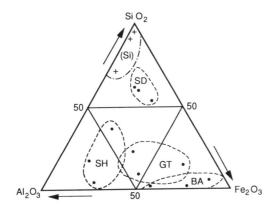

FIGURE 4.5 Ternary diagram to show characteristic compositional ranges for duricrusts (ferricrete) derived from different rocks, and also for some silcretes (Si). SD = sandstone, SH = shale, GT = granite, BA = basalt. Compiled mainly from Goudie (1973) and Schellmann (1981) (after Thomas, 1986; reproduced by permission of Chapman & Hall)

Although mobilisation of Al^{3+} can occur, it is seldom exported on a sufficient scale to leave residues rich in Fe_2O_3 and/or SiO_2 by the process of relative enrichment alone. Small amounts of chromium, titanium (as anatase, TiO_2) and nickel also occur in laterites and may become concentrated as economic ores, as discussed below.

Kaolinite arises from the early weathering of primary silicate minerals and small quantities of gibbsite may form, even at this stage, as the alkali and alkaline earth cations are quite rapidly lost. Schellmann (1981, p. 3) then characterised three later and more intensive stages or pathways of weathering:

(a) *incongruent dissolution* of kaolinite with formation of gibbsite. Because only SiO_2 is removed, Al_2O_3 and Fe_2O_3 are concentrated;

(b) *congruent dissolution* of kaolinite by which Al_2O_3 and SiO_2 are simultaneously removed in a still unknown form; only Fe_2O_3 is accumulated;

(c) dissolution of quartz; Al_2O_3 and Fe_2O_3 are concentrated.

Clearly this list requires an understanding of the conditions leading to the dissolution of kaolinite and quartz. Traditionally quartz had been seen as little affected by these processes, but views have been revised in the light of new evidence (Bennett *et al.*, 1988). The course of weathering is usually thought to be strongly influenced by the mineralogy of the parent rocks; basic and alkaline rocks giving rise to incongruent dissolution of kaolinite and the formation of gibbsite, while congruent dissolution of kaolinite characterises acidic rocks. Similar controls are exercised over the formation of iron hydroxides. These questions have been discussed at some length by McFarlane (1983a, 1987a, 1991), who has questioned the importance of parent rock composition in bauxite formation, and called for quite large-scale mobilisation of aluminium by chelating agents exuded by microorganisms.

The origins of iron in laterites

A majority of common rocks contain only small quantities of Fe, perhaps 1–2%, although this proportion can rise to 10–20% in ultramafic intrusives and in some metavolcanic rocks. It is therefore clear that thick ferricretes are likely to occur only if certain conditions have been satisfied. At its simplest this can be due to one of two circumstances:

(1) where the rock is rich in Fe but low in Al and Si;
(2) where Fe^{2+} is imported into the ferricrete horizon from another site.

Similarly, the formation of alucrete (bauxite) is favoured by:

(1) the presence of rocks rich in Al but low in Si;
(2) where Fe^{2+} is exported to another site, by *chelation* for example.

In both cases the mobilisation and movement of Fe^{2+} is essential to arguments about laterite formation, except in situations where rock composition becomes the overriding factor. The source or provenance of the Fe^{2+} has given rise to vigorous debate. The possibilities are three:

(1) vertical translocation, whereby Fe^{2+} undergoes movement over tens of metres, either downwards by gravitational movement of groundwater, or upwards by *ionic diffusion* (see below);
(2) lateral transfer by convergent flow of shallow groundwater into reception sites, such as valley floors and depressions;

(3) retention and accumulation in surface horizons (in the form of pisoliths or nodules) as the weathering profile is lowered.

All these transfers will be explored in the following discussion.

According to Nahon and Bocquier (1983), there are two pathways for iron hydroxide formation:

(1) directly by re-precipitation of iron as oxyhydroxides at a very short distance (<1 to 100 μm) in the form of a peripheral pseudomorph after the original mineral, or
(2) by co-formation with clay minerals as diffuse segregations, patches, and micronodules.

Nahon (1986) further emphasised that there will be intra- and inter-mineral, small-scale migrations of Fe between crystals that are touching and weathering together. This is seen in thin section as haloes of ferruginous material around Fe-bearing minerals, and by the presence of Fe^{3+} in the lattice of white, Fe-depleted kaolinite adjoining Fe-bearing minerals. During these processes, the red coloration increases, while the dissolution processes described by Schellmann (1981) progress. However, the lithologic structures of the rock can be retained in the case (1) of pseudomorphic (isovolumetric) substitution of Fe_2O_3 for silicate minerals. These processes of iron segregation can continue alongside the leaching of the alkalis and alkali earths and the formation of kaolinite. The loss of SiO_2 will also continue, leading, under favourable conditions, to the breakdown of the kaolinite clays and etching of the quartz grains. At this stage (2), large-scale breakdown of the saprolite can lead to loss of volume. The breakdown of clays within tropical weathering systems has also been seen in connection with podzolisation and the formation of white sands (see Chapters 2 and 8), where Fe^{2+} is also lost in a process of *ferrolysis* (Brinkmann, 1970) or *deferrification* (Sieffermann, 1988). This process is the antithesis of laterite formation and can be responsible for the dissolution of ferricretes, though Butty and Chapallaz (1984) claimed that ferrallitisation can supersede podzolisation if hydromorphic conditions give way to free drainage in the profile.

Stages in the formation of laterite

Nahon (1986) has carefully defined stages in the breakdown of the rock fabric and formation of a new pedological fabric in the formation of laterite (Figure 4.6), the physical reorganisation being a consequence of geochemical changes. His description of a characteristic mineralogical evolution is followed here (Figure 4.7).

First, it is possible to recognise the pseudomorphic replacement of rock minerals by ferruginous septa. This process is mainly important in rocks rich in primary iron bearing minerals. In most profiles, the lateritisation process is first seen in the appearance of Fe-stained patches and segregations of poorly crystalline Fe_2O_3 (soft nodules). The coloration is due to the adsorption of hydrated Fe^{3+} ions, replacing other exchangeable cations, on the kaolinite crystals. Moving upward through the profile the following changes are usually seen in parallel:

(1) corrosion, fissuration and dislocation of quartz as the rock fabric is reorganised in the clay matrix;
(2) comminution of kaolinite crystals with increasing substitution of Fe^{3+} in the octohedral sheet;
(3) increasing adsorption and segregation of Fe_2O_3 hydroxides with increased red mottling of the material and appearance of large segregations (mm^{-1}–cm diameter).

These processes lead to the formation of what has usually been described in the literature as the *mottled clay layer*, and it is frequently argued that it is this which evolves by repeated incremental increases in the Fe_2O_3 content, to form a ferricrete. It is this further evolution that poses many of the most difficult problems in the study of laterites, because the small-scale micro-transfers of iron do not appear sufficient to explain the iron content of many deposits. It is therefore necessary to determine at least two parameters for this system:

(1) the geochemical (or biogeochemical) conditions favouring the inward migration of iron into the developing ferricrete, and

(2) the source or sources of iron capable of supplying the observed Fe_2O_3 content of ferricretes.

Nahon and Bocquier (1983) have shown how Eh values can determine iron migration (Figure 2.5). For example goethite, which is stable at Eh $+0.887$ V, is increased in solubility by a factor of 10^4 if Eh decreases to $+0.2$ V. Such changes occur, along with fluctuations in pH, in the vicinity of the water-table or as a consequence of wetting and drying within the vadose layer.

Transfers of iron are therefore brought about by local reduction of Fe^{3+} that moves into solution as Fe^{2+}, gradually depleting the surrounding saprolite which becomes bleached and may lose kaolinite as well by leaching and dispersion. These changes lead to increases in voids, seen first as a change in microporosity, but developing larger macropores at a later stage. These can become the sites for infilling, often by secondary kaolinite which can become indurated by further fluxes of Fe^{2+} that becomes oxidised to haematite and in the form of indurated nodules. Aluminium which is liberated during the dissolution of kaolinite frequently accumulates *in situ*, often as Al-haematite. But some part of the Al may become mobilised in the presence of chelating compounds.

The iron *segregations* are at first soft and weakly crystalline, but they develop into hard, crystalline *nodules*, that in some circumstances are *pisolithic* (or pisolitic). As the crystalline phase becomes more continuous such pisoliths or more irregular nodules can coalesce to form a massive ferricrete which may be *conglomeratic*, *vermiform* (or vermicular) or *vesicular*, or have a variety of individual features. The continued replacement of kaolinite and quartz by iron in the upper profile thus leads to an increasing continuity of the crystalline phase which is characteristic of hard, ferricrete duricrusts. According to Nahon (1986) the replacement process increases the Al-haematite (13 mole% Al_2O_3 is possible) content and the purple-red coloration. He summarised the result (p. 179):

the juxtaposition and tight cementation of previously formed ferruginous nodules with the

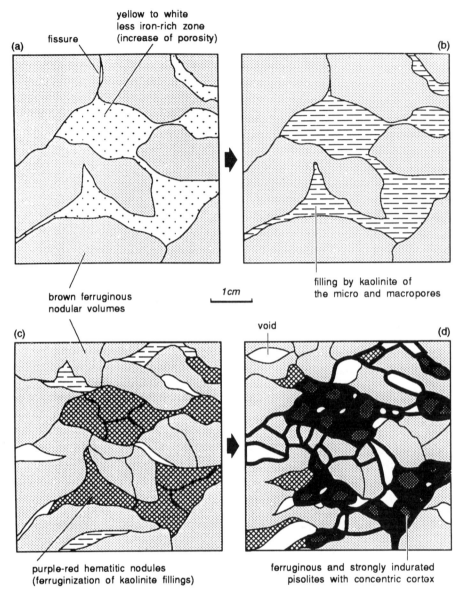

FIGURE 4.6   Sketches of the formation and evolution of successive iron crust facies. (a) Mottled clay layer with ferruginised nodules and bleached zones; (b) secondary fillings of kaolinite in bleaching zones; (c) ferruginisation of kaolinite and the formation of pseudo-conglomeratic iron crust; (d) evolution of haematite nodules into pisolites with the formation of pisolitic crust (after Nahon, 1986; reproduced by permission of Academic Press)

simple gritty facies and the ferruginous purple-red nodules form an indurated iron crust layer with a pseudoconglomeratic facies. This conglomeratic facies is the result of an *in situ* geochemical evolution, contradicting any detrital origin for such iron crusts.

Nahon (1986) argued that the pseudoconglomeratic structure is subsequently replaced by a *pisolitic* structure in the most indurated laterite facies. According to the same author, dissolution cracks may appear in this pisolitic crust, leading to

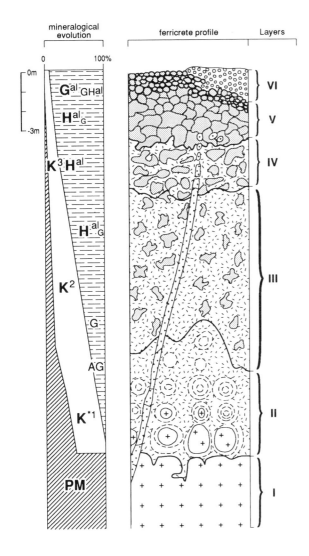

FIGURE 4.7 Sketch of an idealised ferricrete profile and of its mineralogical evolution. Layers: I, unweathered parent rock; II, weathering layer preserving rock textures; III, mottled clay layer with destruction of rock microstructures; IV, transitional layer III–V; V, ferricrete layer with nodular or pisolitic structures; VI, ferruginous pebbly layer from ferricrete dismantling. Mineralogical evolution: PM, parent minerals; K^{*1} well-crystallised kaolinite; from K^1 to K^3 crystallinity diminishes; AG, amorphous to poorly crystallised oxyhydroxides; G, goethite; H^{al}, aluminous haematite; G^{al}, aluminous goethite. The symbol size indicates the relative amount of each iron compound (after Nahon, 1986; reproduced by permission of Academic Press)

dismantling of the laterite into a ferruginous, pebbly layer with an earthy matrix, and this forms the end product of 'iron crust evolution' (Figure 4.7). The nodular zone of developing ferricrete can exhibit some complexity, involving more than one generation of iron enrichment and precipitation, or other features. Some of these problems are discussed below.

McFarlane (1976, 1983a) has taken a different view of an essentially similar progression. She distinguished two classes of laterite: *pedogenetic laterite* that forms in the vadose zone, unaffected by the water-table, and *groundwater laterite* resulting from iron segregation within the narrow range of oscillation of the groundwater table. The pedogenetic laterite forms essentially as described by Nahon (1986, and above), but its development is limited by the supply of iron from within the profile. Two circumstances can increase this supply:

(1) movement of iron into the profile by lateral transfer from an upslope source as Fe^{2+} (Maignien, 1966);
(2) incorporation of additional Fe by bedrock weathering and lowering of the entire profile (McFarlane, 1983a).

Whilst the first of these mechanisms is readily understood (Figure 4.8), the second requires more elaboration. McFarlane (1983a, 1986) has explained a possible sequence of development (Figures 4.9 and 4.10). Instead of regarding the weathering profile as a static column with its different horizons defined by stable water-table relationships, it is viewed as a descending column in which the horizons are continuously transformed as both the ground surface and the weathering front are lowered. This idea was first put forward by Trendall (1962) to explain the thickness of ferricrete crusts which he also thought progressively imposed a condition of flatness on the landscape. In Figure 4.9, the formation of pisoliths and their transformation into a massive 'vermiform' laterite is shown as a series of weathering stages developing through time as the ground surface is lowered. An essentially similar progression was envisaged by Bremer (1981, Figure 30). Because the entire landscape is lowered during

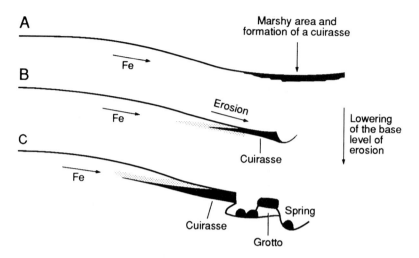

FIGURE 4.8   Lateral migration of sesquioxides and the formation of a ferricrete crust (cuirasse). (A) Migration of Fe in groundwater by organic acids and accumulation in marshy valley floors; (B) and (C) incision of drainage with a fall of water-table leading to induration and erosion of the duricrust (after Maignien, 1966; reproduced by permission of UNESCO, ©1966 UNESCO)

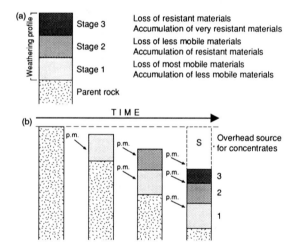

FIGURE 4.9   Stages of weathering and their temporal and spatial relationships, showing accumulation of resistant materials from overhead sources as the ground surface is lowered through time (after McFarlane, 1983a; reproduced by permission of Academic Press)

this evolution, iron may be imported from the underlying rock as a resistant mineral (Fe_2O_3), while mobile constituents are removed in solution. The Fe_2O_3 initially forms soft mottles (Stage 2). As the ground surface is lowered (see Chapter 11), these mottles enter the upper horizon (Stage 3), where they harden into segregations (nodules or pisoliths),

as a result of rapid wet–dry cycles in the upper vadose zone of the profile. The continued lowering of the weathering column by loss of solutes and fine sediment will lead to the accumulation of the iron segregations as 'packed pisoliths' at the surface (Figure 4.10) (though probably covered by topsoil resulting from bioturbation). Continued local transfers of Fe^{2+} can then result in a massive, largely crystalline ferricrete or duricrust, as described above (Nahon, 1986).

There is scope within this hypothesis for the addition of resistant clasts and further iron from upslope sites ('S' in Figure 4.9), but the process is not dependent on these increments. McFarlane's hypothesis on its own is unlikely to satisfy all field situations, but it does couple duricrust formation with the general lowering of the landscape (*etchplanation*, see Chapter 10), and de-couples the process from the requirement for laterites to form under conditions of advanced planation.

In summary, the rival hypotheses for ferricrete formation can be simplified to the following options.

(1) Profile development beneath a stable landsurface of low relief, requiring relative enrichment of iron by leaching of other rock

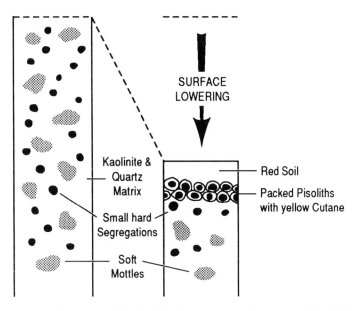

FIGURE 4.10 Accumulation of packed iron oxide pisoliths by removal of quartz and kaolinite matrix as the ground surface is lowered. This process can lead to a continuous skeleton of ferricrete as Fe is re-dissolved and precipitated between the pisoliths (after McFarlane, 1991; reproduced with kind permission from Pergamon Press Ltd, Headington Hill Hall, Oxford OX3 0BW, UK)

constituents, and additions of Fe^{2+} via upward ionic diffusion with minor lateral transfers.

(2) Continuous accumulation of iron within the upper horizon of a descending column of weathered rock, as both ground surface and weathering front are simultaneously lowered.

(3) Inward movement of Fe^{2+} by lateral transfer in groundwater towards reception sites in depressions and valley swamps, that subsequently become positive relief features by subsequent relief inversion.

A number of important issues are not, however, addressed by these processes, and these include:

(1) the link between ferricrete horizons and thick underlying pallid zones;
(2) the role of environmental change in the conversion of laterites to duricrusts;
(3) the conditions for the further leaching of iron to form bauxite (alucrete).

These issues will be taken up below, but it is clearly simplistic to consider the occurrence of thick,

leached (*pallid*) zones in duricrusted profiles as a direct result of ferricrete formation (Bourman, 1993). Such zones may be tens of metres thick and are likely to have deepened *after* the formation of the ferricrete cap, during incision of the landscape (Thomas, 1980, 1989b). It seems that iron is progressively leached to lower levels as the water-table falls. Loss of Fe^{2+} *upwards* as a result of ionic diffusion still remains speculative (Lelong, 1966, 1969; Ollier, 1988a).

Lateritic (residual) bauxites

Bauxites are usually considered the most 'evolved' of the ferrallitic weathering products, and, because Fe^{2+} has usually been lost from the profile, their development is often considered separately. Such residual bauxites are, however, best regarded as *aluminous laterites*. Boulangé and Millot (1988) have proposed the following modes of formation.

(1) Bauxites derived by *direct* or *indirect* alteration of the parent rock:
    (a) *direct alteration* of alkaline rocks to bauxite

is possible, given favourable climatic and drainage conditions; textures and structures of the rock fabric will be preserved in an isovolumetric alteration;

(b) *indirect alteration* involves a kaolinitic (ferrallitic) phase of weathering, and the bauxite is formed as a gibbsite phase due to later desilication; this phase can also involve enrichment of the deposit by Al_2O_3 and Fe_2O_3.

(2) Bauxites derived by *erosion*, and deposited as breciated, nodular or pisolitic deposits (or facies), are also common and may have undergone further geochemical evolution (progressive or regressive).

These broad classes also define petrographic and geomorphic conditions, since those forming by direct alteration neither exhibit nor require any prior development of a weathering profile, and will replace suitable rocks, wherever groundwater conditions provide for the continuous leaching of more mobile constituents. In some cases Al-laterites will reflect the active downwasting of interfluves. This may involve the direct replacement of fresh rock (Thomas, 1980), but can also occur within a deeply weathered (ferrallitic) landscape undergoing dynamic lowering, as found in western Kalimantan (Rodenburg, 1984). In other instances, the deepening of weathering profiles below a pre-formed laterite and bauxite cap (Thomas, 1989b; Plate 3-VI) must be inferred (McFarlane, 1983b). In neither case is the bauxite itself a reflection of the age of formation of any extensive planation surface from which the present relief is derived (McFarlane, 1987a).

Nevertheless, time is also a factor in the evolution of weathering mantles and cannot be discounted from the argument. A two-stage development of bauxite profiles (by indirect alteration—(2) above) has been proposed by Ghosh and McFarlane (1984) and their model is shown in Figure 4.11. In Stage 1, the ferricrete layer has evolved as in Figures 4.9 and 4.10, by lowering of the ground surface. This is accompanied by relief reduction which leads to high water-tables and the congruent leaching of kaolinite, with loss of SiO_2. In Stage 2, incision of

the old landsurface leads to falling water-tables and vadose conditions in the upper profile. Incongruent dissolution of kaolinite leads to relative enrichments of the surface with Al_2O_3 as further SiO_2 is leached. Fe_2O_3 may be leached down profile, as shown in the analysis from the Sula Mountains of Sierra Leone (Figure 4.12), but it is common for any crystalline duricrust to remain as a capping to the bauxite (Figure 4.11, Stage 2 'or', and Plate 3-VI). The dissected remnants of ancient landsurfaces provide ideal geomorphic and relief positions for this type of evolution.

FACTORS AFFECTING LATERITISATION AND DURICRUST FORMATION

It can be seen that the factors which influence the course of lateritisation are those which affect all weathering and pedogenesis (Jenny, 1941), namely:

(1) geological factors, mainly rock mineralogy;
(2) climatic factors, particularly the rainfall regime;
(3) biotic factors, vegetation and biological activity;
(4) hydrologic factors, including soil and groundwater regime;
(5) Eh and pH conditions;
(6) geomorphic and tectonic history, planation and dissection;
(7) the time factor and ageing of duricrusts;
(8) palaeoenvironmental changes affecting factors (2)–(5).

Geological factors

Geological factors affecting laterite/bauxite formation can be very strong, in the sense that the composition and mineralogy of rocks strongly influence the composition of the laterite and duricrust. However, this factor operates in conjunction with more important or general factors such as the regional climate and hydrographic regime, both of which can determine the products of weathering and, according to some authors, override the influence of bedrock.

It has been observed that the progress of weathering into ultramafic and mafic rocks is more rapid than in felsic types, and may be three times

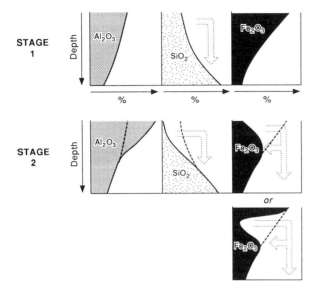

as fast (Nahon, 1986). Since these rocks contain more Fe and less Si than felsic rocks, there will be an enhanced tendency for Fe_2O_3 segregation and for Al_2O_3 to accumulate (often as gibbsite) in preference to kaolinite formation. As a result, aluminous laterites (bauxites) form under favourable environmental conditions. These aluminous crusts are usually less crystalline and remain earthy in texture. Exceptions to this rule include ironstone rocks within greenstone belts and some ultramafic rocks. These can contain up to 20% primary Fe and can indurate rapidly during the early stages of lateritisation by pseudomorphic replacement (Plate 3-VII). In general, rocks containing excess Si or Fe will require a much longer period of ferrallitic weathering to attain the status of bauxites (Butty & Chapallaz, 1984), and for this reason bauxitic residues on such rocks are found on older landsurfaces and where other formational factors are favourable (Figure 4.3). In Figure 4.5, examples of the variable composition of indurated laterites or ferricretes are shown in relation to parent rock. These reveal clearly that bauxites, for example, occur particularly over feldspathic rocks low in silica.

FIGURE 4.11 A suggested two-stage development of bauxite profiles. Stage 1: as relief is lowered the vadose zone narrows and SiO_2 is leached, with a relative accumulation of Fe_2O_3. Stage 2: incision leads to strongly vadose conditions with leaching of SiO_2 and some Fe_2O_3, resulting in a relative accumulation of Al_2O_3 (after Ghosh & McFarlane, 1984a; reproduced by permission of the Hindustan Publishing Corporation)

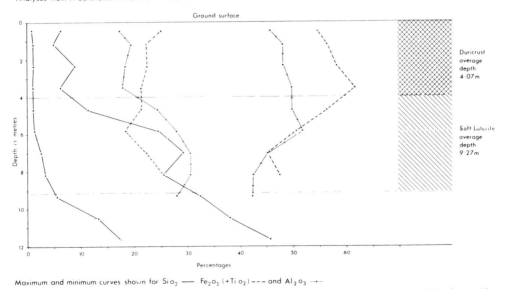

FIGURE 4.12 Mineralogical progressions through a duricrusted, aluminous laterite profile from Sierra Leone, illustrating aspects of the model shown in Figure 4.11 (after Wilson & Marmo, 1958; reproduced by permission of the Government of Sierra Leone)

The above remarks assume an *in situ* process of lateritisation, accompanying the formation of a primary, residual weathering profile in silicate (usually feldspathic) rocks. This, however, ignores the formation of bauxite ores in sediments which have been derived by erosion of such profiles. These sediments were called *laterite derived facies* by Goldbery (1979) and are terrigenous sediments that have accumulated in coastal plains or tectonic basins. They are produced by the re-working of lateritic profiles and constitute a particular type of 'Red Bed' (Pye, 1983). These sediments may contain kaolinised boulders and ferricrete fragments within poorly sorted sequences of sand and clay (which may exhibit white and red hues). They are commonly rich in stable heavy minerals and have a high content of kaolinite, haematite and gibbsite. They may be described as having a low mechanical but a high chemical maturity. They may interfinger with marine or lacustrine deposits that contain evaporites and lignite.

When such sediments are uplifted, a further phase of lateritisation, or perhaps podzolisation, has often occurred with the formation of rich bauxite deposits. These often show lateral gradations into ferruginous laterites (Valeton, 1983). Many would regard these formations as *exolaterites* or false laterites (Bardossy & Aleva, 1990).

The occurrence of *karst bauxites* (Bardossy & Aleva, 1990) trapped within fissures and caverns in limestones and dolostones poses intriguing problems of provenance for the Al and Fe contained within them. Evidence for the residual accumulation of bauxitic residues from the weathering of impure carbonate rocks or from overlying formations during active downwasting and development of karstified terrain, provides one model for the formation of such deposits (Smith, 1971). However, McFarlane (1983a) noted that an allochthonous origin for the bauxitic clay is favoured by many authors, indicating possible transport to their present sites from lateritic hinterlands on silicate rocks.

Climatic factors

Climatic factors controlling laterite formation appear strong, but are not easy to define, partly because most deposits have experienced repeated environmental changes during their development. There is also a difference in optimum conditions, between those leading to progressive ferrallitisation (bauxitisation) of the profile, and those favouring Fe segregation and induration. This distinction not only separates the environments of ferricrete formation from those of bauxitisation, but also suggests that induration *may* be favoured by conditions not conducive to ferrallitic alteration and lateritisation.

Temperature conditions

Limiting temperature conditions for laterisation are difficult to determine. But suggestions that laterisation can occur in temperate environments (Prescott & Pendleton, 1952; Paton & Williams, 1972; Taylor et al., 1992) should be treated with caution for reasons already stated. The association of laterites with ferralsols (oxysols) (Pedro, 1984) makes it clear that these alteration products are formed where seasonal rainfall comes in the hot season (Maignien, 1966). Extensions of these conditions beyond the tropics occur mainly on continental east coasts (southeast Brazil, Natal, eastern Australia, for example). Ferrallitic residues certainly form in the wetter parts of these areas, but ferricrete is not widely found, except in convergent footslope sites. In southeast Brazil the southernmost occurrence of bauxite is at Lages (28 °S), where it occurs on an elevated plateau site (900 m) and is confined to the outcrop of alkaline rocks (Oliveira, 1985, Valeton, et al. 1991).

Annual temperature averages are probably a poor guide to process operation. But, where *soil* water temperatures are maintained close to 27 °C, conditions will be near the optimum, and soil water temperatures > 22 °C probably remain effective. These will occur in areas beyond the tropics which experience summer rains. Many ancient laterites developed in palaeoclimates different from those of today, and there remains considerable uncertainty about this issue. The argument that with the elapse of time cool climate weathering environments will produce laterites and bauxites does not accord well with field occurrences, and this seems to imply that the processes either become so slow as to be ineffective or cease altogether. If as McFarlane (1987a) has argued, microorganisms are involved, then these are likely to be sensitive to temperature variation.

Rainfall regime

Uncertainty about rainfall conditions is also apparent. According to Nahon (1986, p. 185), 'the formation of *iron crusts* [author's italics] is favoured by humid tropical climates with a long dry season that allow successive small-scale redistribution of iron by leaching.' In this view, more humid climates do not favour alternate reducing/oxidising conditions over the very short distances (1–100 μm) involved, while more arid climates lead only to superficial leaching. Nahon (1986) placed the West African iron crusts most frequently in the range 600–1000 mm, but in the author's experience this upper limit is too low. Ferricretes certainly form under higher rainfalls, and Maignien (1966) gave 1250 mm as a *lower* limit over granitic rocks and 1100 mm for basic or felsic rocks. Laterites may possibly become more aluminous where precipitation exceeds > 1200 mm (Bardossy & Aleva, 1990) and if annual rainfall is distributed over 9–11 rainy months, while Butty and Chapallaz (1984) have claimed that full ferrallitisation requires > 1300 mm of rainfall per annum. Young (1976) observed that lateritic formations are found developing within the forested tropics, and periods of drying out in the upper profile are frequent in most tropical climates with a dry season. It is probably not possible to place limiting values around the environments of lateritisation that will be valid for all occurrences. The apparent contradiction between those who require savanna conditions for laterite formation and those favouring a forest climate, can be partly resolved by reference to the following arguments:

(1) that duricrusting may be a separate and later phase following lateritisation, and one that requires a drier climate;
(2) that, irrespective of (1), climate change has affected almost all major laterite deposits, usually involving many wet–dry oscillations;
(3) that mafic rocks with high Fe content will form ferricretes more readily and in drier climates;
(4) that the course of lateritisation depends crucially on other factors of the environment (below).

A new approach to this problem has recently been developed by Bird *et al.* (1989), who carried out stable-isotope analyses of gibbsites from lateritic bauxites from equatorial Brazil (Paragominas) to extra-tropical locations and including the Weipa bauxite of Queensland. Their results show that ^{18}O and D depletion are compatible with results from monsoon rainfall and do not correspond with onshore trade winds or equatorial rains. This interpretation must be considered tentative, but might imply climate change to explain near-equatorial bauxite sites of today.

Biotic factors

Biotic factors can include both the role of the vegetation cover and also biological activity involving organisms and organic materials. All of these strongly influence the course of weathering, and especially the behaviour of Si, Fe and Al in weathering reactions (Chapter 2). However, the contradictions concerning the climates of lateritisation are carried into discussions of the vegetation cover. If laterites are considered as ferricretes and associated with duricrusts then there is a bias towards drier conditions of formation, but if laterites are associated with bauxites as co-formational residues of advanced ferrallitisation then the bias is towards humid and forested conditions of genesis. To resolve this dichotomy it is necessary to return to certain important distinctions.

Al-laterite or bauxite, wherever it is found as a residual deposit, represents a *relative* enrichment in Al, with a corresponding desilicification of the profile (loss of kaolinite and quartz). The behaviour of Fe in the system may vary, but relative enrichment can take place. The apparent stability of the Al in the system implies low acidity, and if both Al and Fe remain stable or are only transported over short distances then a lack of *chelating* agents may also be implied. This condition would be satisfied where free drainage favours rapid oxidation of organic materials.

Where bedrock is rich in Fe, pseudomorphic precipitation of Fe_2O_3 under these conditions can lead to strongly indurated ferricrete, containing significant Al_2O_3. Such duricrusts are found over the gabbroid rocks of the Freetown Peninsula of Sierra Leone and inland over the greenstone schists and ironstones of the Sula Mountains (Thomas, 1974a, 1980). This type of deposit was called *primary laterite* by Harrison (1934) and Bleackley (1964), though the term has fallen into disfavour.

The vegetation cover associated with such deposits today is nearly always forest, though this term includes many woodland types other than true rainforest. The role played by the forest cover in lateritisation processes is seldom specified in detail, but the following influences can be summarised (see also McFarlane, 1976, 1983a; Butty & Chapallaz, 1984; Bardossy & Aleva, 1990).

(1) Forest maintains soil humidity, whilst it prevents waterlogging (through high ET) and enhances infiltration by maintaining high permeability (especially along deep, branching root systems).

(2) Forest protects the ground surface from surface erosion especially by sheet wash and maintains slope stability (on slopes $> 20°$).

(3) The forest environment *either* has a high biological activity that maintains slight acidity (pH 4.5–6) in freely drained soil sites, favouring leaching of SiO_2, but retention of Al_2O_3, while moderate chelation can move Fe^{2+} down profile; *or*, where reduced biological activity follows from greater wetness, it favours strong chelation of Al^{3+} and Fe^{2+}, with Al_2O_3 accumulating in the B horizon and Fe^{2+} migrating within organo-metallic complexes sometimes to form ferricretes downslope.

(4) Forest accelerates silica loss from the saprolite through uptake of SiO_2 by plants within which it is stored in leaves and returned via litter in a more mobile form.

(5) Forest brings about local reduction of iron

around roots where pH is lowered; thus allowing partial leaching of Fe^{2+} from the profile.

(6) The forest environment encourages soil fauna that may play significant roles in homogenisation of the topsoil and concentration of coarse elements including Fe_2O_3 (and Al_2O_3) nodules (such processes leading to the formation of a ferricrete layer appear more effective under dry woodland types such as *Miombo*).

(7) The forest supports a range of microorganisms that play significant roles in the decomposition of alumino-silicate minerals, as agents of dissolution and in catalysing certain reactions (Bardossy & Aleva, 1990), forming complexes with Fe^{2+} and also causing desilicification (McFarlane, 1983a).

The importance of microorganisms

The role of *microorganisms* in the formation of bauxites has received relatively scant attention until recent years, although some early writers suggested its importance. Preliminary investigations by Heydeman *et al.* (1983), McFarlane and Heydeman (1985) and by McFarlane (1987a) point towards the role of *bacilli* and to significant leaching of Al from kaolins in laboratory experiments. *Bacillus circulans* has previously been implicated in SiO_2 leaching (Hess *et al.*, 1960), while the role of bacteria in Fe mobilisation and precipitation is more widely known (Byers, 1974; Chukrov, 1981), possibly mediating in the process of chelation.

The operation of these organic materials and microorganisms substantially alters the terms for mobilisation of the more stable elements in a developing saprolite, in the sense that geochemical principles may be overridden, and silica, iron and aluminium may all go into solution at near neutral pH.

Hydrologic factors

Hydrologic factors are determined by the climatic regime and its interaction with lithology and relief. They combine to create a range of local

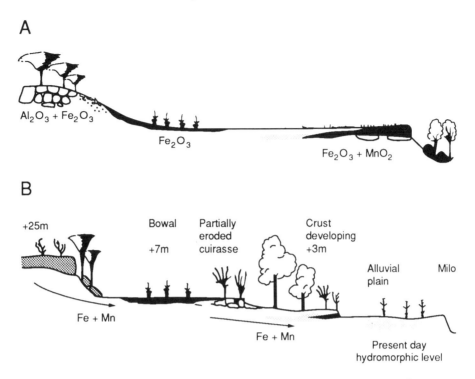

FIGURE 4.13 Association of duricrusts (ferricrete) with relief features: (A) accumulation of Fe_2O_3 on a terrace, near Labé Guinea; (B) Catena showing levels of crust formation on the terraces of the Milo River, Guinea (after Maignien, 1966; reproduced by permission of UNESCO, ©1966 UNESCO)

environments with contrasting conditions for iron mobilisation or precipitation. It has long been realised that catenary relationships exist in lateritised terrains (Maignien, 1966; Ollier, 1976), and these tend to emphasise the possibility of *lateral* transfers of Fe^{2+} in the profile (Maignien, 1966, and Figures 4.8 and 4.13). However, the formation of *groundwater laterite* is also determined by oscillations of the water-table within narrow limits that depend on conditions of low relief and poor vertical drainage.

Many earlier theories about laterite genesis insisted on the importance of extensive planation for the development of laterites (Prescott & Pendleton, 1952). This notion may have resulted from a false premise, that the flatness of many terrains containing duricrust sheets is a cause of their development. More recent studies have increasingly recognised that the low relief of lateritised plains may be a *result* of the duricrust

sheets which impose a certain flatness as landscape is lowered and the iron crust thickens and becomes more continuous. Equally, it can be argued (McFarlane, 1976, 1983a) that many landscapes exhibiting duricrust-capped hills have undergone a relief inversion, whereby former footslope ferricretes, by virtue of their resistance to erosion and dissolution in a dissected landscape, gradually replace the interfluves as hill-top cappings.

The argument is not, however, straightforward, and the following checklist draws attention to some of the issues.

(1) Al-laterites (alucretes or bauxites) require free vertical and/or lateral drainage and are frequently associated with elevated plateau edges or isolated hills, the thickness of the deposit wedging out away from the bounding slopes.

(2) Fe-laterites (ferricretes) commonly require the

addition of Fe from subjacent rock to explain the Fe content and this implies a long-term lowering of the terrain in a state close to dynamic equilibrium (see below and Chapter 8), and requiring sufficient relief to energise the landscape.

(3) Footslope-laterites develop in the region of ion accumulation at base of the catena, and although they are formed on gently sloping sites and experience a narrow oscillation of the water-table, the surrounding slopes may be steep and high, and the wider landscape is not required to have a low relief.

(4) Groundwater laterite (as distinct from (3) above) depends on local sites of low relief, and appears favoured by wide shallow valleys, such as are found on extensive plateaus in central Africa, Brazil and elsewhere on the Gondwana continents; the possibility of long-distance lateral transport of iron should not however be discounted.

Clearly, the hydrologic regime is controlled only in part by the topography, and catenary relationships vary with climatic regime (see Duchaufour, 1982). Thus leaching beneath plateau edges will not readily bring about bauxitisation if the climate is seasonally arid, while Fe deposition on footslopes does not appear common within equatorial rainforest regimes. More subtly, the locus of Fe_2O_3 deposition within a valley-drainage system may be influenced by a combination of relief and climate. In *dambo* areas (see Chapter 8) within central Africa, groundwater laterite can be found beneath the central depression in the upper valleys, where the water-table falls perhaps 3–4 m below the surface, but occurs around the valley edge in downstream and other situations where the water-table is close to the surface in the central depression throughout the year.

Eh and pH conditions

Eh and pH conditions in the saprolite profile most immediately affect the mobility of aluminium and iron within the thermodynamic system. This topic has been explored (Chapter 2) in relation to weathering reactions. It is necessary to note here only that lateritisation depends on the incongruent dissolution of kaolinite, removing SiO_2, while concentrating a gibbsite residue. This can take place when pH > 4 at 25 °C (Garrels & Christ, 1965) (Figures 2.7 and 2.13).

The behaviour of iron is more complex and again has been outlined in Chapter 2. In the present context the mobilisation of Fe^{2+} is inhibited by high pH (>8) values that exist close to the weathering front and by Eh values in excess of $+0.7$ (Figure 2.5). Theoretical Fe^{2+} solubility at pH 5/Eh 0.3 is 56 ppm, but this decreases to 5.6 ppm at pH 7/Eh 0 (Butty & Chapallaz, 1984), yet Fe^{2+} concentrations in groundwater tend to be much lower, in the range 0.01–2.8 ppm according to Tardy (1971). Higher values only occur in the presence of dissolved organic matter (*chelating agents*), and particularly in hydromorphic environments.

Aluminium has a limited mobility within natural environments, ionic forms occurring below pH 4.5 as Al^{3+} and above pH 8 as $Al(OH)^{4-}$. Aluminium ion concentrations in natural groundwater tend to lie within the range 0.005–3.2 ppm. Contrary to some assumptions, aluminium may be chelated as organo-metallic compounds as in the case of iron, and McFarlane and Bowden (1992) have demonstrated higher Al concentrations in groundwaters (from Malawi), by omitting pre-treatments that destroy the metastable organic compounds. Al mobility in this form may be high in podzolic environments. Percolation experiments carried out by Lucas *et al.* (1987) in surface horizons of podzols in Amazonia returned figures of 20–30 ppm during a 10 day experiment, by applying water at the rate of 10 mm day^{-1}. On the other hand, Al mobility is generally considered low in the ferrallitic environment, where aluminium is fixed as kaolinite. However, evidence from perhumid, equatorial areas such as Kalimantan indicates a possible transition from ferrallitic towards podzolic pedogenesis (Chauvel, 1977; Buurman & Subagjo, 1978; Lucas *et al.*, 1987, 1988; and Chapter 8). In seasonal climates, alternating mobilisation and immobilisation of Al leads to the formation of concretions of Al monohydrate and haematite (Butty & Chapallaz, 1984).

The equilibrium conditions between Al^{3+} and Fe^{3+} determine whether ferricrete or bauxite is formed. This is partly dependent on the alumina concentration in the system and Didier *et al.* (1983) have defined the stability fields of gibbsite–goethite–haematite in terms of Al_2O_3 concentration and water activity, showing that haematite formation is inhibited by the presence of Al_2O_3. Relative humidity as a function of water activity determines the haematite–goethite transformation, goethite being stable with RH over 88% according to Didier *et al.* (1983) in the absence of Al in the system, while McFarlane (1983a) reported the boundary at RH 65%.

The complexity of these systems and the lack of correspondence between theoretical solubility and measured concentrations of Al^{3+} and Fe^{2+} ions in groundwater, emphasise the empirical nature of many conclusions regarding the formation of bauxites and laterites.

Geomorphic and tectonic history

The formation, evolution and destruction of lateritic formations (ferricretes and bauxites) is closely associated with the history of the landscape. The time taken for significant laterite formations to develop ensures that they record the changing environments of weathering and pedogenesis over long periods (Bourman, 1993). It has long been recognised that indurated horizons (duricrusts) strongly influence local relief forms, but, more important is the realisation that such surface materials evolve with the total landscape. It is only when this fact is fully acknowledged that the true nature of laterites and some other surface materials such as *white sands* or *stone-lines* (Chapter 8) can be evaluated. Geomorphologists have been slow to take full account of surface mantles in developing theory; equally, many geologists and geochemists have been reluctant to enter into geomorphological discussion. The arguments adduced here for the development of laterites will be discussed again at a later stage when considering theories of landform development in the tropics (see Chapters 9 and 11).

The central difficulty of this topic is the relationship of duricrusts (of different kinds) to world-wide or regional planation; itself a controversial issue. Laterites (duricrusts) in Australia, for example, have been dated back to Cretaceous or early Tertiary times, perhaps earlier, with subsequent formation in the Middle and Upper Tertiary (Späth, 1987), and similar age bands have been suggested for other areas including the Guiana Shield (Aleva, 1984), but McFarlane (1976) warned against the use of duricrusts as markers of particular erosional plains on the basis of detailed studies in Uganda. There are many claims that bauxites and laterites can be arranged in a kind of *chronosequence*; the most evolved (bauxitic) crusts surviving on the oldest landsurfaces; the younger (ferruginous) laterites appearing as ferricretes on lower surfaces. Aleva (1979, 1984) has adduced such a scheme for the Guiana Shield, and Grandin (1976 and Figure 4.3) has claimed a similar sequence for the Ivory Coast which has also been discussed by Boulangé and Millot (1988). However, the evidence presented in these studies is not conclusive and alternative models have been offered for Sierra Leone (Thomas, 1980; Thomas & Summerfield, 1987), and Suriname (Kroonenberg & Melitz, 1983; Pollack, 1983), while McFarlane (1983b, 1987a) has discussed alternative models for general application.

Throughout the Gondwana fragments of Africa, Australia, India and South America, the rupture of the ancient Gondwanaland supercontinent in the late Mesozoic triggered intense erosion which has all but destroyed the ancient landsurfaces that were existing at that time (Partridge & Maud, 1987). Because the regional tectonic picture is varied and the initial relief of the new continents was higher towards the centre of former Gondwanaland, where, according to Partridge and Maud (1987) southern Africa already stood at 2000 m, attempts to define planation surfaces on world-wide scale have remained controversial (King, 1950a, 1956, 1962; Summerfield, 1985, 1988). Nonetheless, although each continent, even sub-divisions within each continent, has evolved independently during the late Mesozoic and Cenozoic, there may have been world-wide periods of optimum conditions for planation (Melhorn & Edgar, 1975). The significance of these surfaces in the present context may have been overstated; many emergent crystalline terranes (mostly cratons) have been exposed to supergene weathering and sub-aerial denudation for much of Phanerozoic time, during which pedimentation and etchplanation may have alternated (Fairbridge & Finkl, 1980), or continued together (Thomas, 1989a,b). By whatever model of planation adopted, elevated plateaus are likely to exhibit the most evolved profiles of weathering and contain locally rich bauxite deposits. It is convenient to group these features according to continent-wide erosional events, but many are determined more by rock structure and lithology (Kroonenberg & Melitz, 1983; Thomas, 1989b).

Nevertheless, it is important to record the widely held view that economic bauxite deposits are associated with older planation surfaces, and in particular those formed during the long, tectonically quiescent, interval from the Palaeocene to the early Miocene. With only minor disagreements over 'age' these embrace the *Sul-Americana* surface in Brazil, and possibly the Guianas (King, 1956, 1962; Valeton *et al.*, 1991); the *African* surface (King, 1962; Partridge & Maud, 1987), and the *Neturhat* surface in India (Evans, 1970, quoted by Ghosh & McFarlane, 1984) (see Tables 11.1 and 11.2).

As indicated above, the weathering profiles evolving on these developing surfaces of low relief would experience progressively more impeded drainage, leading to ferricrete (and possibly silcrete) concentrations, and ultimately to podzolisation in sites with hydromorphic characteristics. Uplift and dissection (which can be long delayed as incision works its way into the drainage basins) will trigger a new evolutionary phase, leading to the incongruent dissolution of kaolinite and progressive concentration of Al_2O_3 in the profile (Figure 4.11). The claim by McFarlane (1987a, p. 337) concerning the Darling Range of Western Australia that, 'the formation of the protore and its bauxitisation appear therefore to be chronologically separate events' probably has wide application.

Valeton *et al.* (1991) has advanced a similar argument for the bauxites found on the uplifted (horst) plateaus of eastern Brazil. A study of the Serra da Mantqueira led the authors to the conclusion that the 10–30 m thick ferrallitic weathering mantle of this region had developed on an undulating 'half

orange' topography with a relief of 100–300 m, during the Cenozoic. Faulting, uplift and dissection during the Quaternary then led to destabilisation of the kaolinite–goethite residue, overprinting it with an Al–goethite–gibbsite-etched quartz, bauxite. This type of bauxite was described as 'syngenetic', having formed during relief dissection, by modification of the pre-existing ferrallitic mantle. The free drainage and oxidising conditions retain the Fe^{3+} with the Al^{3+}, and permit the direct transformation of the rock below any ferrallitic mantle to an impure bauxite. Similar formations are found, on the Darling Ranges east of Perth, Western Australia; on the Eastern Ghats of India; the Usambara Mountains in Tanzania, and widely in West Africa. Most of these bauxites are of low grade and high in Fe_2O_3, except where they have formed on alkali ring complexes as at Lages, Poços de Caldas, and Passa Quatro in southeast Brazil (Plate 4-I).

Valeton *et al.* (1991) also cited the bauxites developed on the 'half orange' hills near Tayan in western Kalimantan, Indonesia, as examples of this type. The syngenetic nature of these deposits is evident from their conformity with the present relief (Figure 4.14). These deposits were described by Rodenburg (1984), who thought they were developing with the present relief in an equatorial climate. The most controversial aspect of Valeton's (1991) discussion is a conclusion that such 'half orange' relief has remained unchanged throughout much of the Tertiary. To a geomorphologist this seems unlikely, although the persistence of such hill forms during the dynamic lowering of the landscape might be postulated. But these convex hills show many signs of active morphogenesis; landslides and colluvium-filled hollows abound. These are mostly of late Quaternary age (Modenesi, 1988) (see Chapter 8), but it is difficult to believe that they were not similarly affected over much longer time periods involving repeated environmental fluctuations.

In contrast to *syngenetic* development of bauxite during dissection, subsidence of marginal platform areas leads to the development of hydromorphic conditions, where reducing conditions (low pH and Eh) lead to lateral or ascending solutions and

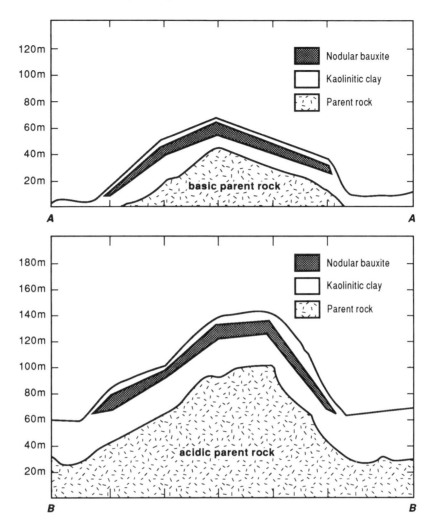

FIGURE 4.14 Dome-shaped bauxite-bearing hills, Tayan area, West Kalimantan, Indonesia. Note how the bauxite horizon follows the configuration of the present topography which forms a multi-convex (*meias laranjas*) relief (after Rodenburg, 1984; reproduced by permission of the American Institute for Mining Engineers)

precipitation of Al_2O_3 within a smectite/kaolinite saprolite, as textured stratified weathering horizons (Valeton *et al.*, 1991).

Relationship of ferricrete to bauxite

Many bauxite deposits occur beneath a ferricrete cap, and the true significance of this observation is sometimes evaded. Where present, this ferricrete capping must record *prior* conditions favouring accumulation and immobilisation of iron in the upper profile, while the underlying bauxite and kaolinite record the progressive deepening of the profile and subsequent leaching of Fe^{2+} from the saprolite as discussed above (Figure 4.11). In these conditions, the crystalline haematite and goethite in the ferricrete may remain relatively immobile and indurate to form a duricrust capping that protects the underlying saprolite from erosion. Uplift and dissection of the terrain then leads to deepening of the weathering profile, and/or to the formation of bauxite by direct alteration of the subjacent rock (Plate 4-II).

Groundwater and footslope laterites

Because of the relative mobility of iron in reducing conditions, and the rapidity with which it can be redeposited, the formation of mottles and segregations within the range of water-table fluctuation is a characteristic early stage in the formation of *groundwater laterite* (McFarlane, 1976), and this tends to occur in areas of convergent drainage (Maignien, 1966). Widespread groundwater laterite is found in seasonal climates, and beneath landsurfaces with a narrow range of water-table fluctuation. The broad planation surfaces of the Gondwana continents can provide these environments (most are savanna or deciduous woodlands), but thickening of the ferricrete will depend on one of two conditions being met:

(1) a steady influx of Fe^{2+} from upslope, by groundwater movement to sites where evaporation leads to oxidation and formation of Fe_2O_3 (Maignien, 1966, and Figure 4.8). This requires a landscape with significant relief, and in the extreme case, these are footslope ferricretes below major hillslopes. Derivation of the Fe from older duricrusts is also common (Figure 4.13).
(2) a dynamic lowering of the landscape by simultaneous weathering and low-energy surface erosion. As new saprolite is formed iron can be added from the subjacent bedrock (below) and also re-cycled from the developing ferricrete (above), as described by McFarlane (1976, 1983b).

Thus in neither case is a state of near planation, persistent over long time periods, favourable to the formation of thick Fe-laterite (ferricrete) deposits. Thin ferricretes will however form in many relatively stable landform positions during shorter episodes ($< 10^4$ years).

The phenomenon of *footslope laterites* is widespread in tropical climates with strong seasonal contrasts. Remarkable in this respect are the ferricrete surfaces around residual hill masses in Sierra Leone (Bowden, 1980, 1987; Thomas, 1980, 1983). These consist mostly of re-cemented ferricrete and bedrock breccia, and are associated with large macropores and cave systems. Where they border isolated hill masses, as in the Kasewe Hills, they take the form of pediments or *glacis*; where they border river valleys, as in the Orugu Valley of the Freetown Peninsula, they take the form of fans or terraces (Figure 4.15). Many have the appearance of *fanglomerate* facies, now strongly cemented by iron. The clear implication is that the ferricrete has formed within a sedimentary deposit dating to an arid or semi-arid period of morphogenesis (see Chapter 8), and the iron appears to have come from the more recent hillslope weathering and groundwater flow within a forested climatic regime (Figure 8.14, Thomas & Thorp, 1992).

The time factor and the age of laterites

The association of many laterites with extensive planation surfaces has led to many suggestions that they share a common 'age', but this simple idea is unfortunately inconsistent with the development of laterites, as discussed above, and with ideas concerning planation. Extensive surfaces of planation can have no single 'age', but develop through long time periods of perhaps 10^7 years (Ahnert, 1970). During such periods, the landsurface will undergo important fluctuations of environment, possibly as characterised within the *cratonic regime* described by Fairbridge and Finkl (1980) (see Chapter 11). Lateritisation and duricrusting of such surfaces are complex processes, which require a variety of conditions of formation that will determine the time and period(s) of formation. Ferricrete may therefore:

(1) commence formation beneath both interfluve and valley floor sites *during* the period of planation by dynamic lowering of the surface;
(2) develop as groundwater or footslope laterite on lower slopes, as a result of lateral transfers of Fe^{2+};
(3) become indurated during dissection of a plateau

FIGURE 4.15   Ferricreted fans or *glacis* slopes in the Orugu Valley, near Freetown, Sierra Leone. The ferricrete has formed in a brecciated sediment

surface, following uplift, with bauxitisation down-profile;

(4) become subject to dissolution and disintegration due to changes in plant cover and water regime (e.g. wetter climates, subsidence);

(5) proceed through different stages of evolution during all of the above (1–4).

Within these constraints some efforts have been made to determine the time necessary for the formation of specific lateritic formations. Thus Nahon and Lappartient (1977) calculated that a nodular ferruginous horizon 0.5–1.0 m thick required 0.3–0.75 Ma to form in the present-day semi-arid zone of Senegal. It was calculated that 6 Ma was necessary to form a complex profile on these rocks (containing 8–10% Fe_2O_3); to allow the geochemical evolution to destroy most of the

kaolinite and for the massive iron crust to break down chemically into a nodular, ferruginous horizon. This calculation was based on the age of the unweathered parent rock (7.6 Ma) and the age of an overlying dolerite (0.8 Ma). But the rate may be more rapid in a wetter climate, and Nahon (1986, p. 186) correctly pointed out the limitations of these calculations, and stated that, 'such a period will depend on the nature of the parent rock, the intensity of climatic changes, the tectonic evolution, and other factors'.

It might be more relevant to refer to the water flow through the profile during its formative period. According to Kronberg *et al.* (1986) this leaching is a major determinant which clearly depends on rainfall input and permeability of the saprolite. They estimated 10–100 Ma for the formation of the Paragominas bauxite deposits of Amazonia, during which time they required sustained high rainfall (2200 mm) and tectonic quiescence. In the Sula Mountains of Sierra Leone, Gaskin (1975) used estimated rainfall infiltration rates and measured silica extraction in stream water to calculate the time taken to form a 115 m weathering profile in the Tonkolili Ironstones. His figures ranged from 43 Ma (infiltration 1524 mm year^{-1}) to 64 Ma (infiltration 1015 mm year^{-1}), and it will be seen that this gives a rate of 1 m in 0.37–0.56 Ma^{-1}. Trendall (1962) calculated that for a 10 m ferricrete duricrust forming over granitoid rocks, 180 m of groundsurface lowering would be required, and that geological evidence in his Ugandan field area indicated that this may have taken place over 35 Ma (5 m Ma^{-1}). In Chapter 2 it was suggested that lateritic residues might form at a rate of 2–9 m Ma^{-1}, and this seems the right order of magnitude, where these are the products of protracted ferrallitic weathering. The direct alteration of rocks to bauxite might require far less time, and Valeton *et al.* (1991) has indicated that the bauxites of southeast Brazil are Quaternary in 'age', though developed in some cases on a much older ferrallitic mantle.

The association of bauxites with older and higher planation surfaces implies either that they have evolved over a long time period on an ancient landsurface, or that the bauxites have resulted from recent uplift and dissection, as already discussed (McFarlane, 1983a,b; Thomas, 1989b; Valeton *et al.*, 1991). The appearance of nodular bauxite within developing weathering profiles in lowland Kalimantan (Rodenburg, 1984) also suggests that dissection and rainfall are the keys, irrespective of the 'age' of the landsurface.

It remains true that many ancient laterites and bauxites survive on high residual relief, where they may mark ancient levels of planation and past incision (Aleva, 1984).

Palaeoenvironments and environmental change

The importance of changing world climates to the study of long-term landform evolution is stressed elsewhere (Chapter 11). In the geological record, bauxites and laterites occur in higher latitudes as a consequence of climatic change and the movements of crustal plates. Bardossy and Aleva (1990), using generalised global temperature and precipitation curves after Frakes (1979), have plotted the frequency of lateritic bauxite deposits against age. Notwithstanding the objections levelled above, this exercise suggests a major extension of lateritic bauxite formation in the late Cretaceous–Palaeogene. The absence of large, earlier deposits could be a function of their destruction by erosion, or possibly the absence of microbial activity (McFarlane, 1983b) and suitable vegetation cover. The warmth and quiescence of the Cretaceous–Palaeogene environments, prior to Alpine tectonism and the formation of the Antarctic ice sheet, may well have led to the onset of lateritisation and other forms of advanced weathering world-wide. The study of supergene mineralisation in Chile by Alpers and Brimhall (1988) showed that hyperarid conditions halted these processes there after 15 Ma.

Clearly many fluctuations of rainfall amount and distribution, as well as of vegetation cover, have affected most areas during the last 100 Ma, during which many laterites have formed. Arguments about 'cool-climate lateritisation' have been discussed, but there is widespread agreement that ferricrete formation requires a seasonally wet climate with high soil-water temperatures during the wet season. Switches between humid and semi-arid conditions seem likely to have been more important than temperature changes within present-day tropical areas. Ferricrete duricrusts are found today in desert marginal climates, where they form ancient and partially disintegrated cap rocks, and also in rainforest environments, where they appear to be undergoing chemical dissolution (Figures 3.13 and 3.14). A study from the equatorial rainforest area of Haut-Mbomou in central Africa (Beauvais & Tardy, 1991) has demonstrated the dissolution of ferricrete under forest, with the formation of goethite- and gibbsite-rich ferrallitic soils. The possible progression of climate changes affecting such areas is explored elsewhere (Chapter 8).

Bauxites of the Amazon Basin

Some of the difficulties and complexities of tropical weathering studies are exemplified by the ferrallitic and bauxitic profiles found in the sedimentary rocks of the Amazon Basin which include terrigenous (fluviatile, lacustrine, estuarine) sediments of probable end-Cretaceous to Pliocene ages, known as the *Barreiras Formation*. These sediments have much in common with Cenozoic sedimentary formations found in the coastal plains of the Guianas and in West Africa. They are derived from ancient weathering mantles, and are widely bauxitised (many contain important reserves of aluminium ore; Valeton, 1983). At several locations in Brazil (Juruti, Paragominas, Trombetas) the bauxites have been the subject of investigation, and different accounts of their origins have been published. The uppermost part of the sequence is a structureless, yellow-brown clay, called the *Belterra Clay*, which reaches a maximum thickness of 15–20 m. This is underlain by bauxite and by ferruginous laterite which in turn grades into a lower mottled clay.

Butty and Chapallaz (1984, p. 141) favoured an application of hydromorphic podzolisation to the 'bauxitisation of continental to shallow marine sediments deposited around the edges of ancient crystalline shields, with particular reference to the presence of a ferritic horizon'. With reference to the Trombetas region, they proposed the following sequence:

podzolisation
  → occasional flooding (deposition of Belterra Clay)
    → uplift
      → ferrallitization

The link between ferrallitic pedogenesis and podzolisation was discussed above in Chapter 2, but in this instance a reversal (podzol→ferrallite) is proposed, and this has required uplift and a new source of Fe, possibly from the Belterra Clay. Exhaustive analyses have, however, failed to determine the origins of the superficial, Belterra Clay, and Truckenbrodt *et al.* (1991) offered three different hypotheses: (1) *in situ* formation of the clay by weathering processes; (2) clay deposition after lateral transport; and (3) vertical transport of saprolitic material by termites. Option (1) is favoured by mineralogical affinities with the underlying saprolite, but option (2) is suggested by the sharp break in the profile beneath the clay horizon, and although option (3) could be attractive, the thickness of the clay (up to 20 m) precludes termite action as the sole process.

Kronberg *et al.* (1982) explained the extensive formation of gibbsitic bauxite in sediments at Paragominas by invoking strong lateral water flow above an impermeable layer over a period of $10–10^2$ Ma. The homogenisation of upper horizons under humid tropical pedogenesis is poorly understood, but clays apparently similar to the Belterra Clay overlie bauxitic horizons developed on an andesitic suite of rocks in Kalimanatan, Indonesia. On the other hand, the upper horizon on alluvial terraces in this area is a white sand. Depositional processes have been argued for all of these horizons, but more authors reason that they are pedogenetic. We may yet be some way off a final resolution of this problem, and the reader must be left unsatisfied by the difficulties of explaining profile reversals and of adjudicating depositional origins (Grubb, 1979; Kotschubey & Truckenbrodt, 1981, 1984; Kotschubey & Lermos, 1985) *versus* pedogenetic origins (Dennen & Norton, 1977; Aleva, 1981; Kronberg *et al.*, 1982; Sugden, 1984; Kobilsek & Lucas, 1988) of superficial sands and sandy clays in these profiles.

LATERITIC ORES

Much of the literature on lateritic ores is based on studies of *gossans*, defined by Govett (1987, p. 101) as 'the weathered expression of rocks that contain sulphides'. This author recognised that these are only a particular case of weathering leading to the formation of ironstones (ferricretes) in tropical weathering systems.

The typical *gossan* profile is illustrative of mineral zonation in weathered mantles (Table 4.1).

Ionic scavengers in soils

As a result of the intensive leaching of many weathering profiles, the upper zones become dominated by Al oxides, while metal concentrations increase with depth and may become concentrated

TABLE 4.1 Zonation and mineralogy of gossan profiles formed by deep weathering

Profile zones	Dominant minerals
(1) Leached Fe-oxide zone	Haematite, goethite, minor quartz
(2) Intermediate Fe-oxide zone with secondary ore minerals	Haematite, goethite, minor quartz with sulphates, carbonates, phosphates, etc.
(3) Supergene sulphide zone	Secondary sulphides, and sulphates, native metals
(4) Unoxidised zone	Primary sulphides

After Butt (1987) and Govett (1987).

in ferruginous horizons. According to Kühnel (1987), finely divided, amorphous and poorly crystalline constituents of weathered mantles carry high ionic charges and act as *scavengers* that catch and fix elements in migrating soil solutions. Subsequently these amorphous constituents undergo ageing which leads to dehydration and crystallisation (e.g. induration of laterite to form ferricrete). During the progression of these processes, in which initially disordered compounds become ordered crystalline structures, some ions will be lost, but others become incorporated into the structures of the neoformed minerals. Kühnel's (1987) summary is shown in Table 4.2.

Both cations and anions can migrate towards scavengers with which they have affinity (Table 4.2), and resultant precipitation may either be simultaneous, leading to complex intergrowths, or sequential, leading to coatings and cements on old nuclei. These processes can be repeated as rhythmic precipitation in alternating environmental conditions and colloform (botryoidal) fabrics can be produced.

Nickeliferous laterites

Nickel is one of the more important economic minerals to enter into the lateritic weathering system

TABLE 4.2 Examples of some ionic scavengers in laterites and soils

Scavenger	Collected cations (anions)	Possible products after recrystallisation
Fe-oxyhydroxides	Ti, Cr, Al	Magnetite, maghemite, haematite
	Cr, Al, Ni, Co, (F)	Goethite and other FeOOH polymorphs
	Co	Goethite + heterogenite
	Cu	Delafossite, goethite + cuprite
	(AsO_4)	Goethite + struvite (?)
Mn-oxyhydroxides	Co, Ni	Todorokite, woodruffite
	Ba, Zn, Al	Psilomelane
	Li, Al	Lithiophorite
	K, Na	Cryptomelane, birnessite
	Zn, Pb, Ag	Hetaerolite, quenselite
	Ca, Mg	Marokite
	Fe, Mg	Bixbyite, jacobsite
	(SiO_4)	Braunite
Silica	Al, Ni, Na, K	Opal, jaspar, quartz
	Cu	Chrysocolla
Al-oxyhydroxides	Ni, Co, Cr	Clay minerals
+ silica	Ni	Smectites
+ magnesia	Cu	Talc, sepiolite, serpentine
	Ni, Fe, Al	Vermiculite
	$Ni(CO)_3$, (SiO_4)	Pyroaurite, takovite

After Kühnel (1987). Reproduced by permission of Elsevier Science Publishers BV.

over ultramafic rocks (dunite, peridotite) (Schellman, 1978). These typically contain 80% SiO_2 and MgO, 10% FeO and 0.25% Ni (Trescases, 1983). The Ni is contained in easily weathered primary minerals such as olivine and serpentine and must be adsorbed by minerals such as goethite and Mn oxides, but may also replace Mg in the lattice of the primary serpentine, forming garnierite under neutral to alkaline pH conditions that are usually found close to the saprolite/rock interface. As the serpentine weathers, Fe and Ni become linked and concentrated in goethite or intermediate smectite clays. All of this transformation takes place in the upper saprolite; any Ni mobilised in the lateritic layer above tends to migrate down-profile into the saprolite. An example of this process has been offered by Aleva (1991 and Figure 11.5). According to Trescases (1983) pre-concentration of Ni takes place with the weathering and formation of flat relief on what are now 'old' landsurfaces, but that the metal may migrate towards lower surfaces following uplift and dissection. In Goias, central Brazil, Melfi *et al.* (1988) described nickeliferous laterites developed by lateritic weathering below early Tertiary silcretes (on the 'Sul Americana' erosion surface), and the subsequent migration of Ni to adjacent lowlands ('Velhas' erosion surface), where enrichment to form ores took place.

At Liberdade, a neighbouring site in central Brazil, Esson and Surcan dos Santos (1978; also Esson, 1983) found a high concentration of Ni (in garnierite) at the base of a lateritic profile 10.7 m deep. Up to 97% of the Ni had been leached from the upper profile (compared with Ti, assumed to be stable), and precipitated to form an enriched layer at the base of the laterite profile in which the weight gain was more than 400% in some samples. These authors also calculated that the 10.7 m of soil had been derived from the weathering and collapse or removal of 241 m of the serpentonite rock; the Ni was thus concentrated by progressive lowering of the whole profile, as hypothesised by McFarlane (1983a) for the formation of thick ferricretes. However, such calculations ignore lateral transfers to and from the sites of mineral concentration and must be considered with great caution.

These findings indicate the importance of the entire weathering profile in geochemical prospecting, and, more particularly, an understanding of the profile history, including possible truncation by erosion. Butt (1987) has noted that erosion commonly removes the leached top of many weathering profiles, exposing mineralised layers to both contemporary erosion and possibly to hydromorphic conditions, where relief has been lowered. In such eroded profiles, ferruginous mottles and concretions often form a lag deposit along with vein quartz and other resistant clasts. Such mottles and concretions can carry Cu, Co and As (Cu and Mo were recorded at 250–1800 ppm from a site in Guyana). In many cases these mineralised mottles become concentrated within stone-lines (Chapter 8).

Auriferous laterites

The presence of gold in lateritic weathering profiles is well known, particularly from Australia (Mann, 1984; Wilson, 1984). This supergene gold is often found in a very pure form (>95% Au). It may be precipitated in cracks of iron-stained quartz and in clays. In non-gossan laterites, gold is often associated with Fe nodules or concretions. Ironstone pebbles may contain fine gold (99.46% Au, at Coolgardie, the purest on record), and very thin layers of supergene gold are found interlayered with limonitic or bauxitic nodules or pisolites.

Some of this gold may come direct from weathered veins, but it is also precipitated from Au-bearing waters, and the conditions under which the mobilisation and precipitation of Au occur is clearly of great interest. The very fine gold found in laterites may be adsorbed on the Al and Fe hydroxides (possibly related to bog-iron deposits). Larger nuggets are often found near the base of pisolitic laterites, where they may be derived from the bedrock source and arise from residual accumulation, but Au is also accreted in gravels around iron oxide nodules (Evans, 1982; Bhaskara Rao, 1987a,b; Nair *et al.*, 1987).

*Transport mechanisms for gold
in lateritic environments*

Considerable uncertainty persists concerning the mechanisms for the transport of gold within lateritic weathering environments. Wilson (1984) observed that gold is often associated with hypogene pyrite, and liberation from the sulphide lattice occurs under very acidic conditions in which the Au would become attached to Fe oxides, sulphates and clays. As the profile deepens, pH and Eh conditions will change and the gold can be taken up by the thiosulphate ion, $Au(S_2O_3)^{3-}$ which is stable and could seep down-profile until it becomes unstable near the water-table, leading to the enrichment of the weathered ore body.

Mann (1983, 1984), on the other hand, has emphasised the role of *ferrolysis* (Brinkman, 1979) in the presence of continental brines, rich in the Cl^- ion. The semi-arid environment of the Yilgarn Block in Western Australia has been the location for these studies, and it must be recognised that similar conditions are not frequently encountered on the global scale. Here, the gold is found intimately associated with the iron oxide in lateritic profiles and may also occur as nuggets in the coarse pisolitic layer of ironstone pebbles. Groundwaters in this area contain high concentrations of Cl^- (in excess of 0.1 M) under acidic conditions (pH 2.8). These conditions probably arise from the oxidation of pyrite, bacterially catalysed by *Thiobacillus ferroxidans*. Gold is solubilised as $AuCl_4^-$, but the increase of Fe^{2+} and decrease of oxygen down-profile would, according to Mann (1984), reduce the solubility of the gold, which becomes precipitated as small crystals of very great purity:

$$AuCl_4^- + 3Fe^{2+} + 6H_2O \rightarrow Au^\circ + 3FeOOH + 4Cl^- + 9H^+ \qquad (14.1)$$

The major source of Cl^- was thought to have been the onshore westerly winds that bring rainfall to southwestern Australia. Mann attributed the production of Fe^{2+} ions to low Eh and near-neutral pH below the water-table at the bedrock interface, leading to an upward diffusion of Fe^{2+}

towards higher Eh and lower pH conditions higher in the profile, an example of the *ionic diffusion* process. However, these conclusions must be tentative and others have found that high pH conditions hinder Fe^{2+} solubilisation at the weathering front.

The problem with the processes involving $AuCl_4^-$ in an inorganic system is the apparent rarity of the postulated conditions. But the difficulty of mobilising gold is eased if the role of humic acids is invoked. Baker (1978) found that Au is mobilised in the presence of humic acids, reaching concentrations 19–33 times greater than those found in pure water. In humid environments these humic acids remain sufficiently stable to allow transportation of the gold vertically and laterally. In very wet environments, rapid oxidation of plant debris takes place in the uppermost part of the profile and continuous leaching of the humic materials, possibly as colloids can give rise to concentrations of the gel-like substance *humate* in the soil. When this becomes concentrated in sandy sediments, it forms a *humicrete* which creates a barrier to vertical water movement and may accumulate to more than 0.5 m in thickness (Chapter 8). The humicrete may contain significant amounts of aluminium as gibbsite or within halloysite clay. This process does not, however, appear to relate directly to lateritic environments, and the soils in which these processes occur are mainly podzols. It should be noted, however, that Dissanayake (1987) found an association of gold with a lateritic peat deposit in Sri Lanka.

The mobilisation of Au in these systems appears possible (Ong & Swanson, 1969; Wilson, 1984), but, since most of the sandy sediments are unconsolidated river or beach sands, the source of the gold remains uncertain. In Kalimantan, very delicate grains of gold were found in Quaternary river sands (Thorp *et al.*, 1990; Teeuw *et al.*, 1991), suggesting that these may have been deposited between and around the quartz grains after deposition, possibly in association with the build-up of humicrete layers. In this case, it would be assumed that the gold had been washed in to the sediments in a complexed form, not deposited with the sediment. However, other evidence from the same deposits also indicates the deposition of placer gold in association with coarser gravels in the Quaternary sediments.

Truncation of weathering profiles

Many of the most important mineral occurrences in regoliths are associated with old landsurfaces, dating to the early Tertiary or before, and these reflect the continued evolution of the weathering profiles. The erosion of these old weathering profiles will remove the leached top, to expose possibly mineralised layers to soil formational processes in the present environment. Such a situation will increase the possibility of *biogeochemical dispersion* of elements in soil and surface waters. On the other hand, where truncation involves the undermining and destruction of a lateritic duricrust, nodules enriched in heavy metals may be washed across the slope pediments, in a process of *mechanical dispersion*, to appear in *stone-lines* or in valley-floor *swamps*. In Guyana, nodules containing 250–1800 ppm of Cu and Mo have been recorded from such swamp environments (Butt, 1987). Lead and gold are also subject to this kind of mechanical transfer in ferruginous nodules and pisoliths, enriching stream sediments 500–1500 m away from sites of formation.

Hydromorphic dispersion

Much interest surrounds the degree to which mineral species are dispersed away from the site of mineralisation in soil and surface waters. Conditions for this dispersion are usually found at the water-table, where redox environments favour solubilisation of many metal cations. In arid areas, water-tables may be deep beneath the landsurface, and thus mineral species move within the weathered profile itself. In areas that are at least seasonally humid, high, or perched, water-tables allow the transfer of elements in the soil water, leading to *dispersion haloes* of mineral enrichment, commonly across distances of 500–1500 m. In mineral exploration programmes such haloes and anomalies in stream sediments provide geochemical signatures involving *pathfinder elements* used to detect the presence of gossans.

The Baluba Ore Body, Zambian Copperbelt

Studies of hydromorphic dispersion, involving both soil catena and stream sediment studies, in the Zambian Copperbelt are revealing (Govett, 1987). The Copperbelt area is situated on the high interior plateau of central Africa and the relief is possibly derived from late Mesozoic to early Cenozoic planation and weathering (the 'African' surface?). This is a deeply weathered landsurface (or *mantled etchplain*) with evidence of a relict duricrust in a landscape of low slopes (<2°) and wide, shallow valleys known as *dambos*, forming the seasonally flooded drainageways. Dambo geomorphology is discussed below (Chapter 8).

Interest in the Baluba area centres on Copper anomalies. In the free draining soils of the local catena, these are only high in the vicinity of the duricrust outcrop. Upslope from the ore body, Cu is only 40–50 ppm, rising to 500 ppm over the duricrust, then declining downslope for more than 600 m to 200–300 ppm. However, when the dambo margin is reached, the total Cu values rise dramatically to 2000 ppm in the B horizon, a major component of which is cold-extractable Cu that has evidently reached the dambo by hydromorphic dispersion and seepage from the surrounding slopes. In addition to this component, there is also strong evidence for mechanical dispersion, in the high Cu content of the top 40 cm of the dambo soils, containing 4% coarser than 1.35 mm diameter clasts rich in Cu. Govett (1987) concluded that the duricrust contains a palaeo-anomaly, while hydromorphic dispersion was taking place today along the water-table, leading to Cu seepages into the dambo itself.

High Fe content in dambo soils has also been noted in Sierra Leone (there called 'bolis'), where 1700 ppm has been recorded (IITA, 1975). Other element transfers are also important in these environments and McFarlane *et al.* (1994) have found chromium anomalies of significance in Malawi. These authors also call for *microbial leaching* to be a mechanism for dispersion of both Cr and Al (McFarlane & Bowden, 1992).

SILCRETE

Silcrete forms one of the silicate-derived duricrusts that also include ferricretes and alucretes. The way in which each end member of the series intergrades with the other two, with mixed 'fersiallitic' crusts forming the central ground, is illustrated by Dury's (1969) classification (Figure 4.2). But this is deceptive, because most silcretes exhibit a much greater 'purity' than ferricretes or alucretes. The percentage of SiO_2 is usually >90%, and typical analyses return values of 96–97% (Wopfner, 1978) from Australia and South Africa (a typical range would be 86–99% SiO_2). Comparable values for Fe_2O_3 in ferricrete would be 20–75%, with many lying in the range 40–60%. However, it must be realised that the parent rocks may contain 60–70% SiO_2 (granite is often close to 70%), while the Fe content of common silicate rocks will be in the range 2–13%, with ironstones in greenschist belts possibly exceeding 20%. The actual quartz content of the parent rocks may range from 30% in granite to 65% in sandstone. The amount of silica enrichment required is therefore no more, and in some cases rather less, than the iron enrichment needed to form ferricrete. A part of the SiO_2 content of common silcretes will in fact be floating quartz grains or larger pebbles. Silcretes usually also show TiO_2 enrichment up to 2%. Many of the mixed species of duricrust in Figure 4.2 would pass as laterite or ferricrete in a hand specimen, but they would not appear as silcrete.

It was pointed out above that Schellmann's (1983) substitution of kaolinite for silica (Figure 4.1) offers a significantly different perspective on the compositional variations, because it relates laterites and bauxites to kaolinitic clay weathering products (saprolite). This poses the question of whether silcrete is also related to kaolinite in the weathering profile and, if so, whether breakdown of the clay is responsible for producing the silica subsequently precipitated in silcretes.

Most discussions of laterite and bauxite ignore the occurrence and formation of silcrete, and it does not therefore appear to be generally regarded as a product of lateritisation processes. Yet strong views are held to the contrary, some authors arguing that the SiO_2 content of the silcrete comes predominantly from within kaolinised weathering profiles (Mountain, 1951; Wopfner, 1978, 1983; Summerfield, 1983a; Butt, 1985). This argument closely associates the formation of silcrete with the *pallid zone* of the 'lateritic' or ferrallitic weathering profile. For this material to provide enough SiO_2 to form a silcrete layer, possibly several metres thick, appears to require the dissolution of SiO_2 without mobilising the Al^{3+}. This question is explored further below, but there must be some doubt about the validity of arguments based on this premise, unless the environments and products of the incongruent dissolution of kaolinite can be shown to differ from those which are thought to give rise to podzols (congruent dissolution of Al and Si).

These introductory remarks are intended to ask whether silcretes are formed as a result of the weathering systems as so far discussed in this book, or whether they have quite separate formative histories, e.g. within drainage systems or around lake margins. To some extent this may be a false opposition, and for two different reasons. First, because there are different categories of silcrete which may have different formational histories. Second, because silcrete is often regarded as a later modification of a weathering profile which has evolved through a long and complex history of changing pedologic environments.

Classification of silcretes

Silcrete classes based on four criteria were advanced by Wopfner (1978, 1983). The criteria he used were: (1) minerological composition of the matrix, (2) macroscopic textural appearance (or habit), (3) retention of primary host-rock textures, and (4) thickness and type of associated profile. The matrix was used as the primary division into groups:

Group I   Well crystallised quartz matrix
Group II   Cryptocrystalline quartz matrix
Group III Amorphous (opaline) matrix

Groups I and II silcretes, according to Wopfner (1983), are always formed in association with a kaolinised profile, and this has been supported from studies in South Africa by Summerfield (1983a,b). Group I silcretes are less common and resemble quartzites, having a framework of clastic quartz grains tightly cemented by a quartz-growth matrix. Wopfner (1983) listed the following characteristics: (1) SiO_2 reaches 97–99%, the main constituent being quartz; TiO_2 makes up the remainder; (2) they are almost entirely depleted of Fe, Al, K, Mg and P. Silcretes of Group II do not retain any host-rock texture. Group III silicates can occur without profile development and the amorphous silica is found as a cement in sediments that may be brecciated, conglomeratic or fine grained. These silcretes perfectly preserve host-rock textures.

The cryptocrystalline silcretes of Group II are the most common and they exhibit a *floating fabric* (Van de Graaff, 1983), in which the grains seldom touch and proportions of grains and matrix approximate 50%. Cones of silcrete were found on pebbles and cobbles by Van de Graaff (1983), working in Western Australia, and he deduced that these signified powerful evaporation. Underlying these deposits are sometimes found silcretes with preserved textures from lower horizons of the weathered profile. Finally, *laminar/pisolitic* silcrete was found occasionally as a surface horizon or filling solution pipes. Further subdivision of the silcrete microfabric has been proposed by Summerfield (1983a), who noted: GS (grain supported) fabric; F (floating) fabric in which skeletal grains >5%; M (matrix) fabric in which skeletal grains <5%, and C (conglomeratic) fabric containing pebbles >4 cm.

Geomorphic and stratigraphic setting of silcretes

The common Group II F-fabric silcretes were found by Van der Graaff (1983) to occur on the higher parts of relict duricrusted landscapes in Western Australia. They were found most notably in Tertiary sands and gravels, but also in Permean sandstones and granitic saprolites. In the Lake Eyre basin Wopfner (1978) found the silcretes formed on

emergent, subdued antiform structures, and seldom carried through beneath basin sediments. More widely, silcrete-capped residual hills today form the positive relief of the landscape, but whether relief inversion has occurred is often uncertain.

Formation of silcrete

Arguments about silcrete formation closely parallel those concerning ferricrete (or laterite), and turn on the provenence of the SiO_2 and the conditions under which it has been mobilised. Extreme positions have been taken on both issues. Thus long distance transport of SiO_2 from the eastern Highlands of Australia into the interior basins was advocated by Stephens (1971), while Wopfner (1978, 1983) has increasingly emphasised that the SiO_2 is drawn mainly from the kaolinised profile within which it is found. These are versions of the *lateral transfer* and *vertical transfer* models of silcrete genesis. Conditions for mobilisation have been considered in terms of semi-arid climates (Bond, 1964; Watkins, 1967; Pettijohn, 1975) with rainfalls of 250–740 mm year^{-1} being suggested. However, although Van de Graaff (1983) advocated a climate with strong evaporation, he favoured rainfalls of 1000 mm year^{-1}, which he thought had occurred during a period of declining rainfall in the Oligocene in Western Australia. But he also attributed some importance to the higher humidity of the pre-Oligocene climates which he thought responsible for prior karstification of silicate rocks in the area. Wopfner (1978) revised his earlier support for arid pedogenesis (Wopfner, 1974) of silcrete and advocated conditions humid enough to support the 'classical fluviatile coal swamp' for Groups I and II silcretes to form. Group III silcretes possibly constitute a separate case.

Some more general observations may be relevant at this point in the argument. If we review the gross climatic character of each of the southern continents, then South America is the most humid and Australia the most arid, with Africa and India lying in-between. The present-day distribution of surface silcretes reflects these differences. They are widespread throughout the centre and west of Australia; patchily developed in Africa, mainly in the Sahel and in the south, and rarely reported from South America. When the distributions of ferricrete and silcrete are compared *within* either southern Africa or central Australia, then a climatic zoning is evident (Grant & Aitcheson, 1970; Stephens, 1971, Goudie, 1973), with major occurrences of silcrete always found in drier areas. Arguments that ignore these distributions appear to the present author to lack commonsense.

Furthermore, if desilication of kaolinite is considered to be the main source of migratory silica, under conditions of impeded drainage and low pH, then the question of Al^{3+} mobility within such a system has to be faced. Silcretes do not appear to be products of a podzolisation process as described above (Chapter 2), yet it has already been observed that hydromorphic podzols are thought to develop from the breakdown of ferrallitic residues in conditions of impeded drainage. The difference in regime possibly hinges on severe seasonal desiccation of the profile in the silcrete-forming environment (Van de Graaff, 1983).

The geomorphic setting of many silcretes is also quite distinct. Within the Great Artesian Basin of Australia, silcretes are found on and within the fluviatile sequence of the Eyre Formation in a conformable manner. Thus according to Wopfner (1978, p. 103), 'the occurrence of this silcrete is closely related to a quasi-planar landsurface of enormous dimensions which existed in the early Tertiary over most of inland Australia.' But Wopfner is equally clear that silcretes of Groups I and II are invariably part of a profile, with silcrete at the top, underlain by brecciated, silicified bedrock, below which a kaolinised zone is always present and may exceed 30 m depth. Wopfner (1983) has summarised possible conditions of formation:

(1) silicification took place at or near the top of a constant groundwater level;
(2) to achieve the observed accumulation of silica, the solubility of Al_2O_3 must have exceeded that of SiO_2, requiring a very low pH-environment that is only likely in paludal conditions;
(3) SiO_2 was provided by the release of silica during kaolinisation of feldspar and other silicates.

In Western Australia, Van de Graaff (1983) found the floating fabric silcretes (Gp II) preserved on the crests of interfluves of palaeodrainage systems, often in terrace gravels and sands. He also noted patchy surficial ferruginisation of the profiles. The partial lateritisation of the silcrete, together with the occurrence of laterite at lower levels suggested that the process of lateritisation had come later. The loss of primary rock textures from the floating fabric silcretes has led most authors to conclude they were formed close to the ground surface, in the upper levels of the weathering profile. Although Van de Graaff (1983) drew attention to the contrast between his well-drained upland sites for silcretisation and the planar riverine swamps referred to by Wopfner (1978), it is fair to add that the latter author stressed that silcretisation took place on gentle antiform structures within these plains.

The conditions for the loss of aluminium from the profiles were considered by Summerfield (1983a), who utilised isovolumetric calculations to estimate a 95% loss of Al, with silica depletion from kaolinite of 40–50%, and accumulation of 30–40% in the overlying silcrete. Titanium enrichment in the silcrete approached 120% (on Dwyka Tillite, from Grahamstown, South Africa). Titanium becomes increasingly soluble at pH < 3.75 (Baes & Mesmer, 1976), while aluminium solubility rises steeply below pH 4, and overtakes the solubility of silica in these conditions. As previously noted, chelating agents can abstract Al at higher pH, but Summerfield (1983a,b) argued for a pH fluctuating between 3.5 and 4. Such conditions may imply impeded drainage, with silcrete formation taking place preferentially within the range of water-table fluctuation. It can be objected, however, that low pH in these conditions would apply only to surface swamp environments with high organic activity. If impeded drainage occurs deeper within kaolinised profiles, then pH is likely to rise as a consequence of solute concentrations approaching equilibrium levels. Under these conditions Al_2O_3 would remain stable but SiO_2 would migrate with increasing mobility at pH values > 8. The answers to these difficulties of interpretation probably come from: (1) pH and Eh changes down profile; (2) fluctuations of pH through time as groundwater conditions change, seasonally or because of climate change; (3) the intervention of organic compounds; and (4) changes in the groundwater regime due to uplift and dissection of the landscape. Different conditions will apply to individual cases and no universal model of silcrete development is applicable. Some possible formational conditions can be suggested:

(1) silicification of an upper (but not surface) horizon of a complete weathering profile due to vertical transfer of SiO_2;
(2) silicification of a lower horizon of a truncated profile;
(3) silicification of sediments or laterite due to lateral transfer of SiO_2.

The silicification of laterites has been noted by a number of authors, and this must signify a reversal of the normal weathering sequence by the introduction of SiO_2 from beyond the site of weathering, probably by lateral transfer. The need for strong evaporation to explain silcrete formation and morphology, and the absence of silcretes from humid tropical areas, where podzolisation takes place, suggests very different environments of formation, although Al^{3+} migration is postulated in both cases. It can be seen, however, that some weathering profile silcretes remain difficult to explain.

For Group III silicates which have no associated profile Wopfner (1983) argued for an evaporitic and alkaline environment, as formerly found around the margins of the palaeolake, Lake Eyre. This interpretation is broadly followed by Summerfield (1983a) in respect of silcretes impregnating sediments in the Kalahari Basin in Botswana, Namibia and South Africa. Here silcretes have formed within clastic sediments, and it is tempting to argue that the SiO_2 has been imported by lateral inflow of rivers and near-surface groundwater, and has also been liberated as a result of the breakdown of smectite clays, a point made by Summerfield (1983a). In semi-arid basins smectites would form a major component of the clay in sediments. Where strong evaporation takes place with rising pH, the

solubility of SiO_2 would rise steeply, and concentrations in sites of impeded drainage would lead to precipitation from saturated solutions.

CALCRETE

Calcareous crusts are characteristic of semi-arid climates and are commonly described as *calcrete*, but are also known as *caliche* and *kunkar*. Like ferricrete, calcrete may form pedogenetically or by precipitation from groundwater (Goudie, 1983b; Cooke *et al.*, 1993), and pedogenetic calcretes are generally confined to areas with rainfalls between 400 and 600 mm year^{-1}. However, groundwater calcretes can occur within the savanna climates, where rainfalls exceed 800 mm year^{-1} (Archer & Mäckel, 1973). These deposits will be considered only briefly here. Some categories of calcrete have high Mg:Ca ratios and merit the term 'dolocrete', but the majority may be considered in terms of Ca^{2+} availability and mobility, and many calcretes approach 80% $CaCO_3$ (Goudie, 1983b).

Pedogenetic calcretes

Calcretes are so widespread within the semi-arid and arid zones that local rock sources of Ca are insufficient to explain their occurrence, and it appears most probable that the Ca comes from wind-blown dust. Proximity to calcic dune sands is thus found to be important. This, however, poses the problem that dryness leads to more dust but little leaching, while relative wetness allows less dust transport but promotes leaching of the calcium. This has led to the conclusion that many, if not most, thick calcretes (up to 3 m) are relict and products of climate change. Some calcretes in Namibia have been dated to the last glacial period (30 000–15 000 BP; Blümel, 1982), when winter rain may have been an important element in the pedogenetic system.

Savanna and forest climates have generally been remote from major sources of Ca from dust, though past dry climates may have promoted the formation of thin calcic horizons in some savanna soils.

Groundwater calcretes

Calcretes formed by precipitation of calcite from groundwater have a much wider distribution and their formation may follow different pathways. The present or former position of the water-table has less significance for $CaCO_3$ deposition than for ferricrete formation, and the important element is groundwater flow with loss of CO_2 by degassing. This can occur with the capillary rise of groundwater, and the evaporation of soil moisture in this way may lead to the formation of calcareous nodules in poorly drained soils. In well drained soils within the sub-humid and humid tropics the Ca^{2+} ions will normally remain in solution and be leached from the soil system.

Throughflow models, therefore, have much wider application in more humid climates, particularly in the context of drainage from and over calcareous bedrock, and within alluvial sediments. In alluvial sediments, Mann and Horwitz (1979) have suggested that the calcite is precipitated due to lateral flow beneath the water-table. Fluctuating hydrological conditions then lead to solution and re-precipitation of the $CaCO_3$ which eventually matures into pods and domes that displace the alluvium.

Archer and Mäckel (1973) described calcrete formation within the *dambo* environment (see Chapter 8) of shallow valleys on the Lusaka Plateau (altitude 1300 m asl) of Zambia. The rainfall here is 800–1000 mm year^{-1} and falls between November and March. The calcretes have formed in shallow depressions within dambos that drain towards a nearby escarpment, and which are underlain by limestones and dolomites of the Katanga System. The groundwater flows across and possibly within the limestone bedrock, beneath the dambo sediments, until the gradient of these headwater valleys increases near the scarp and the carbonated groundwater is forced to the surface. Loss of CO_2 and evaporation, together, produce supersaturation with respect to $CaCO_3$ and calcite is precipitated. Although not mentioned by Mäckel, the broad floodplain of the Kafue River into which these streams ultimately drain, is characterised by discontinuous calcareous deposits forming the pods and domes described above.

These examples serve to illustrate the importance of bedrock composition, and groundwater hydrology in determining the mobility of elements like Ca and Fe within the geomorphic system. The precipitation of $CaCO_3$ on a limestone plain in the dry part of northeastern Brazil was discussed by Branner (1911), and there are probably many examples of calcite solution and precipitation across a wide range of climates in areas of carbonate bedrock.

PLATE 2-I Arched weathering front (or basal weathered surface) on granite in Western Australia. Note the sharp interface between the fresh rock and the saprolite; the ferruginous zone is confined to the upper profile

PLATE 2-II Weathering front on gneiss near Sekondi, southern Ghana. The section is ~30 m depth and is exposed in quarrying operations

PLATE 3-I Very deep weathering (>100 m) below the central African plateau surface, Kolwezi, Shaba, Zaire (copper mine). Note the confinement of the ferrallitic zone to the upper horizon (<20 m)

PLATE 3-II Ferrallitic weathering beneath interfluve with the formation of core-stones in granodiorite, Singapore Island

PLATE 3-III A very deep weathering profile
(>60 m) with core-stones in granite, southeastern
Brazil

PLATE 3-IV Hand specimen of dioritic igneous
rock showing almost complete transformation to
saprolite within 10 cm, Kalimantan, Indonesia

PLATE 3-V Weathered material developing
between pressure release sheets, Nigerian forest zone.
This phenomenon is widespread and can promote
slope instability on the flanks of hills

PLATE 3-VI Laterite–bauxite profile over
feldspathic gneiss, Mokanje, southern Sierra Leone.
Nodular ferricrete forms a surface duricrust
(fragmented), below which is a bauxite-kaolinite
pallid zone

PLATE 3-VII Residual laterite forming directly from weathering of gabbro rock (10–20% Fe total) and preserving its textures, Freetown, Sierra Leone. This is a free-draining site in a forest environment (rainfall > 2500 mm year^{-1})

PLATE 3-VIII Deep weathering in gneiss (a measured 80 m) in southeastern Brazil (rainfall ~ 1500 mm year^{-1}). Much of the lower profile that can be inspected shows transitional or fersiallitic features of sandy weathering

PLATE 3-IX Deeply weathered compartments in multi-convex relief within the Koidu basin area, Sierra Leone

PLATE 4-I Bauxitised summit profile over alkaline igneous rocks, near Lages (28 °S) in southeastern Brazil

PLATE 4-II Bauxite preserving bedrock textures beneath a forested summit site in Sierra Leone

PLATE 5-I Surface water flowing over podzolised white sands in Kalimantan, Indonesia. The dark 'tea water' coloration is due to dissolved organic matter

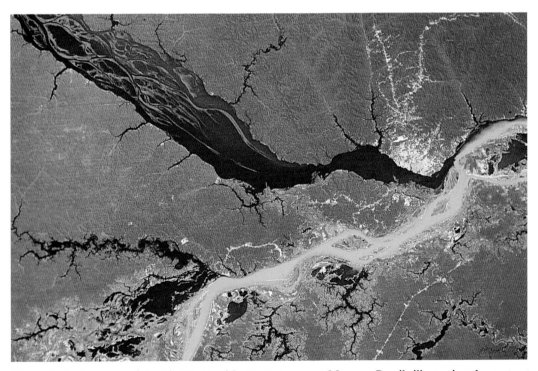

PLATE 5-II Junction of the Rio Negro with Amazonas, near Manaus, Brazil, illustrating the contrast between the blackwater streams draining crystalline rocks and alluvium, with the whitewater Amazonas flowing from the Andes and carrying a high suspended sediment load. Reproduced by permission from Geopic Satellite Corporation (1983)

PLATE 6-I Bottle-shaped gully eroded into thick saprolite in a road cutting on the Rio de Janeiro to São Paulo highway, Brazil

PLATE 8-I Stripped granite hillslope with deposition of colluvium beneath the footslope, Jos Plateau, Nigeria

PLATE 6-II Deep section through weathered landslide deposits beneath a spur site, near Botuvera, southeastern Brazil. The right hand plate shows the existence of slickensided (polished clay surfaces) shear failure planes in the mass which is potentially unstable

PLATE 8-II Undulating stone layer associated with a kaolinised weathering profile, Biano Plateau, Shaba, Zaire. The surface is covered by large termite mounds (10 m diameter, 2–4 ha^{-1}), and the association of the mounds with an undulating stone-line is a matter of debate

PLATE 8-III Coarse stone layer separating a recent soil profile from ferrallitic saprolite, near Curitiba, southeastern Brazil. Professor Bigarella, who is in the picture, has called these 'palaeopavements' and has associated them with Pleistocene semi-arid climates

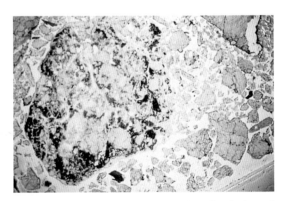

PLATE 9-I Microscope slide showing dissolution of Fe_2O_3 nodule within a swamp sediment beneath a small stream floor in Koidu, Sierra Leone (courtesy R. M. Teeuw)

PLATE 10-I Granite domes and landsliding in saprolite, in the Serra do Mar, on the road to Teresopolis from Rio de Janeiro, Brazil

PART II
DENUDATION PROCESSES

5

Surface Processes in Tropical Climates

RAINFALL IN THE TROPICS

The nature and patterns of tropical rainfall are critical to nearly all discussions of tropical geomorphology. Different aspects of this question can be considered under the following heads:

(1) annual and seasonal water balance,
(2) concentration and seasonality,
(3) intensity and erosivity, and
(4) the occurrence of high-magnitude rainfall events.

Since the last of these topics is closely linked to the question of landsliding, it will also be discussed in that context within Chapter 6.

Water balance studies

The relationship between precipitation and evapotranspiration controls many geomorphic processes in the tropics. Most rainforest climates exhibit a positive water balance over the year, though Davies and Robinson (1969) have shown that near the forest/savanna boundary in Nigeria there is a tolerance of about −200 mm, where total rainfall is 1400 mm. This can arise because of the storage of large water surpluses during the wet season. These water surpluses probably influence more than any other climatic factor the potential for deep weathering in the tropics, since they provide for the infiltration of large quantities of water.

Rainfall surpluses in the savannas occur only briefly during the months of heaviest rainfall and this clearly limits the potential for the formation of a weathered mantle and helps to explain why this is either thin or survives as a relict of past conditions in these areas. It is also important to note that, even in most wet, tropical climates, there will be a short period of water deficit, when soil moisture is utilised and the saprolite may dry out to depths of many tens of metres (Lelong, 1966). The effects of drying out on iron segregation can therefore be effective, even in climates with high annual precipitation. Many such climates are monsoonal and experience a marked dry season of 4–5 months duration, despite rainfalls of 2000–5000 + mm year^{-1}.

Concentration and seasonality of rainfall

Concentration and seasonality of rainfall are reflected in the values given above, but were also used by Fournier (1960, 1962) to predict surface erosion rates for the continent of Africa. In this exercise Fournier used a simple index of seasonality, p^2/P, where p is the mean rainfall for the wettest month and P the mean annual rainfall. He compared this with sediment yields for large catchments and offered maps of contemporary erosion, also incorporating a relief factor. The data he obtained for rates of erosion were probably not representative of continent-wide conditions, and the maps could be misleading (Walling, 1984), but the overall pattern is interesting. Outside of the mountainous zones of Ethiopia and other parts of East Africa, the highest erosion rates were predicted for the wet, monsoonal climates of West Africa, and in the drier savannas north and south of

Latitute 10°, while the lowest rates attach to the equatorial plains of Zaire, with low to moderate values affecting the rest of the humid tropics. The question of sediment yield will be explored further below.

Intensity and erosivity of rainfall

Intensity and erosivity of rainfall are perhaps the most important measures of rainfall as a geomorphic agent, since they relate not to annual or monthly means, but to storm events and the nature of the rainfall itself. Tropical environments experience high rainfall erosivity, and this is a function of rainfall intensity (Lal, 1977; Kowal & Kassam, 1976; Morgan, 1986). According to Kowal and Kassam (1976), erosivity can be measured in terms of:

(1) drop size (diameter, mm);
(2) drop number;
(3) rainfall intensity (mm h^{-1}) and,
(4) energy load (W m^{-2}).

Nearly all measurements attest to the much greater erosivity of tropical rains as compared with temperate rains, but numerical expressions of these properties are not always straightforward when used to predict soil erosion hazard. Many studies seek to express erosivity in terms of kinetic energy ($1/2 Mv^2$) rather than in terms of momentum (Mv). The kinetic energy is a function of both drop size and number, and these measures correlate strongly with rainfall intensity (mm h^{-1}).

Rainfall intensity is more readily measured than kinetic energy and figures are therefore more widely available for a range of tropical climates. These make it clear that tropical rainfall is far more intensive than temperate precipitation. At Ibadan in Nigeria, for example, an analysis of rainfall over a two year period (Lal *et al.*, 1981) showed that maximum intensities over 7.5 min were: 29% <25 mm h^{-1}, 61% 25–75 mm h^{-1}, and with a total of 35% >50 mm h^{-1}.

These values may be compared with those for New Guinea, where 34% of annual rainfall is >25 mm h^{-1}; apparently a much lower value, but

measured over longer time periods (30 min or 60 min), and still much higher than in temperate areas. Hudson (1971) suggested 40% of rainfall >25 mm h^{-1} was characteristic of the tropics and about 5% for temperate regions. The higher percentages obtained for maximum intensities in individual storms implies that there are short periods with cloudburst characteristics (1 mm min^{-1} for 5 min), and these comprise a significant percentage of the total rain which falls. According to Mohr and Van Baren (1959), cloudbursts represent 22% of all rainfall at Bogor in Java (cf. 1.5% for Bavaria). Most tropical storms reach peak intensities within 15–30 min, but Starkel (1976) quoted figures from Kalimpong in the Sikkim Himalayas which show that peak intensities >50 mm h^{-1} often occur towards the end of long periods (2 days) of continuous rainfall. Starkel (1976) suggested that, in monsoon climates, annual daily maxima of 100–500 mm occur, with 1–5 days when rainfalls >50 mm h^{-1}, while in equatorial and mountain regions rainfalls >50 mm h^{-1} occur on as many as 20 days year^{-1}. At Ibadan the return periods for storms of high intensities were found (Lal *et al.*, 1981) to be:

200 mm h^{-1} (7.5 min) return period 100 years
165 mm h^{-1} (7.5 min) return period 10 years
145 mm h^{-1} (7.5 min) return period 2 years
100 mm h^{-1} (30 min) return period 2.8 years.

Such high intensity rainfalls are associated with large drop size and high drop density (Laws & Parsons, 1943). Median (D_{50}) drop sizes of 3 mm are common and drops of 4.9 mm have been reported. However, as drop density increases, larger drops become unstable and form smaller droplets. According to Lal *et al.* (1981), this is apparent when rainfall intensity is >50 mm h^{-1}. Terminal velocity increases with drop size, as shown by Wischmeier and Smith (1958), from 1.0 m s^{-1} for $d = 0.25$ mm, to 4 m s^{-1} at $d = 1.00$ mm, and to 8.1 m s^{-1} at $d = 3.00$ mm. The kinetic energy exerted by the raindrops is related to their terminal velocity, so that this term is important in determining the total energy load of a storm:

TABLE 5.1 Energy load of rainfalls of varying intensity

Rainfall type	Rainfall intensity (mm h^{-1})	Median diameter (mm)	Fall velocity (m s^{-1})	Drop number (m^{-2} s^{-1})	Kinetic energy (W m^{-2})
Light rain	1	1.24	4.8	280	3.3×10^{-3}
Heavy rain	15	2.05	6.7	495	9.4×10^{-2}
Excessive rain	41	2.40	7.3	820	8.9×10^{-1}
Cloudburst	100	2.85	7.9	1214	9.2×10^{-1}
Cloudburst	100	4.00	8.9	440	1.1

After Lull (1959).

$$E_k = \frac{IV^2}{2} \qquad (5.1)$$

where E_k is the energy in W m^{-2}, I is the rainfall intensity in mm s^{-1}, and V is the velocity of the rainfall before impact in m s^{-1} (Lal, 1977, 1981). Examples of energy load have been given by Lull (1959; and Table 5.1).

At Samaru, northern Nigeria, Kowal and Kassam (1976) found that for storms between 40 and 80 mm h^{-1}, 41% of the increase was due to the number of drops and 49% to the increasing size of drops (10% error). They also found that 60% of

drops were >3 mm. Lal *et al.* (1981) observed a very erosive rainstorm with a total energy load of 275 J ha^{-1} mm^{-1} of rain (70–100 J ha^{-1} mm^{-1} is common), associated with a median drop size of 2.4 mm (Figure 5.1).

It is also necessary to take some account of wind velocity in tropical storms, since this factor greatly increases the kinetic energy of the rainfall. According to Lal *et al.* (1981), at 35 km h^{-1} the kinetic energy could be increased by a factor of 3.2. At Ibadan they found that windspeeds >30 km h^{-1} occurred in 30–83% of storms over a 3 year period, but maximum rainfall intensities tend

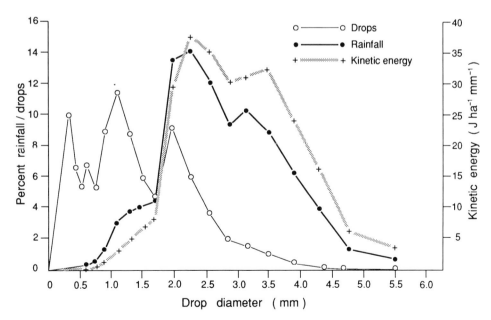

FIGURE 5.1 Drop size distribution and kinetic energy of a rainstorm received at Ibadan, Nigeria (after Lal, 1981; reproduced by permission of IAHS Press)

to lag behind maximum wind velocities, so the relationship is not straightforward.

Wischmeier and Smith (1958) developed an *erosivity index* (EI_{30}), based on the product of the kinetic energy (KE) and maximum intensity during the first 30 min of a storm (I_{30}), and this has been developed by Delwaulle (1973), Lal (1977) and adapted by Roose (1977, 1981), who was able to establish a relationship between I_{30}, and daily rainfalls for the whole of West Africa and to compute an annual *climatic erosion index*, $R(EI_{30})$, that approximates half of the mean annual rainfall. This may seem a surprising result, since it relates erosivity to total rainfall, and indicates that humid tropical climates are more erosive than sub-humid or semi-arid climates. It also serves to emphasise that within the humid (forested) tropics there are enormous and important variations in rainfall and erosivity. For example, the value of R near the forest–savanna boundary in the monsoonal regime of Sierra Leone becomes 1250 which is nearly 80% above that for comparable locations in the more equatorial regimes of Ghana or Nigeria ($R = 700$). In an examination of extreme rainfalls in Nigeria, Ayoade (1976) showed clearly that 24-h totals declined from the wet delta areas (>250 mm day^{-1}) in the southeast to the northern areas where it may be less than 150 mm day^{-1}.

Other methods for estimating erosivity have been devised and they are discussed by Lal (1977) and by Morgan (1986). Lal (1977) reported that EI_{30} was less appropriate for the tropics than for the USA, and that a better correlation was found between the product of total rainfall amount (A) and peak storm intensity (I_{m}). Hudson (1971), on the other hand, used the accumulated kinetic energy of storms with intensities >25 mm h^{-1} (KE>25), derived from studies in Zimbabwe. This index was based on the premise that 25 mm h^{-1} represents a threshold value for soil detachment under tropical conditions, and it has been used successfully by Rapp *et al.* (1972) for Tanzania and by Morgan (1974) in Malaysia. Morgan (1986) has also pointed out that this value may be an order of magnitude higher than the relevant threshold for Western Europe which also implies a major difference in *erodibility* between temperate soils and many

tropical soils. The choice of erosivity index remains unresolved and both Rapp *et al.* (1972) and Morgan (1986) show that EI_{30} and KE>25 produce very different results. There are differences in the database for individual calculations and comparisons of maps from different sources should be avoided. Useful reviews of this topic also include Jackson (1989) and El-Swaify *et al.* (1982).

High magnitude rainfall events

The occurrence of high magnitude rainfall events is a function of the mechanisms of rainfall generation and their synoptic settings within tropical climates. Although the Inter-Tropical Convergence, combined with the monsoonal reversals of airflow in Asia, Africa and Australia, largely determine the seasonal rhythms of precipitation, the occurrence of extreme rainfalls is more commonly associated with specific synoptic situations and with orographic factors. The importance of *tropical cyclones* (hurricanes, typhoons) was stressed in Chapter 1 (Figure 1.1), and these tend to be generated between 5° and 15° N and S of the Equator, thus exempting the inner humid tropics from their depredations. They then move polewards, becoming less vigorous and therefore less frequent beyond 20° N and S. However, very heavy rains are also associated with *easterly waves*, and with *monsoon depressions* in the Indian subcontinent. Severe storms also take the form of *line squalls* ('disturbance lines' in West Africa), particularly over land in Africa (Hastenrath, 1991).

Tropical cyclones are associated with 24-h rainfall totals exceeding 250 mm, and possibly reaching 450 mm or even 600 mm locally on a less frequent basis. Gupta (1988) cited several sources which indicate that such storms have recurrence intervals of 5 years (250–350 mm) to 25 years (350–600 mm) in the Caribbean. Monsoon depressions can give rise to prolonged heavy rain amounting over 3–6 days to >3000 mm (e.g. 3213 mm over 6 days at Cherrapunji, India in 1931). Line squalls can be violent and tend to produce concentrated rainfall over a few hours, often giving individual falls of between 100 and 300 mm.

These rainfall systems not only produce erosive rainfall, but can give rise to major runoff events leading to high floods on tropical rivers (Baker, 1983b; Gupta, 1988).

INFILTRATION AND PIPE FLOW

Infiltration

While soil detachment initially depends on the kinetic energy of the storm (or is related to some surrogate measure), erosion of sediment from the soil surface requires the generation of runoff and this depends crtically on both the rainfall (intensity and amount) and the infiltration capacity of the soil. *Infiltration capacity* is defined as the maximum sustained rate at which a saturated soil can absorb water when lateral flow is unrestricted and is distinguished from *infiltration rate* which refers to the rate of water entry when lateral flow *is* restricted (Greenland, 1977a). These measures are dependent on pore size and are influenced by subsurface soil horizons. These properties in turn reflect particle size distribution, soil aggregate size and stability, and the distribution of macropores (pipes, fissures). Gardner (1977a) has briefly summarised the theoretical basis for studies of infiltration and pointed out that infiltration responds to gravity and the force of attraction between water molecules and the surfaces of soil particles. This decreases during a rainstorm and a number of empirical equations have been proposed to describe the process, including that of Horton (1933):

$$i = i_\infty + (i_0 - i_\infty) \exp^{(-bt)} \qquad (5.2)$$

where i is the infiltration rate at any time t, i_0 is the initial infiltration rate at $t = 0$, and i_∞ is the final infiltration rate at infinite time; b is a constant derived from the properties of the soil. Initially, capillary forces are added to gravity to drive water movement in a non-saturated soil. The initial infiltration rate is a function of these forces plus the *hydraulic conductivity*. As the soil pores become filled, by dislodgement of fine particles and breakdown of soil aggregates, the capillary forces are reduced gradually to zero. As a result the soil infiltration capacity reduces with time under real

rainfall of variable intensity. Gardner (1977) quoted the work of Philip (1957), who defined a coefficient called 'sorptivity' which is proportional to the increase in water content during infiltration and to the square root of the *soil water diffusivity*. This gives the equation:

$$I = St^{1/2} + Kt \qquad (5.3)$$

where K is the hydraulic conductivity of the soil, S is the sorptivity, and t is the time elapsed since the start of the storm.

In addition, Dubreuil (1985) has emphasised the importance of establishing an *antecedent moisture index*, to distinguish the effects of current from previous rainfalls. Stream base flow can be used or some expression of time elapsed since the previous storm event. This leads to the concept of *limiting rainfall* (Dubreuil, 1985), which is the rainfall required to exceed infiltration somewhere in the (small) catchment. The limiting rainfall is found to rise steadily with the duration of the period between rainfalls. The importance of this principle will be evident; for in the absence of other factors, runoff will occur more often when precipitation events recur frequently. Correspondingly, larger storm events will be necessary before runoff occurs in situations where the previous rain was many days before.

Reduction in infiltration rate with time has been discussed in terms of the antecedent rainfall, but it may reflect one or more of several factors:

(1) increase in soil moisture content with time and the filling of some of the pore space initially available, which will reduce the *diffusivity* or rate at which the water penetrates the soil;
(2) breakdown of soil aggregates and downwash of fine particles which clog the natural pores of the soil, sometimes known as *crusting* or *slaking*;
(3) the presence of swelling (2:1) clays that effectively seal off any fissures in smectite clay soils;
(4) the presence of a less-permeable B horizon and slow lateral movement (throughflow) such that deeper percolation of infiltrating water is

prevented, leading to saturated conditions in the surface soil after prolonged, heavy rain.

Many humid tropical soils, dominated by kaolinite clay, are found to have very high infiltration capacities. Thus Nye (1954, 1955) found that soil over a deeply weathered gneiss near Ibadan, Nigeria absorbed water at the rate of $284.5 \, mm \, h^{-1}$ for the first 30 min, $198.0 \, mm \, h^{-1}$ for the next 30 min, reducing to $82.5 \, mm \, h^{-1}$ during the second hour. This took place under experimental conditions beneath an *Imperata cylindrica* grass cover. Slight seepage along the stone-line was observed after 132 mm of rain, but no other runoff was recorded. These results accord with similar experiments conducted in the Ivory Coast (Lafforge quoted by Dubreuil, 1985). However, Morgan (1972) recorded runoff at Kuala Lumpur when rainfall intensities reached 60–75 $mm \, h^{-1}$, with a return period of 60 days for runoff events. He has also noted (Morgan, 1986) that runoff is commonly seen to take place at rainfall intensities which are much lower than experimental results suggest. Despite the factors discussed here, the reason for this is not clearly understood.

Wilkinson and Aina (1976) measured both infiltration capacity and infiltration rate over sandy soils (10% clay) in southern Nigeria and obtained values of $>100 \, cm \, h^{-1}$ and $>20 \, cm \, h^{-1}$ respectively under a bush fallow (secondary forest). These figures are very high and soils with high clay content will return rates perhaps an order of magnitude less. A factor operating where bare soil occurs is the formation of soil *crusts*, a misleading term applied to situations where the soil pores are clogged with fine particles. However, soil crusts involving the precipitation of carbonates will also develop in the semi-arid zone.

A study by Singh *et al.* (1984) of infiltration characteristics in a *transported soil* in Kenya demonstrates much lower infiltration rates ($<20 \, mm \, h^{-1}$ during the first 30 min) than those measured in *residual soils*. But he also recorded a constant rate of infiltration after the first hour and very low rates of percolation through horizons B and C. Such transported soils lose much of their former structure as residual soils, with the collapse

of pipes and other macropores and of the internal structure of micro-aggregates.

It is shown in Chapter 8 that transported soil layers are commonly present in the tropical environment; and for this reason alone very high spatial variabilities in infiltration capacity should be expected, and as indicated here the values may range across two orders of magnitude (from $<20 \, mm \, h^{-1}$ to $>1000 \, mm \, h^{-1}$).

Soil characteristics are genetically linked to slope position in catenary patterns or *toposequences*, and these often determine where runoff will be generated within small basins. Thus ferrallitic soils are commonly restricted to summits and upper slope facets, where they exhibit high infiltration rates, at least into the upper horizon. On summits it is common to find either a nodular lateritic gravel or a sandy topsoil from which the *runoff coefficient*, according to Roose (1973), will reach only 1–2% on freely drained soils with rainfall $\sim 2000 \, mm$ $year^{-1}$, and rise to only 5–15% at higher rainfalls $\sim 3400 \, mm \, year^{-1}$ on less well drained soils. Dubreuil (1985) quoted results from studies by Lafforge on the Ivory Coast, where simulated rainfall of $120 \, mm \, h^{-1}$ failed to produce runoff from well drained soils in the savanna regions (rainfall $2100 \, mm \, year^{-1}$). By contrast, the hydromorphic soils which occur at the footslope position produced runoff at 80% of input when simulated rainfall reached $30 \, mm \, h^{-1}$. This phenomenon is a function of the groundwater rise to the surface in the wet season; the hydromorphic soils are not inherently impermeable.

Although this is a crucial topic for geomorphologists there is a lack of consistent data for infiltration into tropical soils. This is undoubtedly due to a series of factors affecting experimental results. These include:

(1) the inherent high spatial variability of the surface soil, including the frequent presence of fissures and macropores;
(2) common lack of attention to antecedent moisture conditions;
(3) difficulties of observing runoff under deep litter in natural forests.

Pipe flow

The first of the factors listed above leads into the subject of pipe flow. The development of soil pipes and tunnels is not confined in any way to the tropics, where observations of these phenomena are still comparatively few, but they are very common phenomena in both wetter and drier climates in the tropics. Löffler (1974) has correctly pointed out the neglect of subsurface erosional processes in tropical geomorphology and the topic remains comparatively obscure (when compared to the numerous studies of runoff and sediment yield). Some factors contributing to the formation of pipes have been clearly described, but many instances occur where these do not appear to operate.

Formerly, pipes were thought to be characteristic of semi-arid zones and specifically related to the presence of smectite clay minerals (Parker, 1963). These cracking clays form deep fissures in the long dry season and these later admit rainfall, but soon become sealed as the clay minerals swell or deflocculate on taking up water. A few major cracks remain open as a result of large volumes of water entering and causing erosion, and this concentrated flow first moves vertically downwards and then spreads laterally, following the still-open fissures at depth or flowing along textural discontinuities in the soil profile where permeabilities remain high. This set of conditions is common, but not universal, and striking examples can be found in highly seasonal tropical climates. A wide ranging review of piping in different environments has been offered by Jones (1981), who adopted a simple threefold division of pipe formation advanced by Parker and Jenne (1967):

(1) desiccation-stress crack-piping, dominant in many dry climates, and involving water entry into open soil cracks (or other conduits such as root channels and animal burrows) and eroding tunnels down the hydraulic gradient to gully head or side slope;
(2) variable permeability-consolidation piping, developing upslope from gullyheads, where these intersect horizons or beds of higher permeability confined by lower permeability materials above and below;

(3) entrainment piping, where newly created large hydraulic heads cause channelised subsurface flow with entrainment of water-saturated materials, often leading to subsidence.

According to Goldsmith and Smith (1985), macropores that can become pipes may develop from: (1) stress fractures, (2) root networks, (3) biotic activity, (4) texture contrasts leading to marked reductions in permeability at certain depths in the soil, and (5) the presence of a zone of potentially dispersive soil.

In Zimbabwe, Stocking (1981b) monitored pipe development over a 5-year period in soils with a high

FIGURE 5.2 Gullied alluvium and colluvium near St Michael's Mission, Umsweswe River, Zimbabwe. Previous gully floors are marked by pebble layers; black tones denote pipe exits

FIGURE 5.3 The same location as in Figure 5.2, showing how pipe flow can excavate cavernous exits into the gully system, leading to its enlargement. Formed in sodic materials (see text)

exchangeable sodium percentage (ESP). The area occupies the headwaters of the Umsweswe River and contains both weathered Karroo sediments and thick colluvium derived from these rocks. Relief is comparatively gentle though the valleys are inset into a planation surface of very low relief. Rainfall is 750 mm year^{-1}. The smectite-rich soils are deeply fissured, and clay-bound gravel layers represent former gully floors at different levels (Figures 5.2 and 5.3). According to Stocking (1981), in soils with a high ESP (>15–20%) the clay deflocculates in the presence of water and blocks the natural pores, so that throughflow must take place above or below such a horizon which acts as a 'subsurface dam wall'. Thus pipe roof samples gave a median ESP of 40, while large pipes have

ESPs of <5–20. Where very high ESPs occur (~40–60+) the soil appears massive and contains a dense network of micropipes. The piping system was thus attributed to:

(1) bad drainage in the Karroo-derived deposits, allowing build up of Na$^+$ ions (high ESP);
(2) impeded groundwater movement resulting from deflocculation of clays;
(3) proximity to gully heads creating a steep hydraulic gradient;
(4) breaches of subsurface dams during storm flows (initiated by inflow through deep vertical fissures in cracking clays);
(5) extension through self-perpetuating feedback as gully walls retreat.

This association of piping with gullying is very important and is repeated in many different examples. But piping is by no means restricted to these situations. Pipe entrances are found in fissured clays on dambo floors, where no incision is taking place (Figure 5.4). Piping also seems characteristic of high rainfall areas where either textural contrasts occur in the soils or within the parent materials (usually when these are sediments). The formation of *humicrete* layers beneath leached and eluviated podzolic sands in alluvial sediments induces lateral flow and pipe formation in the perhumid equatorial climates of Kalimantan (Figure 5.5) Löffler (1974, 1978) observed pipes in the humid tropics (2500–5000 mm year^{-1}) of Papua New Guinea, where he found them widespread in poorly consolidated sediments, and usually associated with steep slopes or gully walls. He found them apparently linked upslope to sink holes or similar depressions. These were characteristically 5–10 m in diameter and 4–6 m deep. The sediments were usually high in clay but also permeable and dominated by halloysite. Piping in both alluvial and colluvial sediments is common, probably because textural contrasts often occur close to the surface. Piping was also identified as an important process in Sarawak by Baillie (1975).

Very large pipes, really cave systems, have been observed beneath ferricrete duricrusts. These have been described from the Kasewe Hills in Sierra

FIGURE 5.4   Vertical macropore (pipe) entrance in cracking smectite soils in dambo floor, eastern Zambia

FIGURE 5.5   Piping in podzolic 'white sand' alluvial deposits in Kalimantan, Indonesia. Horizontal pipes develop over *humicrete* layers or above clayey sediments

Leone (Bowden, 1980), in an area known to the present author. In some places open pipe entrances plunge steeply many metres into the unconsolidated sediments beneath the duricrust, and cave exits are seen downstream in cliffed gullyheads. These caverns are large enough to be entered and have the form of vadose stream tunnels. A wide range of surface depressions in duricrusted terrain probably attests the efficacy of these pipe systems in removing saprolite and/or sediment, causing subsidence and fracturing of the overlying crust.

The association of pipes with near-vertical slopes may be due mainly to observational bias, since they are not easily seen in other places. Equally, many gully systems are cut into unconsolidated materials and these provide many sites where pipes can be examined. Over crystalline rocks, enclosed depressions may be found on plateaus and ridge tops and this phenomenon is discussed later as *pseudo-karst* (Chapter 11). The occurrence of pipe entrances 1 m in diameter in the bottoms of such surface depressions was observed by Feininger

(1969) in Colombia and the collapse and coalescence of such depressions on sloping ground can initiate gullies. The development of subterranean conduits within otherwise fresh rock is, however, a separate topic from piping in soils and sediments.

RUNOFF GENERATION

The subject of infiltration is closely linked to that of runoff generation, which brings the discussion closer to the processes of flow and erosion on slopes. Runoff can be divided into:

(1) direct surface runoff, occurring immediately, when infiltration capacity of the soil is exceeded (this is 'Hortonian overland flow');
(2) delayed surface runoff, occurring only when the surface or subsurface soil layer(s) has become saturated, and usually starting at the lower end of a plot or small catchment (this is 'saturated overland flow');
(3) subsurface runoff, occurring preferentially within more permeable soil horizons (usually called 'throughflow').

We can expect runoff to be generated by storms of particular intensity, size and frequency. Dubreuil (1985) has tabulated some contrasting data derived from studies in West Africa (Table 5.2). It is clear that in many areas runoff will not occur without a threshold value of antecedent moisture being exceeded. The figures presented here indicate that rainfall of 50 mm will produce runoff on most soils, according to antecedent moisture conditions.

Runoff coefficients in the tropics

Our knowledge of runoff generation associated with tropical rainfall under forests and savanna woodlands, however, remains sketchy. Despite many individual studies of water balance and of runoff under forested conditions, differences in lithology and soil structure, of slope angle and length, as well as of actual vegetation cover all lead to widely varying measurements. The calculation of the annual *runoff coefficient* as a percentage of the rain falling on a slope or catchment area is clearly

an important expression of the dynamics of the morphogenetic system, but the temporal and spatial aspects of this parameter are complex. Individual slopes (measured by erosion plots), small catchments and larger catchments all behave differently, and figures obtained on an annual basis are very different from those recorded during an actual storm event. It is therefore important not to compare figures obtained under different circumstances.

Roche (1981) found that runoff coefficients of 16–27% characterised experimental, forested watersheds in French Guyana (rainfall ~ 3500 mm year^{-1}) and such values are not untypical. However, others range from <10% to >35%, and Ogunkoya *et al.* (1984) and Ogunkoya & Jeje (1987) obtained annual runoff coefficients of 1–40% for 15 third-order drainage basins in Nigeria. But these annualised data from drainage basins aggregate all rainfall events (in time) and all sources of runoff (in depth and in space) and do not reveal the response to individual storms or the behaviour of specific soils and slopes. Higher coefficients come from steeply sloping sites in the humid tropics. For example Bonell & Gilmour (1978) and Bonell *et al.* (1983) monitored two basins in northeast Queensland, where average slopes are 19°, and high rainfalls exceeding 4000 mm year^{-1} are concentrated (63.3%) into the summer months, December–March. Their results showed that 47% of discharge appeared as *quickflow* attributed largely to saturated overland flow. In this very wet climate, rainfall events follow each other so rapidly in the wet season that saturated overland flow is almost instantaneous, and the vegetation cover acts mainly to protect the soil from erosion; not to reduce runoff *per se*.

However, much lower coefficients have been obtained in some studies in humid forested areas. Sharma and Correia (1989) conducted detailed measurements under pine forests on the Shillong Plateau in India (altitude 1000 m, mean annual temperature 26 °C, average rainfall 2077 mm year^{-1}), and found that only 0.52% of the rainfall was lost as overland flow, also citing Pathak *et al.* (1984), who obtained a value of 0.41% under similar vegetation in the Himalaya. The Shillong

TABLE 5.2 Minimum rainfall and antecedent moisture producing runoff at stations in West Africa: (1) minimum rainfall to produce runoff; (2) antecedent moisture index

Structure of the watershed	Beginning of the rainy season (1) (mm) (2)	Middle and end of the rainy season (1) (mm) (2)
(1) Very permeable thick soils Granitic boulders, very thick granitic sand, sandy soils on colluviums (Lhoto, Bénin, 1100 mm year^{-1} with two seasons)	10 if $\Sigma_1^{30} P_a > 150$ mm 20 > 100 mm 40 > 60 mm	10 if $\Sigma_1^{30} P_a > 150$ mm 20 > 100 mm 40 > 60 mm
Thick sandy leached ferruginous soils with low gradient + 15% of gley soils (Tiapalou, Bénin, 1300 mm year^{-1} within 6 months)	50 if $\Sigma_1^{90} P_a > 500$ mm	10 ($t_a = 1$ day) or $\Sigma_1^{30} P_a > 250$ mm 20 ($t_a = 10$ days) or $\Sigma_1^{30} P_a > 150$ mm
(2) Vertisols 55% of hydromorphic soils in the valley floors + very permeable ferrallitic soils (Sakassou, Ivory Coast, 1200 mm within two seasons)	35 with $IH \leqslant 25$ $IH = \Sigma_1^{45} P_a \cdot e^{-0.10\, t_a}$	15 with $IH = 100$
Vertisols on marly shales (Lac Elia, Togo, 1050 mm within 2 seasons)	40 if $\Sigma P_a = 350$ mm	7 ($t_a = 1$ day)
(3) Groundwater rising to surface Lithosols and very thick sandy leached ferruginous soils on sandstones (Sanguéré, Cameroon, 1050 mm per 5 months)	30–50 if $\Sigma P_a \leqslant 400$ mm	25 $400 < \Sigma P_a < 800$ mm 5–15 and $\Sigma P_a > 800$ mm
Gravelly red tableland soils, thick clay sandy ferrallitic soils on granitic sands (Korhogo, Ivory Coast, 1400 mm per 6 months)	30 if $Q_o = 25$ l s^{-1} or $IH = 10$ $IH = \Sigma_1^{14} P_a \cdot e^{-0.18\, t_a}$	15 if $Q_o \geqslant 125$ l s^{-1} or $IH \geqslant 70$

Note: $\Sigma_1^{30} P_a$ = total of antecedent rainfalls on 30 days; t_a = duration from previous rainfall in days; IH = antecedent moisture index; and Q_o = minimum discharge before rainfall. After Dubreuil (1985); reproduced by permission of Elsevier Science Publishers BV.

Plateau sites had slopes of 13.8° and experienced frequent small storms throughout the monsoon period, 73% of the rain falling in events of <20 mm (56.4% at <20 mm h^{-1}), and only 2.9% in events of >60 mm (2.9% at >100 mm h^{-1}). They inferred that much of the flow takes place in the upper layers of the soil, and that overland flow results primarily from the formation of a *perched water-table* and saturated soil conditions as a consequence of frequent rainfall. Nortcliffe *et al.* (1990) monitored three carefully instrumented sites on Maracá Island, northern Brazil, beneath undisturbed rainforest and

obtained runoff coefficients of 7.7% for top slopes, 13.9% for mid slopes (maximum slope 13°), and 11.8% for bottom slopes, over a 300-day period during which the largest daily rainfall was 115.1 mm, and there was also a spell of wet weather over 35 days during which repeated daily falls of 20–40 mm produced 540 mm over the period.

In sub-humid savanna areas, with an open vegetation cover and prevalence of 2:1 lattice clays, high runoff can result from individual storms. Ledger (1964) analysed figures from small West African catchments receiving <750 mm year^{-1}, and found that runoff from individual storms could

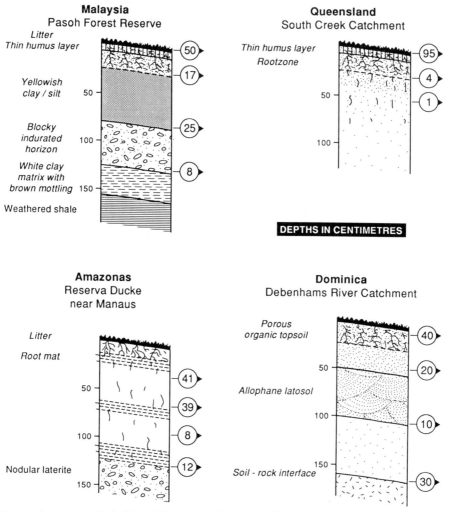

FIGURE 5.6   Proportion of runoff draining different surface and soil horizons in four tropical rainforest sites (after Douglas & Spencer, 1985a; reproduced by permission of Unwin Hyman)

be as high as 80%, while annual figures averaged 30%. Working in Chad (300–500 mm rainfall year⁻¹), Rodier (1975) also found a major difference between runoff coefficients for coarse granitic sands (5–20%) and sandy clay soils of *glacis* and valley floors (up to 75%). Much larger catchments (>1000 km²) gave Ledger values of as low as 6%; these are due to increased evaporation from water surfaces and floodplain soils, and also to floodplain storage and aquifer recharge. It should be noted that Cordery and Pilgrim (1983) analysed 53 basins of varying size in Australia and

concluded that rainfall losses from storm runoff are not correlated with soil or cover characteristics but, 'vary, apparently randomly, within a fairly narrow band.'

More interest, in this context, possibly attaches to the generation of runoff on individual slopes, but again very variable data have been obtained. Douglas and Spencer (1985a) summarised a number of trap experiments from rainforest locations (Figure 5.6), and these illustrate contrasting conditions of substrate, soil and slope. The Amazonas example measured by Nortcliff and

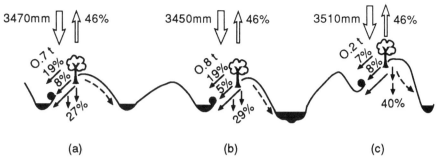

FIGURE 5.7 Distribution of terms in the hydrological balance of three experimental watersheds (ECEREX) under forest in French Guiana. Values refer to soil loss (t ha^{-1} year^{-1}), surface runoff (%), throughflow (%) and infiltration (%), respectively, e.g. (a) 0.7 t ha^{-1} year^{-1} soil loss, 19% surface runoff, 8% throughflow, 27% infiltration (after Roche, 1981; reproduced by permission)

Thornes (1981) indicates that no measurable overland flow occurred, but that 80% of the runoff takes place as throughflow between 15 and 70 cm depth. In contrast, Leigh (1982) found that conditions in Malaysia, towards the base of a 17–22° slope, could produce 50% of runoff as overland flow. This figure rises to 95% at a site in Queensland, due to impermeable soil layers below 10 cm.

The three small (1–1.5 ha) experimental watersheds reported by Roche (1981) from French Guyana, where slopes ranged from 15 to 25%, also provided information on the water balance and sediment yields of the catchments (Figure 5.7). High figures for infiltration to provide groundwater recharge (27–40% or 900–1400 mm) indicate the major source of perennial flow for these small rivers, and also the amount of water available for rock weathering. Surface runoff did not rise above 19% and the sediment yield was 0.2–0.8 t ha^{-1} year^{-1}, with bedload returning slightly higher values than suspended load. About 0.5 t ha^{-1} year^{-1} was lost as dissolved load.

The results quoted by Douglas and Spencer (1985a, and Figure 5.6) indicate the importance of horizon discontinuities in generating lateral flow at different depths, so that few simple statements about lateral water flux in tropical soils are likely to be valid (Nortcliff & Thornes, 1981), although most of this throughflow appears to take place within 1.5 m of the surface. Commonsense tells us, however, that sampling at greater depths is uncommon, and that deeper *interflow* towards stream channels must take place, to supply the base

flow discharge. The importance of piping in this regard has been emphasised.

Base flow discharges

Because much of the rainfall reaching the groundsurface is infiltrated into deep soil and weathering profiles, this water moves towards the stream channels over a long period, and can then maintain a steady base flow. However, Dubreuil (1985) found that, from 40 representative (small) basins in West Africa, only those receiving >1600 mm year^{-1} *consistently* exhibited a steady base flow component of discharge. No steady base flow was measured with rainfalls <1000 mm year^{-1}. However, within the intermediate category (1000–16 000 mm year^{-1}), conditions varied according to bedrock lithology. In general, basins on schists and clays had no steady base flow; some on weathered granites and sandstones were able to maintain a base flow, but catchments had an individual behaviour requiring detailed site analysis.

RUNOFF AND SEDIMENT YIELD

Surface flows and sediment yield

Douglas and Spencer (1985a) collated a large amount of information and numerical data on rates of denudation and of sediment yield in the forested tropics, but their tables show some divergent data. Nevertheless, there is some convergence of results from moderately sloping sites under forest in different parts of the tropics. Roose (1973) studied rainforest plots on the Ivory Coast and obtained values of 0.03, 0.2 and 1.0 t ha^{-1} year^{-1} on slopes of 4°, 12.5° and 24° respectively. He also obtained a value of 0.5 t ha^{-1} year^{-1} for a 5° slope under semi-deciduous forest. Similar rates of soil loss were recorded on slopes of 2–13° under rainforest in Brazil by Nortcliff *et al.* (1990). Sharma and Correia (1989), in their study of the Shillong Plateau discussed above, recorded sediment yields on an event basis over two years, with a mean of 4.71 kg ha^{-1}. Using their data, a sediment yield of 0.18 t ha^{-1} year^{-1} is implied, averaging two years across six sites with a mean slope of 8.67°. The sediment yield did not show a statistical relationship with either storm size or intensity; only with overland flow depth. They deduced from this that sediment was entrained by overland flow and not by rainsplash which was rendered ineffective by the canopy and the litter layers, and that sediment was lost largely due to the preparation of the surface by rodents and other burrowing animals, and possibly by the effects of wetting and drying on soil aggregates.

However, Spencer *et al.* (1990), in a study of small catchments in Sabah, Malaysian Borneo, argued that sediment loss was promoted by natural disturbance of the rainforest due to processes such as tree fall (by death and by windthrow), fire, and landslides. Debris and roots on the forest floor were found to act as traps for litter and sediment entrained on the slope, and this sediment would be released by the effects of high magnitude rainfall events producing sufficient runoff to sweep away branches and other material. Quite large dams were found on small rivers (first- to third-order catchments) and these can persist over some (undefined) period of years, but probably much less than the 20+ years recorded for the temperate forests of northwest USA. The rapid growth of pioneer species on soils exposed by natural disturbance reduces the impact of these openings on runoff and sediment generation from the slope, and many authors argue that, under the undisturbed forest canopy, the kinetic energy of leaf drips from mature trees, combined with the waterflow down tree trunks combine to produce a runoff regime capable of dislodging soil particles and entraining

sediment on the forested slopes. The fact that Nortcliff *et al.* (1990) found that partial clearance of forest had little impact on the soil loss probably reflects these factors.

It should be noted, however, that while Kesel (1977) obtained values of 3.3–7.7 t ha^{-1} year^{-1} in the Rupununi savannas of Guyana, he also recorded figures of 7.3–8.5 t ha^{-1} year^{-1} under rainforest with slopes of 10–15°. Lal (1981, 1986) quoted Brünig (1975) that, while under virgin forest sediment yield may be as low as 0.2 t ha^{-1} year^{-1}, this can rise under cultivation to 600 or even 1200 t ha^{-1} year^{-1} which shows clearly that when vegetation cover is altered at a site, usually by clearance of the forest, runoff and sediment yields increase alarmingly.

These figures offer a guide to the orders of magnitude of sediment transport on undisturbed slopes, but the high variability of figures obtained points to several sources of variation. These may be experimental, e.g. different observers using different sizes or designs of erosion plot, and observing over different intervals or periods of time. The more fundamental reasons for 'conflicting' results include:

(1) variations in rainfall characteristics (totals, distribution, intensities) between stations;
(2) variations in rainfall totals and storm magnitude/frequency during the collection periods;
(3) differences in forest type (composition, structure and canopy), resulting from climatic factors (including former climates) and edaphic conditions;
(4) spatial variation of the canopy cover, due to patterns of disturbance and regeneration at different scales;
(5) spatial variation in soils and soil parent materials creating soil mosaics (in addition to catenas or toposequences) with contrasting permeabilities (surface infiltration capacity; subsurface discontinuities);
(6) differences in topography, especially in slope angle, length and position.

Despite many statements to the contrary, surface wash processes do occur within the rainforest.

Chatterjea (1989) has offered some detailed studies from the Bukit Timah Nature Reserve in Singapore, effectively extending earlier work by Douglas (1967a,b). This area receives 2369 mm year^{-1} in an equatorial rainfall regime. It is hilly, with slopes exceeding 20°, and some evidence of breaks in the 25–34 m high, rainforest canopy. Rainfall intensities of 120 mm h^{-1} over 15-min periods have been measured. The interception of rainfall by the canopy is highly dependent on the total rainfall, and varied from nil to a maximum of 91%, averaging 58% during measurements of 35 storms. In forest-covered areas, much of the throughfall arrives from leafdrips and stemflow, and the former can generate large drops that are capable of reaching near-terminal velocities from larger trees (Young, 1972), and if not intercepted at lower levels in the forest, can exert a high kinetic energy on the forest floor. Chatterjea (1989) described the formation of earth pillars and occurrence of tree-base scour. On steeper slopes, leaf litter may be very thin, only a few millimetres, but on gentler slopes may reach 250 mm according to Ruxton (1967). Both Hortonian and saturated overland flow were observed at Bukit Timah, a perched water-table forming in the top 20 cm of the soil during intense rainstorms. Sediment movement is also generated by small slope failures and by the effects of tree falls, and the existence of soil pipes from 2–8 cm in diameter must add to the total water and sediment transfer downslope. The surface wash observed was invariably diffuse and had a low velocity. According to Chatterjea (1989), its function is mainly to transport sediment dislodged by raindrop impact, but it is strongly affected by leaf litter, broken branches and root mats. Loss of sediment (ground retreat) from two grid plots measured over 405 days was 4.00 and 4.67 mm year^{-1} respectively.

Such studies emphasise the importance of the litter layer and other obstacles to the acceleration of surface water flows, since flow velocity largely determines the erosive power of the water. Velocity, in turn, depends on the hydraulic radius (or flow depth) (r), the roughness of the slope, expressed as the Manning coefficient (n), and the slope (S):

$$v = \frac{r^{2/3} \, S^{1/2}}{n} \tag{5.4}$$

This is known as the *Manning Equation*, and it applies only to fully turbulent flows such as are encountered in river channels. The *Reynolds Number* (Re) gives an index of turbulence:

$$Re = \frac{vr}{u} \tag{5.5}$$

where r is the hydraulic radius, which is taken as the flow depth for unchannelled flows, and u is the kinematic viscosity of the water. When $Re < 500$ flow is laminar; when it exceeds 2000 it is fully turbulent. According to D'Souza and Morgan (1976) soil loss varies with the Reynolds number, and Savat (1977) also used the *Froude Number* to determine subcritical and supercritical flows, but determined a number of empirical relationships for surface flows that do not satisfy equations developed for channel flow. These are quoted by Morgan (1986). Clearly the essential relationship is between discharge or flow rate (Qw), slope (S) and velocity (v) which can be given as:

$$v \propto S^{1/3} \, Qw^{1/3} \tag{5.6}$$

after Meyer (1965). Meyer and Wischmeier (1969) then determined that detachment (Df) and transporting capacity (Tf) would be:

$$Df \propto S^{2/3} \, Qw^{2/3} \tag{5.7}$$

$$Tf \propto S^{5/3} \, Qw^{5/3} \tag{5.8}$$

providing that detachment varies with the square, and the flow capacity with the fifth power of the velocity (see Morgan, 1986). These somewhat difficult relationships will not be explored in detail here. Flow conditions on naturally vegetated slopes are difficult to simulate in the laboratory, and results from experiments must be taken as a guide only to local conditions. However, the principles of erosion are well established, and the common experience with slope flows is that they have low

Reynolds Numbers for turbulence. This means that surface erosion is highly dependent on detachment by raindrop impact. The classical Hjulström curve (1935) illustrates clearly that the erosion velocity of flow greatly exceeds the fall velocity, especially where particles < 0.5 mm predominate, so that fine particles dislodged by high energy rainstorms can remain in suspension within low velocity surface flows. Particles between 0.2 and 0.3 mm are most easily mobilised by flows with low velocities. Morgan (1986) observed that overland flow is generally deficient in particles > 1.0 mm and also carries much finer material. He also considered that much of the transport of sediment by overland flow is as bedload, with particles rolling downslope.

Although much of the work on soil erosion has been directed towards slopes under controlled conditions of plant cover, these principles also apply to many natural slopes (Elwell & Stocking, 1973, 1975, 1976, 1982), and are relevant to discussions about the effects of climate and vegetation changes in the past (Chapter 7). Where strong sheet wash occurs without gullying, there will be a selective removal of medium and fine sand (perhaps < 1.0 mm diameter), silt and clay, such that the upper slopes become progressively dominated by coarse sand and gravel, together with larger clasts of vein quartz, iron nodules and duricrust fragments. The finer particles may be swept into river channels, but will commonly accumulate on lower slopes. The importance of this process is illustrated by the differences in soil properties across slopes, and also in the development of some distinctive landforms such as the *dambos* of central Africa and many other areas which commonly have clay infills and sandy marginal slopes (Thomas & Goudie, 1985; Thomas, 1986) The same argument also underlies many explanations for stone-lines commonly found within soil profiles or buried by finer sediment (Alexandre & Symoens, 1989).

Sediment yield from catchment studies

When sediment yield from catchment studies is substituted for erosion plot data, other processes of erosion become involved and the data are not strictly comparable, although some data from small catchments have been quoted above. Ogunkoya and Jeje (1987) studied sediment yields from 15 third-order basins (2–18.8 km^2) draining metamorphic and igneous rocks in southwestern Nigeria. These areas lie within the present rainforest zone, and receive an annual rainfall of 1400 mm. Only one year of records was used, so the effects of large storms with long return periods will not have been included. Their results, however, were interesting. Suspended sediment yields ranged from 0.4 to 29.5 t km^2 year^{-1} and were less than solute yields by factors of 1.5–6.8. It was found that 82% of the variance in sediment yield can be accounted for by runoff, basin relief and basin shape. Despite a range of forest cover from 14 to 87% among the 15 basins, this factor was not statistically significant in explaining the sediment yields. This may appear to contradict much of what has been stated above, but in this region a combination of low relief and slope values, long-established traditional systems of rotational bush fallow agriculture, and a climate lacking extreme events may all contribute to this result. There was, however, a very marked difference in behaviour between streams with perennial flow and those with periods from 100 to > 200 days without discharge. Perennial streams had a mean sediment yield of 15.3 t km^{-2} year^{-1}; those with more than 100 days without flow had a mean of 3.9 t km^{-2} year^{-1}. This factor is of course closely linked to variations in total runoff, a relationship established for some Kenyan rivers by Dunne (1979) and Dunne *et al.* (1978). Ogunkoye and Jeje (1987) compare their results with those from small basins in eastern Australia reported by Slaymaker (1978), who found suspended sediment yields of 11.4–22.3 t km^{-2} year^{-1}. Such small basin studies are important for the understanding of geomorphorphic processes because they do not aggregate conditions over large and diverse catchment areas. Larger basins in central and southern Nigeria were reported by Oyebande (1981) to return sediment yields of 20–80 t km^{-2} year^{-1}.

In areas of high relief, as in eastern and southern Africa, sediment yields can be one to two orders of magnitude greater, from 100 to 1000 t km^{-2} year^{-1} (Walling, 1984). Records selected from basins of larger but modest size (< 10 000 km^2),

TABLE 5.3 Sediment yield from African catchments under 10 000 km²

River catchment	Area (km²)	Sediment yield (t km^{-2} year^{-1})	Source
Ethiopian highlands	150	1680	Virgo & Munro (1978)
Upper Tana, Kenya	9520	500	Dunne (1979)
Ikowa, Tanzania	640	287	Stromquist (1981)
Orange, South Africa	6362	890	Rooseboom (1978)
Caledon, Lesotho	945	1979	Chakela (1981)

TABLE 5.4 Sediment yield from African catchments over 10 000 km²

River catchment	Area (km²)	Sediment yield (t km^{-2} year^{-1})	Source
Limpopo, Zimbabwe	196 000	17.3	Ward (1980)
Mbam, Cameroon	42 300	85	Olivry (1977)
Sanaga, Cameroon	77 000	28	Olivry (1977)
Chari, Chad	600 000	3.9	ORSTOM[a]
Logone, Chad	85 000	14.9	ORSTOM[a]
Ouham, Central African Republic	44 700	9.4	ORSTOM[a]

[a]Reported by Walling (1984).

some of which drain largely semi-arid areas, are shown in Table 5.3.

It is interesting to see how, when the scale of enquiry increases again to large basins (>10 000 km²), the figures tend to decline (Pilgrim *et al.*, 1982). Ward (1980) reported 38 t km^{-2} year^{-1} for large rivers in Zimbabwe and some individual records reported by Walling (1984) are shown in Table 5.4. The giant Zaire basin is reported to have a sediment yield of only 11.3 t km^{-2} year^{-1}; the Niger, 33.1 t km^{-2} year^{-1} (Milliman & Meade, 1983).

At the continental scale Milliman and Meade (1983) have compared sediment delivery to the oceans world-wide from large catchments and produced some interesting comparisons that update previous calculations made by Holeman (1968). Using their figures for sediment yield, these can be ranked by major regions (Table 5.5).

The data in Table 5.5 do not tell us about individual locations, but by aggregating large regions, other aspects of denudation are revealed. Continental areas which are relatively arid and do not have active mountain ranges (Australia and

Africa) return low values. These increase for South America which is both more humid and more mountainous. The rapid erosion of the monsoon lands of southeast Asia is a function of high and concentrated rainfall on mainly steep catchments of mountainous areas, and these factors increase in magnitude for the islands of Indonesia, Papua New Guinea, Taiwan and Japan, where rates of erosion increase dramatically.

These rates of erosion do not easily relate to values obtained from small plots, for several cogent reasons:

(1) Erosion plot experiments are usually under a single vegetation cover or land use; large catchments combine all land uses within the area drained.

(2) Most erosion plots are on gentle to moderate slopes (<20°); many large catchments originate in mountainous areas that supply the bulk of the sediment load (e.g. Amazon, Ganges).

(3) Rivers derive sediment not only from surface wash processes but also (mainly) by bank erosion (of creep and landslide debris) and also by channel processes.

TABLE 5.5   Sediment yield ranked according to major
regions of the world

Area	Sediment yield	
	t km^{-2} year^{-1}	t ha^{-1} year^{-1}
Australia	28 (41)	2.8
Africa	35 (25)	3.5
South America	97 (57)	9.7
Asia	380 (543)	38.0
Large Pacific Islands	1000 (na)	100.00

Data taken from Milliman & Meade (1983). Values in parentheses
are from Holeman (1968) for comparison.

(4) Sediment reaching the oceans is only a
proportion of that eroded from headwater and
intermediate areas; sediment stores tend to
increase in importance downstream, mainly in
the forms of active floodplains, palaeoplains
and terraces.

Nearly all the large African rivers, for example,
drain into former basins of internal drainage, where
large sediment stores have accumulated throughout
much of the Cenozoic. The Niger, Nile, Zaire
and Zambezi are all geologically and geomorpho-
logically complex in this sense. These factors have
recently been reviewed by Milliman and Syvitski,
(1992; and see Chapter 1).

Another major difficulty in making comparisons
amongst the available data is the problem of data
reliability and accuracy (Walling, 1984). Suspended
sediment is easy to collect, but the location and
frequency of sampling strongly affect the results,
as may the equipment used.

Sediment yield and rate of denudation

Sediment yield recorded as t km^{-2} year^{-1} can be
computed as a rate of denudation (D), as m^3 km^{-2}
year^{-1} or mm year^{-1}:

$$D = \frac{\text{total load (tonnes)}}{\text{area (km}^2) \times \text{specific gravity of rock}} \quad (5.9)$$

where the specific gravity of rock is commonly
taken as 2.64 (or 2.5).

Douglas and Spencer (1985a) collated sediment
yields for tropical rivers in this form (m^3 km^{-2}
year^{-1}). The results range across four orders of
magnitude, from a minimum of 0.3 m^3 km^{-2} year^{-1}
in the Xingu, a lowland tributary of the Amazon,
to 8000 m^3 km^{-2} year^{-1} for the Cilulung, draining
volcanic mountain slopes in Java. These figures,
although they have a much wider range of values,
indicate a similar trend when compared with those
aggregated for large regions by Milliman and Mead
(1983). Aggregating and averaging groups of figures
from Douglas and Spencer (1985a, Table 3.7)
throws some light on conditions in particular
tropical areas:

(1) the large, mountainous islands of Papua New
Guinea and Indonesia return high values
averaging 1463 m^3 km^{-2} year^{-1} (1592 m^3 km^{-2}
year^{-1} if Ei Creek is omitted) and 2056 m^3
km^{-2} year^{-1} respectively;

(2) the Philippines average only 379 m^3 km^{-2}year^{-1},
while Malaysia returns 83 m^3 km^{-2} year^{-1};

(3) less mountainous regions in the humid tropics
return figures around 30 m^3 km^{-2} year^{-1}
(humid Australia, 36.2 m^3 km^{-2} year^{-1};
Cameroon 30.9 m^3 km^{-2} year^{-1}; Kenya
(except Grand Falls at 793) 29.8 m^3 km^{-2}
year^{-1}; Amazonia, 30.7 m^3 km^{-2} year^{-1});

(4) 10 large lowland tributaries of the Amazon
return an average value of only 8.5 m^3 km^{-2}
year^{-1} (forested lowlands in Nigeria also
return low figures—see below);

(5) the semi-arid zones of Nigeria return a higher
average value of 116 m^3 km^{-2} year^{-1};

(6) studies of small catchments concentrated in
areas of serious soil erosion in Tanzania return
an average figure of 428 m^3 km^{-2} year^{-1}.
However, the large Rufiji catchment, also
in Tanzania yields only 60 m^3 km^{-2} year^{-1}
and is probably more representative of the
region.

Using the data provided by Ogunkoya and Jeje
(1987) for small lowland forest catchments in
southern Nigeria, much lower rates of denudation
can be calculated, ranging from 0.16 m^3 km^{-2}
year^{-1} to 11.14 m^3 km^{-2} year^{-1} with perennial

streams averaging $5.45 \, m^3 \, km^{-2} \, year^{-1}$ and seasonal streams $1.24 \, m^3 \, km^{-2} \, year^{-1}$. From these data it is possible to agree with the authors that runoff and basin relief may be important determinants of denudation rate. Thus while mountainous headwaters of major rivers such as the Amazon or the Ganga–Brahmaputra systems have very rapid rates of denudation; the headwaters of rivers draining cratonic areas, in the absence of bounding escarpments, may have very low rates of denudation. These are a consequence of lower rainfalls, lack of groundwater storage, periods of very low flow (Ledger, 1964), and low relief and slope values.

These values can be read as mm per 1000 years (mm $year^3$ or 'B'—Bubnoff units), and compared with rates of denudation for temperate areas given by Ahnert (1970). It is interesting to note that the Alps return a value of 379, which is identical with that for small eroding catchments in Tanzania. The figure of 1000 given for the Himalayas is only 50% of that recorded for Indonesia. Whether such direct comparisons should be made is open to question, but it will be maintained here that the humid tropical environment can produce conditions for very high rates of denudation in mountainous areas. The causes of this situation lie in the high rates of weathering that, in lowland areas can produce very thick mantles of saprolite, but in steep lands promote rapid denudation by mass movement which is discussed in Chapter 6.

On the other hand, if large rivers are compared from different climates the disparities are smaller: 51 for the Mississippi; 38 for the Columbia, but only 29.8 for the Amazon. The Colorado which drains a mountainous and largely semi-arid catchment rises to 165 (mm $year^3$ or $m^3 \, km^{-2} \, year^{-1}$). Rates of denudation calculated by this means remain incomplete, for they neither account for bedload nor do they consider the solute load.

Some rates of accelerated erosion

A comparison of some 'natural' rates of erosion with 'accelerated' rates of erosion in tropical areas (excepting mountain areas with very steep slopes $>25°$), as measured on experimental erosion plots (see Stocking, 1984), are summarised in Table 5.6.

The first thing to be said about these results is that too much reliance should not be placed on their precision. The second is to point out that erosion plot experiments tend to return different values from those obtained for small catchment studies. Obviously most catchments have a variety of land cover types and usually a significant amount of actual land use; sediment yield is therefore likely to be higher than for an erosion plot beneath undisturbed forest or savanna. On the other hand only a few catchments will experience the full effect of total clearance on sediment yield, though some have been indicated in Table 5.2, and rates of erosion can rise to $1500–2000 \, t \, km^{-2} \, year^{-1}$; much higher in badlands. Small erosion plots, when left bare of plant cover, can return sediment yields exceeding $5000 \, t \, km^{-2} \, year^{-1}$ (Table 5.6, data given in ha^{-1} should be multiplied by 10 to compare with those given in km^{-2}) on steep slopes but are generally $<2000 \, t \, km^{-2} \, year^{-1}$ on slopes of $<15°$.

Of course the data used here are selected and it is very evident from the literature that comparability between sites and experimental periods is seldom achieved, though the figures produced by ORSTOM and reported by Roose (1986) come close to this ideal. Temple (1972) also analysed a large number of results from Tanzania. Results from erosion pin exposure have been excluded here, and they are generally thought unrepresentative and unreliable (measurement is at one point on a slope and the pin itself induces turbulence). Some reasons for the differences in rates of sediment production at different scales of enquiry have been discussed by Millington (1981) in relation to his measurements in Sierra Leone.

Very high sediment yields recorded from severely eroded catchments reflect much more than the rate of erosion from the valley slopes. Such rivers will experience very high flood peaks and these short-term discharges will cause rapid bank erosion, whether into saprolite, colluvium or older alluvial sediments. Under these conditions, debris flows may also occur and some of these may reach the permanent stream channel system. It is therefore important to emphasise that so-called 'soil erosion' as measured for erosion plots of a few square

TABLE 5.6   Selected erosion plot results

Location	Site details Vegetation	Slope	Mean annual rainfall (mm)	Erosion (t ha^{-1} year^{-1}) Natural	Bare plot	Source
Freetown, Sierra Leone	High forest 1 year experiment	16°, 3°	3300	0.05 (16°)	30 (3°)	Millington (1981)
Shillong, India	High forest 2.25 years	4.3–13.8°	2079	0.182	—	Sharma & Correia (1989)
Adiopodioumé, Ivory Coast	Secondary rainforest 6 year experiment	4.5% 7% 20%	2100 2100 2100	— 0.017 0.052	60 138 568–746	Roose (1975, 1986)
Bouaké, Ivory Coast	Dense bush savanna	4%	1200	0.01–0.16	11–52	Roose (1975, 1986)
Saria, Burkina Faso	Sudan-sahel bush	0.7%	850	0.2–0.07	14–35	Roose (1975, 1986)
Henderson, Zimbabwe	Grass cover 10 year experiment	4.5%	750	0.7	127	Elwell (1971)
Ethiopian Central Plateau	Suspended sediment yield		500–800	—	165	Virgo & Munro (1978)
Mpwapwa, Tanzania	Bare plot 2 year experiment		620	—	146	Staples (1936)
Ife, Nigeria	6 year experiment bare plot	10%	924–1417 (6 years)	—	1.6	Jeje (1987)

metres, concerns only one part of the landscape at risk from accelerated erosion, and a disproportionate amount of experimental time has been devoted to this question.

Nevertheless, the loss of surface soil remains of prime concern and the data show that on natural slopes rates of erosion can be as low as 0.01 t ha^{-1} year^{-1} and rarely approach 1 t ha^{-1} year^{-1} (0.1–10 t km^{-2} year^{-1}). The rapid increase in sediment yield from plots with degraded plant covers is not surprising and it results from a complex of interrelated factors, starting with the impact of high intensity rainfalls (see Table 5.1). Here, this problem will be pursued only in so far as it throws light on the particular characteristics of tropical soils and their behaviour in disturbed denudation systems. Roose (1986, p. 326) has summarised the position: 'degradation of chemical and physical properties (of soils) following land clearing is usually attributed to the aggressive climatic conditions, soil fragility and the rapid mineralisation of organic matter which increases losses by erosion and drainage.'

The factors which influence these results are not only the rainfall intensity and plant cover. The existence and durability of water stable aggregates $> 200\,\mu$ in diameter is one of the major controls on runoff and erosion of soils in the tropics (De Vleeschauwer *et al.*, 1978). This property of saprolites has been discussed above (Chapter 3, pp. 60, 61), and, together with the associated organic matter strongly influences soil erodibility. Millington (1982) suggested that slope, soil erodibility (especially particle bonding) and rainfall erosivity are the three most important parameters controlling the rate of soil erosion in Weste Africa, while on a landscape scale drainage density also influenced the results.

GULLY EROSION

Piping and gully formation

Gullies are usually divided according to whether the dominant process is overland flow, creating a waterfall headcut, or subsurface flow and seepage, involving pipe flow (Imeson & Kwaad, 1980). Gully development is frequently associated with piping, but the relationships between the two phenomena are not clearly understood. In a recent review, Bocco (1991) suggested that gully initiation might more often be due to saturated overland flow and/or piping than to Hortonian overland flow that results where high rainfall intensity exceeds infiltration capacity. Once gullies have been formed, piping is frequently associated with the headwall and implicated in the extension of the gully system; the two processes of pipe formation and headwall recession being closely linked by the increased hydraulic gradient created by the gully. In some instances discontinuous tunnel gullies form by the collapse of pipe roofs. In these cases the pipe system must predate the formation of the gullies. This was recognised by Morgan (1979) and examples have been cited from the Sudan (Berry, 1970) and Hong Kong (Berry & Ruxton, 1960).

According to De Ploey (1974), gullies studied in Tunisia fall into three categories:

(1) *axial* gullies developed in gravelly soils;
(2) *digitate* gullies formed in clay soils; and
(3) *frontal* gullies formed mainly in loamy sands with a prismatic structure and subject to piping.

In Natal, South Africa, Downing (1968) noted both long dongas with upstream and tributary extensions marked by lines of sink holes, and shorter funnel-shaped gully heads marked by tunnels. Similar links between piping, tunnels and gully development were explored for the western USA by Heede (1974), who argued that diversion of surface drainage into open fissures would initiate piping which would then develop until progressive collapse of the pipes led to the extension of gullies. These in turn might degrade and become vegetated.

It would be premature (and possibly wrong), however, to claim that gully formation always depends on the existence of pipe flow, though this may be more a question of definition and lack of field knowledge. In very many tropical and sub-tropical soils, textural contrasts between A and B horizons in soils are common and these lead to strong subsurface flows. Such flows are often associated with stone-line gravels and with a compact, illuvial clay horizon formed at the base of the S layer. Spatial anisotropy is almost always present and this leads to the channelling of subsurface flows and (potentially) to the formation of pipes. There is a tendency for these phenomena to be developed mainly in seasonal climates, where the downward flux of dissolved and particulate material is counterbalanced by dry season desiccation and flocculation of clays with sesquioxides.

Subsurface flow can also be channelled around corestones lodged in colluvial or landslide debris; and along sheeting planes that are frequently quasi-parallel with valley sides and steep mountain slopes. Gullies can be initiated by landslides on such slopes, and if the slide strips the weathering front, then seepage water is able to flow from the base of the saprolite and possibly wear back along lines of weakness or preferential flow.

A connection exists between the formation of rills and the development of pipe flow on the one hand, and the formation of badland topography on the other. It is common for badlands to occur where rilling leads to a dense network of small surface channels, feeding into more permanent gullies. These may develop, once dissection into the soil profile is deep enough, as surface water in rills is captured by subsurface flow. However, although rilling has been linked to pipe flow, the connection cannot be essential, since rills develop only where the vegetal cover is sparse or becomes breached.

Both pipes and gullies can occur under well vegetated slopes lacking evidence for rill development. According to Morgan (1986), however, the breaching of vegetation cover is a common cause of gully development, but without signs of rill formation. The breaks in the vegetation cover lead to sealing of the soil surface as aggregates are broken down by intense rainfall. Shallow depressions form that concentrate the enhanced runoff and which then develop a small headwall upslope. This in turn creates a small waterfall and plunge pool; the gully headcut is then undermined by scour and retreats by slab failure, while the runoff scours a channel downslope. As the headwall retreats it becomes higher, and the increased hydraulic gradient will encourage pipe flow in suitable materials as described above.

Lavaka gullies in Madagascar

This process sequence was also argued by Wells *et al.* (1991) in explanation of numerous lobate gullies called *lavaka* in Madagascar. These are cut into toe and mid-slopes of weathered Precambrian basement rock hills near Antananarivo. The area receives rainfall in the range 1000–2000 mm year^{-1} and lies between 1000 and 2000 m asl. The soils here are ferrallitic and underlain by thick saprolite. The lavaka are thought to develop from bare patches, as suggested by Morgan (1986), and erode deeply into the saprolite, often striking fresh rock which prevents further incision but encourages widening of the gullied area. As the gully walls retreat (see also Blong, 1985), the colluvium buries the floor and, gradually, the gully becomes overgrown. Wells *et al.* (1991) argued that the end products are *hillslope hollows* that have been shallowed, flattened and smoothed by continuous infilling (see pp. 183, 184).

Gullying in the rampa complexes of Eastern Brazil

The lavaka gullies share many features with the gullies associated with the *rampa complexes* of eastern Brazil. These were analysed by de Oliveira (1990) in relation to hillslope form and with instructive results. His test area was around Bananal, where studies of colluvial/alluvial stratigraphy have been undertaken (Chapter 8 and Figure 8.17). Hillslopes were classified as convex (cx) or concave (cc) in both elevation and plan. The upper slopes are thus convex in both dimensions (cx/cx) and hollows, in contrast, are cc/cc. Although cx/cx slopes are prominent in the landscape (61.5%), only 10.4% of these slopes are dissected by gullies. It was found that 70% of the

gullies cut cc/cc slopes and 67% occur where the underlying sediments are heterogeneous. Most such gullies widen downstream and join the stream network. Narrow gullies forming on upper slopes are usually developed in homogeneous saprolite and narrow downstream, remaining unconnected to the stream network. A few have developed connections between upper and lower slope systems (Figure 5.8). De Oliveira (1990) argued that the lower slope (cc/cc) gullies develop by seepage erosion (which could include pipe flow), while the upper slope gullies arise from concentrated overland flow. Incision rates and depths appear to be greater where there is a superposition of several colluvial sheets (see also Allison, 1991).

Gullies on the Jos Plateau of Nigeria

Most active gullies have responded to disturbance of the vegetation cover and have quite short 'lifetime' histories. Thus when Grove (1952) wrote about soil conservation on the Jos Plateau of Nigeria (10 °N, rainfall 1250 mm year^{-1}) in 1952, he felt able to state (p. 39), 'large gullies are rare on the Plateau where soil-covered slopes are usually gentle', and did not find examples of the spectacular scars now so evident. He did observe some grass-filled gullies which were almost inactive and he warned of the potential for rapid gully growth as population pressure on resources increased. By 1975 Jones was able to measure >8000 km of mostly active gullies, that had excavated an estimated 100 Mt (enough soil and sediment to fill 1000 supertankers each of 100 000 t capacity) (Jones, 1975). He analysed aerial photographs from 1956 and 1972 and found that, in that period, larger gullies had cut back 300 m (20 m year^{-1}). Most of the gullies have been cut into the Bokkos and Rayfield colluvial deposits of late Quaternary age, described by Hill and Rackham (1978) (Figures 5.9, 5.10, 5.11 and 8.26). However, the current wave of land degradation is, in many instances, merely the latest of several cut-and-fill cycles on the same site. Smith (1982) has described two cut-and-fill episodes from gullies west of Zaria, northern Nigeria (11 °N, 1100 mm year^{-1}) and attributed the sediments to the effects of late Quaternary

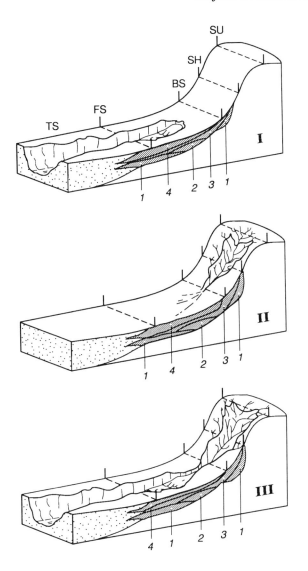

FIGURE 5.8 Types of gully erosion into concave slopes in the Bananal area, Brazil. I, seepage erosion type; II, concentrated overland flow type; III, combination of Types I and II. Note the incision into the multilayer colluvial–alluvial fill (as described in Chapter 8, Figure 8.17) (Bocco, 1991, after de Oliveira, 1990; reproduced by permission of Edward Arnold)

environmental changes. This example recalls the gully development at St Michael's Mission in Zimbabwe discussed above (Figures 5.2 and 5.3). Such gully sites are often the most sensitive parts of the landscape, and can usually be predicted by prior interpretation of the geomorphic landscape.

FLUVIAL EROSION

The efficacy of fluvial erosion

The data for suspended sediment load within river basins do not reveal a great deal about the efficacy of fluvial erosion in the tropics, a subject much debated by geomorphologists (Tricart, 1965; Baker, 1978; Douglas & Spencer, 1985a,b; Gupta, 1988). Studies of bedload transport in tropical rivers have been few and there is a lack of reliable data on almost all aspects of this topic.

This lack of data, and a predominance of early observations in tropical plainlands, have contributed to the perpetuation of generalisations about the load characteristics of tropical rivers that need to be carefully scrutinised. The opinion that the speed and thoroughness of tropical weathering leads to stream sediments lacking abrasive tools for erosion was advanced more than 30 years ago (Büdel, 1957; Birot, 1958, 1968; Louis, 1964; Tricart, 1965), and was held responsible for the apparent lack of valley incision in many areas of the seasonally humid tropics. Thus, although core boulders produced by weathering of crystalline rocks abound in the weathering profiles, many are too large to enter directly into the bedload, while at smaller diameters (0.15–0.25 m) they appear to disintegrate into sand and clay within the saprolite or reach the stream channel in a weakened condition. The size grading of the saprolite mantle has been shown to contain mainly clay and silt-sized particles and sands, and it is these that are thought to dominate the sediment carried by tropical rivers. Such rivers would not, according to the theory, be capable of effective channel cutting and would play a passive role in the denudation of the landsurface. Some support for these ideas, in their appropriate contexts, has come from studies in the Amazon basin (Gibbs, 1967; Tricart, 1965; Baker, 1978).

Büdel (1957, 1982) also thought that this concept was supported by circumstantial field evidence from many seasonal tropical areas, where valley-side slopes are no steeper than the stream thalwegs (1.5–3%). He also observed that river sediments had similar grading characteristics to the saprolite of the slopes. Both these observations led Büdel to the conclusion that such valleys developed not by true

FIGURE 5.9   Advance of 'frontal' gully heads in colluvium on the Jos Plateau, Nigeria. Piping is active

linear erosion but by an extension of wash processes across the valley floors. At high water some lateral erosion was accomplished and at low water superficial gullying of the sediments took place. These observations were combined with the suggestion that waterfalls and rapids along stream thalwegs are due to the outcrop of unaltered rock at the surface, and that these features persist in their locations over long periods of time. Both Büdel (1957) and Bakker (1960) referred to evidence of varietal differences in water snails found above and below cataracts on rivers in Surinam.

A similar argument was advanced to explain what have been described as *sula complexes* in Surinam by Zonneveld (1972). These are rocky zones along stream courses, where channels divide into anastomosing patterns. These were explained in terms of the irregular weathering front, the river branching where this intersects the (non-erosive) channel; the lack of abrasive materials in the stream load; possible inheritance of braided conditions from Quaternary dry periods, and the induration of channel sediments with ferromanganese compounds.

These observations apply to rivers lacking mountain catchments and which flow over high grade metamorphic and igneous rock formations on the Precambrian cratons. They do not apply to mountainous areas, fault scarps and other areas of dissection. However, it would be unwise to dismiss these earlier ideas completely, and quantitative observations of stream behaviour lend support to some aspects of the general thesis, as will be shown.

FIGURE 5.10   Linear gully, cutting colluvial and alluvial deposits beneath shallow slopes on the Jos Plateau, Nigeria. Note the pipe exits exploited by stream scour on the right bank

But what must be refuted is any suggestion that river valleys in the tropics are invariably shallow and lack bedload. This remains a difficult subject for discussion, but some more recent studies are helpful.

Several rivers in Papua New Guinea have now been the subject of semi-quantitative studies: the Angabunga (Speight, 1965), the Fly and the Purari (Pickup *et al.*, 1980; Pickup, 1984; Pickup & Warner, 1984), all of which drain mountainous catchments to the Gulf of Papua and exhibit ridge and ravine topography (*gorge section*) in their headwaters which drain mountains rising above 3000 m, a *valley section* of quite intense dissection and *alluvial plains*. In the gorge sections, the rivers occupy the entire valley floor, enclosed by high walls, and carry sediment from boulders down to fine silts and clays. The valley sections in the Purari and Fly rivers dissect poorly consolidated Miocene and Pliocene basin sediments, and the river channels start to braid. The valley floor is comprised of present and former channels. Downstream the braids decrease until the river adopts a single channel of low sinuosity and alternate bars which are often submerged. The Fly River then crosses 400 km of alluvial plains, before entering its delta.

FIGURE 5.11   Healing of a gully at a site near to those seen in Figures 5.9 and 5.10, showing the collapse of gully walls and the growth of vegetation. Reasons for the cessation of activity are not well understood

Pickup (1984) compared the flow variability on the Fly and Purari basins with a number of rivers from Australia, and was able to demonstrate very clearly the much lower variability of the Papua New Guinea rivers. This was attributed to the high and regular equatorial rainfall over most of the year. Catchment conditions of temperature and antecedent moisture remain similar, so that catchment response is almost the same throughout the year. Tropical cyclones are absent and the large mountainous catchments respond to the aggregate effects of many individual convectional rainfalls. As a consequence of these conditions the 100 year flood is only twice the size of the 2 year flood. In northern and eastern Australia this factor rises to between six and eight.

In Sarawak, Malaysia, Rose (1984a) found confirmatory evidence for bedload, transport, river braiding and fan development in a humid tropical catchment (Melinau River, 4 °N, > 5000 mm year^{-1}), but again this stream drains a very steep catchment, rising in Gunung Mulu at 2377 m asl, and falling 2000 m in about 10 km. Coarse bedload is transported from the mountain course (192 m km^{-1}), through an intermediate section with a gradient of 9.36 m km^{-1}, and a further 10 km downstream along a lowland channel with a gradient of only 1.05 m km^{-1}. Some 95% of the bedload in the > 132-mm size range is derived from the Gunung Mulu headwaters and the river transforms from fan building to braiding within a sinuous channel, to a meandering stream over this distance. This and other examples therefore serve to emphasise the importance of mountain catchments in controlling sediment characteristics in the tropics, as elsewhere.

Downstream changes in fluvial activity

The sediment load of mountain headwaters is very high. Pickup *et al.* (1980) calculated the rate of erosion in the steep forested catchment of the Ok Ningi tributary to the Ok Tedi, at 3–4 mm year^{-1} on the basis of suspended sediment concentrations in gulp samples. In Sarawak, a similar figure (3000 mm ka^{-1}) was quoted by Rose (1984a). These rapid rates of erosion were supported by Löffler (1977), who calculated a rate of sediment accumulation behind a dam in the Bewani Mountains of 2831 t km^2 year^{-1}, and also noted an accumulation of sediment in braided channel sections at the rate of 2–3 m year^{-1}.

In the upper gorge sections of rivers studied by Pickup (1984), most material reaches the river via frequent landslides that carry coarse boulders and gravel from fresh outcrops of igneous and limestone rocks or finer sediments from shales, sandstone and weathered crystallines. In the valley and plains courses, most sediment is derived from the river banks which are subject to small landslips after high water recedes. A steep decline in the grade of sediments found in bank sections downstream from landslides indicates rapid sorting of the sediment entering the river. Sediment rating curves show an immediate response in the gorge sections to increases in discharge, showing that most sediment comes from sources close to the river channel, but in other mountainous tracts there is evidence for the later arrival of sediment from highland sources, and in the valley and plains sections there is progressively less relationship between discharge and suspended sediment load.

Downstream changes in sediment grading result from:

(1) the nature of sediment input to the channel;
(2) sorting by the river (competence), and
(3) particle breakdown and comminution due to abrasion and corrosion.

But in these Papua New Guinea channels, downstream sorting does not increase exponentially, despite the absence of steep slopes to supply bedrock pieces in the lower reaches. The lack of relationship is thought by Pickup (1984) to reflect the hardness of many rock fragments and the short period spent in the river by the clasts. Several lengthy sections of channel were found to have little internal variation in sediment characteristics; each reflecting local conditions of sediment supply. In an earlier study of the Angabunga River in Papua New Guinea, Speight (1965) found that where the river debouched from its mountainous catchment the bedload has a mean diameter of 110 mm, but a few kilometres downstream the load was comprised almost entirely of sand, though this was achieved by fan building as much as by loss of clasts through abrasion and corrosion and the example probably illustrates the same features as described by Pickup (1984) and by Rose (1984a,b).

Two further factors affecting these and many other rivers were identified which must be considered of general importance. First, bank erosion in the valley and plains sections of the rivers sometimes intersects coarse gravel terraces dating from periods of Pleistocene sedimentation. Second, the lower alluvial plains have been affected by sea level changes and particularly by the Holocene sea level rise. These factors operate very widely throughout the tropics, where palaeohydrological conditions were in most areas very different from those of today.

It is not surprising that the mountainous catchments should mainly reflect present-day conditions in which high rainfall and rapid rates of vegetation colonisation develop thin, unstable regoliths on steep slopes which then become subject to shallow translational slides and rockfalls. These materials are then despatched rapidly downstream by consistently high discharges, flowing down steep channels and generating a stream capacity in excess of sediment supply, so that most channels are formed in bedrock and are swept clean of all but a thin veneer of sediment.

This is in great contrast to rivers draining catchments of low relief, but demonstrates the importance of river channel processes in humid tropical mountains. Beyond these mountainous zones, conditions of erosion and sedimentation change, in favour of more widespread saprolite mantles, and accumulation of sediments in floodplains, buried channels and terraces.

It is difficult to deduce too much from comparative data of sediment loads, because they are so sensitive to vegetation changes from human activity. Temple and Sundborg (1972) studied the sediment load of the Rufiji River in Tanzania, a catchment of $156\,000\,km^2$, and calculated an annual sediment discharge of 17 million tons, which equates to $109\,t\,km^{-2}\,year^{-1}$ and returns a denudation rate of $0.06\,mm\,year^{-1}$. Temple and Sundborg noted that a high proportion of the sediment load comes from a small part of the catchment, and this probably accounts for an erosion rate three times that given for Africa as a whole (Table 5.4). Very high sediment loads were associated with flash floods, during the rising limb of the hydrograph, when sediment concentrations rose to a maximum of $5900\,mg\,litre^{-1}$. Much of this was derived from areas of severe soil erosion.

The petrology of river sands

While changes to sediment grading appear sensitive to the conditions in each section of a river, the petrology of river sands may show clear downstream changes. Franzinelli and Potter (1983) studied the river sands of the Amazon system and found that they were moderately to well sorted throughout the 3000 km channel from the Peruvian border to the Atlantic, and remained angular. However, their chemistry and petrology changed markedly from lithic arenites with a ratio of quartz:feldspar:rock fragments (QFR) of 47:8:45 and SiO_2/Al_2O_3 of 7, in the Andean source areas to a QFR of 85:3:11 and SiO_2/Al_2O_3 of 32, near the river mouth. Where the river traverses crystalline rocks, coarser angular quartz arenites are found, indicating the production of 'first cycle' quartz sands by the tropical weathering. Significantly, the rivers draining the Andes and also those draining Tertiary molasse sediments in the lower, western part of the basin are muddy and carry high suspended load, especially in flood. But the rivers

draining the Precambrian rocks from the shield areas north and south of the trunk river are clear except for the 'tea water' staining by organic matter (Plates 5-I and 5-II) which gives the name Rio Negro.

Comparisons with other large rivers in sub-arctic or temperate environments, including the Mississippi over a 1600 km course, indicate (Franzinelli & Potter, 1983; Potter, 1978) that while the sands of such non-tropical, big rivers maintain a similar mineral composition from source to mouth, this is not so in the Amazon, where 'maturation' of the sediment occurs. Equally, rivers draining ancient cratonic landmasses in the tropics produce quartz arenite sands as a consequence of the advanced weathering of the rocks.

Savage *et al.* (1988) studied the mineralogic maturity of sands on a different scale, along the coastal plain southwest of Rio de Janeiro, where Pliocene and modern fans grade into fluvial sands within short catchments draining the coastal granite mountains of the Serra do Mar. These sands are then redistributed along the high energy coast as beach sands. The QFR ratios are given in Table 5.7.

Data were not given for Pliocene fans, but these were declared to be quartz arenites. The loss of rock and feldspar fragments is a clear indication of abrasive destruction of grains weakened by tropical weathering, while the evidence from the Pliocene fans indicates the *in situ* transformation of lithic arenites towards quartz arenites in the weathering environment.

Rivers draining sedimentary terrains recycle sediment already weathered in the source deposits, illustrated by recent studies of the Imo River in southeastern Nigeria, by Amajor (1989), who found

TABLE 5.7 Petrology of sands from coastal plain sediments near Rio de Janeiro, Brazil

Sediment location	QFR ratio
Modern alluvial fans	53:26:21
Fluvial sands	63:27:10
Ancestral beach	69:17:14
Beach ridge (restinga)	100:0:0
Modern beach	97:3:0

After Savage *et al.* (1988). Reproduced with kind permission from Pergamon Press Ltd, Headington Hill Hall, Oxford OX3 0BW, UK

FIGURE 5.12   Granite core boulders filling a valley floor in Matopos Hills, Zimbabwe. The river runs deep below the surface accumulation

that the feldspar content was nil at all locations, reflecting the quartz arenite status of the Cretaceous bedrock. However, rock fragments comprised up to 30% of the bedload samples in the upper reaches, declining downstream to zero over a distance of ~ 80 km, until replenished by influxes from tributary valleys. He found coarse sand and gravel more or less randomly distributed throughout the length of the river.

There are too few detailed studies available to make valid generalisations about stream loads in tropical climates, but it is clear that rock fragments abound in steep headwater reaches and persist in the rivers for variable distances, dependent on the competence of the river rather than the speed of weathering of the clasts. However, once deposited and abandoned by the river, rock fragments weather rapidly and any feldspars are transformed to secondary minerals. Over long distances the QFR ratio in sands will change systematically downstream in humid tropical catchments, and rivers flowing from deeply weathered basement terrains are likely to carry only quartz arenites.

In erosional terms, the performance of tropical rivers remains difficult to evaluate, and much will depend on the nature of the catchment geology and relief. In steep terrain, rock weathering may be rapid, but the result is to detach rock masses of varying size from their roots. These may reach the river channel, usually by mass movement, before further decomposition destroys their fabric, and they may be utilised as abrasive tools in the channel. On the other hand, rivers draining from planate or moderately undulating terrain may be supplied overwhelmingly with quartz sand and fine gravel, together with a suspended load of alteration products, mainly clays.

Dissection into deeply altered mantles is almost certainly rapid, if stream gradient and discharge are sufficient to remove the non-cohesive saprolite. Further dissection into bedrock could be retarded if the channel is formed in partially altered bedrock, but in many cases resistant quartz veins, bands of quartzite, or conglomeratic sediments will provide local inputs of bedload which becomes available for abrasion.

Another phenomenon of stream activity is characteristic of granite areas producing abundant corestones, and this is the formation of valleys along structurally aligned zones of presumed (former) deep weathering, such that corestones not only remain in place along the stream channel itself, but may collapse inwards from surrounding slopes to choke the channel which is effectively buried and the river continues to flow many metres underground (Figure 5.12).

FLUVIAL DEPOSITION AND FLOODPLAIN FORMATION

A full appreciation of the great variety of floodplain architectures has been slow to emerge from studies of fluvial geomorphology and sedimentology, and it remains uncertain whether tropical floodplains differ in essential elements from those in temperate and cold environments. Several problems afflict this discussion and one is that the study of some very large systems such as the Amazon and the Congo/Zaire rivers is extremely difficult, and each is unique. Another is that very few tropical floodplains have been closely examined, so that all we have are some possibly unrepresentative examples. Notwithstanding these difficulties, some information is available, and a few general pointers can be provided for rivers outside of mountainous zones:

(1) most rivers have sand-bed channels;
(2) lowland rainforest rivers have very low suspended loads;
(3) rivers experiencing regular high floods may switch between modes of channel formation and sedimentation between high and low flows;
(4) the frequency of high magnitude flood events and disparity between the competence of large floods and post-flood flows tends to perpetuate flood effects (Gupta, 1988);
(5) predominantly sandy floodplain facies are non-cohesive and subject to re-organisation during successive floods, as well as encouraging lateral channel migration;
(6) fine calibre loads may build cohesive channel banks by vertical accretion, limiting the free migration of meanders;
(7) coarse gravel bars are usually due to palaeo-hydrological conditions during the late

Pleistocene or early Holocene, and may have become cemented, resisting channel migration;
(8) many catchments now reflect the impact of extensive deforestation and serious soil erosion, leading to sporadic inputs of high sediment volumes from tributaries and other sediment stores.

Floodplain depositional processes

A useful categorisation of floodplains has been advanced by Nanson and Croke (1992), based on force–resistance relationships, whereby the amount and texture of sediment are viewed as functions of stream power. They point out that many climates oscillate wildly about present-day means, reflecting multidecadial variations rather than secular climatic change. This can lead to the almost complete reorganisation of sand-bed rivers between drought periods and flood periods.

One of the main issues regarding floodplain development is the fundamental distinction between *lateral accretion*, which results from channel migration and the successive formation of point bars (Wolman & Leopold, 1957) and *vertical accretion*, which results from repeated overbank flooding with the deposition of levee and slackwater deposits. Nanson and Young (1981) pointed out that the view that floodplains usually develop by channel migration was inadequate without a proper recognition of the role of vertical accretion, and cited this process as dominant in a number of streams in southeastern Australia. In 1986, Nanson advanced a model of episodic floodplain accretion and catastrophic stripping, brought about by the confinement of high discharges within the channel, as vertical accretion built higher levees after each major flood (Nanson, 1986). Eventually, a very large and rare flood event would overtop the levee and cause avulsion, accompanied by widespread removal of floodplain sediment. This model may well apply to some small West African rivers discussed below. In 1992 Nanson and Croke distinguished three energy-based systems:

(1) high-energy ($>300\,\mathrm{W\,m^{-2}}$) non-cohesive floodplains, where very coarse alluvium and rock channels inhibit migration;

(2) medium-energy (10–300 W m^{-2}) non-cohesive floodplains, with many variants;

(3) low-energy cohesive floodplains (<10 W m^{-2}), built of silt and clay, and associated with single thread or anastomosing channels.

Within these categories it is possible to recognise braid accretion, point-bar accretion, vertical accretion, scroll-bar formation, and anabranching, together with cut-and-fill episodes. In some cases the deposition is confined by bedrock or older terrace fragments.

The complexity of sedimentation processes has been demonstrated on an annual/seasonal basis for the Auranga River which drains the Lohardaga Plateau in India (Gupta & Dutt, 1989). The catchment experiences a monsoonal climate (1200–1500 + mm year^{-1}, with 80% of rain in 4 months from June to September) and occasionally extreme storm intensities of 200 mm h^{-1}. The catchment is seriously eroded though widely replanted with young trees. In the dry season, the river is braided from bank to bank in predominantly sandy sediments; the braid bars are flat (<15 cm high) and intervening channels shallow (<30 cm deep). The channel bed represents a mobile sand sheet <1 m thick in most places. During high flows the braids are destroyed and the meandering main channel builds point bars, and there is also a higher flood level with a limited number of coarse sand and gravel flood bars. These very high floods are responsible for the width of the channel and for reorganising the sediments.

In a study of the Yom River in north-central Thailand, Bishop (1987) recognised the arguments about lateral and vertical accretion, pointing out that, with vertical accretion, the recurrence interval for overbank flooding must increase, unless of course the levees are breached. This can occur by *avulsion* which is marked by sand splays into backswamps. The Yom River was described as having an upper and a lower floodplain, both inset within older terraces. The upper floodplain was overtopped only rarely and exhibited 8 m of alluvium. Channel sands are marked by cross bedding, while levee deposits show repeated upward fining sequences <30 cm thick. Over these

sediments there is around 2 m of uniform loams and clays, structureless and with a 'bland uniform appearance'. Such overbank deposits, which show signs of multiple sub-units, comprise 50–67% of the upper floodplain sediment. In the view of Bishop, the Yom River had, initially, freely migrated, depositing channel sands. But at some point the channel had become fixed and there was a switch to vertical accretion of the upper floodplain which is not a terrace but is visited rarely by very high floods.

Many lowland tropical rivers have relatively stable channels confined between high levees and flanked by thick overbank deposits which are further stabilised by the tropical vegetation (Savat, 1975). This is encouraged in the humid tropics by the advanced weathering of the sediment reaching the river, and where this is derived from fine-grained rocks or from clay-rich saprolites, comparatively little sand is entrained. On the other hand, over granitoid rocks or sandstones, much less fine material is available and more mobile scroll-bar formation takes place.

The middle Birim Valley in Ghana (6 °N, rainfall 1750 mm year^{-1}) drains a forested area of weathered phyllites, and flows in a relatively straight, single thread, box-like channel 30 m wide and 7 m deep. Its overbank vertical accretion deposits can be 5–6 m thick; the channel appears remarkably stable, and late Pleistocene low terrace deposits also retain overbank fine sediments (Hall *et al.*, 1985, and Figure 5.13). There is little evidence here of the dominance of extreme flood events over the activity of the river, probably because the river drains a forested catchment without high mountains and, although experiencing heavy rains associated with line squalls, the area is not subject to the frequency of extreme events experienced in monsoonal and cyclonic areas of the tropics.

According to Gupta (1988), the physiographic indicators of the impact of large flood discharges include: (1) a sharply differentiated terrace, submerged by high floods; (2) large bars, flood channels and chutes in the channel, on the floodplain and terrace; (3) sedimentary structures associated with high velocity flow; (4) young

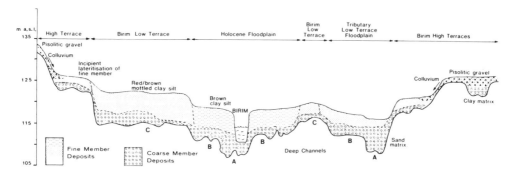

FIGURE 5.13   Cross-section through the floodplain of the middle Birim River, Ghana, illustrating thick clay/silt vertical accretion deposits on both the present floodplain and a low terrace; also the box-shaped narrow channel confined between cohesive and vegetated banks (after Hall *et al.*, 1985; reproduced by permission of the Geological Society Publishing House)

vegetation indicative of periodic flood destruction; (5) a high width–depth ratio in the channel cross-section; (6) abundant coarse alluvium stored in lower valley flat; (7) finer material indicating low-magnitude flows confined to lower parts of the channel. Many of Gupta's (1988) examples come from the Himalaya, including the Darjeeling Hills studied by Starkel (1976), where large core boulders, 3–5 m in diameter are carried in small channels (>10 m in diameter in larger channels). These conditions are an important example of efficient stream erosion in the tropics, but do not offer insight into the behaviour of many lowland catchments.

The importance of palaeofloods and palaeoenvironments

The question of the palaeoflood hydrology in prehistoric and historic times is a topic of increasing interest, and develops from an appreciation of the role of extreme events in shaping landforms (Wolman & Miller, 1960; Starkel, 1976, 1983, Gupta, 1988). Eybergen and Imeson (1989, p. 312) have pointed out that, 'important effects of climatic change with regard to landform will be mainly due to the change in frequency and magnitude of so-called extreme events.' The role of such events in the restructuring of stream channels and in the development and destruction of floodplains is

considered decisive, and a sedimentary history that stretches across 10^3–10^4 years will be a record of the clustering of these events (Starkel, 1983). Down-cutting, avulsion episodes, deposition of coarse sediments and periods of rapid sedimentation all offer a record of the palaeoactivity and flood history of a river.

Evidence for exceptional floods is preserved in high-level silt deposits and flood debris which can be dated. Not surprisingly much of this evidence comes from the arid and semi-arid zones, where the imprint of rare flooding is both clear and a fundamental aspect of the fluvial system (Baker, 1983a,b, 1988; Starkel, 1983; Pickup, 1991; Smith, 1992). Some findings concerning palaeoflood analysis are indicated in Chapter 7 (Table 7.11).

Floodplain evolution

It is also important to emphasise the course of *floodplain evolution* over time. In the examples given above, there is ample evidence for the existence of older channel deposits that signify a different regime (Figure 5.8). The chronology of these sediments is discussed subsequently (Chapter 8). They contain coarse gravel bars, absent from the present river channel except where they are (randomly) intersected, and these tend to be inset into bedrock channels in the middle Birim (Hall *et al.* 1985; Thorp & Thomas, 1992; Thomas & Thorp, 1993). Large, coarse gravel bars are also found along the banks of the Luangwa River, in Zambia, where they form benches of cemented cobbles.

Many rivers such as the middle Birim in Ghana, appear to have experienced major floods during the Pleistocene–Holocene transition, and to have been deposited mainly by lateral accretion of point-bar and braid-bar sediments at that time. These early Holocene floodplains fall into either the high or medium energy non-cohesive categories. But during the Holocene, energy conditions declined with the establishment of forests. Channels became stabilised within banks formed from cohesive silts and clays and anchored by vegetation. Vertical accretion became dominant, and today these are low energy, cohesive floodplains. However, such conditions may be susceptible to change as deforestation leads to higher flood peaks and an increased supply of largely sandy sediment. Bank erosion, channel widening and braid-bar formation may all follow, as appears to be the case in the Auranga River (Gupta & Dutt, 1989).

DISSOLVED LOAD AND CHEMICAL DENUDATION BY RIVERS

Perhaps one of the most interesting functions of the river system is its ability to transport a large dissolved load, and this amounts to a form of chemical denudation that has only quite recently received much comment.

Solute concentrations in tropical rivers

Solute concentrations in tropical rivers are not necessarily high, despite the inferred rapidity of tropical weathering. The reasons for this situation, however, are not difficult to find and will be discussed below. Some of the lowest recorded figures (Walling & Webb, 1986) have come from lowland rivers in Amazonia, where the Negro, Xingu and Tapajos average solute concentrations in the range 5–8 mg litre^{-1}. Such figures compare, on a world-wide basis, with a mean concentration of 255.6 mg litre^{-1}. According to Walling and Webb (1986), this figure reduces to 120 mg litre^{-1} when weighted according to total load (total load from 496 rivers divided by total runoff), and

equates to 39.5 t km^{-2} year^{-1}. Some of the highest solute concentrations are found in rivers draining saline (evaporite) rock strata, where solute loads can range from 2000 to 60 000 mg litre^{-1}.

Studies from the Amazon Basin

In their seminal studies of the geochemistry of the Amazon, Stallard and Edmond (1981, 1983, 1987) explored the weathering environments, products and river loads within the basin, and their data can also be compared with those obtained by Furch (1984). These authors all confirm the low electrolyte levels in the lowland rivers which drain the cratonic part of the basin or which flow over the basin sediments. There is an important distinction to be made between catchments where denudation is *weathering limited* (mainly steep catchments in the Andes and their foothills) and those which are *transport limited* (areas of deep soils on cratons and basin sediments) following Carson and Kirkby (1972). According to Stallard and Edmond (1983) the Amazonian rivers fall into four principal groupings based on total cation charge (TZ+) and geology:

(1) Rivers with TZ < 200 μeq litre^{-1} drain the intensely weathered Tertiary sediments of the Negro basin, and other highly weathered areas, and regimes are transport-limited. Their solute load is characterised by relatively high levels of Fe, Al, H^{+}, and they are enriched in Si. According to Furch (1984) their solute loads represent only 5% of the 'world average freshwater' load, and trace elements (Fe and Al) amount to 24% of the total in this category.

(2) Rivers with TZ 200–450 μeq litre^{-1} drain siliceous terrains, and higher values come from catchments that are weathering-limited. These show some enrichment in Na and Ca relative to K and Mg, when compared with bedrock.

(3) Rivers with TZ 450–3000 + μeq litre^{-1} drain marine sediments in the Peruvian and Bolivian Andes, where there are large areas of carbonate rock and minor evaporites, and water is enriched in Ca and Mg.

(4) Rivers with TZ > 3000 μeq litre^{-1} drain massive evaporite deposits also in the Andes.

Furch (1984) has adapted a fourfold division of Amazonia into geochemical regions proposed by Fittkau (1971, in Furch, 1984). This recognises the western periphery (Andean) that includes the categories (3) and (4) above together with parts of (2), the northern and southern peripheral areas that drain the cratonic areas and platform sediments of the basin margins, and the central Amazon which corresponds to category (1). His sampling all came from a small area in the vicinity of Manaus, but included the Solimões-Amazon, that branch which drains from the western Andean periphery. When measured against world averages, *all* the Amazon waters proved relatively poor (at Manaus), and even those of western peripheral origin did not reach a 50% value. However, when compared with other catchments, the Rio Negro and local forest streams not only have very low specific conductance (~ 5.2) when compared with the Solimões (31), but also are relatively enriched in trace elements including Fe and Al (Rio Negro: 29% of world average; Solimões: 21%). The proportions of the alkalis Na and K, relative to Ca and Mg, are higher in the electrolyte-poor central Amazonian streams. The Rio Negro also has twice the content of Ca when compared with the forest streams, in which this ion is below the average for rainwater. Thus the 'electrolyte poor' waters form a separate class from the normally carbonated waters containing significant Ca^{2+} and HCO_3^-. In these waters, 80% of the solute load is comprised of HCO_3^-, SO_4^{2-}, Ca^{2+}, and SiO_2. But in the blackwaters of the Rio Negro and waters of the forest streams, the acidity (pH 4.5–5.1) leads to a cation–anion imbalance with H^+ ions making up a significant proportion of the specific conductance, along with Fe and Al, the ionic forms of which were not studied by Furch (1984).

According to Walling and Webb (1986) as much as 40% of the solutes recorded in river water may come from external sources such as rainfall and windblown dust, plus atmospheric CO_2 (P_{CO_2}). But in the humid tropics vegetation also acts as a long-term sink for nutrients and this may mean that stream loads seriously underestimate the level of solutes present in surface runoff.

Bruijnzeel (1983) studied the chemical budget of a small drainage basin in central Java and showed that calculations based on solute loads can underestimate the rate of chemical weathering for major ions, if the uptake of nutrients by the vegetation is not accounted for. The values he obtained were: Ca^{2+}, 75%; Mg^{2+}, 33%; Na^+, 17%; K^+, 70%; SiO_2, 12% (i.e. percentage by which the rate of weathering is underestimated).

Suspended and solute loads

When the suspended and solute loads of rivers are compared on a world-wide basis, an average value of 3.5:1 has been obtained (Walling & Webb, 1986). But this conceals very wide variations from <1 in Europe to >10 in Oceania. When individual large catchments are compared, similar differences are apparent, and these are shown in Table 5.8.

Walling and Webb (1986) found a wide scatter of points when data for dissolved and suspended sediment load were plotted for 302 world rivers, and they deduced that three major factors influenced the data. Higher *basin relief* tends to increase both the particulate sediment load and the dissolved load, while increasing *annual runoff* has the effect of diluting the suspended sediment in relation to the solute load. Increasing *basin area* also causes a decline in particulate load as the sediment delivery ratio declines. This is a reflection of the size of sediment stores in large floodplains.

Inspection of the data in Table 5.8 shows clearly the effect of large mountain catchments on the Ganga system in returning a ratio (S:D) of 11.0:1. By contrast, the Congo or Zaire catchment has only a very small mountain component in the east; a catchment twice the area of the Ganga, and a vast forested central basin containing an enormous sediment store. As a consequence both sediment and solute loads are low and very similar in magnitude (S:D = 0.9:1). In the Amazon, which is by far the largest catchment in the world, loads are higher than in the Zaire (Congo) (Moukolo *et al.*, 1993), largely as a consequence of the major Andean tributaries and much higher runoff. The much higher loads (particulate and dissolved) in rivers with headwaters in orogenic zones (Ganga, Amazon) is strongly influenced by the presence of thick sedimentary strata including carbonate rocks and

TABLE 5.8 Ratio of suspended sediment to dissolved load for selected tropical rivers

River	Area $(km^2 \times 10^6)$	Load $(t \times 10^6\ year^{-1})$		Ratio S:D
		Suspended	Dissolved	
Amazon	6.15	900	290	3.1:1
Zaire	3.82	43	47	0.9:1
Ganga/Brahmaputra	1.48	1670	151	11.0:1
Niger	1.21	40	10	4.0:1
Zambezi	1.20	48	15.4	3.1:1
Orange	1.02	17	12	1.4:1
Orinoco	0.99	210	50	4.2:1

Adapted from Walling & Webb (1986), mainly after Meybeck (1976) and Milliman & Meade (1983).

evaporites. According to Walling and Webb (1986), dissolved loads in rivers draining igneous rocks average only 20% of equivalent rivers draining sedimentary rocks. However, in a study of the Krishna River which drains the Western Ghats to the Bay of Bengal in peninsular India (Ramesh & Subramanian, 1986), it was found that tributaries and trunk streams consistently returned dissolved loads in excess of suspended loads in ratios (S:D) from 0.9 down to 0.4 (less at the outfall). This was considered by Ramesh and Subramanian to typify conditions in peninsular India, in contrast to the mechanical dominance of rivers draining the Himalaya.

Problems of data interpretation

There are many problems with these figures. First amongst these is the unreliability of the data. Most large catchments have impoundments and these can affect sediment delivery rates. Figures for the same river system can vary widely according to author, and this probably reflects wide variation in the stream loads with time, and differences in sampling design. Douglas and Spencer (1985a) have pointed out that as flood peaks occur solute loads not only fluctuate (generally falling as the peak flow arrives), but the balance between different ions in transport changes. Thus Na^+ and Ca^{2+} may decline rapidly, while K^+ declines less and later. In one instance from northern Queensland (Douglas, 1973), SiO_2 fell sharply as discharge increased, but then rose again, suggesting some interaction between the suspended sediment and stream water, leading to SiO_2 dissolution.

It has already been established (Chapter 2) that the dissolution of rocks in the humid tropics usually involves the formation of organo-metallic complexes in the presence of both bacteria and organic acids. Such metals which include Fe and Al will not then be present in ionic form within the water. Meybeck (1979) considered dissolved organic carbon (DOC) for different climatic regions and found that DOC values were highest under *taiga* forests ($10\ mg\ litre^{-1}$), and lower in the humid tropics ($6\ mg\ litre^{-1}$), with lowest figures of all coming from temperate and semi-arid areas ($3\ mg\ litre^{-1}$). When the much higher runoff from tropical catchments, as compared with those of the northern taiga forests, is taken into account, it is remarkable how high this figure ($6\ mg\ litre^{-1}$) remains, and if metals are also being transported by this means, then the denudational effects may be significant.

The organo-metallic complexes involved are usually metastable and transported metals are readily precipitated. Sites for these precipitates occur in the floodplains, where the rise and fall of water-tables can lead to Fe_2O_3 precipitation to form ferruginous or ferricreted layers. In drier seasonal areas $CaCO_3$ may be deposited as calcrete nodules or more massive lenses. Al^{3+} and SiO_2 together with other cations including Fe^{3+} also recombine to form new clay minerals in the floodplain environment.

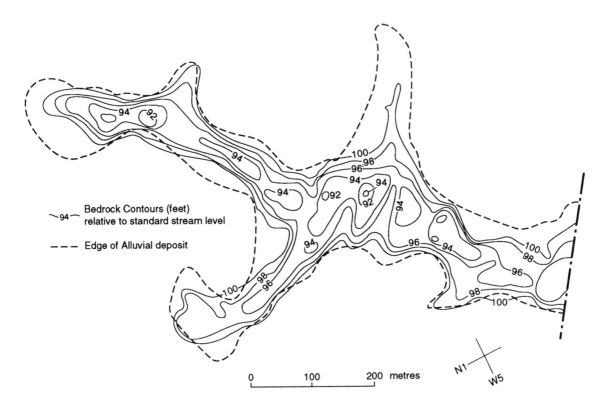

FIGURE 5.14   Bedrock configuration below finger tip tributaries of Neuyi Valley, in the Koidu basin area of Sierra Leone. These are channelless valleys ('bolis'), largely immune from mechanical scour; the bedrock surface has been etched by chemical weathering. (Reproduced by permission of Unwin Hyman from Thomas & Thorp, 1985, after unpublished data by R. W. Crofts)

In summary, the factors that can lead to comparatively low solute loads in humid tropical catchments include the following.

(1) Drainage from already leached terrains, typically underlain by deep kaolinised mantles or by quartz sands. Such old landsurfaces predominate on the Gondwana continents and strongly influence the data for highly integrated, large drainage basins that are associated with these terrains (e.g. Amazon, Orinoco, Zaire, Zambezi). The 'electrolyte poor' *blackwater* rivers which usually drain *white sand* (Chapter 7) regions are typical of this situation.

(2) Elsewhere the relief may be 'stabilised by a layer of indurated cation-poor material' according to Stallard and Edmond (1983).

(3) Long-term storage of both particulate and solute loads in floodplain sediments, especially in the vast basinal tracts of alluviated land that occur on the Gondwana cratons which have been tectonically warped into large basins and swells (Niger, Zaire, Amazon). Only where hillslopes directly adjoin the stream channel, usually in mountainous areas, can the ionic flux from the slope enter directly into the river.

(4) Long-term storage in vegetation which has significant capacity to absorb silica and other cations (Lovering, 1959) and which recycles a high proportion of the nutrients from fallen and decayed trees. Although this does not form a closed system, solutes are more or less permanently isolated from the river systems by this means. Stark and Jordan (1978) estimated

FIGURE 5.15   Unrolled and etched quartz fragments (sized >8 mm) from swamp ('boli') gravels in Koidu, Sierra Leone

that >99% of Ca and P is assimilated into the biomass, while amounts of K, Ca and Mg in the total biomass are typically three to five times those found in soil A and B horizons from which leaching into rivers takes place.

(5) High peak and annual discharges can be accounted for if data are available, but without these data crude figures for solute loads may be misleading.

(6) Drainage from cystalline rocks which are widely exposed on the upswells of the cratonic regions will be poorer in solutes than drainage from orogenic regions, where carbonate and evaporite rocks, together with fissile greywackes and schists, plus volcanic suites will all weather rapidly to supply ions into the river systems.

Chemical denudation rates

Catchment studies

The rate of chemical denudation as computed from total dissolved solids in river waters can simply be expressed as $t\,km^{-2}\,year^{-1}$ (Sarin *et al.*, 1989) or it can be converted to a rate of groundsurface lowering in $mm\,ka^{-1}$ (or $m^3\,km^{-2}\,year^{-1}$) which can, in turn, be converted to 'Bubnoff Units' ($1\,B = 1\,mm\,ka^{-1}$ or $1\,m\,Ma^{-1}$; Saunders & Young, 1983). If the average sediment has a mass of sg 2.65, then $1\,m^3 = 2.65\,t\,km^{-2}$. On this basis, rates of lowering resulting from chemical denudation, according to Sarin *et al.* (1989), range from a maximum of $143\,t\,km^{-2}\,year^{-1}$ (54 B) in

FIGURE 5.16   Semi-enclosed valley head etched into
Koidu etchplain, Sierra Leone

the Yamuna River, India, a tributary of the Ganga, down to $9\,t\,km^{-2}\,year^{-1}$ (3.4 B) for the Zaire River. From these data the Amazon River erodes $35\,t\,km^{-2}\,year^{-1}$ (13.2 B). Values given earlier by Saunders and Young (1983) suggest that chemical denudation of crystalline rocks falls into the range 0–15 B, without any clear climatic contrast between temperate and tropical savanna areas. Rates for limestone denudation are much higher, commonly in the range 30–100 B. On mixed, mainly sedimentary, lithologies values of 20–60 B predominate.

Unfortunately such methods of estimating chemical denudation have very limited value; first, because the groundsurface lowering implied does not take place evenly over the catchment area and partitioning the data is generally not possible, and, second, because ion loss does not equate with volume loss where hydration of the residual compounds takes place. Much weathering is considered to be *isovolumetric*, leading to saprolite mantled landscapes without surface lowering. Only where *congruent dissolution* (of Si, Al, Fe, for example) takes place can immediate groundsurface lowering be implied. This will occur on limestones which are virtually monomineralic, and dolomites. However, as will be shown in the discussion of groundsurface lowering (Chapter 11), the transformation of saprolite into soil can involve

considerable volume loss. This can reach 75% according to Pavich (1986), who also found, in the Appalachian Piedmont, that base-flow stream chemistry indicated a minimum rate of lowering/weathering of 4 B, but consideration of other factors in the soil and weathering system increased this rate by at least five times to around 20 B.

These values are not adequate but they do converge around characteristic ranges and they suggest that chemical denudation is very significant, and may exceed mechanical denudation across a wide range of mainly humid to sub-humid climates. That congruent dissolution of silicate rocks is slow and confined to special environments (such as swamps) is evidenced by the formation and preservation of very thick saprolites. Where the rate of denudation is low overall, the chemical component of that erosion may dominate, but when large catchments are compared such areas are usually subsumed within large heterogenous regions such as Amazonia. The Zaire River is interesting in this respect with its S:D ratio of 0.91.

Morphological evidence, the etching of valley floors

Many headwater valleys in relatively planate terrain appear to be notched or etched into the bedrock by chemical weathering processes, accompanied by the removal of solutes and fines in very low energy (often channelless) systems. These processes are exemplified by the *bolis* of Sierra Leone and by the *dambos* of central Africa, but analogues occur in most continents and climates (Thomas & Goudie, 1985). Etching patterns in the bedrock floor suggest non-fluvial processes of weathering beneath a sedimentary fill (Figure 5.14), while the basal gravels found in these bolis are mainly sharp, angular vein-quartz lag deposits (Figure 5.15), remaining after fines have been washed from the upper levels of a saprolite. That is to say the clasts are not moved by the normally low-energy swamp system, but accumulate until they are flushed through the system by a major flood event. The valley floors are also punctuated by large corestones and they exhibit spring sapping along structural lines penetrating the bounding slopes (Figures 5.16

FIGURE 5.17   Channelless valleyheads (bolis) etched into the granite corestone zone in the same area as that shown in Figure 5.16

and 5.17). There is therefore evidence for both mechanical channel processes and for chemical denudation in these areas. Studies of dambos in Zimbabwe have come to similar conclusions (McFarlane & Bowden, 1992) and aerial views of the drainage system suggest the gradual etching of the drainage network into the basement rocks rather than by a conventional extension of finger-tip tributaries (Figure 5.18). This type of development can be inferred for many granite catchments within which are often developed a series of linked basinal widenings of the valley floors. The importance of sapping processes in sandstone terrain under temperate conditions has been stressed by Laity and Malin (1985) from studies of the Colorado Plateau.

Chemical denudation on slopes

The examination of these relationships at the small catchment or slope scale has been neglected. Douglas and Spencer (1985a) noted that studies from Pasoh in Malaysia showed that solute transport by throughflow is particularly important, accounting for up to 27.4% of the total sediment and dissolved load transported by all types of flow. Douglas and Spencer pointed out that contact between the water and soil particles is important in raising solute levels in throughflow, and this is one process involved in the leaching and eluviation (*lessivage*) of soils, bringing about volume loss and collapse of structure. This process has its ultimate

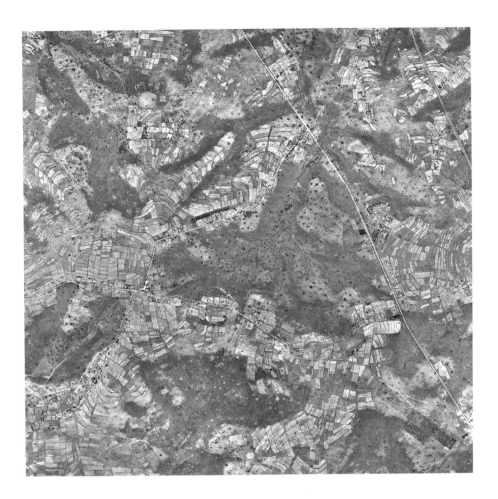

FIGURE 5.18   Dambo headwaters displaying etching patterns into the plateau surface, near Harare, Zimbabwe

expression in the formation of *hydromorphic podzols* or *white sands* (Chapter 2).

Once the loss of material from hillslopes by solution, affecting all rocks to some degree, is acknowledged then there are important inferences for slope denudation. There is, in fact, a direct loss from all points on the hillslope (Carson & Kirkby, 1972; Young & Saunders, 1986). Ions in transit only flow directly into the rivers where streams are actively eroding bedrock channels and possess no true floodplain; in other instances the flux is dissipated at the margins of the valley, some distance from the stream channel. It is often here

that calcretes and ferricretes can be found, or where 2:1 clays are formed. The well known 'black cotton' soils are nearly always associated with areas of major alluviation along river courses. These soils, as described above (Chapter 2) are dominated by montmorillonite. The clay lattice thus 'locks in' the former solutes which fail to reach the stream channels. In this way we can begin to see how chemical denudation can affect the tropical landsurface in fundamental ways, without leading to high solute loads in major rivers. Unfortunately it is difficult to quantify these effects, and for this reason alone they are frequently neglected.

6

Landsliding and Other Mass Movement

LANDSLIDES IN TROPICAL ENVIRONMENTS

Taiwan is a mountainous island bisected by the Tropic of Cancer and it experiences high-intensity tropical rainfalls. In an area of 104 km², containing 22 catchments, 7810 slope failures involving 151 million m³ of rock and soil, occurred between 1965 and 1977 (Lee, 1981). This is but one illustration of the importance of landsliding in tropical and sub-tropical environments, yet the subject of landsliding in the tropics is still poorly served by the major literature on the subject. The topic is unaccountably neglected in symposia on problems of land stability and instability (Brunsden & Prior, 1984; Anderson & Richards, 1987), engineering geomorphology (Fookes & Vaughan, 1986), and tropical geomorphology (Douglas & Spencer, 1985b). Chorley *et al.* (1984) acknowledge that conditions in the humid tropics are particularly favourable to catastrophic slope failures, but examples do not appear in their systematic study of mass movement. Similarly, Carson (1976) was unable to tackle this issue fully in his study of mass-wasting and climate. These omissions are odd, since some of the most landslide-prone areas of the world are situated in the humid tropics, and Rio de Janeiro, a city of more than six million people, has a long recorded history of landslide hazard (Branner, 1896; De Meis & Da Silva, 1968; Jones, 1973; Smith & de Sanchez, 1992), and Hong Kong is scarcely less well known in this context (So, 1971; Lumb, 1975, 1983). Serious studies of landslide activity in Hawaii were undertaken more than 40 years ago (Wentworth, 1943; White, 1949), while

detailed studies of landsliding in Tanzania were published by Temple and Rapp (1972), who also referred to earlier accounts for Brazil (Freise, 1932; Sternberg, 1949; Birot, 1968).

One reason for this neglect of the subject is that landslides are difficult to study and expensive to monitor, and there has been a concentration of research in areas of rapid urban development faced with earthquake hazards, such as California and Japan. Catastrophic land failure is seen as a major threat from earthquakes, even though return periods for prolonged heavy rainfall, likely to cause extensive sliding, are almost always much shorter.

Landslide scars are quickly concealed by renewed vegetation growth in the humid tropics and sub-tropics, and this may have led to serious underestimates of their frequency of occurrence and importance within the morphogenetic system. On the other hand, engineers encounter palaeo-landslides during excavation and construction, and contractors often trigger new or renewed activity by removing support from the base of metastable slopes. Engineers have for a long time recognised the importance of deep residual soils in tropical areas in this context (Deere & Patton, 1971). However, the much greater volume of construction undertaken in temperate areas has led to a greater emphasis in soil mechanics on conditions in these climates, and conferences dealing specifically with the geomechanics of tropical soils are of recent date (ISSMFE, 1985, 1988, 1990).

Landslides and earthquake shocks

The importance of earthquake shocks in triggering catastrophic slope failures has often overshadowed discussion of the climate and environment of these events (Simonett, 1967, Harp *et al.*, 1981). One example is the Huascarán catastrophe in Peru, where a rock avalanche and debris flow killed 70 000 people in 1970, as a result of an earthquake of 7.7 Richter magnitude, which detached 25×10^6 m^3 of mountainside that struck lowland towns at an estimated speed of 400 km h^{-1}. Such events heighten awareness of the destructive power of mass movement, but they are not a direct response to climatic controls or meteorological events. However, climate may still be implicated. The Huascarán slide involved a failure in partially weathered granodiorite, and Hsü (1975) has suggested that the granite blocks were dispersed in a dense interstitial mud and flowed down the valleys.

In a recent study of a large landslide (180×10^6 m^3) in the Bairaman Valley, Papua New Guinea, which followed an earthquake of 7.1 Richter magnitude, King *et al.* (1989) referred to the weakening of the rock mass by weathering and deep karstification, and while it is difficult to define the specific influence to these factors, there can be little doubt that the nature of landsliding in the humid tropics reflects not only high magnitude events (climatic or tectonic) that act as triggers for the movement, but also the preparatory rock weathering that weaken rock fabrics and produce susceptible mantles of non-cohesive material.

Another example from Papua New Guinea has been examined by Peart (1991), who described a major planar slide containing 2 km^3 of rock, covering an area of 11.4 km^2, and leaving a back scarp 1600 m high! This slide killed 74 people, and dammed local rivers for a period of 8 months. These rivers were then supplied with copious sediment which has led to widespread aggradation. The point of interest in this landslide was that there was no known trigger event, no earthquake recorded or heavy rainfall within weeks of the failure. Peart (1991) attributed this slide to repeated seismic loading over time; the existence of a dense network of rock discontinuities consequent upon uplift and unloading of the steep terrain, and to the prolonged high rainfalls of the region.

In regions such as Papua New Guinea, where there are convergent plate boundaries and frequent earthquakes, slides are common. The humid tropical and montane climates contribute to the hazard through weathering, karstification and the reduction of shearing resistance in saturated rocks. In the Philippines, a massive limestone block 18 km $\times$ 25 km $\times$ 300 m high glided a distance of 5 km over a slope of only 0.6°, and Wolfe (1977) has argued that this was largely due to these factors, but triggered by seismic events during the Holocene. It is the existence of weathered mantles, however, that is the focus of attention of engineers working in tropical environments (Deere & Patton, 1971; Brazilian Society for Soil Mechanics, 1985; Geological Society, 1990).

Tropical rainfall and landsliding

It has been shown in Chapter 5 that tropical climates are typified by more intense rainfalls than most temperate regions, and this intensity is accompanied by larger storm sizes, and often by much higher annual rainfall totals. These factors were considered important in determining the erosivity of rainfall, and there has been much research directed at elucidating such relationships. However, rainfall events that are known to have triggered major landslides, seldom occur on an annual basis and may have return periods of 10, 50 or even 100 years.

When rainfall intensity is considered alone, it is seen that high intensities are not unusual. Ibadan, Nigeria, for example, receives 100 mm h^{-1} for at least 30 min once every 2.8 years, and 35% of all storms exceed 50 mm h^{-1} for short periods (Lal *et al.*, 1981). Other tropical locations experience comparable storm intensities, but it is clear that landslides are usually triggered by still larger but less frequent rainfall events, and that deep-seated landslides in particular, are strongly influenced by the effects of prolonged or antecedent rainfall over a period of days or weeks.

Landslides in eastern Brazil

It is interesting to consider the rainfall records associated with major landslide events in eastern

Brazil. In the Serra des Araras escarpment zone (a part of the Serra do Mar) 50 km west of Rio de Janeiro, Jones (1973) noted that 218–275 mm of rain fell on 22–23 January 1967, causing >10 000 slope failures over a small area of 180 km². Rainfall intensities reached 114 mm h^{-1} during this period. The results of this storm were catastrophic, and 1700 deaths were recorded; the main power-generating plants for Rio de Janeiro were overwhelmed by mudflows, and countless failures affected the road system. Public services including water supply, power and transport were seriously disrupted for 2–3 weeks and damage to property was enormous. According to Jones (1973) many of the slides occurred beneath undisturbed, forested slopes. The major characteristic of this storm was its intensity and the small area affected. It did not register as a major event in the city of Rio de Janeiro itself, though the city suffered a major storm 4 weeks later when 300 mm fell on 19–20 February. Jones (1973, p. 22) was sufficiently impressed by the Serra des Araras event to comment: 'Beginning at about 11:00 pm, an electrical storm and cloudburst of 3½ hours duration laid waste by landslide and fierce erosion a greater land mass than any ever recorded in the geological literature.'

In the previous year, a larger but less intense, storm had hit the city of Rio de Janeiro itself, when 484 mm of rain fell over a 3-day period, under the influence of a stationary cold front. The resulting landslides produced >300 000 m³ of rubble in the steets of Rio and more than 1000 people died. In this event, forested areas beyond Rio de Janeiro itself were little affected, and the impact of human interference with the natural systems was most apparent. Many cut-and-fill slopes failed, and drainage outlets were impeded by infill material used to reclaim swamps in Guanabara Bay in the 19th century. This type of event was not without precedent, and major flooding and sliding had occurred in 1942 which led to decrees to prevent dangerous construction (1955) and to promote forest conservation and land-use control (1959). However, the storm of 1966 exceeded in magnitude all previous records for intensity, and storm size (484 mm in 3 days; previously 340 mm). De Meis and Da Silva (1968) analysed the frequency of heavy rainfall in Rio de Janeiro between 1955 and 1967. All monthly falls >200 mm (14 in 12 years) occurred during the December to March summer period, and on five occasions totals exceeded 300 mm, with a maximum of 620 mm in January 1966.

A similar catastrophic event afflicted the Tubarau area of Santa Caterina State in southeast Brazil in 1974, when nearly 50% of the mean annual rainfall of 1558 mm fell during 16 days of March (742 mm), culminating in 392 mm in the 3-day period 23–25 March, of which 240 mm fell during the final 24 h. Catastrophic landslide flows took place, mainly on deforested hillslopes, and serious flooding occurred in the rivers; 25 people were killed in one slide alone (Bigarella & Becker, 1975). In February 1988, 200 were killed and 20 000 made homeless by debris avalanches in Rio de Janeiro (Smith & de Sanchez, 1992). All along the Serra do Mar region of Brazil similar conditions exist, with lowland rainfall averages of 1500–1800 mm year^{-1}, rising to >5000 mm year^{-1} along the higher mountainous zones of the plateau edge (>1500 m asl). Prolonged, heavy rains are associated with stationary cold fronts, but the extreme events do not always strike the same areas.

The importance of antecedent moisture conditions is not sufficiently stressed in these studies, though the build-up to the Tubarau slides is made clear. Ab'Saber (1988), however, gave a detailed account of previous rainfall, leading to serious debris avalanching near Cubatao, Brazil, on 26–27 January 1985. The rainfall records at Paranapiacaba show abnormally wet conditions persisting through the months of October–December 1984, prior to the trigger event in January 1985, when 872 mm fell in one month, and 111 mm fell on 1 and 2 January; 80 mm on 7 January; 101 mm on 13 January; 94 mm on 17 January and, finally, the two days of 22 January and 23 January received a total of 344 mm (246 mm in 24 h on 22 January). The inexorable build-up of this event is well demonstrated, and the Serra de Paranapiacaba were devastated by debris avalanches.

Landslides in Sierra Leone

A further example from Freetown, Sierra Leone (Thomas, 1983) serves to illustrate still higher magnitude rainfall leading to landslides under forest in an area with mean annual totals of 2500–4000 + mm year^{-1}. Between 4 and 9 August 1945 there occurred the wettest 5-day period since records began in 1876, with rainfall in August 100% above the mean for that month. In the 5 days preceding the slides, 1121 mm of rain fell on Freetown and on 10 August, the day of the slides, there was a fall of 401.8 mm. This led to a massive debris slide and to slumps in the Orugu Valley that claimed the lives of 13 people, but we have no records of other events in the area on that date. Interpretation of 1:12 500 scale aerial photographs, combined with visits on the ground, demonstrated a high spatial frequency of landslides on the main ridge of the Freetown Peninsula, where maximum slopes commonly range from 28° to 38°. Most of the slides visited were 'recent' in the sense that the debris was little altered and there was no lateritisation or dissection. But villages are built on the toe areas of certain slides, signifying ages exceeding 150 years, and it is concluded that the event of 10 August 1955 has a return period of around 100 years (but not to the same place), with the implication that the characteristic landslide morphology of the area represents the cumulative effect of events over a 10^3 year period.

Rainfall magnitude and frequency of landsliding

Landslides in Hong Kong

Parallels in other tropical regions of the world are ready to hand. Hong Kong lies in a climatic zone in some respects comparable with southeast Brazil and has aspects of relief and geology comparable with Rio de Janeiro. Rainfall averages around 2000 mm year^{-1}, and heavy rains come either from tropical cyclones or from slow-moving low-pressure systems. The short-lived cyclonic storms appear to be less important than the prolonged rains brought by the troughs. The incidence of slides and their relationship with rainfall were examined by Lumb (1975). Landslide events were classified as: *disastrous*, when >50 slips occurred throughout Hong Kong in 1 day; *severe*, when 10–50 slips took place; *minor*, when activity was localised and there were <10 slips; and *isolated*, when only one slip

occurred. Records were available from 1950 to 1973 and Lumb (1975) was able to demonstrate the importance of cumulative rainfall leading up to the events, by plotting cumulative rainfall from 1 January in each year against the cumulative average for the same period. Almost all landslides occur during the period May–September, so this gives an indication of the antecedent wetness. The influence of rainfall in the previous year was shown to be less important, though very wet conditions following a dry year produced significantly more severe and disastrous events. Detailed analysis of rainfall records led Lumb (1975) to conclude that the total 24-h rainfall on the day of the event together with the rainfall over the previous 15 days best predicted the severe events (disasters after 350 mm + >100 mm on the day of the event) (Figure 6.1). However, a more recent analysis by Brand *et al.* (1985) has contested this view, arguing that landsliding is triggered mainly by the 24-h rainfall amount and peak hourly rainfall intensity. A statistical relationship with antecedent rainfall can be found because large storms and prolonged rainfall are associated. If 24-h rainfall does not exceed 100 mm very few slides occur, but when it exceeds 270 mm almost every rainfall 'event' will cause slides to take place. When these events involve intensities exceeding 70 mm h^{-1} significant landsliding occurs. These authors cited a 24-h storm of 380 mm in October 1978, but maximum rainfall intensity was only 38 mm h^{-1} and it produced only one slide, whereas a 24-h period of 200 mm of rain in April 1966 was accompanied by intensities up to 68 mm h^{-1}, resulting in a large number of slides. Many other data were cited by Brand *et al.* (1985), but it would be unwise to discount the importance of antecedent rainfall from these data, and much may depend on whether the slides are shallow features or are deep seated.

It is interesting that these values are lower than many from other tropical areas and one can speculate that this is because most of the slides in Hong Kong come from construction sites and hills already affected by previous engineering works. Under natural conditions more rain may be necessary to produce comparable results. However, So (1971), who examined landslides associated with

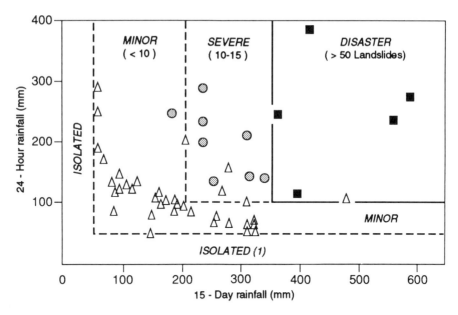

FIGURE 6.1   Relationship between rainfall and the incidence of landslides in Hong Kong, established by Lumb (1975). Reproduced by permission of the Geological Society Publishing House

a rainstorm in Hong Kong in June 1966, found that 35% of the slips occurred under woodland which covered only 8.4% of the land area of Hong Kong at that time. Tree roots are not observed to have stabilising effects on deep-seated slides.

In sub-tropical central Japan, Fukuoka (1980) found a relationship between landslide activity, intensity and amount of rainfall, based on empirical data from the Boso Peninsula. Here, in one typhoon event, more than 1000 slides within 1 km^2 occurred after 350 mm of rainfall. When less intense rainfalls were analysed Fukuoka predicted that five slides per 1 km^2 would result after preceding rainfall of 60 mm and an intensity of 50 mm h^{-1}, or after 250 mm and intensities of less than 20 mm h^{-1}. Fukuoka (1980) plotted landslide probability against rainfall intensity and preceding total rainfall over a 36-h period (Figure 6.2). These empirical data apply to one small area but the principle can be employed elsewhere.

In the San Francisco area, Mark and Newman (1988), recorded >250 mm of rain during the two-day storm of January 1982, representing 30% of the mean annual amount. In this area, Cannon and Ellen (1988) found that abundant debris flows

occurred after 500–760 mm of pre-storm rain (seasonal, July–January) and >8 h of rain with intensities 10–20 mm h^{-1}, where mean annual rainfall was >660 mm, but that 381–483 mm of pre-storm rain and 17 h at 2.5–6.4 mm h^{-1} triggered similar activity in the drier areas.

Shallow slides (mainly debris flows) can, nevertheless, be triggered by short, intense rainfalls, particularly where there has been extensive deforestation. This was demonstrated in the Mgeta area of Tanzania, where Temple and Rapp (1972) found that damaging rainfall intensities of 75 mm h^{-1} have a return period of 1.5–6 years, according to station, and have a close relationship with daily rainfalls >75 mm. Extreme rainfalls recorded for this area exceed 100 mm day^{-1} for all stations considered and reach a maximum of 195 mm day^{-1}. One 3-h storm of 100.7 mm on 23 February 1970 caused more than 1000 new landslide scars and reactivated some older features in the Mgeta area.

Landslide recurrence intervals

Lumb (1975) and De Mello (1972) both referred to the impact of major events in one year on the

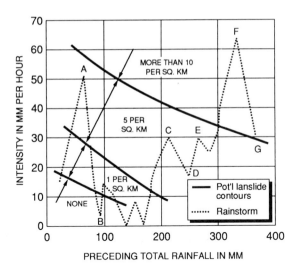

FIGURE 6.2 Probability of the occurrence of landslides according to Fukuoka (1980), based on conditions in Japan. The broken line plots the intensity of rainfall against the amount of preceding rainfall during a storm; the solid lines (contours) divide the graph into four regions according to the risk of landsliding (no. km^{-2}). Thus at B, the preceding rainfall has been >80 mm but the intensity is <5 mm h^{-1}, and there are no landslides; at F, however, the preceding rainfall was 330 mm and intensity 65 mm h^{-1}, and there were more than 10 landslides km^{-2}. Reproduced by permission of the South East Asian Geotechnical Society

Janeiro, while deforestation has affected other areas to different degrees. However, attempts to relate the incidence of landsliding to deforestation have been only partially successful, and it is clear that exceptional storms are capable of triggering landslides on steep, forested slopes (Temple & Rapp, 1972; Jones, 1973; De Ploey & Cruz, 1979). Nevertheless, it is probable that more frequent landsliding will occur on deforested slopes (Lumb, 1975), in response to events of lesser magnitude.

The two salient points to emerge from a consideration of rainfall events in relation to landsliding in the tropics are:

(1) the antecedent rainfall over a period of days, weeks or even months may be as important as the precipitation during the critical 24 h containing the slides. However, in most cases >100 mm falling in 24 h is necessary before significant landsliding occurs;

(2) extreme storm events lasting a few hours can produce widespread devastation, resulting from intense runoff causing debris flows and avalanches. In these cases the peak intensity of the storm may be as important as the rainfall total, with >100 mm h^{-1}, and total storm sizes 200–400 + mm indicated.

According to Temple and Rapp (1972), rather similar conclusions will apply to temperate landslides which mainly occur in alpine areas receiving heavy orographic rainfall or in areas subject to tropical storms moving into temperate latitudes. Caine (1980) has analysed 73 published records of shallow landslides and debris flows world-wide and derived an empirical relationship between intensity and duration of rainfall which he expressed:

$$I = 14.82 D^{-0.39} \qquad (6.1)$$

in which I is the rainfall intensity (mm h^{-1}), and D is the duration of rainfall (h). A similar relationship was found with total storm rainfall (depth, d):

$$d = 14.82 D^{0.61} \qquad (6.2)$$

possibility of repeated sliding at the same site in the following year in Hong Kong. A series of major slides occurred in 1972 but were not repeated during the wetter year of 1973; a period of 'ripening' apparently being required to produce comparable disasters. The question of recurrence intervals for landsliding has been investigated in the Pacific northwest of the USA, where Benda and Dunne (1987) found that dating of basal colluvium in bedrock hollows suggested a recurrence interval of 6000 years. This period was reduced when all slopes within first- or second-order basins were examined, and the comparable intervals became 1500 and 750 years respectively.

Some of the events documented here (see Table 6.1) come from areas which have suffered extensive deforestation and varying amounts of engineering modification to the natural slopes. The latter is certainly true of both Hong Kong and Rio de

TABLE 6.1 Some examples of landslide events in the humid tropics (excluding landslide events associated with earthquake shocks)

Location, date (source)	Mean annual rainfall	Previous rain	24 h rainfall	Comment
Hong Kong, 1972 (Lumb, 1975)	2000	>350 (15 days)	>100	182 slides 138 deaths
Hong Kong, 1966 (So, 1971)	2000		401	702 slides
Java, 1983/84 (Heath & Saroso, 1988)	2340 2753	(5–7 days >100 mm year^{-1})	168 246 (48 h)	slides slides
Puerto Rico, 1970 (Deere & Patton, 1971)	3000	900 (different events)	150	slides
Cubatau, Brazil, 1985 (Ab'Saber, 1988)	4121	407.2 (30 days)	246	avalanches
Tubarau, Brazil, 1974 (Bigarella & Becker, 1975)	2000	742 (16 days)	240	avalanches
Serra des Araras, Brazil, 1966 (Jones, 1973)	2000	Isolated storm	218–275	10 000 slides
Rio de Janeiro, Brazil, 1967 (Jones, 1973)	1700		484 (3 days)	1000 deaths
Caraguatuba, Brazil, 1967 (De Ploey & Cruz, 1979)	1560	260 (previous day)	420	slopes >20 °
Rungwe, Tanzania, 1955 (Haldemann, 1956)			425	
Mgeta, Tanzania, 1970 (Temple & Rapp, 1972)	1058		100.7 (3 h)	>1000 slides
Freetown, Sierra Leone, 1945 (Thomas, 1983)	3000	1121 (5 days)	401	13 deaths
Nilgiri, India, 1978/79 (Seshagiri *et al.*, 1982)		unprecedented		>300 slides
Darjeeling, India, 1968 (Starkel, 1972)	4000	>50 mm day^{-1} (12 days)	465	25% area slides

However, Caine's (1980) database included only a few tropical examples, and may be compared with findings from Japan (Fukuoka, 1980; and Figure 6.2).

VEGETATION AND SLOPE STABILITY

Particular attention needs to be paid to the role of the vegetation cover in either promoting or inhibiting mass movement on slopes, and there has been much unsubstantiated argument concerning this topic. Whilst there is abundant experimental evidence to show the increased rate of surface erosion resulting from the removal of natural vegetation cover, the effects of such changes on the incidence of mass movement are less clear cut.

On the one hand, there is good evidence for increased landsliding in areas where *construction* has altered natural slope profiles, and studies from cities such as Hong Kong (Lumb, 1975) and Rio de Janeiro (Jones, 1973) emphasise this point. On the other hand, some other studies have indicated that slope failures occur under both forested and deforested slopes in humid regions. So (1971), for example, noted that 35% of 700 landslides resulting from a single storm in June 1966 occurred under woodland that occupied only 8.4% of the land area. Most other landslide studies from areas where forest cover survives in part have shown that landslides do occur under forest, particularly in response to high magnitude events (Branner, 1896; De Ploey & Cruz, 1979; Jones, 1973; Thomas, 1983; Greenway

TABLE 6.2   Effects of vegetation on slope stability

	Effect on stability
Hydrological mechanisms	
(1) Foliage intercepts rainfall, causing absorptive and evaporative losses that reduce rainfall available for infiltration.	Beneficial
(2) Roots and stems increase the roughness of the ground surface and the permeability of the soil, leading to increased infiltration capacity.	Adverse
(3) Roots extract moisture from the soil which is lost to the atmosphere via transpiration, leading to lower pore-water pressures.	Beneficial
(4) Depletion of the soil moisture may accentuate desiccation cracking in the soil, resulting in higher infiltration capacity.	Adverse
Mechanical mechanisms	
(5) Roots reinforce the soil, increasing soil shear strength.	Beneficial
(6) Tree roots may anchor into firm strata, providing support to the upslope soil mantle through buttressing and arching.	Beneficial
(7) Weight of trees surcharges the slope, increasing normal and downhill force components.	Adverse/ beneficial
(8) Vegetation exposed to the wind transmits dynamic forces into the slope.	Adverse
(9) Roots bind soil particles at the ground surface, reducing their susceptibility to erosion.	Beneficial

After Greenway (1987); reproduced by permission.

et al., 1985). It is therefore necessary to enquire further into the role of the vegetation cover in stabilising slopes.

Landsliding on the steep, forested slopes of the Serra do Mar in eastern Brazil is well documented, and although Jones (1973) described mainly superficial debris slides and flows, De Ploey and Cruz (1979) noted large-scale features. They also found that the root mat was generally confined to a depth of 3 m, roots of larger trees penetrating to 5 m. The root mat had the function of increasing infiltration and throughflow. These processes increase the likelihood of slope failure by promoting a rise in pore-water pressure and a consequent decrease in cohesion; they also induce a downslope, seepage stress. Shear stress stengths appear to decrease towards the base of the slope and De Ploey and Cruz argued for more attention to be paid to the initiation of landsliding at this position. This is in contrast to most debris slides which start high on the slope, close to the interfluve, and are often initiated by runoff rather than by throughflow. Debris slides can be favoured by lack of forest cover, although many originate in pre-existing water courses that become overwhelmed by floodwater and sediment. Nevertheless, most writers have

tended to emphasise the net stabilising effect of forest growth and have pointed to the reinforcing role of tree roots. Tree planting to arrest erosion, including landsliding, is also widely practised.

Mechanical and hydrological effects of vegetation

More recently, Greenway (1987) reviewed a large number of studies in this field and classified vegetation influences on slope stability as either *mechanical* or *hydrological* in effect, and assessed whether these operate to reinforce stability or to undermine it. His conclusions are reproduced here as Table 6.2.

It is clear that, under moderate rainfall conditions, the hydrological effects of the forest will operate more to stabilise the slope than to destabilise it. Surface runoff is reduced by increasing infiltration, and soil moisture is decreased by enhanced evapotranspiration. However, during extreme rainfall events, such as have been exemplified above, rapid infiltration may build up pore-water pressures to dangerous levels. In addition, throughflow may be encouraged by piping along root channels, increasing the downslope drag. In highly seasonal climates, prior conditions of desiccation may have opened up deep cracks, that promote water entry to the soil which exaggerates these effects.

On the other hand, most of the mechanical effects of the tree cover act positively to reinforce stability. The roots possess tensile strength and frictional properties, and can anchor into firm strata, buttressing the soil upslope from the tree. The weight of the biomass, however, acts as a (small) surcharge on the slope adding to the normal and downhill components of force. On balance, this surcharge is thought to act positively towards stability.

Yet the relationships within these factors become complex during storm events and a listing of the possible adverse effects of tree cover on slope stability is illustrative of the problem:

(1) enhanced infiltration raises pore-water pressures;
(2) increased throughflow produces seepage stress;
(3) rapid throughflow from upslope can induce instability at lower slope positions as pore-water pressures rise faster;
(4) confinement of a majority of roots to a superficial mat can promote sliding across the lower surface;
(5) in high winds trees transmit dynamic forces to the slope;
(6) tree fall with uprooting will increase both erosion and infiltration and produces a down-slope drag.

The well-known effects of vegetation cover in promoting *interception losses* from incoming rainfall must reduce erosion risk on slopes, but this effect is partly overcome in large storm events. The soil moisture extraction from roots and the high *evapotranspiration* rates from broadleaved forests combine to reduce soil moisture and lower ground-water tables over annual cycles of climate. These positive effects are again rendered ineffective during prolonged, heavy rainfall, because they operate at a slower rate when compared with high rainfall intensities.

The depth of rooting can in certain circumstances attain 30 m (Greenway, 1987), but a high proportion of roots (perhaps 80%) remain within 3 m of the ground surface, spreading to a radius of >50 m. Roots possess tensile strength and undoubtedly reinforce soil strength, but their effects on slope stability depend on the depth of potential slip surfaces within the slope (Tsukamoto & Kusakabe, 1984). On deep, tropical ferrallitic soils roots are mostly superficial, being involved in the recycling of nutrients within the biomass. They therefore can have little effect on deeper, slump-earthflow type slides.

In a study of landslides under temperate coniferous forests from northwest USA, Orme (1990) suggested conclusions that might be valid elsewhere, namely that, in steep catchments with rainy climates, shallow soils and a liability to high-magnitude intense rainstorms, landsliding (debris avalanches and flows in this study) is a recurrent process under both older and more recent forest cover. Orme's (1990) record stretched over 3400 years and he concluded that major events inducing landslides have a recurrence interval of 60–70 years in this area. Unfortunately no comparable records from tropical climates are known to the present author. Intuitively, however, the conclusions, in general terms, accord with observations from many tropical forest areas. The high spatial incidence of landslides in the Freetown Peninsula has been noted, and the existence of slump-earthflow slides under forest in the Owen Stanley Mountains of Papua New Guinea can be clearly seen (Figure 6.3). Aerial reconnaissance over parts of Kalimantan, Indonesia, also indicates the occurrence of some quite large slides that have produced debris fans around some of the larger residual (forested) hills.

LANDSLIDE TYPES

Some forms of mass movement may not be regarded as landslides *sensu stricto*. Skempton and Hutchinson (1969) argued that: 'the generic term, landslide, embraces those down-slope movements of soil or rock masses as a result of shear failure at the boundaries of the moving mass.' Crozier (1986) pointed out that this will include many movements where the displacement occurs by flow rather than slide, and this latitude will be accepted

FIGURE 6.3   Small slump scars and earthflows on slopes cleared of rainforest, over weathered schists in the Owen Stanley Mountains, Papua New Guinea. Similar, if less distinct, forms can be identified within the forest, where they may be associated with tree falls. The alluviated surface in the foreground may be partly due to solutional subsidence or to tectonics

for the present discussion. There have been many classifications of landslides and other mass movements (Sharpe, 1938; Varnes, 1958, 1978; Carson, 1976; Coates, 1977; Hutchinson, 1977; Hansen, 1984; Crozier, 1986), and this topic will not be discussed in detail here. All types of slides may occur in deeply weathered terrain, but some are more common. If we exclude *creep* processes which are considered separately, landslides can be grouped into three broad categories: falls, slides, and flows.

(1) *Falls* involve the free-fall, usually of rock, but may involve soil materials (more likely as topples where a pivotal movement occurs) and are not particularly common in tropical environments.

(2) *Slides* always involve a definable shear surface and are often categorised as rotational, when the shear surface is concave upwards, or planar (alternatively, translational). Slumps typify the former category; debris slides, the latter.

(3) *Flows* involve material moving as a viscous mass which is dominated by intergranular movement, in contrast to moving over a shear surface. Some flows may be dry, and usually involve sand or loess-type materials, but most are wet and are typified by mudflows. There is probably more difficulty with this group than with others, because the water content and clastic material are very variable, and the morphological expression may be less clear. However, some authors distinguish between

earth flows and *debris avalanches* (less water; more debris), and *mudflows* and *debris flows* (more water; less debris).

The rate of movement can vary from slow (<0.06 m year^{-1}–1.5 m year^{-1}), to moderate (1.5 m year^{-1}–1.5 m day^{-1}), and rapid (1.5 m day^{-1}–3 m s^{-1}), according to Hansen (1984). It is tempting to suggest different rates amongst the landslide types listed above, but in reality, many landslides are composite. Shear failures (slides) very commonly lead to earthflows or mudflows downslope. The nature of the previous argument gave emphasis to triggering events liable to produce rapid movements, and although forest disturbance by moderate or rapid landsliding is important (Garwood *et al.*, 1979), too great an emphasis on these events may underestimate the geomorphological role of slow movements that can be partially accommodated by the vegetation cover.

According to Deere and Patton (1971) four types of slide are most common in weathered rock, residual soil and colluvium. These are:

(1) shallow slides in upper residual soil, resulting from either a rise in pore-water pressure above profile discontinuities, or flows resulting from the collapse of piping systems;
(2) block or wedge slides along relict joints, or other planes of weakness in the weathering profile (saprolite); these slides are perhaps less commonly encountered on natural slopes than in excavations;
(3) slides in shallow colluvium involving the 'ravelling' back of a thin layer above residual material;
(4) deep-seated slides in colluvium, often moving across buried, B horizon planes of weakness in the residual soil beneath.

However, this list is probably less appropriate for the study of slides on natural slopes than for the study of failures in excavations. According to Jones (1973), the landslides near Rio de Janeiro fall into four categories, and these can be widely recognised throughout the tropics.

(1) *Slump-earthflow landslides* developing initially from a concave shear plane, the moving mass becoming a viscous liquid downslope. These depend on the infiltration of water, probably over a period of days or weeks, leading to a rise in pore-water pressure to the point when the angle of shearing resistance is reduced to a value close to that of the slope itself.
(2) *Debris slides and avalanches* which are shallow and move across a shear plane parallel with the underlying bedrock surface. Debris slides commonly have slump blocks at their heads and tend to break up as they move slowly downslope. In debris avalanches, progressive failure is more rapid and the material moves in a single mass.
(3) *Debris flows and mudflows* form in the surficial deposits on slopes; debris flows involving coarse material; mudflows, sand, silt and clay. These flows almost invariably result from intense precipitation leading to torrential runoff, and commonly originate in hillslope hollows and follow existing water courses on steep mountain slopes. They may be more common on deforested slopes but they also occur within the forest. The erosive runoff detaches soil materials, and receives additional sediment from sloughing, rapidly becoming a high density flow (60%–70% solids by weight) capable of achieving considerable destruction. As runoff from lower slopes increases the water content and contribution of finer material, the debris flows become transformed to mudflows that follow river courses down valley. Jones (1973) documented many hundreds of such events following the intense rainstorm in the Serra das Araras region in 1967.
(4) *Rockfalls and rockslides* may be associated with slides involving saprolite. These are usually rock slides involving slabs detached from rock masses by pressure release, and forming part of the saprolite profile, which has been shown to contain large numbers of corestones over jointed granites and granodiorites.

Other writers on southeast Brazil describe landslide movements as *massive planar slides*

(De Ploey & Cruz, 1979; Ab' Saber, 1988), *rock slides* and *debris slides* (De Meis & Da Silva, 1968). In Tanzania, Temple and Rapp (1972) distinguished *sheet slides* as a shallow form of debris slide, often only 2 m deep, 5–20 m wide and up to 50 m long, and occurring on steep slopes between 28° and 44°. These developed into shallow mudflows downslope. In addition, they identified *bottle slides* as a type of slump–debris flow combination, where a broad retrogressive slump headwall (< 10 m high) leads to a channel narrowing downslope, where it accommodates a saturated mass or debris flow. Similar features were called *lavakas* by Tricart (1965), and these have been studied in the island of Madagascar (Wells *et al.*, 1991).

The problems with many such observations is that they have mainly been concerned with the immediate results of intense rainstorms, and have described relatively superficial features a few metres in depth. Some of the areas described so graphically from southeast Brazil now show few obvious signs of the great storms of 30 years ago. Slowly moving slides are widespread and have tended to receive much less attention in the literature. In the Nilgiri Hills, India (Seshagiri *et al.*, 1982) landslides were described as occurring mainly in thick soils, with 51.8% in weathered material > 30 m deep and a further 38.3% in mantles 10–30 m thick. Many of the larger slides are slump-earthflow features with clear evidence of long-term movement. However, debris slides are also common.

In Sierra Leone, there is widespread evidence of slumping in deeply weathered material; many slumps taking place beneath thick duricrust cappings, where dissecting valleys have carved out steep bounding slopes. The hydrology of sub-duricrust saprolites can be complex, with the formation of hollows and subterranean cave-stream systems (Bowden, 1980). Along steep valley sides and escarpments the effect of these systems and the flushing out of clays beneath the duricrust capping can lead to *toppling failures*, similar to rock falls. But, in addition, deep-seated slides have developed, incorporating the duricrust and underlying saprolite in rotational slump-earthflow slides of large dimensions (Gaskin, 1975; and Figure 3.12(D)). Some of these are of unknown or indeterminate age,

but the oldest show signs of recementation by iron and are possibly early or pre-Quaternary.

The Freetown Peninsula landslides include slumps in deep saprolite, one having been triggered by dam construction, and several planar, debris slides that have been strongly influenced by steeply dipping (28–38°) joint structures. However, there are some large slides incorporating rock and saprolite, that can be recognised only from aerial photographs and these appear to be complex in form involving important rock slides, if not actual rock falls. In certain cases blocks weighing an estimated 10^2 t have been transported > 200 m across gently sloping ground at the foot of the slides. This implies a rapid movement and fluidisation of the saprolite matrix. The precise cause of such movements is not known and the influence of small earthquake shocks, in combination with the effects of heavy monsoonal rains, cannot be excluded.

Slump-earthflows and mudslides may take place quite slowly over periods of many years, experiencing occasional surges of movement after prolonged and heavy rainfall (Brunsden, 1984). It is also clear that most types of slide occur widely in different climatic environments, and their particular importance in tropical areas has been more often asserted than demonstrated. Nevertheless, the incidence of rainfall events liable to trigger landslides appears to be much higher in tropical and some sub-tropical regions, particularly if we exclude high mountain areas from any comparison.

LANDSLIDE PROCESSES

Detailed discussions of landslide processes and mechanisms are well illustrated by the work of Terzaghi (1950) and Varnes (1958, 1978), while discussions of slope stability are found in Deere and Patton (1971), Brunsden and Prior (1984), Crozier (1986), and Anderson and Richards (1987). Useful introductions are also given by Chorley *et al.* (1984), Gerrard (1988), and Selby (1982).

Behaviour of slopes in residual soils

According to Deere and Patton (1971, p. 87), 'the key to understanding the stability of slopes in residual soils lies in recognising the roles of weathering profiles, groundwater, and relict structures.' They also offer the following conclusions concerning the behaviour of natural slopes in areas of residual soils (pp. 99, 100):

(1) landslides are a common and perhaps the predominant method of slope development in areas of deep residual soils;
(2) they are associated with characteristics of the weathering profile;
(3) they are especially common during periods of heavy rainfall and/or earthquakes;
(4) shallow slides are the most common on the steeper slopes but deeper slides which incorporate weathered rock and both shallow and deep layers of colluvium are also frequently encountered;
(5) rapid erosion is common on slopes where the silty sands and sandy silts of the saprolite . . . become exposed. . . .'

(See Plate 6-I). If correct, then these conclusions should provide a major input to all discussions concerning tropical landform development, where *transport-limited* as opposed to *weathering-limited* (Carson & Kirkby, 1972) slopes are widely encountered. However, to take the argument further the landslide processes themselves need definition.

Varnes (1958, p. 42) pointed out that, 'all true slides (excluding falls) involve the failure of earth materials under shear stress. The initiation of the process can therefore be reviewed according to (a) the factors that contribute to high shear stress and (b) the factors that contribute to low shear strength.' It is, however, necessary to recognise that a single factor may operate to affect both shear stress and shear strength simultaneously.

According to some engineering usage most saprolites and residual soils come under the heading of 'mudrocks' (see Gerrard, 1988), but this term will be avoided here. These materials obey the *Coulomb equation* regarding shear strength:

$$S = C + \sigma \tan \Phi \qquad (6.3)$$

where S is the strength, C is the cohesion, σ is the normal stress and Φ is the angle of internal friction or shearing resistance. If pore-water pressure U is taken into account then this will become:

$$S = C' + (\sigma - U) \tan \Phi' \qquad (6.4)$$

This simple formula thus takes into account *cohesion* or the resistance (strength) of the material to shear stress, irrespective of any weight imposed normal to the surface of potential movement. The angle of internal friction (Φ) reflects the weight of material inducing friction along the shear plane as a normal stress. This angle, for any given material, is approximated by the angle of repose, which is governed by the density, shape, size and sorting of granular materials, and on the cohesion and moisture content of clay soils (mudrocks).

Under strain, the shear strength will increase to a *peak strength*, at which point failure will occur and the remoulded material is found to have a lower *residual strength*. This arises especially from orientation of clay platelets at the point of failure and from an increased water content.

According to Blight (1990), most natural slopes in the tropics are initially stable except after severe events, but the stability depends on the unsaturated state of the regolith, within which the capillary tensions enhance the strength of the material. He observed that there are records of 70° slopes 10 m high, and 45° slopes 20 m high, remaining stable for at least 40 years. These figures are accountable only by invoking capillary stresses on pore water, above a relatively stable water-table.

During heavy rain a *wetting front* advances rapidly into soil at a rate determined:

$$V = \frac{k}{(1 - S)n} \qquad (6.5)$$

where V is the velocity of wetting front advance, k is permeability, S is initial degree of saturation, and n is porosity. The build-up of water within the soil causing pore-water pressures to rise, gradually reduces the *factor of safety*. This is the 'ratio of average shear strength and average shear stress along the most critical potential failure surface' (Brand, 1985, p. 23), and when this falls below 1, failure is likely. Figure 6.4(a) and (b) illustrate this

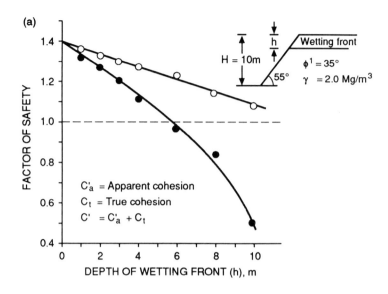

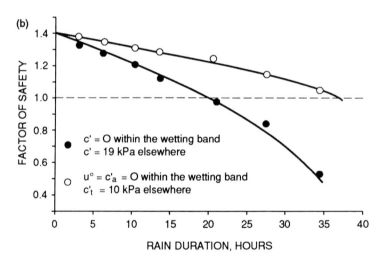

FIGURE 6.4   (a) Effect of rainwater infiltration on the factor of safety. (b) Effect of rainfall duration on the factor of safety of a slope. (Reproduced by permission of Balkema from Blight, 1990)

result. According to Skempton and Delory (1957), the factor of safety may be estimated:

$$F = \frac{\dfrac{C_r'}{\cos\alpha} + (\gamma_t\, Z\cos\alpha - Zm\cos a\gamma_w)\tan\Phi_r'}{\gamma_t\, Z\sin\alpha}$$ (6.6)

where C_r' = apparent residual cohesion,
        α = slope of surface and failure plane parallel to it,

Z = the vertical depth of potential failure plane,

$m = Z$ divided by the height of the phreatic surface above the potential failure plane,

γ_t = bulk density of soil + water,

γ_w = density of water, and

Φ_r' = apparent residual angle of internal friction.

More simply this relationship can be generalised as:

$$F_s = \frac{\text{Shearing resistance or strength}}{\text{Applied stress}} \quad (6.7)$$

When $F_s < 1.0$, failure will occur. According to Deere and Patton (1971), many slopes in residual saprolite and soil will have a factor of safety close to 1.0 after heavy rain.

Anderson and Richards (1987) have drawn attention to the possible importance of soil suction (negative pore-water pressure) in tropical soils which can alter the parameters for determining stability by introducing *pore-air pressure* into the equation. Their discussion includes a possible technique for incorporating this term.

Causes of slope failure

Much of the interest in mass movement concerns *stability analysis* of earth materials and the identification of factors leading to slope failure. In this section emphasis will be given to the latter, and a check list of factors, taken from Varnes (1978) is shown in Table 6.3 (after Chorley *et al.*, 1984). However, it will be seen that important factors can contribute to both increased shear stress and to reduced shear strength, making the distinction blurred. Most important in the tropical environments are the effects of weathering, and the results of heavy rainfall.

LANDSLIDING IN THE WEATHERED MANTLE

The effects of weathering are fundamental and pervasive in this context. Varnes (1958) drew attention to the *initial* state of the earth materials under the headings: (1) composition, (2) texture, and (3) gross structure. Thus the composition of decomposed rocks, sedimentary clays and shales, volcanic tuffs and some schistose rocks either imparts an inherent weakness to the materials or a liability to become weakened by changes in water content or by further weathering. The texture of loose materials such as sands and sensitive clays and the gross structure that includes rock discontinuities, plus the arrangement of rocks all come under this heading. The characteristics of the weathered mantle, its texture and clay mineralogy, and the presence in it of relict discontinuities are specifically very important in tropical environments and lead to low shear strength on many natural slopes (see Chapter 3).

There is some evidence to show that the stages of weathering in a granite can influence the incidence and type of landsliding. Durgin (1977) recognised fresh rock (<15% weathered rock), corestones (15–85% weathered rock), decomposed granite (>85% weathered rock), and saprolite (fine grained, possibly lateritic).

(1) *Corestone slopes* are subject to rockfall avalanches in which the cores themselves can roll downslope causing serious damage, or become entrained in fluidised mud.

(2) *Decomposed granite* in which biotite and feldspar have rotted, exhibits an increase in microfractures and porosity, and a decrease in bulk density and shear strength. When saturated the material is virtually cohesionless, and the transition to fresh rock is often abrupt. The water becomes perched above this weathering front and moves through the regolith creating significant *seepage forces* (Hartsog & Martin, 1974, as reported by Durgin, 1977), capable of inducing planar sliding (Figure 6.5). Decomposed granite, however, is virtually cohesionless in the dry state and *dry debris slides* are known to occur on steep slopes. On the island of Honshu in Japan, landslide disasters indicate a concentration of slides in decomposed granite and saprolite, while repeated sliding is largely confined to saprolitic profiles (Oyagi, 1968, reported by Durgin, 1977).

(3) *Deep saprolites* behave as 'mudrocks' (above) and possess cohesion as high as $2000 \, \text{g cm}^{-2}$ when unsaturated, reducing to zero when saturated (Lumb, 1965). In these materials the bedrock interface is less important and rupture is characterised by a circular failure plane, leading to rotational slumping. Water continues to flow along the relict shear planes and

TABLE 6.3   Causes of mass movement

I   *Factors that contribute to increased shear stress:*
   (1) Removal of lateral support (undercutting–steepening of slope)
      (a) erosion by rivers
      (b) erosion by glaciers
      (c) wave action
      (d) weathering
      (e) previous rockfall or slide, subsidence, or faulting (steepens slope)
      (f) construction of quarries, pits, canals and roads
      (g) alteration of water levels in lakes and reservoirs
   (2) Surcharge (loading of slope)
      (a) weight of rain, snow, water from pipelines, sewers, canals
      (b) accumulation of talus
      (c) vegetation, trees
      (d) seepage pressures of percolating water
      (e) construction of fill, waste piles, buildings
   (3) Transitory earth stress
      (a) earthquakes
      (b) vibrations, blasting, traffic
      (c) swaying of trees in wind
   (4) Removal of underlying support
      (a) undercutting by rivers and waves
      (b) solution at depth, mining
      (c) loss of strength of underlying sediments
      (d) squeezing out of underlying plastic sediments
   (5) Lateral pressure
      (a) water in cracks
      (b) freezing of water in cracks
      (c) swelling (hydration of clay or anhydrite)
      (d) mobilisation of residual stress (pressure release)

II   *Factors that contribute to reduced shear strength:*
   (1) Weathering and other physiochemical reactions
      (a) softening of fissured clays
      (b) physical disintegration of granular rocks (frost action, thermal expansion, etc.)
      (c) hydration of clay minerals causing decreased cohesion, swelling
      (d) base exchange (changes physical properties)
      (e) drying of clays and shales — racking, loss of cohesion — entrance of water
      (f) removal of cement by solution
   (2) Changes in intergranular forces due to water content (pore-water pressure)
      (a) saturation — buoyance, decreased intergranular pressure and friction, and capillary cohesion destroyed
      (b) softening of material
   (3) Changes of structure
      (a) fissuring of shale and consolidated clays
      (b) remoulding of loess, sand and sensitive clay
   (4) Organic
      (a) burrowing animals
      (b) decay of roots

From Chorley *et al.* (1984) after Varnes (1978); reproduced by permission of Methuen & Co.

FIGURE 6.5 Translational landslide in gabbroid intrusive rocks, Freetown Peninsula, Sierra Leone. The slide took place across the sharp weathering front during the wettest 5-day event on record, in August 1945, and killed 15 people

precipitates Fe and Mn. Shear strength along the rupture surface is usually 50–65% that of the saprolite, and thus repeated failures involving the same rupture are likely to occur.

Further loss of shear strength can arise in different ways:

(1) physical disintegration of granular rocks due to thermal expansion (or frost), leading to loss of cohesion;
(2) absorption of water by clay minerals (hydration) decreasing cohesion of saprolites and soils behaving as clays (with high clay content), and particularly the loss of cohesion with swelling of smectites;

(3) base exchanges in clays affecting their physical properties;
(4) removal of cements, especially from sandstones, by solution;
(5) loss of water from clays will also give rise to loss of cohesion due to the formation of cracks, which will subsequently admit water to the interior of the material.

The effects of water entering rock and saprolite will be considered following an examination of some of the above factors in a tropical context. A number of parameters of the saprolite relevant to this enquiry have already been identified (Chapter 3), including: mineralogy, particle size composition, void ratio, structure and inter-particle bonding, permeability and the presence of low strength discontinuities. A moment's reflection will make it clear that in all these categories tropical saprolites may behave differently from those formed under temperate conditions.

Gidigasu (1988) has stressed that the rate of chemical weathering in the tropics, and the nature of the products of that weathering, place severe limitations upon the application of classical rock and soil classification and rating systems for foundation engineering, and he noted that numerous earth failures in the tropics had resulted from the misapplication of those principles. This problem must equally apply to the failure of natural slopes in the tropics. Gidigasu (1988) particularly stressed the following priorities:

(1) the need to consider the whole profile of residual materials;
(2) the need to pay particular attention to clay mineralogy (especially the presence of allophane, halloysite and montmorillonite);
(3) the need to recognise the complex interrelationships between rocks, environmental factors, chemical processes and profile development and patterns.

This high variability which results from the interplay of different factors in the formation of weathering profiles has also been stressed by Irfan (1988) in terms of the fabric variability in granite saprolites. Baynes and Dearman (1978) related the microfabric differences within a granite saprolite mainly to: (1) the degree to which the feldspars are weathered; (2) the proportion of clay produced; (3) the degree of eluviation of fine particles.

However, these changes are evident in other properties such as, the microcrack density, the modal analysis of the percentage of decomposed minerals, and weight percentages of quartz and feldspar. Some weathering indices were previously discussed in Chapter 2. Baynes and Dearman (1978) also proposed a void ratio–degree of decomposition index, and considered the changes in engineering properties induced by weathering to reflect: (1) microfracturing; (2) opening of grain boundaries; (3) development of intergranular porosity. Vaughan and Kwan (1984) concluded that the weakening process in weathering could ultimately be seen as due to the formation of a large number of holes in the parent rock.

Saprolite microfabric was discussed in Chapter 3 (see Figures 3.7 and 3.8) and Collins (1985) has further stressed that: (1) 'complex multi-level pore systems will be found' in tropical saprolites, representing a wide variety of microstructural forms; (2) microstructures will be heterogeneous, reflecting variations in degrees of decomposition and the presence of relict grains or sesquioxide nodules, for example; and (3) the links between microstructure and behaviour are important but little researched and therefore poorly understood.

It is also widely recognised that the clay percentage is difficult to define in the presence of sesquioxides (Fe_2O_3 and Al_2O_3), the figure obtained being highly dependent on pretreatment and on the method of determination. It is, as a result, often a poor guide to engineering behaviour. Terzaghi (1962) and Ruddock (1967) both found clusters or aggregates of clay minerals in larger particle sizes, 'porous aggregate textures' (Collins, 1985). These seem to be associated with sesquioxide coatings on mineral grains that bind them together. Such clusters appear to act as though they were grains with rough surfaces, and impart high void ratios, low compressibility, high permeability, and high angles of shearing resistance to clay-rich saprolites. These properties are very common in ferrallitic and ferruginous (lateritic) soils, as found by Smart (1973), Smart and Tovey (1982) and by Tuncer and Lohnes (1977).

The presence of low strength discontinuities in saprolites (see Chapter 3) is an additional, and very important, factor in landslide activity. Such discontinuities include, planes of schistosity, joint planes including dilatation joints parallel with many hillslopes, horizon discontinuities including the weathering front itself, and old shear planes which mark the occurrence of previous landslides at the same site (see below). Water can move along many of these features and shearing resistance is reduced as a result of polished surfaces, oriented clay platelets, and precipitation of low friction Fe and Mn compounds. Not all of these discontinuities promote unidirectional movement, but they contribute to block failures on very steep slopes and to the occurrence of translational slides in many locations (Franks & Woods, 1993).

The different factors discussed above combine in varying ways to cause considerable heterogeneity within saprolite mantles, and this can be considered on different scales of enquiry, each leading to the recognition of different features:

(1) microfabric studies lead to conclusions about mineral alteration and assemblages, pore sizes and distribution;
(2) profile description recognises differences in rock mass weathering and properties, including the formation of soil horizons, the existence of relict structures, grains, and rock cores, and petrographically determined variations in degrees of weathering;
(3) catenary slope profiles reveal the effects of site drainage, dissection into pre-weathered materials, and formation of both residual and colluvial soil horizons.

Despite the high variability of tropical residual soils, certain properties are widely encountered and have a profound influence on their behaviour.

Repeated failures of old landslides

The occurrence of repeated failures along pre-existing shear planes and affecting old landslides was stressed by Nilson and Turner (1975) from studies in California. Some such instances may reflect the effects of construction (loading, removal of support) in triggering fresh movement on old slides. But the phenomenon appears likely to occur on natural slopes underlain by deep saprolite, where close inspection of cuttings can reveal the size and importance of some of these features (Plate 6-II). It is also important to recognise that many landslides, especially deep-seated slides, continue to move over many centuries, and once rupture has created a shear plane, then movement along this is likely to recur during extreme rainfall events.

The role of colluvium in landslide activity

The term 'colluvium' is imprecise, in both its reference to materials and to processes of formation. The term has been used to describe a broad range of deposits resulting from mass movement and non-channelised flows of water and sediment. Reference was made above to the occurrence of both deep and shallow landslides in colluvium.

Colluvial sediments have frequently resulted from periods of rapid climate change or have accumulated in conditions of drier (or wetter) climates of the past. However, they also result both from high magnitude events within the broadly contemporary climatic system (over 10^2 years rather than 10^3 years), and may in fact represent previous landslide or earthflow debris.

Hillslope hollows and debris flows

Colluvial material concentrated in *hillslope hollows* is frequently involved in, mainly shallow, landslides. Such hollows, and related *swales* in less steep terrain, represent channelless extensions of the present stream channel networks and have been described as 'zero order basins' (Dietrich *et al.*, 1987). According to Reneau and Dietrich (1987), 'most soil-mantled hillslopes in humid and semi-arid landscapes are finely divided into small unchannelled valleys which, because they focus runoff and sediment transport, are the dominant sites of saturated overland flow, periodic gullying and debris flow initiation.'

Attention was drawn to the presence of streamless hollows in the landscape by Hack and Goodlett (1960), but the significance of these features for landslide activity has been emphasised more recently by Dietrich and Dunne (1978), Dietrich and Dorn (1984), Dietrich *et al.* (1986, 1987), Reneau and Dietrich (1987), Reneau *et al.* (1986), Alger and Ellen (1987), Wilson and Dietrich (1987) and in Ellen and Wieczorek (1988), in a series of studies from the Pacific northwest of the USA (Ellen, 1988; Ellen *et al.*, 1988). The role of colluvium in landslide activity was also considered by Iida and Okunishi (1983) in their studies from Japan. Work on colluvial landscapes in Brazil (see Chapter 8) has many links to these findings.

The studies from the Pacific northwest of the USA demonstrate the importance of colluvium-filled hollows as sites for shallow landslides. After prolonged rainfall, rapid debris flows are generated within the colluvium and scour first- and second-order stream channels downslope. Not all of these events scour to the bedrock interface, and pollen analysis has dated charcoal samples from the basal sediments in many hollows to the latest Pleistocene period (typically 14 500–9000 years BP), which led Reneau *et al.* (1986) to suggest that rapid climate change at the end of the last glacial period may have triggered a large number of the more deep-seated slides. Shallow slides, however, can be triggered by rainfall events within the present-day climate, and Dietrich *et al.* (1987) have suggested that on steep slopes ($>50°$) on pumice and ash, the recurrence interval may be only 20–30 years. This probably rises to about 1000 years on 20–40° slopes on granite.

In parallel studies, undertaken following a catastrophic storm on 3–5 January 1982, Ellen *et al.* (1988) suggested that debris flow activity in Marin County had a recurrence interval of 20–100 years, while recurrence at a particular site might range from <33 to >1950 years. In this study 80% of 4600 documented flows occurred on slopes >27.5°; 50% originated in amphitheatres above drainage heads, and 26% were contiguous with previous landslide scars.

The system of hollows functions by concentrating water and sediment flows into concavities on hillsides, and failure occurs only after the build-up of sediment in the hollow has reached a critical thickness. This depth will vary with slope steepness and vegetation cover, but the forms under discussion are all quite small, averaging 7–10 m in width, 10–20 m in length and 0.7–1.1 m in depth of flow, although some larger slides, >2.5 m in depth, accounted for 21% of the volume in the central Californian Coast Ranges (Reneau & Dietrich, 1987). According to Dietrich and Dunne (1978), as soils thicken, root penetration to bedrock becomes progressively more difficult until this stabilisation mechanism is eliminated. However, Tsukamoto and Minematsu (1987) have suggested that vertical roots are deeper on dry 'nose' slopes and lateral roots are more numerous. The excess moisture concentrated in swales and hollows leads to shallow rooting and to a relatively sparse network of lateral roots.

The mechanisms of water flow include *saturated overland flow* which is strongly influenced by exflow from bedrock fractures, and *subsurface pipeflow* (Chapter 5) which, according to Ziemer and Albright (1987), can account for up to 95% of all flow in granite terrain in Japan. Their measurements found flows of 8 m s^{-1} at 2 m depth. This activity is certainly common in tropical environments and it accelerates most forms of denudation. Chemical or solutional removal is probably augmented by biological processes in pipeflow systems. Physiochemical erosion (*suffosion*) develops as chemical processes increase soil macroporosity, allowing water flowing through the soil to detach and move colloidal particles. The importance of subsurface flow relates to the presence of voids and discontinuities in the soil mass and profile.

The particular relevance of this work for tropical geomorphology lies in the widespread availability of thick saprolite layers that furnish colluvium under conditions of accelerated surface erosion. Infilled hollows and gullies are commonly encountered on tropical hillslopes, and there seems to be little doubt that the mechanisms outlined for temperate USA and Japan have also operated in the tropics. This inference can be illustrated from work in Puerto Rico (Jibson, 1989), where multiple slides (debris flows) were triggered by heavy rainfall during 5–8 October 1985. During this time the 24-h rainfall exceeded 400 mm (maximum >560 mm with 4-day rainfall >750 mm), and fell with intensities of 70 mm h^{-1}. This 24-h total represented around 30% of the mean annual rainfall (1400 mm year^{-1}). Jibson (1989) found that the slides concentrated in areas of steep slopes from 38°–50°, and that most (though not all) occurred in pre-existing gullies and depressions. Most slides affected <2 m of colluvium, yet some were quite large, >10 000 m^3, and 600 m long (Figures 6.6 and 6.7).

The wider significance of colluvial slope materials

Colluvium is very widespread in tropical landscapes, and the figure of 50% for the proportion of land in São Paulo State (Brazil) covered by colluvium indicates its importance (Ferreira & Monteiro, 1985). In Hong Kong, where great emphasis has been placed upon the nature of granite weathering profiles, it is nevertheless the case that many slopes which fail are underlain by colluvium which is extensive and may exceed 30 m in thickness (Brand *et al.*, 1985; Franks & Woods, 1993). This material has a high boulder content (of dislodged corestones) and can induce perched water-table conditions that during very heavy rainfalls may trigger, mainly shallow, landslides.

In mountainous terrain, exceptional thicknesses of colluvium may be encountered. Murray and Olsen (1990) described a 200 m thick colluvial accumulation derived from a fluidised slide arising from the catastrophic failure of a ridge in Daria limestone at the Ok Menga hydroelectric scheme, near Tabubil in Papua New Guinea. The authors suggested that a *lower colluvium* at the site is a result of a continuous process of weathering and colluviation that was interrupted (*c.* 8000 BP) by the *upper colluvium* event. The upper colluvium is non-plastic, has a low (<10%) water content, and virtually no clay. Its bulk density is around 2.0 and angles of shearing resistance reach 39–43 °. However, the lower colluvium has a plastic limit of

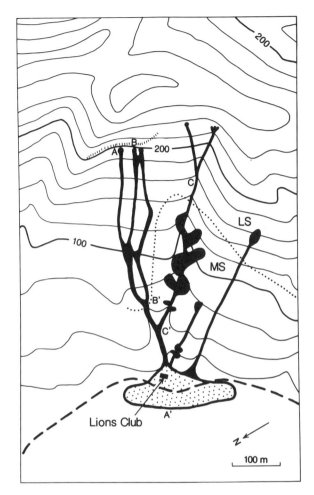

FIGURE 6.6 Debris flows on steep (40°) slopes in limestone, southern Puerto Rico, resulting from cumulative rainfalls exceeding 400 mm in 24 h (6–7 October 1985). (Reproduced by permission from Jibson, 1989)

FIGURE 6.7 Debris flows in the rainforest zone of Puerto Rico, following intense rainfall (photo from Jibson, 1989)

17–25%, with a water content of 20–30% and variable clay. It is a stiff silty clay containing rafted limestone blocks. The upper material tends to break away by slab failure on the upper slopes, probably by stress release but assisted by high rainfalls, while the lower colluvium is subject to shallow slides turning into mudflows after rain.

Complex slope sequences of weathered volcanics and derived colluvium have been identified widely in Java, wherever road traces have been made (Heath & Saroso, 1988). Landslides and groundwater releases are common and creep failure

resulting from saturation of the lowest deposits is widespread. It has been calculated that there are 1.4 slope failures for every 1 km of road in Java (Brand, 1985). In these circumstances, the role of impermeable layers causing a build-up of groundwater and saturated conditions during heavy rain is crucial, and the model used by Whipkey and Kirkby (1978) can be applied. Most of the material is porous, with voids ratios of 0.8–1.2, and also contains fissures and pipes creating a high permeability. The more permeable deposits receive almost continuous groundwater flows from the higher volcanic mountains, and, when 24-h rainfalls approach or exceed 100 mm, the collapse of fissures and pipes quickly leads to translational slides that develop into mudflows towards the base of the slopes, where groundwater flow has become obstructed. Along many road sections the removal of support has led to rotational slumping.

These processes are aided by the presence of smectite clays that comprise 60% of the volcanic ash derived materials and 35% of the brecciated and shale derived colluvium. The total picture is therefore one of rapid weathering of susceptible volcanic and volcano-sedimentary deposits in a climate with equatorial rainfall that can reach 5000 mm year^{-1}. Past periods of slope activation and colluviation were interrupted by new volcanic episodes, and the deposits are now subject to rainfall-induced shallow landslides and mudflows, that are closely related to the stratigraphy and properties of the colluvium.

This suggests, but does not prove, that the stability of regolith on slopes is conditioned by the climatic regime, probably through its effects on soil properties and vegetation canopy. The humid tropical landscape with its abundant vegetation cover can therefore withstand storm sizes and rainfall intensities that are much higher than those which trigger widespread landsliding in temperate areas.

In reporting on hillslope colluvium, gullying and landslides in eastern Brazil, the term 'rampa complex' has been coined to describe colluvium-filled, concave hollows and their detrital aprons and alluvial fans (De Meis & Monteiro, 1979; De Meis & De Moura, 1984; De Oliveira, 1990). The initiation of large hollows in deeply weathered, multi-convex or 'demi orange' topography (Modenesi, 1988) is widespread in the humid tropics. These features are discussed in Chapter 8.

The importance of pore-water pressure

Residual soils usually exist in a state of partial saturation with a continuous air phase in their voids (Geological Society, 1990). This is a reflection of the high void ratio and permeability of many saprolites. The pore-air pressure will approximate to atmospheric pressure, while pore-water pressure will be sub-atmospheric as a result of the capillary effects in the smaller pores. This leads to a

component of effective stress due to *pore-water suction*, and it increases the strength of the material. There is no agreement concerning the importance of this suction in maintaining the stability of natural slopes, but Vaughan (in Geological Society, 1990) suggested that it has a significant if limited effect, the increase in cohesion reflecting about 36% of the pore-water suction value.

The loss of pore-water suction following heavy rainfall may therefore predispose many slopes to failure. But eventually it is the rise of *pore-water pressures* above atmospheric pressure that is the most potent cause of landsliding in the tropics (as elsewhere). The rise in pore-water pressure causes a loss of cohesion, because it reduces normal stress in proportion to the height of the water-table above a point.

Changes in the water content of the saprolite are obviously usually a result of rainfall and they are the most frequent cause of slope failure in the tropics. According to Crozier (1986), the effects of these changes can be summarised:

(1) increasing interstitial *pore-water pressure* decreases *cohesion*, by reducing the capillary and electro-molecular forces binding the material (these can drop from 200 kPa to 0); this takes place when percolation is insufficient and the pore water forms a continuous film, reducing cohesion and adding positive pore-water pressure;

(2) causing *cleft-water pressure* along discontinuities, which is shown to reduce the effect of the angle of internal friction on cohesion;

(3) inducing *seepage pressure* which is a drag stress brought about by water percolation downslope; this may also lead to subterranean erosion, removing support from the overlying mass;

(4) increasing *weight* or loading of the slope is difficult to separate from reductions of normal stress and cohesion resulting from pore-water pressures, but in expansive clays very large amounts of water may be absorbed and then the additional weight becomes a trigger mechanism for failure.

The influence of *lubrication* by water films has been contradicted by most workers in this field, who have shown that water does not act as a lubricant between particles. However, slides which have moved over major discontinuities, where the underlying material is impermeable may have been influenced by water moving across this surface. Also, if saprolite absorbs sufficient water to reach its *liquid limit*, it may convert to a slurry in which water does act as a lubricant.

These principles help to explain why *cumulative precipitation* is so important as a mechanism for triggering landslides, since this leads to a continuous filling of void spaces and to the development of seepage drag as the water moves downslope within the regolith. In the tropics, the very high permeability of the (kaolinitic) mantle with its aggregate structure, also requires very intense precipitation before failure will occur.

The role of slope in landslide activity

In the foregoing section, the role of slope was implied by reference to the incidence of landslides, particularly debris flows, on slopes of different inclination. It is perhaps necessary to comment more generally on this factor, because very few landslides are generated on shallow slopes under natural conditions in warm, humid climates. Apparently stable slopes well in excess of 20° are widely encountered in tropical areas with deep kaolinitic soils, and most of the self-evident slope instability occurs on slopes above 26–28°. The higher of these two figures has been widely observed, in Tanzania (Temple & Rapp, 1972), Japan (Iida & Okunishi, 1983), Sierra Leone (Thomas, 1983), and also in northwestern USA (Ellen *et al.*, 1988). On the other hand, Jibson (1989) found a concentration of debris flows on slopes of 38–50° in Puerto Rico, and the maximum incidence of landsliding in central Japan was found by Iida and Okunishi (1983) to lie between 35° and 40°. The importance of landsliding on steep slopes within the Gunung Mulu National Park, Sabah has been stressed by Day (1982), who found that most large, planar slides on shale bedrock occur on slopes of 40–50°.

The frequency of landslides involving highly weathered rocks probably diminishes on slopes over 45–50°, however, because slopes become *weathering limited* and regolith is removed downslope under gravitational forces before the mantle can 'ripen' to a thickness within which shearing failures can take place. On a global scale such very steep slopes are also less common, but of course they are liable to sudden failure involving rock falls, often as a result of seismic shocks. Slopes of less than 20° cannot always be considered 'safe' during extreme events, and specific instances of slope failure on slopes of lesser value probably result from localised site factors such as bedrock fractures or other planes of weakness combined with subsurface water flows (bedrock springs). In San Mateo County, California, Smith (1988) mapped all areas with concave contours (including all hillslope hollows) with slopes in excess of 20°, prior to a major storm of 3–5 January 1982, and found after the event that this map had a 98% predictive accuracy for the incidence of debris flows.

However, because debris flows normally mobilise only the surface 1–3 m of colluvium, they are especially sensitive to slope conditions. Deep-seated, slump-earthflow slides may respond differently. These features depend more on bedrock conditions or the existence of a deep saprolite, and cannot therefore be favoured by slope class alone. However, most slopes >28° are not underlain by deep saprolite, except in multi-convex, 'demi orange' terrain, where short steep slopes are commonly found. Work in Brazil, detailed in Chapter 8, has demonstrated the incidence of slumping in such terrain.

Many slump earthflow slides occur along plateau edges, escarpments, and in deep valleys, the walls of which become scalloped by this activity. Usually the rock formation is implicated in the landsliding, commonly because of an impermeable layer, above which pore-water pressures are caused to rise after prolonged rainfall. In such cases rainfall intensity is not *per se* likely to be decisive.

Earthquakes as a trigger mechanism

Earthquake zones are of course not specifically related to tropical environments, but the two coincide in important areas of the world, as in Central America and around much of the Pacific Rim. According to Garwood *et al.* (1979), some 18% of the world's rainforests occupy earthquake-prone areas, and these authors invoke earthquake-caused landslides as a major mechanism leading to disturbance of tropical forests.

Simonett (1967) drew attention to the importance of seismicity in causing landslides in Papua New Guinea, where he studied the frequency of landslides within and beyond the seismic zone of the Toricelli Mountains. He found that the volume and number of slides decreased logarithmically with the distance from an epicentre, and that the denudational effect represented a landsurface lowering of approximately 1 mm year^{-1} in the shock zone, whereas beyond the seismic area, rainfall-induced slides effected less than one-quarter the amount of erosion (0.22 mm year^{-1}). Garwood *et al.* (1979) compared events in Papua New Guinea and Panama and found that, in the former, 8–16% of the area was affected by landslides every 100 years (year^2), while in the latter this fell to only 2% year^2 which was matched by erosional slides (3% year^2). Their conclusion, that recovery times (of the rainforest) following major landslides would be 200–500 years, also has important implications for geomorphology, though these figures cannot be used in relation to all or perhaps the most common, shallow slides.

King *et al.* (1989) more recently described the effects of a 7.1 Richter magnitude earthquake on highly weathered rocks on the island of New Britain, Papua New Guinea. Widespread landsliding occurred over a radius of 20–30 km in steep, forested terrain. Shallow translational slides dominated the volcanic rocks to a distance of about 20 km, while large block slides affected the limestones which had been dissected into rugged relief. One slide (the Bairaman Landslide) occurred in weathered limestone and was initiated as a blockslide that immediately broke up into a debris avalanche. A back scarp 200 m high was formed and the debris was estimated at 180×10^6 m^3, accumulating to form a dam 200 m high in the Bairaman River. This allowed a lake to form, eventually containing 50×10^6 m^3 of water.

The dam was breached in a controlled manner after 16 months, producing a large debris flow that caused extensive destruction. The failure was confined to one rock type, a biosparrudite limestone, 'weakened by weathering, faulting and dissection by karstic processes' (King *et al.*, 1989). The humid tropical, mountain environment in this example has probably accelerated deep karstification (Jennings & Bik, 1962), and a more general weakening of an already sheared and brecciated rock by weathering.

Some conclusions about landsliding in the tropics

Several conclusions seem inescapable:

(1) that landslides (*sensu lato*) are an integral and major component of the humid tropical morphogenetic system;
(2) that the properties of tropical residual soils, and especially their high void ratios and infiltration capacities can promote deep landsliding under certain conditions;
(3) that prolonged, intense rainfall events in many tropical climates are the triggers for most landslide occurrences;
(4) that slopes on most lithologies remain stable to >20° even during these events and most slides occur on slopes of 28–45°, possibly with a maximum incidence between 35° and 40°;
(5) that the spatial and temporal frequency of landsliding in the humid tropics must have profound implications for slope evolution in the medium and longer timescales, and for the exposure of rock slabs on residual hills (*inselberge*).

If high mountainous areas (>3000 m) are excluded, then the recurrence interval for major storms appears to be shorter in most humid tropical regions than in non-tropical areas. This certainly appears to be the case in eastern Brazil as well as in cyclone-prone areas such as Hong Kong. Measures of the *intensity* of tropical rainfall (in mm h^{-1}) are probably less important than storm size in explaining the incidence of landsliding under forest, but they can explain dramatic increases in erosion rates and in superficial sliding (mainly debris flows) once forest is cleared.

The incidence of landsliding in seasonal, savanna and deciduous woodland climates appears to be low, and there are relatively few records of events in such areas. This may be because high-magnitude storm events are rare and tend to fall on soils with low antecedent moisture content. This also illustrates the limited effectiveness of short-term, intense rainfalls that are common throughout most tropical areas.

SOIL CREEP

Soil creep is virtually a universal process on all slopes, where unconsolidated materials move downslope in response to small changes in the stress–strain relationships between individual particles or aggregates. Textbook descriptions always allude to wetting and drying or to freeze–thaw cycles as major reasons why soil materials move downslope under the influence of gravity, but a more continuous, 'viscous' creep has been described by Van Asch *et al.* (1989). This process does not require any shrink–swell mechanism for its initiation. Creep processes are, however, certainly rapid in cold, montane climates as a consequence of frequent freeze–thaw cycles (Saunders & Young, 1983), and where they often merge into solifluction or soil flowage.

In tropical and sub-tropical climates frost is largely absent, but variations in moisture within the near-surface soil layers may be frequent, due to the pattern of short intense storms interspersed with dry weather. However, under a forest cover, surface soil moisture levels may be maintained by high infiltration, reduced air flow and evaporation, natural mulching, high clay content (>20%), and by capillary forces. Most clays on well-drained hillslopes will also be dominated by non-swelling kaolinite or halloysite. The 2:1 clays of the smectite group will occur mainly in semi-arid regions or in footslope sites.

Rates of soil creep in the tropics

Rates of soil creep measured in the tropics remain few, and certainly not sufficient to offer representative figures. Nevertheless, some relatively high figures have been obtained. Eyles and Ho (1970) measured 12.4 cm^3 cm^{-1} year^{-1} for a 10° slope under rainforest in West Malaysia, and they quoted a mean rate for studies in Northern Australia of 4.4 cm^3 cm^{-1} year^{-1} on a 15° slope. This is compatible with the figure obtained by Williams (1973) of 7.5 ± 5 cm^3 cm^{-1} year^{-1} from the same region. Lewis (1976) has provided figures from Puerto Rico, where, for 17–20° slopes, creep ranges from 7–9 cm^3 cm^{-1} year^{-1} under forest, to 3–4.6 cm^3 cm^{-1} year^{-1} under savanna, and only 1.5 cm^3 cm^{-1} year^{-1} in the semi-arid zone, but the slope here was only 8° (Lewis, 1975). These figures may not be strictly representative, but they do indicate, along with some other observations, the comparatively rapid rates of creep in humid tropical climates.

In terms of surface movement, the volumetric figures translate to <3–4.5 mm + year^{-1}. Initially, the rate of creep declined with depth and at 15–25 cm, most rates have reduced to 20–25% of the above figures. Under forest, however, creep then accelerated to a second peak at around 24 cm depth, where the rate of soil movement was 40–80+% of the surface creep. This pattern was not replicated on dry, savanna sites, where soil movement declined sharply below 10 cm. Lewis (1976) recognised that a different process must be responsible for increased movement at depth, and he considered that it indicated the effects of throughflow and slow, subsurface shearing.

In Rwanda, Moeyersons (1989a) was able to corroborate these observations and he obtained figures equivalent to 2.8–7.1 mm year^{-1} in surface soil on steep (40°) slopes, but a maximum of 44.3 mm year^{-1} beneath the surface. Moeyersons (1981, 1989a,b) noted the association of soil creep with terracette formation and shallow slides on Rwaza Hill in Rwanda, and pointed out that the terracettes represent superficial slips and accelerated creep movements. Slope failure in this location takes place mainly at the end of the rains (1100 mm year^{-1}), with the formation of 'bottle slides' surrounded by terracettes. These terracettes appear to comprise soil pillars bent forward and separated by curved fissures (shear surfaces) marked by thin zones of crumbled earth (attributed to rapid creep). On slopes >25°, the angle of internal friction (or shearing resistance) may start at 40° with a moisture content of only 10%, leaving the slope stable. As the moisture content advances to 40% this angle is reduced to 30°, but slopes at lower angles (23–28°) also failed. Moeyersons (1989a,b, 1990) attributed this behaviour to the saturation of a gravelly layer at depths of 10–50 cm. Movement along this surface was partly by shearing between blocks or slabs of soil material and partly between aggregates and particles, indicating renewed creep at this depth.

Simon *et al.* (1990) similarly observed many shallow slope failures in the humid areas (2500–4500 mm year^{-1}) of Puerto Rico, and referred to the tensile stresses occurring at around 0.5 m depth in soils subject to creep. But these authors also recorded discontinuous density contrasts at 3–7 m depth in saprolites >20 m deep. They reasoned that perched water-tables occur at 0.5 m, corresponding with a soil horizon containing maximum clay (translocated from the surface), and at 3–7 m at the saprolite–weathered rock boundary (Zone IV/III boundary in Figure 2.1). The shear strength of the soils reached a maximum in the root zone around 25 cm depth, falling to a minimum above the clay horizon. In its unsaturated state the soil experiences negative pore-water pressures and tension cracks 1–2 m deep develop. During rain storms (>50 mm) infiltration is enhanced by the tension cracks and pore-water pressures rise to peak levels after 12–96 h in soils over diorite. These factors led to shallow slides and soil failures concentrated in mid-slope hollows (called 'cove' slopes by Simon *et al.*, 1990) where moisture is concentrated.

Because creep processes (but not the shallow slides with which they may also be associated) are generalised and respond to low magnitude events or may even be continuous, they are less noticeable in moulding the landscape. But, though few specific landforms are associated with creep processes, they are none the less very important in soil and slope evolution, and provide a non-disruptive conveyor system for residual materials, departing from interfluves and ultimately entering the valley floors and stream systems.

PART III

QUATERNARY ENVIRONMENTAL CHANGE AND LANDFORM DEVELOPMENT

7

Quaternary Environmental Change in the Tropics

THE EVIDENCE FOR ENVIRONMENTAL CHANGE

There are few areas of study affecting the geomorphology of the tropics that have advanced so fast during the last 20 years as research into Quaternary climate change. Yet this topic still abounds with uncertainty and controversy, and much of our understanding is limited to the period of the last interglacial–glacial cycle (150 000 years) and rests on evidence from polar ice sheets, as well as from ocean cores in both tropical and non-tropical latitudes. This situation creates a gap between the models that determine thinking in this field and the evidence on land, within the tropics, for the impact of major climate changes. Discrepancies persist between model predictions of temperature and precipitation changes for tropical latitudes, and the empirical evidence that is available. Nevertheless, the level of corroboration between different strands of evidence: from ice cores, from ocean cores and from lakes and mires on land remains remarkable, and when individual studies claim that local temperatures or rainfalls did not reflect the supposed global pattern of change, such observations should be placed against the wider framework before deciding on their significance. However, some of the evidence for environmental changes derived from sediments or landforms in scattered sites throughout the tropics may not necessarily imply *climate* change, and the reader must guard against circularity of argument.

A much greater limitation on our understanding arises from the lack of accurately dated evidence beyond the range of radiocarbon dating (40 000 BP), for the impact of earlier oscillations of climate on landforms and sediments in the tropics. Even within the span of the major glaciations of the northern hemisphere (c. 700 000 years), we have very little detailed information beyond the last glacial cycle (115 000 years), and most of what is available comes from d^{18}O analyses of deep sea cores, uranium series analyses of coral reefs, and isolated lake sediments. The preponderance of information comes from the last 20 000–40 000 years of radiocarbon datable sequences (a few long range AMS ^{14}C dates extend to c. 60 000 BP). At the time of writing, a new understanding of sedimentation, especially in fluvial environments, is beginning to appear as a result of thermo-luminescence dating, but it is not yet possible to provide a new framework from these results. None the less, Nanson *et al.* (1991) have dated a major pluvial episode in arid northern Australia to 120 000–85 000 BP, using both TL and uranium series dating, which indicates enormous promise for this technique.

An impression will therefore be conveyed of landscapes containing many older features that are largely undated and fragmentary in occurrence, but also a plethora of Late Quaternary forms and deposits. This cannot be the full picture and our perspectives may be distorted by how little we really know of the earlier periods. Nevertheless, the tropical landscape is a mosaic of forms and deposits, many of which are fragile and ephemeral when viewed on longer timescales and they belong predominantly to the last 20 000–40 000 years of

TABLE 7.1 Sources of evidence for palaeoclimatic reconstruction

Continental	Marine
1. *Palaeobiological evidence (from cores and sections)*	
fossil pollen and spores	diatoms
fresh-water diatoms	foraminifera
plant macrofossils	reef corals
vertebrate fossils	
invertebrate fossils (mollusca, ostracods)	
2. *Biogeographical evidence (from present-day distributions)*	
plant and animal distributions	
species diversity in ecosystems	
3. *Palaeo-sedimentology*	
lacustrine sediments	
alluvial sediments	
colluvial sediments (e.g. landslides, mudflows)	
eolian sediments (dunes, sand plains, loess)	
stone-lines and surface gravels	
soil stratigraphy	
evaporites and tufas	
glacial and periglacial sediments	
tephra (volcanic ash falls)	
	continental dust
	fluviatile inputs
	beach deposits, shorelines
	biogenic dust (as above, 1)
4. *Palaeo-landforms*	
palaeo-karst	
glacis and stone pavements	
alluvial fans	
valley systems and gullies	
river terraces	
weathering phenomena (tafoni, pseudo-karst, cores)	
5. *Pedogenic phenomena*	
duricrusts (calcrete, silcrete, ferricrete)	
palaeosols (fossil soils)	
pedogenic phenomena (clays, clay cutans, etc.)	
6. *Archaeological materials*	
plant and animal remains	
hominid remains	
artefacts (bone, stone, wood, shell, leather)	
hearths, dwellings, workshops	
rock art	
7. *Isotope geochemistry*	
oxygen isotopes (ice, marine foraminifera, carbonate)	
carbon isotopes (plants, bones, carbonate)	
hydrogen isotopes (groundwater, peat)	

Modified from Williams (1985).

earth history. Other parts of this mosaic, however, are much older and the superficial deposits and soils are superimposed or imprinted on to weathered mantles and other deposits that record not only the rest of the Quaternary but also reach back through the Tertiary to the Mesozoic. Discussion of these longer term changes will be taken up in Chapter 11.

There are many different categories of evidence for environmental change and some sources for palaeoclimatic reconstruction are listed in Table 7.1, which has been modified from Williams (1985). A good review of our current knowledge in this area is provided by Crowley and North (1991).

Many of these phenomena are not strictly diagnostic of any single climatic condition and most are highly sensitive to local site conditions or to the *context* of the site, which means, for example, the location of a feature in relation to surrounding hills; its catenary position and its groundwater regime. Williams (1985) warned strongly against attaching climatic signals to the presence of calcrete, ferricrete and other pedogenetic phenomena that are strongly influenced by groundwater conditions, which may be controlled by other factors. Nevertheless, when studied in context, such phenomena may be useful. Similarly, alluvial fans may form in mountain areas under humid conditions, but will represent signs of aridity elsewhere. Pitfalls are attached to interpretations of almost all the features listed in Table 7.1 and some problems of palynology are touched on below. Confidence in an interpretation grows when a *consistent record* emerges from many different sites where a particular phenomenon is found, and when *corroboration* is provided from several different strands of evidence. However, in the tropics, the problems of dealing with isolated and geographically distant sites remain very serious (Colinvaux, 1987). Because palaeoenvironmental reconstruction is inherently interdisciplinary, and is pursued by specialists who may be climate modellers or palynologists, marine biologists, sedimentologists, pedologists or geomorphologists, there will always be difficulties in matching the evidence and understanding the subtleties of argument amongst the contributors to a debate.

Geomorphology and environmental change

In many respects the geomorphologist is at the end of a line of argument, beginning with the calculation of orbital forcing mechanisms, and the formulation of global climate models (GCMs), and then supported by evidence from ocean sediment and ice-cores providing continuous records of sedimentation. Comparable long-term records are rare on land, and, where cores are taken from lake beds and swamps, the integration of knowledge about the sedimentary history of a site with the erosional history of its catchment is often inadequate. However, it is this aim, which can be expressed as the attempt to provide models for the behaviour of morphogenetic (and pedogenetic) systems during oscillations of climate and changes in the vegetation cover, that is a major concern of the geomorphologist. This connection, between erosion and sedimentation, is essential for the sciences of geomorphology and pedology and also for the successful management of natural environmental systems, yet it is often neglected or offered as a footnote to ever more detailed constructions of new chronologies.

The recognition that climates within the tropics underwent changes during the Quaternary which may have been comparable in magnitude with those that affected the formerly glaciated regions of the world came slowly, and many doubts remain concerning the geographical patterns of change on the one hand and the severity of the changes (reductions of temperature and rainfall) on the other. It has to be admitted that much more progress has been made on the *timing* of climate changes than on their impact on surface processes.

There was an early appreciation of the importance of Pleistocene palaeoclimates in the desert and semi-arid regions of the world, but many of the earlier studies assumed that the equatorial climates remained stable, forming a core of rainforest environments throughout the glacial/interglacial events of the higher latitudes. Such a view is understandable, because of the awe inspired by the complexity and majesty of the tropical rainforests. Surely these must have been in place for millions of years? On a more prosaic level, these forests obscured the geomorphology of the humid tropics; few scientists ventured to study their soils and surface deposits, except on their drier margins, and for a long time speculation remained free from confrontation by uncomfortable facts.

DEVELOPMENT OF RESEARCH ON THE TROPICAL QUATERNARY

Although there was discussion concerning the relationship of northern glaciations to the formation of pluvial lakes in east Africa in the early decades of the 20th Century, this was centred on the idea that these phenomena were in phase (Wayland, 1929, 1952; Nilsson, 1931). Zeuner (1959) questioned this assumption and both Cooke (1958) and Flint (1959a,b), challenged the idea that tropical pluvial and glacial conditions corresponded.

The possibility that the humid tropics had been subjected to dry episodes during the Quaternary was implied by studies of superficial deposits in Ghana undertaken by Brückner (1955, 1957), while Grove (1958) drew attention to the widespread evidence for past desert conditions in the Sahelian and Sudanian zones of west Africa, and pioneered the study of both humid and arid climatic oscillations as they had affected the Lake Chad basin. Evidence for Pleistocene aridity in the rainforest areas of the Congo/Zaire basin was advanced by De Heinzelin (1952) and De Ploey (1965). In eastern Brazil, Ab'Saber (1957), Bigarella and Salamuni (1958, 1962) and Bigarella and Andrade (1965) all concluded from their detailed work on pediments, palaeopavements and sediments, that profound aridity had affected this humid part of South America during the Pleistocene. However, Cailleux and Tricart (see Tricart, 1985), initially, did not think that the Amazon basin had been subjected to arid conditions, and Tricart (1974) concentrated on the presence of ancient dunes in the Llanos of Venezuela and in the Sao-Francisco Valley of Bahia State in the present-day dry zone of northeast Brazil. Subsequently, he has altered his views in favour of dry conditions in central Amazonia during the Pleistocene (Tricart, 1975, 1985). Journaux (1975) has also drawn attention to the importance of climate change in determining slope forms in this region.

Aubreville, in 1962, reviewed the ecological evidence for anomalous distributions of forest and savanna, and concluded that widespread *savannisation* had occurred in the humid tropics during the glacial maxima. The details of his model for climate change have been superseded, but this study remains an important marker for the development of our understanding of climate change in the tropics. Support for important vegetational changes in northern South America also came from palynological studies by Van der Hammen and Gonzalez (1960), Wijmstra and Van de Hammen (1966) and Van der Hammen (1972, 1974).

By the mid 1960s a thorough reappraisal of late Pleistocene and Holocene climates in Africa and other tropical regions was under way (Coetzee, 1967; Van Zinderen Bakker, 1967), and the extent and severity of African ice-age aridity was clearly stated by Fairbridge (1964) and documented for the western Congo basin by De Ploey (1965), and also by, the now carefully dated, lacustrine sequences in east Africa (Butzer *et al.*, 1972). In 1969, Haffer first advanced the influential but controversial *refugia theory*, to explain disjunct distributions of modern taxa in Amazonia, arguing for widespread semi-arid conditions with the survival of small enclaves of rainforest during Quaternary cool/dry periods. This conclusion was supported by independent recognition that Atlantic deep-sea arkosic sands off the northern coast of South America, represent the influx of feldspathic material during glacial cold/dry conditions affecting the shield areas north and south of the Amazon River (Damuth & Fairbridge, 1970). Similar conclusions regarding central Africa were offered by Hamilton (1976). Dry conditions during glacial periods were deduced from similar evidence for the central Caribbean by Bonatti and Gartner (1973).

The Quaternary of southeast Asia received less attention at this time, although interest in the former land bridges resulting from exposure of the Sunda and Sahul shelves, linking the islands of Sumatra, Borneo and Java with the Malaysian peninsula during glacial low sea levels, was discussed by Mayr (1944) and Fairbridge (1961). A wider discussion has only ensued following work by Haile (1971), Aleva (1973) and the regional appraisals by Verstappen (1975, 1980).

During the last 20 years, there has been an explosion of research and publication on Quaternary climate change. But the problems confronting palaeoenvironmental interpretations in the tropics remain serious. The number of sites for which long-term chronological records are available is small, and few of these lie in the humid tropics. There is still an over-dependence on evidence drawn from lake basins of interior drainage in the semi-arid zone (Street-Perrott *et al.*, 1985). Inland lakes are relatively rare in the humid tropics and studies of sediments and pollen from humid tropical lakes and mires remain few. Even recent work on deep sediment sea pollen, diatom and phytolith records (Diester-Haas, 1976; Pokras & Mix, 1985; Jansen & Van Iperen, 1991; Lézine, 1991) depends on comparisons between isolated cores from widely different locations and environments. Furthermore, an adequate theoretical model (GCM) to explain or predict apparently synchronous *global* shifts in temperature has proved difficult to construct, and climate simulations using existing GCM data sometimes appear to contradict the empirical evidence (Prell & Kutzbach, 1987), particularly with regard to temperature depression in the intertropical zone. In the absence of such a model, good evidence for the depression of snowlines on tropical mountains (Flenley, 1979, 1985; Mahaney, 1988, 1989) has not always led to agreement about the changes of climate in tropical lowlands (Colinvaux, 1987). Clapperton (1993a,b) has recently reviewed much of the evidence from South America, while Hamilton and Taylor (1991) have discussed the history of climate and forests in tropical Africa. Williams (1985) has pointed to one difficulty that confronts all palaeoclimatological reconstruction, namely that geologists and biologists who produce the evidence usually lack an understanding of climatology, while the climate modellers are often unable to assess the evidence from studies of palynology or sedimentology.

Amongst the more contentious issues which remain is the status and interpretation of the forest refugia theory of Haffer (1969, 1987) which has been the subject of much discussion (Livingstone, 1982; Whitmore, 1987; Maley, 1991; Kershaw, 1992; Hamilton and Taylor, 1990). From a geomorphological standpoint the transition from forest towards more open woodland or savanna conditions has great significance for the course of weathering and erosional processes (Thomas & Thorp, 1992), and therefore the changing pattern of such transitions during the oscillations of late Quaternary climates in the tropics should have left a durable imprint on soils, landforms and deposits. In central Africa, Colyn *et al.* (1991) have claimed confirmation

of a major refuge area in the central Zaire basin which has long been postulated (Livingstone, 1975). However Runge (1993) has also found evidence for major environmental disturbance in the eastern Zaire basin which suggests that the extent and boundaries of possible refugia, as well as the severity and duration of forest degradation elsewhere remain important topics for future investigation.

Notwithstanding these problems, a consensus is emerging, with agreement that the lowland tropics experienced temperature reductions during glacial maxima of perhaps $4 \pm 2\,°C$. Concomitant with these changes were widespread reductions in rainfall, in the order of 30%, but opinions still differ concerning the severity of the aridity, its persistence and geographical extent. In the absence of lacustrine deposits over wide areas, evidence for aridity must come from the few available lake sites, from peat deposits, and from geomorphological and sedimentological features. Before detailing this evidence, the inferences made from modelling exercises are considered.

Glacial cycles and tropical climates— inferences from models

The detailed study of glacial cycles in earth history is beyond the scope of this book, but the main outlines of the argument are relevant. In 1941, Milankovich postulated that summer insolation anomalies at 65 °N controlled ice-sheet oscillations, and were a consequence of orbital variations. These showed strongest power at the 41 000 year period *tilt* and 23 000/19 000 year *precession*, but much less power for the 94 000 and 125 000 year orbital *eccentricity* periods. In 1955, Emiliani established a link between the earth's orbital parameters and the timing of glacial periods. Since then, examinations of the oxygen-isotope records from ocean sediment cores have confirmed the relationship of ice volume changes to oxygen-isotope variations (Shackleton & Opdyke, 1973, 1976; Shackleton, 1987), though the signal is also complex. These records contain clear evidence for the 41 000-year and 23 000/19 000-year cyclicity required by orbital forcing of the earth's climate system (Hays *et al.* 1976, Imbrie, 1985). However, Broecker and

Denton (1990) have pointed out that, although the 100 000-year periodicity is dominant in the record, this dominance is not predicted by orbital forcing.

Each 100 000-year cycle reveals a slow build-up of $\delta^{18}O$, followed by an abrupt decrease, described as a *termination*, and marked by sudden de-glaciations. A range of evidence shows that the last termination commenced at around 14 000 BP. Such terminations appear to have world-wide significance on a 100 000-year cycle corresponding with the eccentricity maxima in the earth's orbit. However, to translate this into a mechanism to drive climate change, many, so far untestable, assumptions must be made, and modellers disagree on some fundamental aspects of this problem (Crowley & North, 1991). In most models, the insolation spectra are first used to predict an ice-sheet model involving glacial isostatic lag, and then the impact of ice-sheet growth and decay is transmitted via the atmosphere to account for known variations in sea-surface temperatures. However, Broecker and Denton (1990) have argued that the Milankovitch summer insolation variations at 65 °N cannot by this means be translated into a global climate signal, and quote Menabe and Broccoli (1985) to the effect that climate models based on this premise restrict change to the northern hemisphere, and fail to predict an adequate growth of Antarctic ice. The synchroneity of de-glaciations world-wide at the 14 000 BP termination also precludes any lag effect between hemispheres, while the similar behaviour of high- and mid-latitude mountain glaciers cannot be explained by polar insolation anomalies.

The 23 000/19 000-year periodicity is confirmed by sea-level maxima recorded in coral reefs in Barbados, where peaks occurred at 124 000, 107 000 and 84 000 BP (Mesolella *et al.*, 1969). Similar evidence from the Huon Peninsula in Papua New Guinea (Bloom *et al.*, 1974; Chappell, 1974) confirmed these peaks and also lesser ones at around 62 000, 47 000 and 30 000 BP. However, these maxima are not directly attributable to temperature changes in ocean water, which were considered by the CLIMAP (1981) reconstruction to lie within the range 0–2 °C, a figure corroborated by oxygen-isotope studies of marine foraminifera by Broecker (1986). Broecker and Denton (1990)

have pointed out that this figure is not consistent with much larger reductions in glacial temperature proposed for tropical high mountains (Flenley, 1979; Hamilton, 1982), which are typically in the range 5–9 °C, because such figures would require a greatly increased lapse rate for which there is no theoretical justification. Moreover, important changes in ocean water characteristics (salinity, temperature, biological productivity) have been dated to the last termination, between 14 000 BP and 12 000 BP, corresponding with the rapid meltback of high latitude glaciers after 14 000 BP.

Broecker and Denton (1990) also drew attention to the importance of the *Younger Dryas Cold Event* (11 000–10 000 BP) which, though short-lived, was of a similar magnitude to the previous glacial, but cannot be explained by orbital forcing. Furthermore, the much earlier *Dansgaard-Oeschger Events*, recorded in Greenland between 55 000 and 25 000 BP may have been of comparable magnitude. Broecker and Denton (1990, p. 325) summarised their findings: 'these events suggest that, rather than smoothly following the orbital drive, climate changes in sharp jumps'. They argued that the last termination at 14 000 BP which was synchronous in both hemispheres and in mid-latitudes must have resulted, not from lagged responses in the North Polar ice sheets resulting from isostasy, but from a background climatic event. According to their analysis this depended on the reorganisation of ocean water circulation, and the role of North Atlantic Deep Water (NADW) that acts as a conveyor belt for heat transfer towards the poles, supplementing annual insolation by 25%. Evidence of much lower North Atlantic temperatures in glacial time suggests the weakening or elimination of this heat conveyor which also generates water vapour transfer to both Indian and Pacific Oceans and controls salinity differences between ocean basins. For the Younger Dryas event (11 000–10 000 BP) it is argued that diversion of meltwater from the Laurentide Ice Sheet from the Mississippi to the St Lawrence would have created fresh cold water in the North Atlantic, preventing the formation of NADW and halting the conveyor. Whatever the mechanisms involved, Broecker and Denton (1990) argued for rapid reorganisation of the ocean–atmosphere system at each 100 000-year termination. Not all perturbations in ocean sedimentation, however, can be traced to climatic oscillations and the lithic sediments in the northern Atlantic, thought to result from greatly increased ice rafting in the North Atlantic after 60 000 BP and known as *Heinrich events* (Heinrich, 1988), may have resulted from surges in the Laurentide ice sheet rather than from climate changes (Broecker *et al.*, 1992).

Although cause and effect remain ambiguous, ice ages appear to involve reductions in ocean heat transport and in the greenhouse gases, CO_2 and CH_4, and also an enhanced reflectivity of the atmosphere, resulting from increases in atmospheric dust in both northern (Tundra) and mid-latitude (desert) regions. For these to coincide, a change in the mode of ocean–atmosphere operation may be required (Broecker & Denton, 1990). Kutzbach (1976) has also pointed out that orbital parameters are unable to explain all the conditions of climate observed at 18 000 years BP, a position supported by Peters and Tetzlaff (1990).

The relevance of this debate for geomorphology lies not only in the rhythm of warm/cold oscillations of climate, but also in the changes taking place to other environmental parameters and in their geographic distribution. Global, ice-age depressions of temperature and reductions in rainfall appear well established (Crowley & North, 1991), but the proposed changes in ocean–atmosphere circulation are also required to explain the severe aridity of the lands bordering the Atlantic and Indian Oceans, the apparently increased severity of Northeast Trades in West Africa and other secondary mechanisms leading to climate change in tropical areas. Lack of agreement between predicted sea surface temperatures (SSTs) in the South Atlantic during glacial periods ($-1\,°C$ to $-2\,°C$) and larger temperature depressions on land ($4 \pm 2\,°C$) could be a result of deep upwelling (Benguela Current) that affected the Atlantic Ocean as far as $30\,°W$ longitude. Vincent *et al.* (1989) also suggested the invasion of Antarctic bottom water into the Gulf of Guinea, leading to SSTs of -3 to $-5\,°C$. The Atlantic monsoon that brings rainfall to west and central Africa would have had to traverse this cold

water, and it has been suggested that evaporation losses could have been reduced by 70% in these circumstances.

Lautenschlager (1991), working with a GCM known as T21, also found discrepancies in tropical latitudes and noted that the model did not include the effects of tropical mountain glaciations. But the model does predict an increase in the Northeast Trade Winds in coastal regions by 30–50%, leading to deep upwelling and cooling of SSTs. This in turn could have reduced evaporation in the Eastern Pacific by 20–30% and rainfall by a similar amount over land. Lautenschlager (1991) calculated that precipitation reductions over the Congo/Zaire Basin could have been 29%, but perhaps only 10% in the Amazon Basin. Different views have been expressed on the basis of empirical data (below).

There has been considerable discussion of the variability of the African–Indian Monsoon, and Prell and Kutzbach (1987, p. 8411) have asserted that, 'in the tropics the seasonal monsoons dominate climatic variability on time scales from the annual cycle to glacial–interglacial cycles.' They maintained that variations in the strength of the monsoon, understood to define 'large-scale, seasonal changes of wind and precipitation patterns', are largely responsible for the climate change signals coming from the study of palaeolakes, pollen profiles and deep sea cores. GCM simulations appear also to predict a weakened monsoon during the last glacial maximum (LGM) at around 18 000 BP, and a strengthened monsoon during periods of increased solar radiation, as at 9000 BP.

The conclusion that an increase in the strength of the Northeast Trade Winds will result in a weakened monsoonal circulation (in the northern hemisphere) may not appear immediately obvious. However, Menabe and Hahn (1977) argued that, while predicted changes to SSTs during an ice age could not explain the necessary changes to the penetration of monsoon rains in India (and West Africa), increased continental *albedo* could. Extensive ice sheets, snow, and meagre vegetation, especially over higher latitude plateaus and mountains, would reduce tropospheric air temperatures and create a strong pressure gradient from land to sea (see Sirocko *et al.*, 1993).

Strong and persistent anticyclonic conditions over the mid-latitude deserts, such as the Sahara, during the summer period, also effectively blocked the advance of the monsoonal rains. According to several other writers (Maley, 1976; Rognon & Williams, 1977; Nicholson & Flohn, 1980) this may have resulted from influxes of dry tropospheric air, derived from the cold, ice-covered high latitudes. These arguments may not apply so readily to South America, where the ITCZ, if evident, crosses the Guiana Highlands and Venezuela. Colinvaux (1987) pointed out that southward displacement of the ITCZ in a glacial climate should *increase* the precipitation in the northern Amazon Basin. However, the information which we have about the Amazon Basin appears to contradict this proposition, while evidence for dry glacial climates along the east coast of Brazil must be a response to important changes in ocean circulation (Ab'Saber, 1982).

Street-Perrott and Perrott (1990) concluded that the many abrupt climate fluctuations which have occurred in the tropics during the last 14 000 years (see Table 7.3) require a mechanism capable of responding over periods as short as 10^1–10^2 years. Their extensive review concluded that fluctuations in the production of the NADW which controls the Indian–Atlantic Ocean heat conveyor holds the key to this phenomenon, and that during the late Glacial this was controlled by influxes of fresh meltwater either into the North Atlantic, halting NADW production, or into the Gulf of Mexico, restarting the system. But whilst the correlations between cold SST anomalies in the North Atlantic and low rainfalls in the African tropics appear sound, the meltwater mechanism cannot explain the mid and late Holocene fluctuations in rainfall.

It can therefore be seen that tropical climates were affected by complex circulatory mechanisms that are not yet well understood, and while the changes to these circulations were probably synchronous, their effects were differentiated according to the configuration of ocean basins and continents, altitude and distance from the sea. It would be wrong in these circumstances to expect a uniformity of response in the ecological and geomorphological systems from all parts of the tropics. It is, therefore,

all the more remarkable that striking similarities are found in widely scattered locations.

Glacial cycles and tropical climates— empirical evidence

Data on the abundance of freshwater diatoms in three deep-sea, Atlantic cores from the equatorial Atlantic were compiled by Pokras and Mix (1985), who concluded that high numbers of diatoms in the cores are an indication of deflation from dry African lake beds during periods of aridity. These periods were identified as occurring 115 000–110 000, 95 000–84 000, 68 000–60 000, and 23 000–15 000 BP. Jansen and Van Iperen (1991) examined cores from the Zaire/Congo River deep-sea fan and used an index of aridity based on the relative abundance of opal phytoliths (Ph) (siliceous clasts from vascular plants, mainly grasses), wind transported and freshwater diatoms (FD). They defined a *PhFD Index* (Ph/(Ph + FD)) as a signal of wind versus fluvial transport (high PhFD = reduced diatom production and/or increased wind transport) and found that it mirrors aridity in the equatorial tropics. Their climatic interpretation is shown in Figure 7.1 and in Table 7.2.

Similar conclusions, derived from the analysis of pollen spectra over 140 000 years, were advanced by Hooghiemstra (1989), Hooghiemstra and Agwu (1989), and these have been largely confirmed from the study of a deep core off the northwest African coast (Latitude 13°50′ N) by Lézine (1991). Lézine used an Arid Pollen Index (API) and a Humid Pollen Index (HPI) to define periods of drier or wetter climates in the sensitive (Sudanian) zone lying between the forests and the open Sahelian scrublands (Table 7.2).

A further important line of evidence has come from the *sapropel muds* found in the eastern Mediterranean. These are considered to result from influxes of fresh water and fine sediment from the Nile, during periods of enhanced discharge. Such periods are assumed to reflect the climates in the headwaters of the Nile River system, in Ethiopia and in Uganda. The sapropels have been dated to 129 000, 104 000, 85 000, and a double peak at 11 000 and 9000 BP (Rossignol-Strick *et al.*, 1982; Rossignol-Strick, 1985) (Table 7.2).

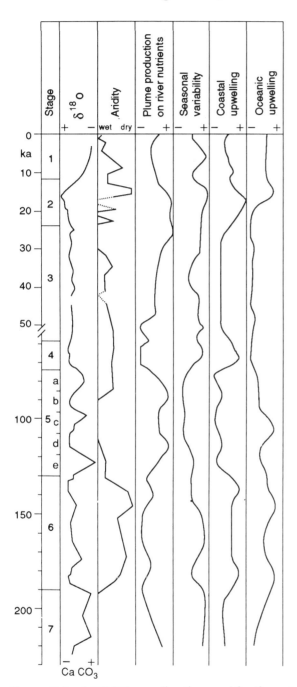

FIGURE 7.1 A 222 000 year climatic record for the east equatorial Atlantic Ocean and equatorial Africa. Evidence from diatoms and opal phytoliths in the Zaire (Congo) deep sea fan, showing oxygen isotope stages, indices of aridity, seasonality and coastal upwelling (after Jansen & Van Iperen, 1991, copyright by the American Geophysical Union)

During the Last Glacial Cycle, interglacial warmth gave way to glacial conditions in high latitudes after 150 000 years BP, but there were warmer and wetter punctuations in the tropical latitudes at 125 000/129 000, 103 000/104 000, 80 000/85 000 and possibly *c.* 47 000 BP, though no sapropel was identified of this age. With the end of the last glacial, humid conditions returned after 12 000 BP. These different constructions are detailed in Table 7.2.

It should be noted that radiation maxima in the northern hemisphere occurred at 10 000, 82 000, 104 000, and 126 000 BP, and that between 75 000 and 15 000 BP there is less coherent variability. Subdued peaks occur at *c.* 60 000 and 35 000 BP. When oxygen isotope ($\delta^{18}O$) temperature and CO_2 records (Barnola *et al.*, 1987) based on ice cores and ocean sediments are compared (Jouzel *et al.*, 1987, 1993; Sundqist, 1987 and Figure 7.2), there is some correspondence with summer solar radiation curves, showing peaks at 58 000/55 000 BP and 40 000 BP, for example. Lézine (1991) compared her humid pollen index (HPI) with Arabian Sea records (Prell & Van Campo, 1986) of cold water upwelling in response to the monsoon (MUI), demonstrating a correspondence of HPI and MUI maxima at 52 000/44 000 BP, with a peak centred on 47 000 BP. This signal may be important in the geo-morphological record, following as it does a marked trough in solar radiation with strengthened Trade Winds at 68 000/60 000 BP. Long range ^{14}C dates from Amazonia and Kalimantan suggest important sedimentation events during the period 56 000 BP onwards (Thorp *et al.*, 1990; Van der Hammen *et al.*, 1992a,b), while at a coastal site in Perak, Malaysia, continuous sedimentation from before 67 000 BP is implied by AMS and TL dating (Kamaludin *et al.*, 1993).

There is evidence for an onset of late Pleistocene cooling by 40 000 BP, and cool moist climates probably prevailed widely after 26 000/25 000 BP (Gasse *et al.*, 1989; Bonnefille *et al.*, 1990; Van der Hammen *et al.*, 1992a,b; Vincens, 1991), but the impact of full glacial conditions on tropical climates is most apparent between 22 000/20 000 BP and 14 000/12 000 BP, when widespread cooling

TABLE 7.2 Ocean sediment signals of Quaternary climatic changes

API Max. HPI Max. (Lézine, 1991)		Fresh - water diatom max. (Pokras & Mix, 1985)	PhFD Index (Jansen & Van Iperen, 1991)	Trade - wind Max. (Hooghiemstra 1989)	Sapropels (Rossignol - Strick, 1985)
			145,000	140,000	
130,000				130,000	129,000
	125,000			120,000	
120,000		115,000			
		110,000	110,000		
105,000	103,000			106,000	104,000
94,000		95,000		96,000	
		84,000	85,000	84,000	85,000
	80,000			76,000	
		68,000	72,000		
60,000		60,000	62,000	58,000	
	47,000			40,000	
			35,000		
32,000					
25,000		23,000			
17,000		15,000			
9,000?	<12,000		12,000	12,000	11,000
			6,000		

*API = Arid Pollen Index; HPI = Humid Pollen Index (after Lézine, 1991); PhFD = opal phytolith/freshwater diatom index; open stipple = arid phases; dense stipple = humid phases (after Jansen & Van Iperen, 1989).

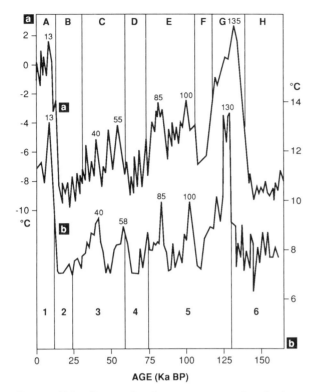

FIGURE 7.2 Curve of palaeotemperature for the last glacial cycle, based on the Vostok isotope temperature record from Antarctic ice, and estimated Indian Ocean sea surface temperatures from drilling log RC11-120 (simplified after Jouzel *et al.*, 1987). Values superimposed on the temperature peaks indicate the approximate timing of possible energy pulses affecting world river systems

and drying of climates is recorded from both mountain and lowland environments (Flenley, 1979, 1985; Street-Perrott *et al.*, 1985; Talbot *et al.*, 1984; Talbot & Johannessen, 1992 and many others). Coolest and driest conditions attach to a period during and following the last glacial maximum centred at 18 000 BP. The rapid change towards post-glacial conditions may have begun before 13 000 BP and was certainly evident by 12 500 BP at most recorded sites. There followed everywhere a short but significant return to cooler and drier conditions during the Younger Dryas Cold Event, 11 000–10 000 BP, after which conditions favouring a return to rainforests in the humid tropics were established by 9000 BP, when inter-tropical lake

levels were also high once more. During the Holocene there were further short-lived climatic oscillations of importance, but these will be dealt with separately. There is some evidence to suggest that post-glacial humidity, in Africa, increased first in low latitudes and moved northwards into the Saharan margins over a period of some 2 000 years. A summary of these findings is given in Table 7.3, and subsequent sections will discuss these climate changes.

THE PATTERN OF QUATERNARY CLIMATE CHANGE IN THE TROPICS

The details of climate change are still poorly understood for rainforest areas, and many reviews of the Quaternary in the tropics depend on field studies from the arid zones (Littmann, 1989a) or from mountain environments (Flenley, 1979). But during the past few years several significant studies of lowland sites in the humid tropics have been published (e.g. Absy *et al.* 1989, 1991; Talbot & Johannessen, 1992; Van der Hammen *et al.* 1992a,b). Results from these studies must be compared with the more abundant evidence from lake sediments and pollen (Street-Perrott *et al.*, 1985; Street & Grove, 1979) from the Sahelian and other semi-arid environments, and from the study of high mountains and the depression of the snow line, tree line and other vegetational belts (Flenley, 1979, 1985; Hamilton, 1982; Mahaney, 1989).

Important sites within the humid tropics include Lynch's Crater on the Atherton Tableland of northern Queensland, where Kershaw (1976, 1978) recorded the pollen spectra from a 40 m core considered to represent 130 000 years of deposition (Figure 7.3), later extended to other sites with some indications of still earlier climatic oscillations (Kershaw, 1992). This long-term record presents a number of problems. It is situated in a zone of very strong E–W rainfall gradient, and also much of the dating is based on estimation. Nevertheless, a major vegetational change from rainforest to a mixed *Araucaria* forest after 79 000 BP is very marked, suggesting a decline in rainfall from >2500 to <1500 mm year[-1]. Warmer conditions appear to have led to a revival of rainforest species between

TABLE 7.3   Tentative chronology of late Quaternary environmental change

Radiocarbon BP	Probable environmental conditions (recorded examples)
c. 80 000–c. 63 000	Indications of cooler, drier conditions (northern Queensland, Australia)
63 000 (?)–c. 52 000	Scattered records of sedimentation suggest rapid warming at end of isotopic zone 3 — dating insecure (Venezeula, Brazil, Kalimantan)
c. 40 000 BP	Cooling of upland climates evident, possibly becoming drier (Amazonia, Uganda) (these earlier stages and dates are very tentative)
>32 000–22 000/20 000	Cooler, probably humid to sub-humid (Ghana, Sierra Leone, Uganda, Brazil)
Post 22 000–13 000/12 000	Becoming cold in uplands and dry in most lowlands; by 18 000 years BP the tree-line is depressed to 1000 m, rainfall is possibly reduced by 50% (most sites in tropical South America, Africa and India and southeast Asia)
12 500–c.11 000	Rapid warming with unstable climates and prolonged heavy rains in tropical Africa, very high lake levels (world-wide)
11 000–10 500	Dry, cool interval in many areas; low lake levels
10 500–8000	Second humid period with high lake levels and discharges; re-establishment of forest by 9000–8500
7800–7000	Reduced lake levels and river discharges in West Africa (Ghana, Brazil)
7000–5000	Increased humidity and modest rise in lake levels
5000/4500–3000	Mid Holocene dry phase, probably quite severe (West Africa)
Post-3000	Increased humidity in forested tropics with rising discharges, but concomitant with deforestation and human occupation

Main sources: Kershaw (1978, 1992); Street & Grove (1976, 1979); Adamson *et al.* (1980); Butzer (1980); Thomas & Thorp (1980, 1985); Talbot *et al.* (1984); Schubert (1988); Absy *et al.* (1989, 1991); Street-Perrott & Perrott (1990); Thorp *et al.* (1990)

c. 63 000 and 50 000 BP, after which drier conditions are evident. Sclerophyll woodland then progressively replaced the rainforest species from c. 38 000 BP onwards (minimum rainfall <800 mm year^{-1} at the LGM), until a marked recovery of the broadleaved forest occurred after 10 000 BP. The trends in this core appear to correspond well with the signals from ocean sediments and pollen, although some of the changes are difficult to interpret (Flenley, 1985). A site in southeast Amazonia has provided further evidence older than 50 000 BP (Absy *et al.*, 1991), but few land sites in the lowland, humid tropics so far offer a dated record much beyond the last 40 000 years of radiocarbon dating, and their details will be considered in that context.

Great emphasis has been placed upon palynology in the study of past climates, but there are many pitfalls in the use of this technique, as with most others. The provenance of pollen at many sites must often be questioned, particularly in steep terrain, where pollen from different elevations may become mixed in a single deposit. There will also be problems of interpretation where vegetation mosaics exist in an area, as a response either to stable edaphic contrasts (Riezebos, 1983, 1984) or to preferential colonisation or survival during periods of rapid ecologic change (Fölster, 1992). The relative productivity of different taxa must also be considered. Lake sites are particularly prone to individual behaviour, and the simple issue of whether there is standing water or not in a lake basin may pose difficult problems. Thus, a lake fed by groundwater or from a distant watershed may persist long after a smaller basin, dependent only on local Precipitation:Evapotranspiration ratio

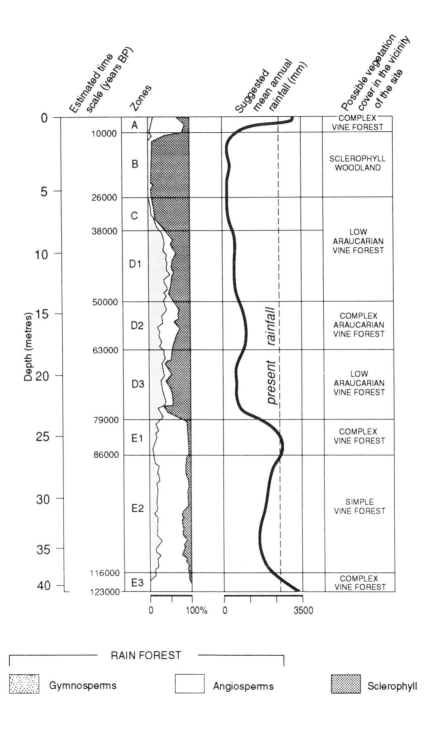

FIGURE 7.3   Pollen record over 125 000 years for Lynch's Crater, northeastern Queensland, Australia (after Kershaw, 1976; reproduced by permission of Unwin Hyman)

relationships, has dried out. Shallow lakes formed along the desert margins may therefore have different histories and return confusing signals about past climatic conditions, when other factors are unknown or ignored.

The glaciation of tropical mountains, and the corresponding shift of the vegetation belts, has been a major source of inference about climate change in the tropics, but, even if it is accepted that important changes to the lapse rate did not occur at times of cooler climate, the presence of snow and ice do create local climates significantly colder than those of neighbouring non-snow areas, and downslope drainage of cold air will have affected neighbouring lowlands. It is also not often clear whether cooler climates have resulted from a colder atmosphere or from greater cloudiness and reduced net radiation.

All of these factors make quantification of changes in temperature or precipitation quite difficult. Nevertheless, the following parameters of change are all important:

(1) the *direction* of change, whether towards aridification or pluviation; cooler or warmer conditions;
(2) the *magnitude* of the change, including the intensity of aridification or pluviation and involving both the annual means of climate variables and the character of individual weather events;
(3) the *rate* of change and whether any vegetational response was concomitant or lagged;
(4) the *duration* of modified conditions; whether sufficient to allow readjustment of environmental parameters to changed conditions; and
(5) the spatial *pattern* and geographical extent of particular climate changes.

These questions are, where possible, considered in context below; the question of cooling and warming being discussed separately from that of rainfall reductions and increases.

Upland and lowland temperatures at the last glacial maximum (LGM)

Modelling exercises to show global changes in temperature frequently only display sea surface temperature (SST) for which microfossil assemblages (foraminifera) and $\delta^{18}O$ data can be used (CLIMAP, 1976, 1981; Menabe & Hahn, 1977). On this basis temperature depression (SST) at the last glacial maximum (LGM) is thought to have been 6–10 °C in northern regions affected by sea ice. But in tropical areas the SST depression shows a range of 0–2 °C over most of the lower latitudes. These conclusions, however, appear to conflict with evidence on land for temperature depression in tropical highlands and the supposed causes of aridity in tropical lowlands (Rind & Peteet, 1985; Crowley & North, 1991).

Syntheses by Flenley (1979) and Hamilton (1982), based largely on palynological studies of montane sites, indicate the depression of vegetation zones in high tropical mountain ranges by 1000–1500 m (Figure 7.4), and this requires temperature reductions of 5–9 °C. In a recent study of equatorial mountain sites in Burundi, Bonnefille and Riollet (1988) estimated a temperature fall of 4–6 °C, between 30 000 and 13 000 BP, but in a more recent statistical study of pollen time-series data from the same area this estimate was modified to 4 ± 2 °C (Bonnefille *et al.*, 1990, and Figure 7.5), and at the lowland site of Lake Bosumtwi in southern Ghana glacial cooling was estimated by Maley and Livingstone (1983) at only 2–3 °C under an increased cloud cover at *c.* 15 000 BP. Verstappen (1980) argued that a depression of temperature of at least 3–5 °C was necessary to explain vegetation changes in southeast Asia and quoted Kraus (1973) in support of an increased lapse rate during glacial conditions. However, Stuijts *et al.* (1988) have suggested a lower figure of 2–3 °C, based on the palynology of upland sites in Sumatra and Java. Their pollen studies lacked evidence for drier climates, but the occurrence of coarser sediments after 18 000 BP indicated a change in landscape dynamics, and possibly drier or more seasonal conditions.

Figures in the higher range are supported by recent reviews of the evidence from central and south America (Markgraf, 1989; Van der Hammen, 1991). Such larger temperature reductions on land may have been magnified regionally by altered patterns of ocean current upwelling (Coetzee &

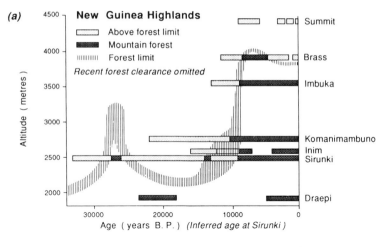

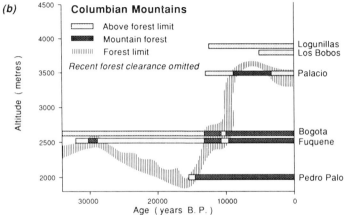

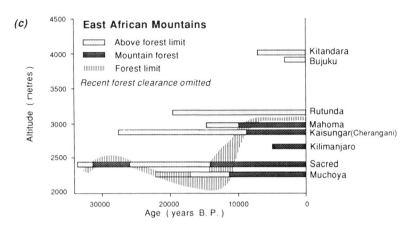

FIGURE 7.4 Summary diagrams of late Quaternary vegetational changes in three tropical mountain areas; (a) New Guinea, (b) Columbia, South America, (c) East Africa (from Flenley, 1985, after Flenley, 1979; reproduced by permission of Unwin Hyman)

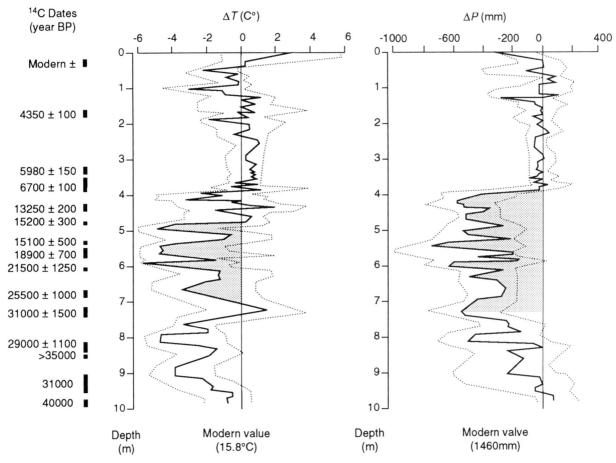

FIGURE 7.5 Reconstructed mean annual temperature and rainfall at Kashiru, (3°28'S), Burundi, expressed as a deviation from the modern value. The dotted lines represent the mean of upper and lower standard deviations (after Bonnefille *et al.*, 1990; copyright 1990 Macmillan Magazines Limited)

Van Zinderen Bakker, 1989; Littmann, 1989b; Lautenschlager, 1991), but were most likely to have been caused by changes in atmospheric conditions, involving the outflow of cold tropospheric air from higher latitudes.

Oceanic evaporation losses may have been reduced by 18% globally and by much more in the tropics (variously estimated at 30–70%), and the weakening of both the ITCZ and the monsoonal circulation would have led to widespread rainfall reductions during the last glacial maximum (22 000–14 000 BP). Van Campo (1986) has adduced evidence for the aridity of southwest India from 22 000 to 18 000 BP, associated with the cooling of

the Indian Ocean and weakening of the monsoon which gradually extended its influence after 16 500 BP, culminating in very humid conditions by around 11 000 BP. Loss of radiative warming due to increases in *albedo* in drier areas may have been a major factor leading to cooling of continental climates.

Available evidence thus suggests that global cooling before and during the last glacial maximum affected all tropical sites so far analysed in terms of their pollen and phytolith records. The magnitude of that cooling may have been regionally variable, and some sites may have only local significance, but an increasing number of records indicate significant cooling of 4±2 °C (Vincens *et al.*,

1993). The timing of the onset of cooler conditions in the tropics remains a matter of debate (Bush *et al.*, 1990), and temperatures may have been depressed for much of the last glacial cycle. However, they probably dipped seriously after 40 000 BP (Bonnefille *et al.*, 1990, and Figure 7.5), and reached their minimum values either side of the LGM, around 18 000 BP.

The question of ice-age aridity

From the pollen record alone, some doubts have been voiced concerning the comparative aridity of the cooler climates particularly in Amazonia (Colinvaux, 1987, 1991), but these doubts have been voiced mainly from hilly areas such as Sumatra (Stuijts *et al.*, 1988) and the Andean foothills of Ecuador (Bush *et al.*, 1990), which may be unrepresentative of the wider lowland and plateau landscapes of the continents. Colinvaux (1987) has expressed serious doubts about ice-age aridity in the Amazon lowlands, emphasising the possibility of regional variations in climate change over such a vast area and the lack of reliably dated features attributable to dry climatic conditions. However, new evidence of important climatic changes in Amazonia has subsequently been published (Absy, 1982; Absy *et al.*, 1989; Soubiès *et al.*, 1989; Van der Hammen *et al.*, 1992a,b; Ledru, 1993) and this area is discussed in more detail below. Similar doubts have been voiced by Walker (1982) concerning the late glacial climates of southeast Asia, despite the claims by Verstappen (1980) that rainfalls declined by more than 30%. However, the latter view has been supported by a more recent interpretation by Heaney (1991). The complexity of records from southern Africa, and particularly from the Kalahari, has led to much disagreement and to two significant warnings; first, by Butzer (1984) not to jump too soon into continent-wide models of climate change, and second, by Shaw and Cooke (1986) that it should not be assumed that the southern hemisphere patterns of climatic oscillation were identical to those of the northern hemisphere tropics.

There are good theoretical reasons for associating cooling with lowered rainfalls at lowland sites, and these include the increased strength of the Northeast Trades and reduced evaporation from the oceans (Williams, 1985), leading to a weakening of the monsoonal circulations. Regional variations will have been important and the apparently extreme glacial aridity of parts of West Africa may have been partly a function of the steep rainfall gradient towards the Sahara desert. This is presently as high as 2.6 mm km^{-1} from south to north, suggesting that a 400–500 km shift of the rain belt could involve a loss of precipitation of >1000 mm (40–50%) in the humid areas of Sierra Leone. But conditions were equally severe in the dune fields of the Venezuelan and Columbian Llanos (Schubert, 1988), and possibly along the eastern coastlands of Brazil (Bigarella & Andrade, 1965; Bigarella *et al.*, 1965), where changes in the ocean currents have been postulated (Ab'Saber, 1982; Bigarella & Andrade-Lima, 1982). Increased continental aridity during ice-age global circulation patterns (18 000 BP) is also predicted by global climate models (Menabe and Hahn, 1977). The T21 model used by Lautenschlager (1991) indicates a loss of precipitation of 29% over the Congo/Zaire basin, but only 10% over the Amazon. This latter figure would, however, not be sufficient to account for observed changes in vegetation and fluvial activity. In a recent review of the palaeoenvironmental history of east Africa Colyn *et al.* (1991) concluded that a major 'fluvial refuge' survived within the Zaire basin but there is evidence for important environmental changes in eastern Zaire (Runge, 1993).

Evidence from closed lake basins and mires

The fluctuation of lake levels during past climatic changes forms one of the major strands of evidence for palaeoenvironmental reconstruction, but there are some problems which have been touched upon already; namely that most lakes are in semi-arid areas, and palaeolakes were often in the deserts themselves; also that lakes with large and diverse catchments do not accurately reflect the climate of the lake surface itself. The water balance for *closed* lakes (after Street, 1980):

$$P + R + G_I = E + O + G_O \pm \Delta S \qquad (7.1)$$

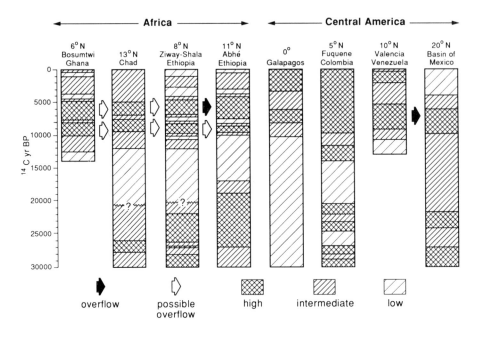

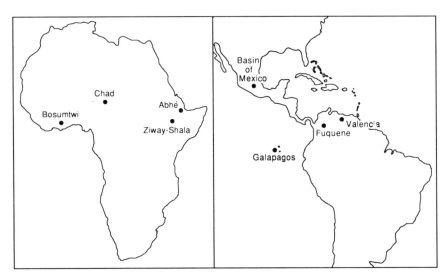

FIGURE 7.6   Detailed lake-level sequences from Africa and Central America (after Street-Perrott *et al.*, 1985; reproduced by permission of Unwin Hyman)

where P is precipitation onto the lake surface, R is runoff from its catchment, G_I is groundwater inflow, E is evaporation from the lake, O is surface outflow, G_O is groundwater outflow and ΔS is change in storage. In order to take account of the size of the catchment in relation to the lake, a ratio, z, has been defined:

$$z = \frac{A_L}{A_B} = \frac{P_B - E_B}{E_L - P_L} \qquad (7.2)$$

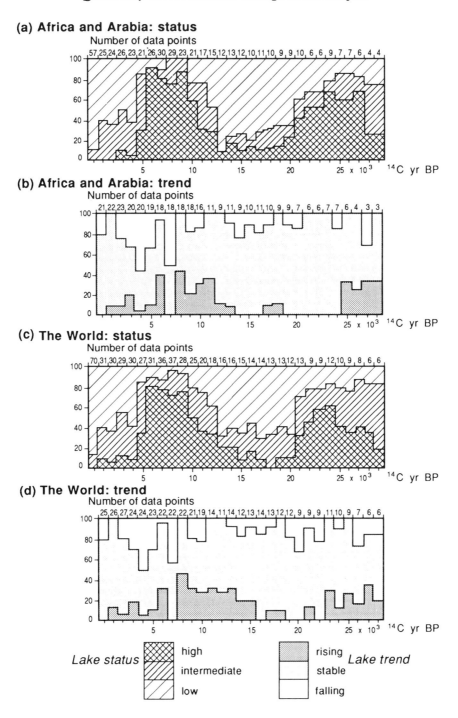

FIGURE 7.7   The variations in lake status and trend in the northern hemisphere tropics over the last 30 000 years for the (a, b) Afro-Arabian, and (c, d) global data sets (after Street-Perrott *et al.*, 1985; reproduced by permission of Unwin Hyman)

TABLE 7.4   Standard lake level sequence based on records from Africa and Arabia

F	*c.* 30 000–23 000 BP	Some lake levels rising and remaining high
E	*c.* 23 000–17 000 BP	Lake levels falling, intermediate or low
D	17 000–12 500 BP	Prolonged, intense aridity with most lakes low
C	12 500–10 000 BP	Transitional period of rising lakes but with recession after 10 800 BP in some areas (Younger Dryas event)
B	10 000–5000 BP	Lake levels rising, from Equator first, towards Sahara, with most high by 9000 BP
A	5000–0 BP	Sharp fall in lake status, with severe desiccation north of 13 °N

Mainly after Street-Perrott *et al.* (1985).

where A is area and the subscripts L and B refer to the lake and its drainage basin respectively. P and E are as in Equation 7.1.

However, the parameters required to solve this equation are not easily determined, and inaccuracies of measurement for palaeolake margins, and catchment hydrological characteristics, led Street-Perrott *et al.* (1985) to use *relative water depth* as a surrogate for lake area. These data came from ^{14}C dated stratigraphic sequences, from which it was determined whether lakes were *low* (0–15% of total fluctuation or dry), *high* (70–100% of fluctuation and overflowing) or *intermediate* (15–70% of fluctuation).

A selection of eight carefully dated sequences from Africa and America reveals the major trends (Figure 7.6), while a larger data set from Africa and Arabia demonstrates that from 27 000 to 21 000 BP and from 12 500 to 5 000 BP, lake levels were relatively high (Figure 7.7). Street-Perrott *et al.* (1985) then proposed a *standard sequence* for the last 18 000 years. Although these authors did not include earlier indications of humidity, this (phase F) has been added here in Table 7.4 for clarity and completeness.

These results confirm what has been known for many years about Lake Chad in West Africa. Lake Chad is interesting because it is a large lake basin receiving tributaries from the south. Its palaeoclimatic signal will therefore represent both local and broad regional trends in climate. A widespread lacustrine transgression affected the Lake Chad basin around 12 000 BP (Servant, 1973; Servant & Servant-Vildary, 1980), and the lake-level record also clearly signals the Younger Dryas (dry) event, following which the level peaked again at *c.* 9000 BP,

with a subsequent regression to low levels *c.* 7500 BP, rising once more to high levels until progressive desiccation set in *c.* 5000 year BP (Figure 7.6).

In East Africa, Lakes Malawi and Tanganyika are thought to have been lowered by 250–500 m during the LGM (Scholz & Rosendahl, 1988; Gasse *et al.*, 1989; Baltzer, 1991; Vincens, 1991; Vincens *et al.*, 1993), and there is much evidence to indicate high lake levels in the early Holocene. An overflow of Lake Victoria into the White Nile is known to have taken place around 12 500 BP (Livingstone, 1980). In their study of Lake Tanganyika (the earth's second deepest lake, at 1470 m), Gasse *et al.* (1989, p. 59) concluded that the, 'abruptness of the major hydrological events recorded at 13 000 BP and *c.* 10 600 BP makes the fluctuations of the lake water balance synchronous with the deglaciation steps recorded in the Atlantic Ocean.' But although the lakes of East Africa (Butzer *et al.*, 1972), which had featured previously in mistaken attempts to link 'pluvials' to successive northern glaciations, also began to fill after 12 500 BP, Lakes Rudolf, Nakuru, Naivasha, and Magadi did not reach high levels until between *c.* 10 000 and *c.* 7000/6000 BP (Lakes Rudolf and Nakuru are known to have been high prior to 20 000 BP). This delay in the filling of the lakes in the East African area may have been due to a lag in the post-glacial warming of the Indian Ocean (Vincent, 1972).

Lake Bosumtwi in southern Ghana provides confirmation of these trends, though this crater lake appears to be subject to quite sudden and short-term variations of water level (data from Talbot & Delibrias, 1980; Talbot *et al.* 1984, and Figures 7.6 and 8.5).

TABLE 7.5   Tentative correlation of the late Quaternary climatic and environmental changes in Africa with world-wide climatic events

KBP	Large-Scale Events [1]	European Climate	Sahara North [1]	Sahara South [1]	West [2][7]	Cameroon [3]	Congo [4][8]	E. Africa [5]	Kalahari North [6]	Kalahari South [6]	KBP
0		SA	Arid	Arid	Humid (Forest)	Humid (Forest)	Humid (Forest)	Semiarid	Sl. Wetter	Semiarid	0
2		SA	Arid	Arid	Humid (Forest)	Humid (Forest)	Humid (Forest)	Semiarid	Semiarid	Semiarid	2
4	Opt Arctic / Disap. Laur. Ice	SB	Desiccation	Hyperarid	Dry	Humid (Forest)	Humid (Forest)	Wet	Wet	Arid	4
6	Opt Europe / Disintegr. Hudson B. Ice	AT	Wet, Warm	Wet, Warm	More Humid (Forest)	More Humid (Forest)	More Humid (Forest)	Semiarid	Semiarid	Arid	6
8	Disap. Scand. Ice	BOR / PB	? / Dry, Wet (?)	Dry Interval / Lacustrine	(transition)			Wet / Lacustrine	Wet	?	8
10	Opt Antarctic	PB / YD	Dry, Wet (?)	Lacustrine / Wetter with Fluctuations	Transition	Transition	Transition	Transition → Wetter	Semiarid	?	10
12	Ice Retreat	AL / OD / BO	Semiarid with Fluctuations	Wetter with Fluctuations	Transition	Transition	Transition	Transition → Wetter	Lacustrine	Transition	12
14		OLD	Semiarid with Fluctuations	Hyperarid	Subhumid-Semiarid (Savanna)	Subhumid-Semiarid (Savanna)	Subhumid-Semiarid (Savanna-Steppe)	Semiarid	Hyperarid	Wetter	14
16	Max. Ext. N Ice-Sheets	Last Glacial	Semiarid Cold	Hyperarid	Subhumid-Semiarid (Savanna)	(Savanna)	(Savanna-Steppe)	Semiarid	Hyperarid	Wetter	16
18	Max. Ext. Ice Drift both Hemisphers	Last Glacial	Transition → More Humid	Transition → Hyperarid							18
20		Last Glacial	Transition → More Humid	Transition → Hyperarid	Humid (Forest)				Wetter	Dry, Cool	20
22		Last Glacial			Humid (Forest)				Wetter	Dry, Cool	22
24		Last Glacial			Humid (Forest)				Lacustrine	Dry, Cool	24
26		Last Glacial	* Highlands Wet, Cool		** Excluding Humid Tropics		*** All-Season Rains		Lacustrine	?	26

* Highlands Wet, Cool   ** Excluding Humid Tropics   *** All-Season Rains

After Kadomura (1986a) (sources: (1) Flohn & Nicholson, 1980; (2) Thomas & Thorp, 1980; (3) Kadomura, 1982; (4) De Ploey, 1968; (5) Butzer *et al.*, 1972; (6) Heine, 1982; (7) Talbot *et al.*, 1984; (8) Peyrot & Lanfranchi, 1984).

There is still some degree of doubt concerning many local and regional details, and recent surveys by Littmann (1989a,b) have noted some of the inconsistencies and problems amongst the data. He recognised the three major phases as: 40 000–22 000 BP, 22 000–14 000 BP and 14 000–8000 BP. Given the imprecision of long-range ^{14}C dates, differences in threshold ages for so-called 'abrupt' changes of climate must be accepted as inevitable, but the variations in the number and magnitude of the changes since *c.* 40 000 BP probably reflect the spacing and the reliability of dates, and the interpretation of pollen spectra in the absence of other indicators of environmental change. Another summary of African conditions during the late Quaternary is given by Kadomura (1986) and his table (Table 7.5) makes clear both the correspondence and the inconsistencies within the existing data sets.

One of the important features of the lake level sequence is the abrupt transition from one climatic phase to another, too sharp to be resolved easily with ^{14}C dates (Street-Perrott, *et al.*, 1985). At the same time, it should be noted that there was a lagged response in the climates of the Sahara and its southern margins, where inter-dunal lakes first appeared *c.* 10 000 BP, and as far north as Latitude 21 °N in Mali, a short lacustrine phase appeared after 9000 BP. A total of 88 ^{14}C dates were analysed by Fabre and Petit-Maire (1988), indicating a lacustrine optimum between 8300 and 6700 BP and followed by a climatic deterioration, culminating in total desiccation by 4500 BP. These dates reflect the time-transgressive nature of the post-glacial increases in rainfall which advanced progressively from the humid tropical low latitudes after 13 000 BP.

There is also apparent evidence for a lagged response of vegetation to changed conditions, so

TABLE 7.6   Water level record for Lake Bosumtwi, Ghana

27 500–24 500 BP	Lake low and saline
24 500–20 000 BP	Lake dilute but sediments disturbed
(*c.* 22 000 BP)	(Dry excursion)
20 000–16 500 BP	Lake saline but depth uncertain
(*c.* 18 500 BP)	(Dry excursion)
16 500–12 500 BP	Low lake, some fluviatile facies
(*c.* 14 700 BP)	(Dry excursion)
12 500–9000 BP	High lake level (+ 45 m)
(*c.* 12 000 BP)	(Dry excursion)
(*c.* 10 500 BP)	(Dry excursion)
9000–3750 BP	High or overflowing lake (185 m depth ?) with thick sapropel
(*c.* 8000 BP)	(Brief fall in level)
3750–present	Dramatic decline in level to − 20 m, followed by revival to + 25 m; cessation of sapropel formation.

After Talbot & Delibrias (1980); Talbot *et al.*, (1984); Talbot & Johannessen (1992).

TABLE 7.7   Water level record for Lake Valencia, Venezuela

13 000–10 500 BP	Dry marsh-like conditions in the lake basin, xeric vegetation
10 500–8700 BP	Increasing lake level to overflow, arboreal vegetation
8700–present	Fluctuating levels and salinities: 7000–6000 BP, low; post-3000 BP, evidence of desiccation

Based on Bradbury *et al.* (1981).

that pollen studies from several sites indicate that rainforest was not re-established until *c.* 9000 BP, despite the rise in rainfall after 12 500 BP. The 3500-year interval implied by these figures was almost certainly one of great geomorphic activity and change, and this will be explored further. One reason for this situation was probably the renewed aridity *c.* 10 800 BP (Younger Dryas event), lasting 500–1000 years in most locations.

Three examples of lakes from the humid tropics provide details to show that important fluctuations in humidity have occurred in these low latitudes. *Lake Bosumtwi* (Latitude 6 °30′ N, 100 m asl, *P* = 1500 mm) is an impact crater with a diameter of about 8 km, situated in the rainforest zone of southern Ghana. Sediment cores have been studied in some detail (Talbot & Delibrias, 1980; Talbot, 1982; Talbot *et al.*, 1984; Talbot & Johannessen, 1992), and the interpreted results are summarised in Table 7.6.

There is some ambiguity in the interpretation of detail in the sediment cores, but the general trends have been established for a lake which must have

responded directly to local climate and vegetational changes, since it has no wider catchment. The River Birim in the same vicinity displays a synchronous response in fluvial activity (Chapter 8).

Lake Valencia in Venezuela lies in a tectonic depression at 470 m asl and provides an interesting comparison for the last 13 000 years (Schubert, 1980; Schubert & Valastro, 1980) (Table 7.7). In this case the earlier pluvial period from 12 500 BP appears to be missing.

Some depressions must count as mires rather than lakes, and may contain thick sedimentary deposits. Cores taken from a small swamp in the Brazilian *Serra dos Carajas* (700–800 m asl, 1500–2000 mm), 500 km south of Belem in Amazonia, confirm the major outlines of previous findings (Soubiés *et al.*, 1989; Absy *et al.*, 1989, 1991). Three organic-rich layers were found divided by two sandy clay horizons rich in siderite and underlain by a third. The lowest silty organic layer has provided a finite ^{14}C date of 51 200 (+ 1700/ − 1400) BP, while the intermediate layer spans either side of 28 660 BP, and the top organic horizon returned dates from

TABLE 7.8   Sedimentation and palaeoenvironments in the *Serra dos Carajas*, Brazil

>60 000 BP	Clastic sand with siderite, dry conditions
>51 200 BP	Organic-rich silts, humid conditions
>31 000 BP	Siderite, sand and clay, reduced tree pollen, dry conditions
31 000–23 000 BP	Black clays with carbonised woods, abundant tree pollen, humid
<23 000–>10 500 BP	Brown sandy clays (siderite), reduced tree pollen, dry
10 500–8000 BP	Black organic clays, abundant tree pollen, humid
c. 8000–3000 BP	Black clays with plant debris, reduced tree pollen, somewhat drier
c. 3000 BP–present	Dark brown sediments with plant debris, sharp rise in tree pollen more humid

After Absy *et al.* (1991).

TABLE 7.9   Vegetation change and palaeoenvironments recorded from the Rukiga Highlands, Uganda

Pre *c.* 42 000 BP	Moist montane forest as today
>42 000–*c.*32 000 BP	Ericaceous vegetation denotes 1000 m depression of tree line with cooler, drier conditions
32 000–*c.* 21 000 BP	Upper montane forest indicating warmer, wetter conditions at site
21 000–*c.*14 000 BP	Ericaceous vegetation, evidence of increased runoff, cool, dry
post *c.* 14 000 BP	Rising temperatures especially to 11 000 BP with forest
post *c.* 4000 BP	Drier conditions
post 2200 BP	Forest clearance evident

After Taylor (1990), reproduced by permission of Elsevier Science Publishers BV.

10 460 BP to recent. A tentative chronology is shown in Table 7.8.

A long pollen record from two Ugandan mires in the *Rukiga Highlands* (Taylor, 1990) has also provided useful corroborative evidence for the major periods of climate change, derived from sites above 2000 m asl. The environmental interpretation proposed is shown in Table 7.9.

Evidence from southern hemisphere transitional climates

Southern Africa

Southern Africa is illustrative of the changing patterns that occur towards higher latitudes, beyond the tropics, and some confusion has arisen over the interpretation of evidence from scattered sites in the Kalahari, Namib and adjacent areas. In fact, there has been quite serious disagreement amongst writers on this area. In a series of studies from Lake Makgadikgadi and the Okavango Basin, Heine (1978, 1979, 1982) recognised a 'tropical' rhythm with high lake levels from *c.* 30 000–19 000 BP,

but much drier conditions between 15 000 and 13 000 BP, then becoming very wet again by *c.* 12 000 BP. However, parallel, and more detailed studies by Grey and Cooke (1977), Cooke (1983), Cooke and Verstappen (1984), Brook (1982) and Shaw and Cooke (1986) have claimed a more complex picture, while Lancaster (1979) and D. Thomas (1989) have cited widespread evidence for late glacial humidity of climate throughout much of the Kalahari and into Namibia.

Fluctuations in the dynamic circulation over southern Africa appear to have led to alternating influences of the tropical (summer rain) climates to the north and the temperate (winter rain) cyclonic climates of the south (Heine, 1977, 1978; Deacon & Lancaster, 1988) across much of this region. The combination of these factors has led to contrasts not only between north and south but also between east and west in southern Africa. This is a vast region, stretching from 19 °S in the Okavango delta to nearly 30 °S at the Orange River. Before attempting to summarise the findings from this area, some of the factors involved can be listed.

TABLE 7.10   Tentative outline of palaeoenvironmental changes in the Kalahari and its margins

>40 000–35 000 BP	Humid climate in northern Kalahari with Lake Makgadikgadi high (945 m)[a]
35 000–33 000 BP	Dry with dune formation in southwest and mid-Kalahari
33 000/30 000–25 000 BP	Fluctuating wetter and drier periods; 33 000–28 000 lakes in the southwest; Namib desert more humid
25 000–20 000 BP	Generally more humid with Lake Makgadikgadi at 920 m and lakes in the southwest, 23 000–21 000
19 000–17 000 BP	Drying of climates after 20 000; lakes dry and dunes in the southwest[b]; dry conditions extended to most areas around LGM
17 000–15 000 BP	Lake Makgadikgadi remained dry but stromatolites in Urwi Pan plus fluvial activity suggest local moisture in Kalahari, and Namib
15 000–12 000 BP	Declining rain in southwest and mid Kalahari; wet conditions in the south (Molopo Valley)
12 000–10 000 BP	Wetter in southwest and mid Kalahari; Etosha Pan lacustrine
10 000–6000 BP	Declining moisture with dunes in the southwest; climate as today in the mid Kalahari
6000 BP–Present	Lesser local fluctuations
(6000–5000	Wetter interval in the mid Kalahari)
(4000–3000	Dry with dune activity in the southwest)

[a]Volume of the 945 m palaeolake Makgadikgadi may have required inflow from the Zambezi system, later followed by overflow to the north across Victoria Falls.
[b]After a dry phase around 19 000 BP, the southern margins of the area, in the Molopo Valley, appear to have received enhanced rainfall between 17 000 and 12 000 BP.
Compiled mainly from Cooke & Verstappen (1984); Heine (1978, 1979); Shaw & Cooke (1986); Lancaster (1989); Nugent (1990); Scott (1989); Vogel (1989).

(1) Local climates and rainfalls of the Okavango and neighbouring Makgadikgadi basins differ from those of their northern catchments which drain from the well watered Angolan highlands (12 °S), that form the watershed between the Okavango (Cubango R.) and the Congo/Zaire systems.

(2) The northern Kalahari including the Okavango and Makgadikgadi basins experiences summer rainfall (95%) associated with southerly movement of the ITCZ, while the southern margins, along with the mountains of the Cape, receive winter rainfall from cyclonic westerly circulation. Over some areas these regimes overlap today, as they must have done in the past.

(3) A general northward shift of the wind belts could have brought the temperate westerly regime (wetter) to the central and northern Kalahari during the LGM, when the incoming rivers from the north were affected by glacial 'dry' climates to the north, producing potentially conflicting signals.

(4) In addition to these factors, neotectonism may have affected the Okavango delta region, blurring the palaeoclimatic signals.

It is difficult to piece together a coherent picture for this area, though the maps produced by Heine (1979) clearly illustrate the problem. Table 7.10 offers a tentative outline of our present understanding of late Quaternary climatic change for the Kalahari and its margins, compiled from various sources.

It is clear from Table 7.10 that as the sub-tropical zone, beyond latitudes 30°N and 30°S, is approached, different directions of climate change are encountered, due principally to the influences of the westerly cyclonic circulation governed by the position of the Polar Front.

South America

This reversal of trends in sub-tropical latitudes is also evident from Argentina and Chile, where Heusser (1990) has elucidated a pollen record from a 10.7 m core in the *Laguna de Tagua Tagua*, in the interior central valley of sub-tropical Chile (34°30′ S). Present rainfall is around 800 mm year^{-1}, and with summer temperatures around 20 °C and winter levels about 8 °C, the area supports a semi-arid sclerophyllous vegetation. For a considerable period bridging the LGM (25 000–14 000 BP) the pollen record shows a temperate, semi-humid woodland of southern beech (*Nothofagus dombeyi*) and podocarp (*Prumnopitys andina*). This vegetation was also possibly present at an earlier stage (>43 000–34 000 BP). According to Heusser (1990) the appearance of these taxa implies a temperature reduction of 7 °C (a similar figure for temperature depression has been suggested for Cape Province, South Africa, from oxygen isotope studies of stalagmites; Talma & Vogel, 1992) and an increase in precipitation of 1200 mm year^{-1}. This large increase in rainfall at the LGM conforms to climate models involving an equatorward shift (possibly by 5° Latitude) in the position of the Polar Front, combined with an intensified westerly circulation bringing cyclonic storms northward from their present-day position. These humid intervals also coincided with glacier advances in the sub-tropical Argentinian Andes dated 50 000–35 500 BP and 28 500–14 500 BP.

Further evidence from neighbouring Argentina comes from high levels in a lake at 33° 20′S (*Salina del Bebedero*), between 17 500 and 13 200 BP (Gonzales, 1981, cited by Heusser, 1990), as well as by abundant evidence from higher latitudes not considered here. Correspondingly, relatively warm and dry conditions returned to these sub-tropical latitudes between 10 000 and 8000 BP, when the polar front shifted to the south (Heusser, 1984). At some time after 6000 BP the pollen assemblage points to very low rainfall, perhaps only 100 mm year^{-1}. There is a reassuring symmetry about this example, reversing as it does the tropical pattern of humid–arid oscillation. The open terrain of southern Africa, lacking barriers to air flow has clearly led to a less well defined transition from tropical to sub-tropical conditions, as illustrated above.

In Central and North America, the lakes of northern Mexico and southwestern USA were high at 18 000 BP (25°N–35°N), and these moist conditions moved northwards during the meltback of the northern ice-sheets, leading to drier conditions farther south, similar to the present day, by 5000 BP (Street-Perrott *et al.* 1985).

Sub-tropical Australia

The Quaternary of sub-tropical Australia is also conditioned by different factors. The bulk of Australia is not truly tropical, and only 7% of the landmass is situated equatorwards of Latitude 15 °N. Except for these northern peninsulas (Cape York, Arnhemland, Kimberley), the continent is either arid or temperate, and the riverine basin of the southeast has been strongly influenced by the climates of the eastern Dividing Range (Snowy Mountains). No major rivers drain from the wet tropical areas in the north as they do in the Kalahari.

There have been several attempts to summarise the Late Quaternary climates for all or parts of Australia (Bowler *et al.*, 1976; Rognon & Williams, 1977; Williams, 1985; Chappell, 1991; Kershaw *et al.*, 1991; Harrison, 1993), but the picture remains less clear than for Africa and there are very few dated sequences for the tropical north. The long pollen record from Lynch's Crater implies relatively dry conditions from perhaps 79 000 to post 10 000 BP but with a more severe climate after 26 000 BP (Kershaw, 1976), and this probably reflects the effects of sea-level depression and exposure of land between Australia and New Guinea during this long period (Chappell, 1983).

It is interesting that Nanson *et al.* (1988, 1991) have successfully used thermoluminescence (TL) and uranium–thorium methods of dating to identify a major fluvial episode in northern Australia (the Gilbert fan-delta) which they date to 120 000–85 000 BP, a period when the Lynch Crater record indicates broadleaved rainforest, and which agrees with data from the Congo/Zaire fan (Jansen & Van Iperen, 1991, Table 7.1). Nanson and Young (1988) have also identified a major pluvial epsiode in the southeast near Sydney, ending around 45 000/40 000 BP. *Lake Amadeus* in central Australia (25°S) has recorded high shoreline dunes at 60 000/50 000 BP (by TL dating; Chen *et al.*, 1990). Several lakes in the south (around 35°S), including lakes George, Mungo and Willandra remained high from before 40 000 BP until around 25 000 BP, during what Chappell (1991) described as the 'lacustral period'.

The LGM in Australia appears to have been relatively dry everywhere with lake levels falling after 25 000 BP, and becoming very low between 17 000 and 15 000 BP. A volcanic crater lake at Tower Hill, Victoria has been shown to have been at its driest from 15 000 to 11 800 BP (D'Costa *et al.*, 1989). Rapid warming after 13 000 BP marked the end of glacial and periglacial conditions in the Snowy Mountains, and increased evaporation rates over interior lake basins across Australia. Consequently lake levels initially fell, but rose again as rainfalls appear to have increased. Lake Keilembete was low at 10 000 BP but high again between 8000 and 5000 BP. Since that time desiccation has been widespread throughout much of Australia (Rognon & Williams, 1977; Chappell, 1991).

The record of dune building in Australia matches the lake level data, though there can be some overlap during periods of transition. Dune formation appears to have been active from 30 000 BP, climaxing between 20 000 and 15 000 BP, then declining rapidly after this latter date nearly to zero by 8000 BP; activity resuming again after 5000 BP.

Although differences between Australia and tropical Africa and America have been stressed, Rognon and Williams (1977) drew interesting comparisons between Australia and *north* Africa which has a similar configuration of great meridional extent and a large central desert area. On the temperate margins of northern Africa there appears to have been a similar 'lacustral period' from 38 000 to 18 000 BP, exemplified by the Ahmet depression, while minimum lake levels may have been reached later than in Australia, at around 15 000 BP. Less arid conditions were not established widely in the Sahara before 9000 BP and these persisted until *c.* 4000 BP (Le Houérou, 1992). However, both dunes and evaporites dating from 9500–7900 BP are found near Ouargla (32 °N), while windblown sand covers pediments at Laghout (24°N) (Conrad, 1969). The climate has remained mainly dry since *c.* 4000 BP.

An important observation relating to the rivers entering these lakes relates again to the source areas for runoff. Thus in the Riverine Plain of southeast Australia, dunes were forming in the western plains when discharges through the main channels of the Murray and Goulburn Rivers were as much as six times those of today, between 17 500 and 15 000 BP (Bowler, 1978a,b). Those high discharges, however, had their source areas in the Eastern Highlands (Snowy Mountains), fed by rivers draining glacial and periglacial environments. In a similar way the great Wadi Saoura in Morocco received high discharges and sediment loads from the Atlas Mountains prior to 14 000 BP, but with a loss of load and discharge degraded its bed from 14 000 to 6000 BP, since when surface flows have become ephemeral.

Evidence from deltaic and marine deposits

The Niger Delta

Continuous records of sedimentation by large rivers are usually best preserved in deltaic sediments offshore. Pastouret *et al.* (1978) studied oxygen isotope ratios of both benthic and planktonic foraminifera in a 15 m core from the outer Niger delta, along with an analysis of the sediments, and established through ^{14}C dating the essential outlines of the environmental history of the Niger basin during the later Quaternary. A comparable record was obtained for the Congo/Zaire estuary by Giresse and Lanfranchi (1984), and both match well with the ages of sapropel muds, recording enhanced Nile floodwater entering the eastern Mediterranean, discussed below. These results are compared in Figure 7.8, and it can be seen that sedimentation rates in the Niger delta remained low at 30–35 cm ka^{-1} until *c.* 11 500 BP, when they rose suddenly to 640 cm ka^{-1}, an increase of 18 times. This rate of sediment delivery almost certainly reflected overflows into the Niger catchment from Lake Chad and from internal lakes in the Middle Niger. Important but less dramatic changes occurred at similar times in the Congo estuary, while the Nile floods peaked around 11 000 and 9000 BP.

A change in the clay minerals reaching the Niger delta, from smectites (pre-13 600 BP) to kaolinite during the subsequent period must reflect changing conditions of pedogenesis and morphogenesis. Surface soils will have responded to fluctuations in

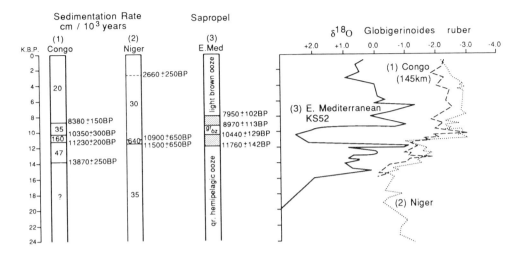

FIGURE 7.8   Sedimentation rates in the off Congo estuary core (Giresse & Lanfranchi, 1984) and the off Niger delta core (Pastouret *et al.*, 1978); the sapropel ages in the eastern Mediterranean KS52 core (Rossignol-Strick *et al.*, 1982), and oxygen isotope records in the three cores (after Kadomura, 1986a)

humidity of climate over 10^3 years, with smectites being in equilibrium with drier conditions and kaolinites evolving in humid periods. The rapid fall in the rate of sedimentation after the Pleistocene–Holocene transition (Figure 7.8) must have been the result of a recovery of the forest and deciduous woodland vegetation and the consequent reduction of sediment yields from slopes and tributary catchments.

The River Nile

This story is corroborated by studies of the River Nile and its discharges into the eastern Mediterranean Sea. Adamson *et al.* (1980) established a prolonged dry period of braided river conditions from *c.* 20 000–12 500 BP, followed by greatly increased discharges, leading to major floods in Egypt. These floods produced a significant dilution of the eastern Mediterranean, leading to a stable density stratification of fresh or brackish water over the older, saline water. This in turn led to the gradual settling of fine mud known as *sapropel*. It has already been indicated that these sapropel muds have formed during periods of peak solar radiation (Rossignol-Strick, 1985), but more detailed study of the Pleistocene–Holocene transition (Termination I)

reveals a two-layer upper sapropel dated at 11 800– 10 400 BP and 9000–8000 BP (Rossignol-Strick *et al.*, 1982, and Figure 7.8).

The climate of the southeastern Mediterranean coast was not wet at that time, and the sapropels must have resulted from heavy rainfall in equatorial Africa after 12 500 BP, interrupted by the Younger Dryas cold/dry event, and then culminating between 10 000 and 8000 BP. Catastrophic floods in the Nile Valley were characterised by Butzer (1980) as the 'Wild Nile', and Paulissen (1989) has dated associated deposits at +30 m above the floodplain at Wadi Halfa to *c.* 12 500 BP. This date corresponds with the first overflow from Lake Victoria in Uganda (*c.* 12 500 BP). Although older than the date given for the first sapropel, these phenomena all fall into the same transitional phase of wetter climates. The conclusion, that major floods were generated during the onset of the late Pleistocene–early Holocene wet phase in eastern Africa, also finds support from studies of river sediments in west Africa, as discussed in Chapter 8. Adamson *et al.* (1980) recorded the dating of mainly Ethiopian silts in the Nile to periods corresponding closely with the sapropel dates: 11 200–9300 BP, 8000–7000 BP and also 5800–5400 BP. A great deal of additional information on the Nile and its catchments is available in Williams and Faure (1980).

The Congo and Amazon estuaries

Studies of the sediments derived from the Zaire (Congo) and Amazon rivers also indicate drier ice-age conditions, with influxes of feldspar and illite, characteristic of less intense weathering under dry climatic conditions (Damuth & Fairbridge, 1970; Damuth, 1977; Jansen *et al.*, 1984; Jansen & Van Iperen, 1991). A deep-sea core 145 km off the Congo estuary revealed the warming of Guinea waters *c.* 13 000 BP, with semipelagic sedimentation reaching a maximum between 11 230 and 10 350 BP, when sedimentation rates reached 160 cm ka^{-1}, a sixfold increase over the former rate *c.* 13 000 BP (Giresse & Lanfranchi, 1984) (Figure 7.8).

Windblown sediments in the North Atlantic

The study of marine sediments from the Atlantic off northwest Africa indicates that the maximum influx of windblown sand shifted 8 ° southwards (from 23°N to 15°N) during the last glacial maximum (Kolla *et al.*, 1979). This must reflect the strengthening of the Northeast Trades, and almost certainly the concomitant weakening of the monsoon over West Africa. This evidence also implies that greatly reduced rainfalls will have affected forested areas which probably deteriorated to open savanna woodlands or scrub at that time. Increased Trade Wind activity alone, however, does not appear to account for this phenomenon (Hare, 1983). Sarnthein (1978) also used deep-sediment cores to calculate that global deserts covered 50% of the land between 30°N and 30°S at 18 000 BP, compared with only 10% today, though his predicted latitudinal shift was only equivalent to 4°.

Degrees of aridity during the last glacial dry phase

It does not seem too extreme today to suggest that presently forested areas in southern Ghana at Latitude 6°N then experienced a climate corresponding with 14°N, and that, in central Sierra Leone between 8° and 9°N, a climate similar to 16–17°N, where today rainfall is ~700 mm and the vegetation is of a Sudano–Sahelian transitional type, may have prevailed for at least 2000–3000 years.

During this 'glacial' dry phase, the degree of aridity remains speculative. Crowley and North (1991) have pointed out that, 'with few exceptions much of the planet seems to have been drier during the ice age'. The major exceptions appear to be those mid-latitude regions to which the storm tracks were displaced, including northwest Africa and the Middle East, southern Australia and southern South America, where conditions became more moist.

Widespread aridity in the tropics has already been established. The suggested 8° southward shift of the rainbelts in West Africa converges with the views of Rossignol-Strick and Duzer (1979) that the ITCZ may not have penetrated far beyond 8°N for 4000–7000 years around the last glacial maximum. If true, this would have desiccated areas of Africa north of Latitude 7° 30′N, and left many areas to the south with rainfalls diminished by more than 50%. A recent model of the palaeosynoptic patterns of West Africa at the LGM by Peters and Tetzlaff (1990) produced a Sahelian precipitation (at 12°N) roughly one-third of the present value. This was based on a reduction of 50% in the number of squall lines propagated across West Africa, combined with a rainfall intensity in squall lines of only two-thirds of the present-day figure.

Rainfall reductions in southeast Asia of 30–50% were proposed by Verstappen (1980) and backed by Heaney (1991), but some recent analyses from southeast Kalimantan (Borneo) (Caratini & Tissot, 1988) and from the Java and Sumatra highlands (Morley, 1982; Stuijts *et al.*, 1988) have suggested only small temperature decreases and minor drying out (see also Morley & Flenley, 1987). Crowley and North (1991) have suggested that, because of the biological feedback mechanism of transpiration from the tropical rainforest which leads to enhanced precipitation, an initial 15–25% decrease in rainfall may be enough to produce sufficient change within the plant cover to become amplified to the 30–50% required to account for some of the features already described from this zone.

The dating of stream sediments within several small catchments in Sierra Leone, which today receive >2000 mm year^{-1} of rainfall (Thomas & Thorp, 1980), points to an hiatus in stream activity between *c.* 20 000 BP and 12 500 BP, and this must reflect a major loss of runoff to rivers, even in the wet season. Similar results have now been recorded from western Amazonia (Van der Hammen *et al.*, 1992a,b). If runoff coefficients rise with rainfall decreases sufficient to degrade forest into open woodland or savanna, then such evidence might imply a much greater reduction in rainfall than would be needed to cause the transition to savanna, and would almost certainly exceed 50%.

Bigarella and Andrade (1965) also deduced that semi-arid conditions had affected eastern Brazil from 3°S to 36°S during the LGM, in areas which today receive 1400–3000 mm year^{-1} precipitation, and more recent research has confirmed and extended these conclusions (Ab'Saber, 1982; Bigarella & Andrade-Lima, 1982).

The evidence from palaeodunes

Perhaps the most striking impact of climate change in the tropics is the formation of dunes, and these have been described from sub-Saharan Africa and from northern South America. Many of these ancient ergs were probably first formed long before the late Pleistocene and were then reactivated during the last glacial dry phase.

The question of an 'age' for the vast area of the Kalahari Sands in central and southern Africa illustrates the problem but the answer remains elusive (D. Thomas, 1984). These sandy beds extend northwards to the Congo/Zaire River and cover large parts of Angola, where rainfall exceeds 1000 mm year^{-1}. They comprise the uppermost strata of the Kalahari Beds that accumulated in a vast, shallow, continental basin that was formed by downwarping between the uplifted rims of southern Africa as a consequence of plate-tectonic movement and the opening of the Atlantic and Indian Oceans. The Sands have been attributed to periods ranging from the Miocene to Plio-Pleistocene, but no reliable dates have been obtained. The fixed dune systems discussed by D. Thomas (1984) were,

apparently, formed on this more ancient sand cover, during the later Quaternary. Heine (1982) was of the opinion that the dune features seen today can be attributed to the dry climates of the LGM, between 18 000 and 13 000 BP. D. Thomas (1989) has recently reviewed much of the evidence for arid zone extensions during the Quaternary and it is clear from his account that a perspective from the mid-latitude deserts towards both polar and equatorial margins straddles climatic fluctuations that have often been out of phase, so that while dunes were forming on the southern margins of the deserts, the northern margins received increased rainfall and vice versa.

THE NATURE OF EARLY HOLOCENE CLIMATES

It has been shown that the early Holocene brought a major extension of humid climates in tropical Africa, South America and in most other tropical regions. In west Africa, Lézine (1988) and Lézine and Casanova (1989, p. 49) concluded that the, 'general palaeogeographic pattern between 6° 30′ and 21° 22′N corresponds to a 400–500 km northward extension of the modern latitudinal vegetation zonation'. This extension of humidity appears to have begun soon after 13 000 BP in low latitudes, but did not reach the Sahara until *c.* 9000 BP (Le Houérou, 1992).

Such swings of the climate belts appear very large, and from the literature, an 8° Latitude 'arid' shift plus a 4° Latitude 'humid' shift—equal to a total effective latitudinal change for west Africa—, from the last glacial maximum to the climatic optimum of 12° of latitude could be deduced, though it would not be widely accepted. It is also not the correct way in which to understand the climatic changes of the Quaternary, since these did not result from simple swings of the zonal climatic belts. Nevertheless, changes of climate equivalent to just half of this value (or 6° Latitude) would have had a major impact on the distributions of forest and savanna.

Evidence already cited for the rise and overflow of African lakes, for increased discharges in major rivers, combined with the pollen evidence for changes towards more humid conditions, offer a consistent picture of increased rainfalls across most of Africa and in many other parts of the tropics. Yet, some disagreement has emerged concerning early Holocene climates in west Africa. Lézine and Casanova (1989) have argued for evenly spread light rains ('drizzle') from *c.* 12 000 to 7700 BP across a wide zone of west Africa, although the evidence cited in support of this view is capable of alternative interpretations, and different conclusions arise from studies of deltaic and marine sedimentation along the central/west African coast (Pastouret *et al.*, 1978, Giresse, 1978, Giresse *et al.* 1982). Field evidence for deep channel cutting in west African rivers at this time (Thomas & Thorp, 1980; Hall *et al.*, 1985) suggests very high peak discharges, and that prolonged heavy rains were a feature of this period. In addition, there is evidence of very high floods in the Nile valley around 12 500 BP (Rossignol-Strick *et al.*, 1982; Paulissen, 1989). Talbot *et al.* (1984) also found evidence at lake Bosumtwi in Ghana for a period of highly seasonal climates, leading to laminated sediments prior to 9000 BP.

However, although the details remain to be elucidated, the evidence given already concerning the Mediterranean sapropels, and the high to overflowing lake levels world-wide (Figures 7.7 and 7.8), combined with information concerning off-shore sedimentation rates and the behaviour of tropical rivers, leaves little room for doubt about the reality of the Holocene 'pluvial'. Many of the data have been summarised by Kadomura (1986a; and Table 7.5).

The onset of warming and the melt-back of northern glaciers commenced before 13 000 BP, and became well established by 12 500 BP, continuing until 11 000 BP, when it was interrupted by the Younger Dryas cold event (11 000–10 000 BP). A similar pattern has been established for tropical mountain glaciations, though enhanced precipitation has led to glacial advances during periods of climate warming. The Andean *Guantiva* interstadial is dated to 12 600–11 000 BP; the *El Abra* stadial to 11 000–10 000 BP (for details, see Clapperton, 1993a).

A 'two-step' warming of climate is confirmed by studies of cores from the tropical Atlantic (Berger *et al.*, 1985), as well as from the Mediterranean sapropels (Rossignol-Strick *et al.*, 1982) and the meteorite crater containing Lake Bosumtwi in Ghana (Talbot & Johannessen, 1992). The warming maxima are usually given as *c.* 12 000 BP and post-9000 BP, when humid conditions were experienced repeatedly until *c.* 6000/5000 BP, after which most climates became drier (Figures 7.5, 7.6 and 7.7, Table 7.5). Exceptions to this trend were probably the inner equatorial tropics and the sub-tropical margins of areas such as the Kalahari (Table 7.5).

The acceleration of sedimentation rates off the Congo/Zaire and Niger rivers indicates greatly enhanced erosion within their catchments, although, in the case of the Niger, the river became integrated with its middle course, which was formerly an enclosed basin of internal drainage, and was also subject to an overflow from Lake Chad into the Benue River, its largest tributary. Nevertheless, the geomorphic implications of this sudden and major change of climate towards wetter conditions are profound and will be explored below.

There is, however, good evidence to suggest that the earlier phase of warming, although it produced catastrophic floods and a major acceleration of landward erosion rates, did not fill the African lakes to levels attained during the second phase of warming post-9500 BP. There is equally good evidence to show that rainforests were not fully re-established until perhaps 9000–8500 BP (Talbot & Delibrias, 1980); even 8000 BP in eastern Brazil (Servant, *et al.*, 1989). The dry phase lasting 500–1000 years after 11 000 BP, following the first pluvial, was probably a major cause of this delay.

Beyond the limits of the tropics *sensu stricto*, a reversal of climatic trend can be detected from records of increased aridity in the early Holocene. This pattern is seen in southwest USA and northern Mexico, the southern Kalahari and Namibia, and also in southern South America and Australia.

Mid-Holocene fluctuations of climate

The Holocene 'pluvial' may have culminated between 9000 and 8000 BP in many inner tropical areas, but the climate did not remain stable on the scale

TABLE 7.11 Holocene climatic signals from palaeoflood analysis

Temperate/sub-tropics (after Smith, 1992) Stream erosion after Brakenridge, 1980 [ ]		Dry periods in the Tropics* (after Van der Hammen *et al.*, 1992b)
Flood periods	(yr BP)	Dry periods
Late 19th Century		
400–1100—[AD 1300–1600]	'Little Ice Age'	400–700
(400–500)[a]		
—[AD 900–1200]		
(1500–1600)[a]		
1800–3000		2000/1900–2700
		(3300–5000/6000 end)
(2500–2900)[a]		3800–4000
4000–5000		4800
(4900–5000)[a]		5400
5900–6500 (Europe only)		(3300–5000/6000 start)[a]
(6000–8000)[a]		(7500–8000)[a]
7000–9000		
10 250–10 900	'Younger Dryas'	(10 000–11 000)[a]

*Dated sequences for Holocene palaeofloods in the tropics are not available.
[a]Dry periods from other sources.

of millennia. Also, the broad regional correlations break down into a number of local sequences that reveal important phase differences with respect to wet/dry episodes. In part this is an artifact of our dating techniques which become more precise as we approach the present day. These more detailed investigations reveal differences of timing or response in local and regional climate/sedimentation systems which remain obscured for earlier periods. We have yet to determine whether there is a clear and consistent synchroneity of short-lived climatic changes (typically 100–1000 years) during the Holocene, when compared with the major glacial/interglacial cycles and their stadia and interstadia.

Nevertheless, by around 8000/7500 BP a dry period lasting more than 500 years occurred in many areas, including west Africa, and this was as severe as the previous Younger Dryas event. Wetter conditions then returned (c. 6000 BP), only to decline again after 5000 BP, when the desiccation of the Sahara set in, with a short revival of moist conditions occurring around 3000 BP. In western Amazonia (Caquéta River), Van der Hammen *et al.* (1992a,b) proposed dry periods around 5400 BP,

4800 BP, and between 4000–3800 and 3300–3200 BP, and between 2700–2000/1900 BP and 700–400 BP. However, after c. 3000 BP the impact of human deforestation on pollen spectra and on sedimentation rates, makes the separate detection of any palaeoclimatic signal very difficult. In South Africa, Brook (1982) detected cooler and wetter phases between 9700 and 7600, 4300 and 2200, +1700 and 400 BP, based on speleothems in Echo Cave, South Africa. This suggests that the Pleistocene reversal of trends as compared with tropical regions to the north continued into the Holocene.

Evidence from large palaeofloods

Recent claims by Smith (1992) for 'global' correlations of large palaeofloods were based exclusively on temperate and sub-tropical data which match the earlier reconstruction derived mainly from North American examples by Brakenridge (1980). When these data are compared with recent findings from Amazonia (Van der Hammen *et al.*, 1992a,b), together with data from West Africa and elsewhere, it is clear that opposite trends between low latitude and higher latitude stations continued through the Holocene (Table 7.11).

SOME CONCLUSIONS AND PROBLEMS

In summary, it can be stated with some confidence that the broad outlines of a chronology for climate change in tropical latitudes during the last 30 000 years are agreed, though the timing of individual events suggests a number of caveats.

(1) Over vast areas of the Amazon and Congo/Zaire basins and the island of Borneo, preliminary conclusions are based on very few dated sequences (these sites may not be representative and may not tell us much about intervening areas).

(2) The interpretation of individual strands of evidence (e.g. water levels, pollen spectra, sediment fluxes) may ignore alternative explanations and/or peculiar site conditions (e.g. local water-tables, complex response of rivers).

(3) The search for global patterns may overlook local and regional contrasts in environmental oscillation (e.g. altitudinal variations, orographic effects on rainfall).

(4) The signals from isolated, localised events (e.g. impacts of single storms, landslides) can obscure or be confused with the effects of longer term or regional trends.

(5) The impact of human population on the natural environment has ultimately exceeded in severity the effects of climate changes, and a question remains as to the period during which this impact may have been significant. The use of 3000 BP is quite arbitrary, and much earlier dates possibly apply in some areas.

This last point is a major one. In the Baliem Valley, in montane Irian Jaya (island of New Guinea) people have been present in significant numbers for at least 25 000 years, and an initial date for the onset of burning and slope degradation has been suggested as 26 000 BP (Haberle *et al.*, 1991). The infilling of swamps in this area started around 7000 BP, and major changes to the plant cover and extent of occupation date from *c* 2900 BP. In northern Queensland Kershaw (1992) has noted the presence of human communities by 5000 BP. The human impact in other parts of the tropics may have been equally significant, though there are good reasons to believe that many areas of Africa and Amazonia experienced very little impact during these earlier periods of human pre-history.

Important sources of variation also probably include differences in behaviour between the Atlantic Ocean and the Indian Ocean in respect of ocean currents and SST. In addition, there is also some variation due to differences in monsoonal circulation and response, as between northern Africa and eastern Asia, and a lagged amelioration of climate post-13 000 BP, as between near equatorial and Saharan marginal latitudes, should be anticipated.

When many palaeoclimatic signals are due to changes in the rate of surface denudation as revealed in lacustrine, fluviatile, deltaic and ocean sediments, it must be recognised that the response of the denudation system to climate change will not everywhere be the same (Fairbridge, 1976). The present-day pattern of climates is very diverse and even when the *direction* of change is similar in all areas and broadly synchronous, this does not necessarily imply an identical response from all catchments and slopes. Nevertheless, the theme of environmental change must be central to any study of geomorphology in the tropics (as elsewhere) and in the next chapter the landforms and sediments produced by these changes are discussed in some detail.

8

Landforms and Deposits Associated with Environmental Change

In this chapter, the development of landforms resulting from the processes and environmental changes outlined in previous chapters is discussed. Consideration of longer term evolutionary trends in the tropical landscape is offered in Part IV. Most landscapes are faceted by recent erosional scars and also present a mosaic of surface deposits that form the parent materials for soils and the foundations of the entire land resource. These forms and deposits are of different ages and origins, and are inherited by the present-day environmental systems within which they may be potentially unstable and liable to rapid readjustment. It is important, therefore, to focus attention on these elements of the landscape.

THE NATURE OF THE FIELD EVIDENCE

The evidence for environmental change has been derived in part from continental deposits, particularly those beneath lakes and mires. But the testimony of the landscape is both more widespread and more diverse than might be inferred from these dated sequences (see Table 7.1). In fact, the very diversity of forms and deposits that are encountered in landscapes mantled (or formerly mantled) by saprolite lends emphasis to the view that these materials are particularly sensitive to changes in the controlling environmental factors. Notwithstanding the importance of deep residual saprolites, surficial deposits that have been transported from neighbouring or distant sites comprise the greater part of the surface materials in many areas of the

tropics. The depth of such deposits is highly variable and may reach tens of metres. But, except where major landslides have occurred, thicknesses of 1–15 m are most common. Equally, the sediments associated with river floodplains in the tropics (as in the temperate zone) record the effects of more than 40 000 years of environmental fluctuations, and in many instances provide clues concerning the river regimes and wider landscape dynamics of the time.

The present state of theory in geomorphology does not, however, allow the dynamic responses of water-, sediment- and nutrient-fluxes to patterns of climate change to be unambigously defined (Eybergen & Imeson, 1989), and some widely used models predicting sediment yield variations in response to climate change (Schumm, 1968) may be difficult to apply in tropical environments. It is therefore necessary to be aware that what we infer from field evidence may spring from faulty deduction, based on partial or inadequate theory, and it is instructive to search for alternative explanations for some of the phenomena described in the literature as the consequences of climatic oscillations. Notwithstanding this caveat, there is overwhelming evidence in the landscape for major changes in dynamic systems, and many of these can be dated. The search for a palaeoclimatic explanation is therefore fully justified.

RIVER REGIMES AND CLIMATE CHANGE

Fluvial landforms and sediments, particularly those of the river floodplain and its terraces, often span

10^4 years of the later Quaternary. Some terraces are clearly much older. The alluvial stratigraphy and the records of downcutting and/or aggradation, must logically span the important climatic and other environmental changes that affected much of the inter-tropical zone during this time. Arguably, these deposits are of greater interest to the geomorphologist than is the sedimentation record of enclosed lakes or mires. Unfortunately, however, the data are more difficult to obtain and the record is less easy to interpret. It is rare for a complete sequence to be encountered at any one section or borehole, because the river channels shift within the floodplain, eroding older deposits and forming new ones from the same sediment.

Problems of interpretation

These and other actions of the river have inhibited attempts to construct Quaternary stratigraphies and histories from floodplain deposits. Some of the problems encountered in the dating and interpretation of alluvial sediments are listed below.

(1) Complex cut-and-fill structures require attention to detailed stratigraphy and close spacing of samples for dating; few natural sections or boring programmes yield sufficient subsurface data over a large enough area to give a clear picture.
(2) Organic material suitable for dating; mainly tree logs, but also leaf litter, small twigs and seed pods, may record a ^{14}C date much older than the sediment containing the fragments. This comes about if sediments are re-worked and are redeposited in successively younger sediments over an extended period.
(3) In this process younger and older organic fragments may become mixed and samples will then become 'contaminated', returning dates that have no significance.
(4) Water constantly moving laterally through sandy sediments may also introduce finely dispersed or dissolved organic matter as later contaminants.

As a consequence of these objections, the reconstruction of environmental history on the basis of a few radiocarbon dates from specific sections is seldom justified. And, even if it is possible to overcome the problems listed above, the interpretation of the results may pose additional questions.

(5) Was the stratigraphic unit laid down over a long period of time as a consequence of significant environmental changes affecting sediment yield and discharge within the basin, or might it have arisen as the consequence of a single major flood event lasting just a few hours or days?
(6) Could the observed stratigraphy have arisen as a result of the *complex response* of the river (Schumm, 1977, 1979, 1981) to the crossing of internal thresholds, as opposed to external forcing factors such as climate change?

To all of these questions there are some partial answers that allow for useful, if limited, palaeo-environmental interpretations to be made. Again, a list of qualifying cicumstances can be offered.

(1) A consistent record from a large number of dated samples from more than one reach of a single river or from more than one river in a local area, should avoid the major problems identified as (1) and (2) above.
(2) By using mainly large, single pieces of wood, the problems of contamination will be minimised and accurate dating facilitated.
(3) If samples in a single section show no *inversion* of dates, then reworking is less likely to have occurred.
(4) Samples derived from backswamp deposits containing thick organic clays are less likely to have been brought in by vigorous erosion of older sediments.
(5) Sedimentological and other morphological observations can be highly significant to a correct interpretation.

Bearing in mind these strictures, some studies undertaken during the 1980s in Ghana and in Sierra Leone will be used to demonstrate the reality of changing environmental controls over stream sedimentation in West Africa.

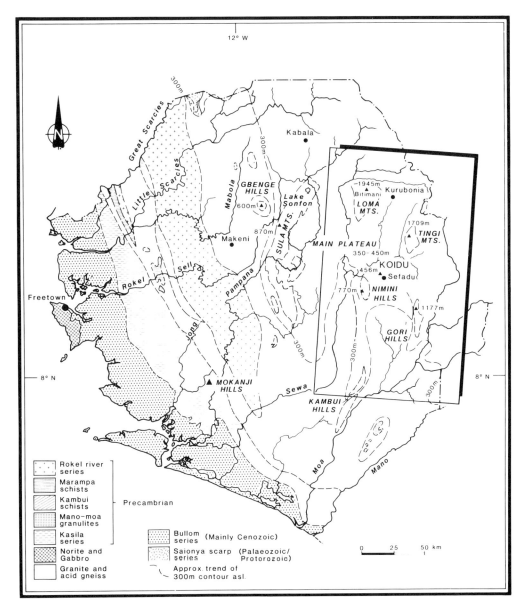

FIGURE 8.1 Geological map of Sierra Leone, showing some locations mentioned in the text. The Koidu basin area in Figure 8.2 is shown by the box (after Thomas, 1989a; reproduced by permission of Gebrüder Borntraeger)

A further problem arises from the limited range of radiocarbon dating, often taken to be < 40 000 BP but now extended by new AMS (Accelerator Mass Spectrometer) techniques to > 60 000 BP with suitable samples. However, new research, based mainly in Australia at the time of writing, and using thermoluminescence dating (TL) techniques, seems set to revolutionise our appreciation of environmental change through the Quaternary (Nanson & Young, 1988; Nanson *et al.*, 1988) with extensions

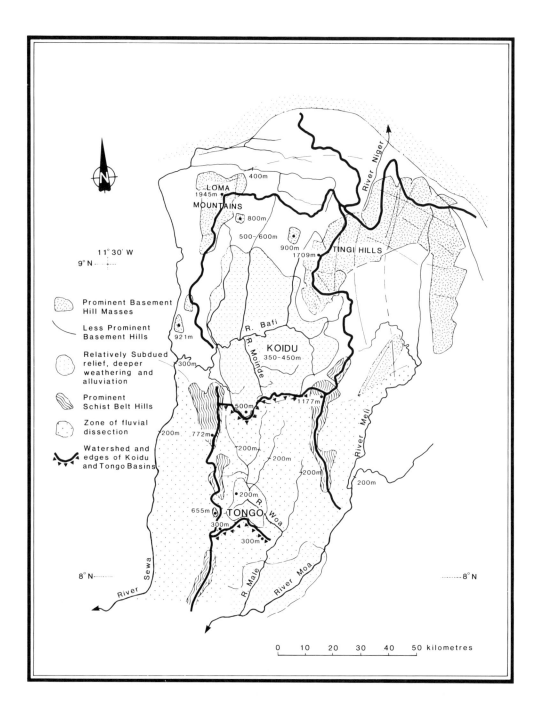

FIGURE 8.2   Geology and geomorphology of the Koidu area of northeastern Sierra Leone, interpreted from Landsat and Sir A radar images (after Thomas, 1989a; reproduced by permission of Gebrüder Borntraeger)

FIGURE 8.3   Alluvial sands and gravels embedded between corestones in a late Pleistocene river channel, in Upper Moinde River, Koidu, Sierra Leone. The hand cleaning of gravels around boulders is undertaken in search for gem quality diamonds

of the datable timescale for sedimentation to > 250 000 BP. Many other issues related to palaeo-hydrological investigation are reviewed in Gregory (1983).

Case studies

Central Sierra Leone—the River Bafi headwaters

In central Sierra Leone (Latitude 8°–9 °N, monsoonal climate with rainfall 2000–3000 mm year^{-1}) there are ferruginised terraces bordering the larger rivers, but they remain undated. Radio-carbon dating has, however, provided a framework for understanding more recent sediments. The

records come from the diamondfields situated in Koidu (Figure 8.1) and relate to sediments deposited by tributaries draining into the Bafi–Sewa river system (Thomas & Thorp, 1980; Thomas *et al.* 1985). The *Koidu Basin* in east-central Sierra Leone is in the forest zone but within 50–100 km of the present forest/savanna boundary. The area has been widely deforested this century as a consequence of population influx to the diamond mines. The area occupies a structural and topographic depression, possibly an ancient graben, developed mainly on Archaean basement migmatites and granites, and it is partially enclosed by duricrusted hills on supracrustal (Schist Belt) rocks (Figure 8.2). The altitude is ~ 350 m asl and the internal relief

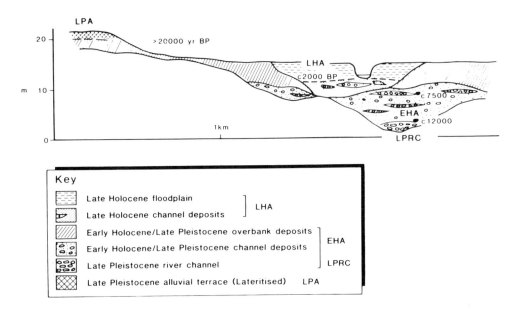

FIGURE 8.4   Schematic section across typical floodplains of small rivers in Koidu, Sierra Leone, indicating periods of late Quaternary deposition. (Reproduced by permission of the Royal Scottish Geographical Society from Thomas, 1988)

amounts to 50–100 m; many interfluves are broad and little dissected.

Stratigraphic data from floodplain sections, soil pits, and drill logs have been examined for this area, and radiocarbon dating of wood and analysis of sediments from several headwater streams have established beyond reasonable doubt that major fluctuations in stream activity have occurred during the last 30 000 years, while analysis of surface deposits has shown a complexity of morphogenesis that must correspond in some way with events in the stream valleys (Thomas & Thorp, 1980, 1985; Thomas *et al.*, 1985; Teeuw, 1989). Samples of black clay, found in thick lenses at the margins of the present floodplains, yielded five dates between 36 000 BP and 20 500 BP. These also correspond with a low terrace. They must have formed under conditions of sufficient humidity to maintain swampy conditions even in quite small floodplains. These floodplains were wider than at the present time though evidence suggests marked oscillations in discharge and sediment loads.

After 20 500 BP no datable material older than 12 500 years was collected (from a total of >40

radiocarbon dated samples). However, the renewed fluvial activity that commenced around 12 500 BP corresponds with timbers jammed under large boulders exposed in bedrock channels, excavated 3–4 m below previous levels and not reached by present streams (Figure 8.3). In larger river channels, pot holes more than 5 m deep were formed, and both these and the deep channels were filled with coarse gravels and clays (Thomas & Thorp, 1980, 1993; and Figure 8.4).

There was more than one phase of this activity and several cut-and-fill episodes took place during the Holocene. It can readily be seen (Figures 8.5, and 7.8) that these events correspond with fluctuating lake levels, extraordinary Nile floods, and spectacular increases in the sediment load of rivers such as the Niger (Pastouret *et al.*, 1978; Street, 1981; Rossignol-Strick *et al.*, 1982; Paullissen, 1989). The bulk of this evidence appears to contradict the view that late Pleistocene climates were typified by low intensity rainfall (Lézine & Casanova, 1989).

Many headwater valleys in this area (Koidu) became infilled with colluvial 'head' deposits that

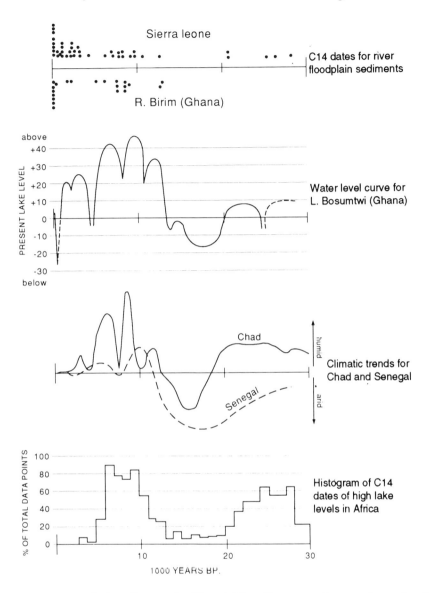

FIGURE 8.5 Late Quaternary correlation diagram for West Africa. The upper diagram shows the distribution of radiocarbon dates from floodplain sediments in Ghana and Sierra Leone, indicating periods of erosion/deposition. These dates can be compared with water level records from Lake Bosumtwi, Ghana (after Maley, 1987); Lake Chad (after Michel, 1980), and a summary of lake levels in Africa (after Street-Perrott *et al.*, 1985) (Reproduced by permission of Chapman & Hall from Thomas & Thorp, 1985)

bury earlier low terrace gravels. In one valley, two phases of colluvial activity are recorded in a situation of minimal groundsurface slope above the valley head (Thomas & Thorp, 1985, and Figure 8.40), and similar infills of colluvium are widespread in this and other areas. On interfluves, surface gravels have become clustered in shallow swales and in places the underlying rock surface has become patinated with iron oxide, possibly indicating strong surface flows of water and removal of finer sediment.

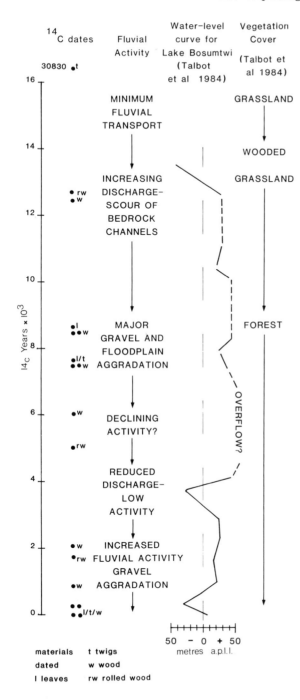

FIGURE 8.6 Suggested correlation between radiocarbon-dated samples, fluvial activity in the Middle Birim River, water level in Lake Bosumtwi and regional vegetation during the late Pleistocene and Holocene in southern Ghana (after Hall *et al.*, 1985; reproduced by permission of the Geological Society Publishing House)

The Middle Birim River, Ghana

A similar record has been obtained from the middle Birim River and its tributaries in southern Ghana (Latitude 6 °N, equatorial regime, 1750–2000 mm year^{-1}). This area is located within 80 km of Lake Bosumtwi which has yielded valuable independent evidence for climate change in this part of west Africa (Talbot *et al.*, 1984). This record has been detailed in Chapter 7 (Table 7.6). The results from the Birim River match those from Lake Bosumtwi and from Koidu in most respects (Hall *et al.*, 1985; Thomas & Thorp, 1992). There is a buried channel, infilled with coarse gravels and dated to 12 500 BP, and the older terrace deposits all pre-date 20 000 BP (Figure 5.8).

What appear to be *mudflow deposits* are found in shallow tributary valleys of the Birim River (Junner, 1943; Hall *et al.*, 1985). The Apiansu (near Akwatia) has been mined for diamonds, revealing several intriguing features. The diamonds were found distributed through a clayey, matrix-supported gravel 1–2 m thick, quite unlike conventional stream gravels in the main river and forming a tongue extending along the valley floor. The weathered phyllite bedrock below exhibits polygonal cracks more than 2 m deep, infilled with pale, coarse quartz sand and fine gravel (Figure 8.27). These features suggest a period of strong desiccation followed by mudflow events, that may have marked the early stages of climate warming after 13 000 BP (at Latitude 6 °N). They were first recognised by Junner in 1943, who found these features rare near the centres of the valley floors but common around the margins, where they, 'suggest dry conditions and very little vegetation at the time they were formed'. No dates for these deposits are available.

All these observations have been placed together for interpretation (Figures 8.5 and 8.6), and can be considered in relation to concepts concerning the response of rivers to environmental changes developed by Knox (1972, 1975) and Bull (1991). The amount of water available for overland flow and stream activity appears to have been insufficient to supply significant discharge to the trunk streams in their upper reaches from 20 500 to 12 500 BP,

TABLE 8.1 Birim valley and Koidu basin chronology

Pre-36 000–21 000 BP	Low terrace, humid/sub-humid oscillations, cool
21 000–13 500 BP	Minimal sedimentation, dry with low-magnitude flash floods
13 500–*c.* 11 000 BP	Increasing discharges and large floods, channel cutting and deposition of coarse gravels
c. 11 000 BP	Possible drier interval lasting around 500 years
10 500–*c.* 7000 BP	Re-establishment of forest with some shallow bedrock channels and coarse sediments
7000 BP–*c.* 4500 BP	Reduced discharges and limited reworking of sediments
4500 BP–3000 BP	Mid-Holocene arid phase, minimal discharges
3000 BP–Present	Increasing discharges and deforestation, sediment reworking and deposition

since timber and other organic trash were not entrained, or were not deposited and preserved during this period. The survival of older timbers, however, suggests that sediment reworking was also minimal during this period. It is possible that the gallery forests largely disappeared, and/or that bank erosion was minimal. Prior to this prolonged dry period (probably at its most arid, 18 000–15 000 BP), conditions may have been cool and moist, and rainfall possibly less intense than today. At the start of the Holocene, conditions became torrential, and although large quantities of sediment became available to the rivers, discharges at high floods were sufficient to cause channel cutting into weathered bedrock. The chronology tentatively put forward is shown in Table 8.1.

Although of little direct interest to palynologists, these results (see also Table 7.6), when placed alongside palynological studies of lake sediments, can be used to address many wider questions concerning past climatic and hydrologic regimes (Thomas & Thorp, 1985; Thorp & Thomas, 1992). The need for geomorphological, pedological and sedimentological evidence for landscape change is reinforced by the quite serious problems encountered when attempts are made to interpret the behaviour of slopes and streams from lacustrine sediments alone. The contrast in approach to the later Quaternary of Central and South America, between Markgraf (1989), who used pollen and lake-level records only, and both Schubert (1988) and Clapperton (1993a,b), who has assembled a wide range of geomorphological, pedological and sedimentological data, is revealing in this context.

Alluvial sediments in Kalimantan, Indonesian Borneo

In western Kalimantan, Indonesia (Latitude 0°, equatorial, 3200–5000 mm year^{-1}), there are large accumulations of Pleistocene alluvial sediments, deposited in the coastal plains (Figures 8.7 and 8.8), and which fall into at least two groups: an older terrace containing abundant timbers dated >40 000 BP at 6–12 m depth (two finite dates 51 000 and 54 000 BP, plus six more >40 ka), and younger floodplain sediments for which there are two near-basal dates of 10 500 BP (Thorp *et al.*, 1990). The older suite is intriguing and it is interesting that similar dates have come from the 'Old Alluvium' (Stauffer, 1973) in a coastal site in Perak, Malaysia (Kamaludin *et al.*, 1993). These are >67 000 years old at the base which is below sea level and <29 000 years old at the top in sediments ~15 m thick. The sediments may relate to prolonged late Pleistocene low sea levels (Figure 8.7), combined with drier climates. It is tempting to suggest that particular episodes of sedimentation occurred at each of the warming peaks on the curve of palaeotemperatures for the last glacial cycle. On the Indian Ocean marine oxygen-isotope record there are clear peaks at 85 000, 58 000 and 40 000 BP (Jouzel *et al.*, 1987, and Figure 7.2), all possibly relevant to the formation of the Older Alluvium of southeast Asia. These and older peaks dated as *c.* 100 000 BP and 130 000 BP are reflected in other ocean signals (Table 7.2), and we should expect some sign on land of these environmental changes.

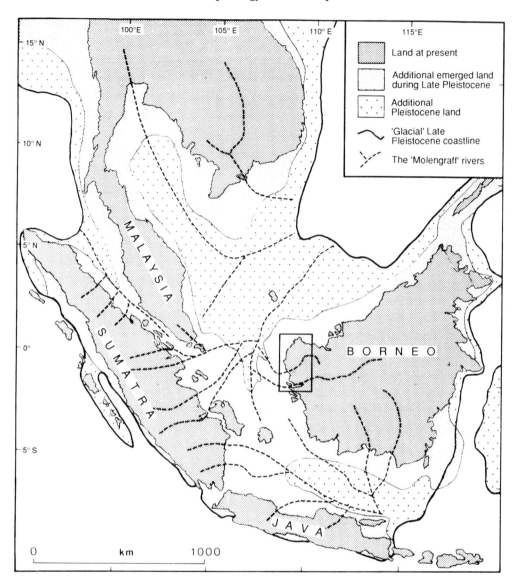

FIGURE 8.7   Sundaland, southeast Asia, during periods of Pleistocene low sea level (after Gupta *et al.*, 1987; reproduced by permission), showing extensions of present drainage. The study area in Figure 8.8 is shown as an inset box

It is sufficient here to point out that the large accumulations of sediment in Kalimantan appear to have been deposited as a result of rapid erosion in small catchments, presently in an equatorial forest environment. The sediments approximate to fans or shallow, braided valley fills and contain little-worn gold and angular quartz shards. They appear today as podzolised *white sands* forming terrace-like features that fall more steeply towards the coastline (4 m km^{-1}) than the gently inclined floodplains. Sudden floods, and rapid deposition appear to have been the conditions of sedimentation, and whilst neotectonic movements in this area cannot be ruled out (Nishimura *et al.*, 1986), they would not adequately account for the sedimentology of the coastal plain deposits (Thorp *et al.*, 1990).

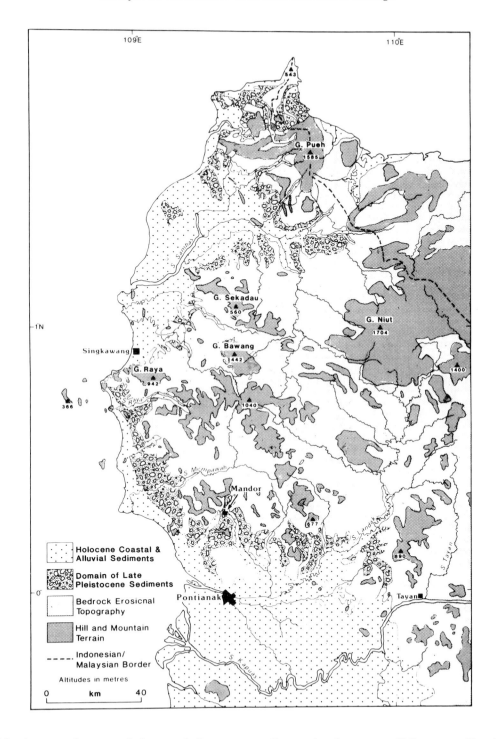

FIGURE 8.8 Aspects of geomorphology and Quaternary sedimentation in western Kalimantan. (Reprinted by permission of Kluwer Academic Publishers from Thorp *et al.*, 1990)

These catchments drain from hilly terrain under 900 m altitude. The wet, equatorial climate is probably optimal for evergreen rainforest *sensu stricto*, and a savanna/forest boundary is difficult to conceive. However, the lowered sea levels of the LGM would have greatly extended the land area, across the Sunda Shelf between Borneo, Sumatra, and peninsular Malaysia, and a drier, seasonal climate has been proposed by Verstappen (1980). Heaney (1991) has also suggested a corridor of low rainfall in the late Pleistocene, extending from southern Thailand through peninsular Malaysia, to include southern Borneo. Walker (1982), however, has expressed doubts about inferred climate change and pointed out that the Sahul and Sunda shelf areas would have been exposed as land but not colonised by forest during the relatively short periods of low sea level at the LGM. The late Pleistocene sediments in this area are almost entirely fluvial in origin and form distinct terraces. The indications of rapid deposition, and aggradation are clear and it is difficult to avoid the conclusion that catchment conditions must have exposed the fragile saprolites to high runoff and severe erosion under an open vegetation cover (Thorp *et al.*, 1990).

Pre-40 000 BP sedimentation in South America.

Schubert (1988) recorded arid conditions of sedimentation for coarse terrace deposits in the northern Venezuelan Andes during a comparable period (*c.* 53 000 BP), and Riccomini *et al.* (1989) have indicated the formation of a palaeopavement with stone-lines *c.* 52 000 years old in eastern Brazil. More recently, Van der Hammen *et al.* (1992a) have dated the sediments of a low terrace on the Caquéta River in Colombian Amazonia, 10–15 m above the river, to the period 56 000–30 000 BP. The lower sediments are sandy with gravels and these pass upwards into grey clays with shallow channel fills, organic remains and some peat. After 30 000 BP the river cut down 10 m and subsequently, after 12 600 BP, built its present floodplain. Van der Hammen *et al.* (1992a) suggested that this sedimentation was influenced by fluvioglacial outflows from the Andean Cordillera during the Middle Pleniglacial. On the other hand, this earlier climatic signal has

also been picked up in the swamp deposits of the Serra dos Carajas (Chapter 7, and Absy *et al.*, 1991). However, intriguing as these parallels are, their true relationships are not yet established.

In assessing the significance of these sedimentary records, Sutherland's (1985) observation that there have been 23 major glacial–interglacial cycles during the last 2.3 Ma, each of which must have caused dramatic environmental changes in the tropics, is clearly relevant. In other words, we should expect the sedimentary record to exhibit the effects of each major palaeotemperature fluctuation, if it was matched by alterations to the rainfall patterns and amounts. The ocean sediment signals for Quaternary climatic change have been discussed (Table 7.2), and cold/dry climate indicators are seen to predominate at 120 000–110 000 BP, 96 000–84 000 BP, 76 000–60 000 BP, perhaps *c.* 40 000 BP, and 23 000–14 000 BP. There is good evidence to suggest that rapid erosion and sedimentation occurred some 1000–2000 years after the last termination, reflecting the instability of the changing climate, together with the lag in re-establishment of the forest. The rather scanty evidence which we have now suggests a similar pattern of events post-58 000 BP (Pollen Zone C, Figure 7.2), with rapid erosion and sedimentation after *c.* 56 000 BP, though the exact timing of these events must remain uncertain. Cooling again set in after 40 000 BP, though conditions were clearly moist across much of the tropics until after 23 000–22 000 BP.

The fluvial sediments thus add to the picture already established from other sources, and confirm that all parts of the landscape were affected by these climatic changes. The behaviour of hillslopes and of small, headwater valleys is clearly a further stage in this discussion.

FORMATION OF PLACER DEPOSITS IN THE QUATERNARY

According to Slingerland and Smith (1986), 'A placer is a deposit of residual or detrital mineral grains in which a valuable mineral has been concentrated by a mechanical agent.' Notable examples of common placer minerals of great economic importance are diamond, gold, ilmenite, rutile and cassiterite—all 'heavy' minerals denser than quartz. More than 50% of the gold ever mined has come from placer deposits (Pretorius, 1976), and major diamond deposits have included valuable placers in Namibia and Sierra Leone. The alluvial cassiterite deposits of southeast Asia are currently the world's largest source of tin (Toh, 1978).

Most placers are water laid and they are usually stream or beach deposits, where the local conditions have acted as a 'dressing mill' for the heavy mineral grains (Slingerland & Smith, 1986). Although the hydraulic conditions for placer concentration are the most important controls on their immediate location, the availability and supply of the heavy mineral grains are decisive factors, and it is in this context that tropical weathering conditions once again prove to be important. This is because the target mineral must first be released from its host rock before it can enter the fluvial system in quantities sufficient to form placers. Many heavy mineral grains are found dispersed throughout large volumes of rock at very low concentrations (e.g. diamond in kimberlite, cassiterite in granite, gold in epithermal rocks) which frequently renders mining of the fresh rock uneconomic.

Climate and placer formation

The sub-humid to humid tropical environments appear to approach optimal conditions for placer formation. Resistant heavy minerals released from the rock fabric by deep tropical weathering enter rivers via mass movement and runoff, and become part of a bedload dominated by quartz and depleted of all feldspars and Fe–Mg minerals. By contrast, arid and semi-arid climates produce only ephemeral water flows that may become so loaded with sediment that they behave as viscous fluids, and fail to concentrate grains according to their natural settling velocities. In extreme cases deposits take the form of mudflows. In some tributary valleys in southern Ghana rich diamond placers are actually associated with local mudflows, but this has been due to a previous cycle of diamond concentration in the Precambrian greywackes from which the grains are derived.

Humid temperate environments have generally led to thin arenaceous weathering mantles, many of which have been wholly or partially removed by glaciation and re-depositited as tills and moraines. Deposition from ice and meltwater has generally failed to concentrate heavy minerals into viable placer deposits.

Placer formation in the Quaternary

The nature of the Quaternary climate oscillations in the humid and sub-humid areas of the tropics tended to confirm the inherent advantages of these areas for placer development. Following the prolonged aridity spanning the last glacial maximum, widespread reworking of floodplain and channel sediments took place, and discontinuous bedrock channels were cut to depths of at least 5 m. The coarse sediments that accumulated in these channels have often been left undisturbed during more recent reworking and channel migration, and it is in these buried gravel bars that rich diamond placers have been found, as in the Bafi headwaters in Sierra Leone, and in the middle Birim river in Ghana (Thomas & Thorp, 1980, 1993; Hall *et al.*, 1985; Thomas *et al.*, 1985; Teeuw *et al.*, 1991). The importance of these late Pleistocene and early Holocene sediments for placer formation is a function of:

(1) reworking of earlier diamondiferous sediments;
(2) deposition of coarser gravels under high energy conditions favouring the entrapment of small diamonds between larger quartz clasts;
(3) repetition of conditions for placer concentration intermittently over 10^3 years;
(4) lower stream power and therefore finer sediment deposition leading to floodplain building during the middle and later Holocene, thus protecting the placers from subsequent erosion.

However, these events were the final phases of diamond concentration that had been continuing over much longer time periods (Sutherland, 1982, 1984). In Sierra Leone, the diamond-bearing kimberlites are Cretaceous in age (*c.* 92 Ma), and

238

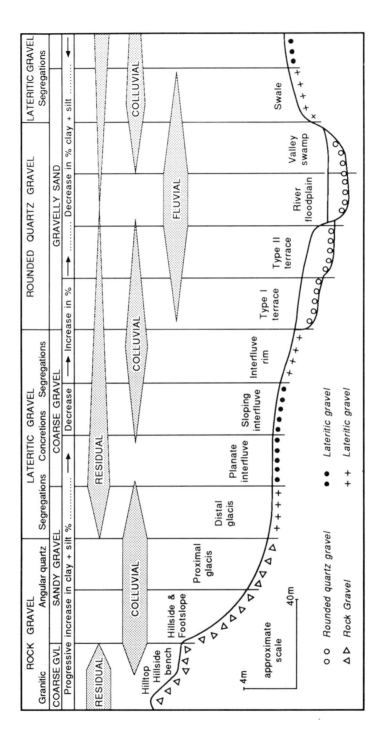

FIGURE 8.9 Process domains and variation in gravel layer composition in the Koidu landscape (after Teeuw, 1989; reproduced by permission of the Royal Academy of Overseas Sciences and Geo-Eco-Trop). Read in conjunction with Figure 8.10

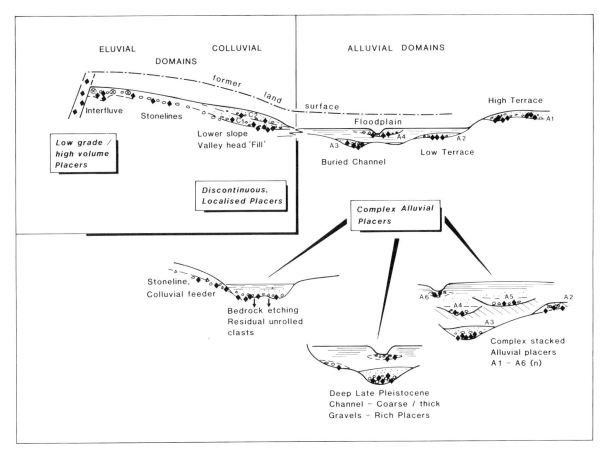

FIGURE 8.10 Process domains within which diamond placers can form in Koidu, Sierra Leone. (Reproduced by permission of Gebrüder Borntraeger from Thomas & Thorp, 1993)

there is evidence for deep erosion into the former pipes, possibly to depths exceeding 1000 m, but certainly to more than 500 m. Much of this erosion took place into a deeply weathered landsurface of possible late Cretaceous to early Tertiary age (Thomas, 1980), so that diamond release was possible over a long period. The Koidu basin, where the placers are found, is also thought to be a downfaulted graben structure (A. MacFarlane *et al.*, 1981), traversed today by rivers with low gradients (Figures 8.1, 8.2 and 9.3).

Although the main placer deposits occur as channel gravels of late Pleistocene age, the Koidu landscape contains hillslopes, glacis and older terrace fragments. Gravel deposits which have been shown to contain diamonds are found at many different levels and belong to *residual, colluvial* or *alluvial* domains (Figures 8.9 and 8.10), and it appears that the diamonds, along with coarse quartz and other clasts, accumulate in etched valley floors and on interfluves as the landscape is lowered (by etchplanation). They may then be transferred from interfluves within stone-lines, to become further concentrated in the river valleys by episodes of enhanced flow.

Optimal conditions for placer formation

Comparisons of this area with other regions of placer formation indicate many similarities (Teeuw *et al.*, 1991; Thomas & Thorp, 1993), and these combine to indicate some of the optimal conditions favouring placer formation (Table 8.2).

TABLE 8.2 Optimal conditions favouring placer formation

(1) Deeply weathered/eroded source rock
(2) Proximity to source or host rocks (<20 km)
(3) Long-term lowering of uplifted, pre-weathered landsurface
(4) Downwarped or faulted (graben) structural setting
(5) Presence of low gradient reaches on saprolites or fine sediments
(6) Occurrence of etched bedrock hollows in both silicate and carbonate terrain (karst depressions)
(7) Coarse clasts in bedload from quartz veins, quartzites or conglomerates
(8) Sediment storage and floodplain re-grading (terrace formation) throughout the Quaternary
(9) Sediments reworked in and deposited in late Pleistocene or early Holocene bedrock channels
(10) Channel sediments containing placers buried by vertical accretion deposits and seldom trenched in Holocene

Mainly after Thomas & Thorp (1993); reproduced by permission of Gebrüder Borntraeger.

The cassiterite (tin) placers of Malaysia and Indonesia occur in similar situations. A thick weathered mantle formed on the granites and andesites is the source of colluvial and fan-like deposits of the locally named 'kaksa' placers (off-shore the Tin Islands Bangka and Belitung) in Indonesia (Aleva, 1973, 1984, 1985; Batchelor, 1979a). These sediments are also found on the Malaysian Peninsula, where coarse boulder and proximal fan facies contain rich placers within karst hollows on the limestone that is peripheral to the granites within which the primary cassiterite was formed. They are probably associated with widespread braided river deposition in the region (Gupta *et al.*, 1987; Kamaludin *et al.*, 1993). Although some terraces contain tektites that have been tentatively dated by their reversed palaeo-magnetism to the Pliocene–middle Pleistocene period (>0.73 Ma) (Batchelor, 1988), confirmatory evidence of a late Pleistocene age for much of the Older Alluvium is now emerging.

However, not all placers are of late Quaternary age. Many are Tertiary or have accumulated throughout the later Cenozoic, and strata-bound placers occur in much older rocks that are not the concern of this study (Minter, 1978). Diamond placers in the Transvaal, South Africa have been described by Helgren (1979) and by Marshall (1988) and many deposits are found in lateritised sediments. Although these have been regarded as older terraces, they may in fact be colluvial in origin and derived from still older gravels in the Vaal basin (Partridge & Brink, 1969; Marshall, 1988), and derived from lateritised deposits on the 'African Surface' of possible Cretaceous age.

HILLSLOPE EROSION AND COLLUVIAL DEPOSITION

A considerable proportion of the superficial deposits in the landscape fall under the heading of 'colluvium' (see Chapter 6). These sediments are important, because they can cover more than 50% of the landscape in humid regions as well as in the drier areas of the tropics, and they attain considerable thickness, especially on lower slopes. They also acquire physical, and sometimes chemical, properties that differ from the sedentary saprolites from which they have been derived.

The formation of colluvial aprons on lower hillslopes depends on the effectiveness of erosional events higher up, and these events may reflect the frequency of high magnitude rainfall events, the degradation or removal of plant cover, or some combination of climate and vegetation change that leads to slope instability.

The notion of stable conditions being interrupted by phases of instability was embodied in the ideas of *biostasy* and *rhexistasy* as advanced by Erhart (1955). Biostasy depends on a biological equilibrium, typically associated with forest, that permits the deepening of weathered profiles and limits denudation to a very low rate of material transfer, mostly in solution. Only chemical sediments will result from these conditions, mainly carbonates and silica. Rhexistasy represents the disruption of equilibria, from tectonic or climatic causes, and leads to the removal of detrital material along with the products of chemical weathering.

This *stripping* of the weathered mantled may be partial or nearly complete, and tends to follow a sequence defined by Millot (1970), starting with chemical sediments under conditions of biostasy and followed perhaps by coal as the forest is destroyed, and then by clays (mainly Al and Fe sesquioxides and kaolin) and quartz sands, before feldspathic sands (arkoses) are derived by erosion of partially altered or fresh rock at the base of the old weathering profiles. This analysis was applied to the great basins of continental sedimentation in Africa (and also in Europe) and to an explanation of the characteristic sediments of the *Continental Terminal*.

Similar principles can be applied to the behaviour of individual hillslopes, and were embodied in the concept of the '*K-cycle*' developed by Butler (1959, 1967) in Australia, though similar ideas were expressed by Ruhe (1956, 1960) from research in Africa. According to Butler (1959, p. 7) the K-cycle may be defined as, 'the interval of time covering the formation by erosion and/or deposition, of a new landscape surface, the period of development of soils on that surface, and ending with the renewal of erosion and/or deposition on that surface'. The periodicity of these events and their impact on similar slopes within regional landscapes (or land systems) was discussed by Butler (1967) and Churchward (1969) in Australia, and similar events in Africa were analysed by Fölster (1969). More recent studies are discussed below.

In eastern Brazil, Modenesi (1988) described a series of superposed colluvia with the most weathered materials at the base and progressively less altered sediments towards the top. This represents an inverted weathering sequence, indicating that the most highly weathered parts of the profile had been stripped first and re-deposited downslope, to be followed by materials from successively lower, and therefore less altered, zones in the weathering profile which were deposited on top.

Many hillslopes in marginal areas of the humid tropics, notably over granitoid rocks, are widely stripped of saprolite, despite the stability of regoliths on present-day slopes of >20° under rainforest. Boulder-controlled slopes and exposed rock slabs are common and extend on to the piedmont or pediment slopes that lead towards the valley floors in areas such as Koidu in Sierra Leone. These features point to the possibility that parts of the humid tropics were subjected to semi-arid conditions for long periods.

Problems of interpretation

Quite complex slope stratigraphies can be built up by such processes, but the interpretation of the formative events poses a number of problems. These can be summarised in terms of: (i) the dynamics of climate change, (ii) neotectonic activity, and (iii) the response of landscape units to change.

(i) *The dynamics of climate change* involve the amounts, distribution and intensities of rainfall, as well as depressions and elevations of temperature means and extremes. Problems associated with these include the following.

(1) The rates of surface processes leading to sediment transfer are not fully understood in terms of their sensitivity to changes in climatic or meteorological parameters; partly because changes to vegetation cover are often unknown.

(2) Key parameters affecting morphogenesis and pedogenesis lack quantification, including relationships of slope to erosion under different vegetation covers.

(3) Intrinsic and extrinsic thresholds within the process systems are imperfectly known, including regolith thickness and slope instability.

(4) The nature of past climates, in terms of the quantification of key parameters such as storm size and frequency, is largely speculative.

(5) The role of extreme events (which may recur in 10^2 years) in triggering erosion and sedimentation is therefore difficult to build into palaeoclimatic reconstructions (requiring periods of 10^3–10^4 years).

(ii) *Neotectonic activity* during the Quaternary has continued in many areas, but is often neglected in palaeoenvironmental reconstructions. Some pitfalls are apparent.

(1) Alluvial channels that control local base levels may be strongly affected by continuing (neotectonic) crustal movement; where movement is positive this leads to dissection of deposited sediment.

(2) Seismicity associated with neotectonic movement can produce significant earthquake-induced events on hillslopes, including rockfalls and avalanches of core boulders.

(iii) *Landscape responses to external forcing mechanisms* will be spatially and temporally complex.

(1) Slope and soil conditions in a landscape will be highly variable and not all *slope facets* will react to a changing environment simultaneously or in the same manner.

(2) Repetition of events at the same locations during successive environmental oscillations is common and may obliterate the evidence of earlier erosional/depositional phases.

(3) Slope conditions are linked (in many landscapes) to local stream channels and slope instability may be triggered by lateral or vertical stream erosion, and not by bioclimatic changes.

(4) Readjustments in the slope system may be due to the *complex response* of the landscape to external factors or to internal re-organisation, and some deposits may have no palaeo-environmental significance.

(5) A significant time-lag in readjustment to new environmental conditions may influence correlation with external factors.

Notwithstanding these theoretical problems, there are good reasons for thinking that many sequences of hillslope deposits have resulted from changes to the external environment, but for the reasons given above, many anomalies are likely to remain.

The importance of colluvial deposition was emphasised in Chapter 6; its significance as an indicator of environmental change is the question explored here. Watson *et al.* (1983) observed widespread colluviation throughout the southern African savannas, where rainfall lies between 600 and 800 mm year^{-1}. They attributed much of this colluvium to the effects of late Pleistocene aridity (20 000–13 000 BP), while Goudie and Bull (1984) referred to two phases of colluviation in Swaziland. Colluvial deposits are not, however, confined to a narrow band of tropical climates. Hill and Rackham (1978) called attention to the importance of colluvial processes linked to alluviation on the Jos Plateau of Nigeria, where rainfall is 1000 to >1200 mm year^{-1}, while detailed studies by a Japanese team in Cameroon (Kadomura, 1977, 1986b) established the importance of colluvial deposits in the forested areas (>1500 mm year^{-1}). Thus Hori (1977, p. 66) commented that, 'in the forest areas (of Cameroon) colluvial deposits almost evenly mantle the hillslopes from the summit to the lower parts of the low-relief hilly terrain including demi-orange hills.'

These statements from studies in Africa can be matched by observations from South America and other tropical landmasses, where rainfalls range from 1200 to >3000 mm year^{-1}. Detailed stratigraphic studies of colluvia have been carried out near Bananal, inland from Rio de Janeiro in Brazil (De Meis & Machado, 1978; De Meis & Monteiro, 1979; De Meis & De Moura, 1984; De Moura *et al.*, 1989). These authors refer to the formation of *rampa complexes* containing bench-like surfaces produced by erosion in saprolites and by colluvial/alluvial deposition. Their findings are discussed below.

Hillslope forms and colluvial deposits

Colluvial deposits are commonly associated with particular slope forms, but can merge with adjacent relief in an indistinct manner. The diffuse nature of water flow and mass movement contributing to the formation of colluvium leads to the construction of undissected slope segments or facets. The conditions of climate and vegetation under which it is thought that these deposits can accumulate involve highly seasonal rainfall regimes with an open vegetation cover. Sufficient rainfall to cause runoff must be combined with sediment yields from slopes high enough to avoid channel cutting and trenching of the slope.

FIGURE 8.11   Rampa complex in southeastern Brazil, comprising landslide scars and colluvium from the erosion of a deeply weathered convex hill, with alluvial deposition forming the foreground surface

These circumstances can promote seasonal transfers of thin layers of sediment, that appear in section as a distinct bedding, with sorting between coarse and fine materials (Figures 8.11 and 8.12). Different conditions can lead to mudflows of heterogeneous material carried downslope in a viscous state, and later deposited as a *polymict* sediment lacking visible sorting or bedding planes. Although both processes can occur together and affect different parts of a landscape, mudflows are usually associated with more arid or more torrential conditions. Mudflows are more often found in valley floors or occupying stream channels (also typical of upper fan environments, especially in semi-arid areas), and their sedimentology and morphology are quite distinct. Water-laid sediments from sheet floods, on the other hand, share some

features with stream deposits and may interfinger with these towards the channel network. Goudie and Bull (1984) adduced evidence for the abrasion of grains due to channelling around corestones exposed by progressive stripping of saprolite during the later phases of stripping and colluviation in Swaziland.

The presence of colluvium-filled *hollows* in humid temperate areas and their significance in landsliding was explored in Chapter 6. Described by Dietrich and Dorn (1984) as 'colluvial wedges', these materials appear to have been particularly active during the climatic changes at the end of the Pleistocene. Reneau *et al.* (1986) found that 9 out of 11 sites in the Pacific northwest of the USA yielded basal dates between 9000 and 15 000 BP which were broadly in agreement with sites in California.

FIGURE 8.12   Colluvial (largely sheet wash) sediments in a valley floor, near Bananal, southeastern Brazil

They concluded that, although climate change was not needed to induce erosion and sedimentation, there was a clear climatic signal in the deposits. This may record the occurrence of more intense storms during the late Pleistocene and early Holocene (Knox, 1984). This region experienced a 'temperate' rhythm of climate change involving increased rainfalls with lowered temperatures (Adam *et al.*, 1981) during the LGM, with a transition to warmer and drier conditions in the early Holocene. This sequence is the reverse of that encountered in tropical latitudes (see Chapter 7), where similar forms and deposits are also common (Jibson, 1989).

Some features of hillslopes in Sierra Leone, West Africa

Dramatic changes in the balance of surface processes affecting hillslopes in the humid tropics are illustrated by features examined in the Freetown Peninsula and several inland mountain groups in Sierra Leone (Thomas, 1983, and Figure 8.13). The Peninsular mountains rise abruptly from the coast to nearly 900 m asl. They are comprised of basic igneous rocks that form steep ridges with flanking slopes >28° (in a monsoon climate receiving 2500 to >3000 mm year^{-1}). The hillslope features (Figures 8.13 and 8.14) include:

(1) rocky outcrops, and boulder slopes, associated with sheeting slabs;
(2) saprolite and colluvium covered slopes;
(3) deep hillslope gullies, infilled with coarse slope deposits of a strong red (5YR) hue, containing boulders;
(4) landslide scars and associated debris that occupy more than 20% of parts of the main hill range of the Peninsula;
(5) ferricreted hillslope colluvium that is becoming fragmented under forested conditions, on slopes >15°;
(6) piedmont glacis, strongly ferricreted and formed of mostly coarse angular debris, suggesting some may be fanglomerates.

There are similar sequences around inland hills. Gaskin (1975) described two periods of landsliding and duricrust formation on the flanks of the Sula Mountains in central Sierra Leone which are formed from supracrustal metavolcanic rocks (Figures 9.6 and 10.16).

Pediments and other piedmont slopes

The term 'pediment' is almost universally used in geomorphology to describe gently inclined, piedmont slopes, from <0.5° to >7° in inclination (more usually 2° to 4°), that connect eroding slopes or scarps with areas of sediment deposition at lower levels (Oberlander, 1989). The potential usage arising from such a definition, however, is very broad, and subject to disagreement over terms and controversy about origins and formative processes (Oberlander, 1989). It is necessary, therefore, to make certain distinctions.

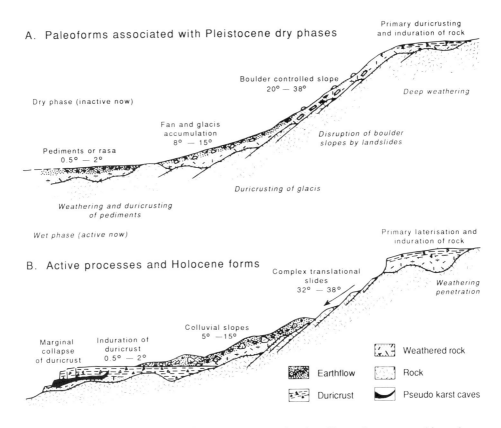

FIGURE 8.13 Hillslope deposits found in the Freetown Peninsula, Sierra Leone, marking changes to slope dynamics from (A) dry periods to (B) humid periods. (Reproduced by permission of Chapman & Hall from Thomas & Thorp, 1992)

Rock pediments have been associated with theories of scarp retreat in resistant rocks, and this will be discussed elsewhere (Chapters 9 and 11). At this point in the argument it will be observed that these features adjoin hillslopes either at an abrupt *piedmont angle* or via transitional slopes of intermediate inclination. Rock pediments are sometimes characterised as carrying very little sediment save for a scattering of coarse clasts and boulders. However, this situation is unusual and most carry an irregular, thin veneer of sediment, perhaps 2 m thick, but exceptionally reaching depths of 10 m. This has the effect of obscuring the uneven rock surface beneath. On granitoid rocks the pediment may be partially mantled by *in situ* regolith, while elsewhere bare-rock pavements emerge (Twidale, 1981b). Many granite pediments

are uneven forms displaying groups of core boulders (Oberlander, 1972, 1974, 1989), and this belies the simple idea of a plane surface swept periodically by unimpeded sheetfloods.

The role of pediments in landform evolution will be considered in Chapters 9 and 11. Here, their significance in relation to climate and climate change will be emphasised. The alternate mantling and stripping of pediments is a theme of several key studies in this field (Ruhe, 1956; Mabbutt, 1966; Busche, 1976; Oberlander, 1974, 1989; Twidale, 1981a). On granite pediments, weathered mantles can be renewed and then later stripped during a sequence of climatic oscillations associated with ecological change. In this context Oberlander (1974, 1989) has demonstrated the importance of more humid climates in pediment formation and

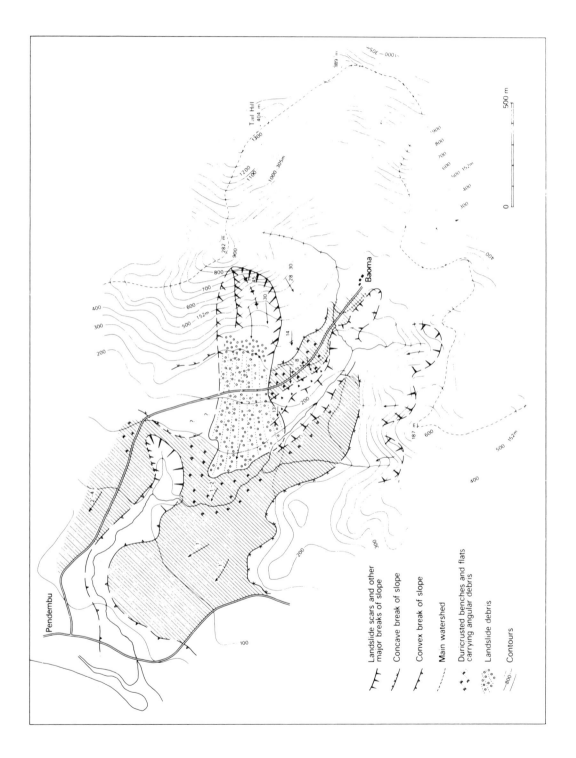

FIGURE 8.14 Hillslope and footslope deposits in the Pendembu Valley near Freetown, Sierra Leone, showing recent landslide debris overlying ferricrete benches (as in Figure 8.13)

extension in the Mojave Desert, USA. Similarly Busche (1976) identified several cycles of deep weathering and stripping, as well as phases of alluvial deposition on a series of pediment surfaces in the Tibesti Mountains of Chad.

Thus within arguments about climate change, pediments occupy an interesting position, being widely regarded as markers of more humid conditions when found in the arid zones (Busche, 1976; Oberlander, 1974, 1989), and as indicators of drier conditions, where they occur within the humid tropics (Thomas & Thorp, 1985). There is a measure of agreement that the processes of pediment extension and maintenance are favoured by highly seasonal, wet–dry, climates. These may be associated with thorn forests (Birot, 1968) or *chaparral* in southwestern USA (Oberlander, 1989), or simply with 'savanna' conditions. If true, this suggests optimum rainfalls for pediment formation of 400–600 mm year^{-1} in the temperate zone and perhaps 600–800 mm year^{-1} in the tropics, though it would be unwise to set too much store by these figures.

King (1953, 1957) insisted that pediments were formed in all climatic environments. In this he has been supported by Twidale (1981a), though these writers differ in their understanding of the role of pediments in landscape evolution. Neither author accords great significance to climate change. The subject is frought with difficulty at this level because the rock pediment retains so little evidence of its age or origins. 'Pediments' that carry significant deposited sediment are, however, important in this context.

Many writers have used the term 'pediment' to describe depositional surfaces fronting mountain ranges and other hills, when these may actually take the form of *fans*, often coalescing to form a *bahada*. The term 'pedisediment' has also been used to describe the alluvial apron beyond the erosional pediment. More helpfully, Bigarella and Becker (1975), in their review of the Brazilian Quaternary, distinguished clearly between *rock pediments* and *detrital pediments*, and presented a model to show how each erosional pediment could be traced to correlative (basin) deposits (Figure 8.16). These, however, were subsequently referred to as *terraces*.

Rohdenburg (1969/1970) employed the term 'slope pediment' to describe slopes resulting from the retreat of low scarps in unconsolidated materials, usually saprolite, and based on studies from the humid tropics of southeastern Nigeria. He also ascribed this process to the effects of climate change, towards less humid conditions in the forest zones. This work has been extended by De Dapper (1989).

This usage is close to the French term '*glacis d'érosion*' (Birot & Dresch, 1966) which can be used to describe any gently inclined surface truncating weak rocks, unconsolidated sediments or saprolite. In Brazil the term '*rampa*' (Portuguese) has been used to describe hillslope hollows fronted by ramp-like slopes of both erosion and accumulation. French writers sometimes recognise '*glacis d'accumulation*' as a separate category of constructional slope. A list of some equivalent and alternative terms to describe mainly piedmont slopes, together with the preferred usage in this study is offered in Table 8.3.

THE INTERPRETATION OF SLOPE FORMS AND DEPOSITS

A problem with many of these landforms and deposits is their indeterminate age, and in the present context this becomes a serious difficulty. It is also important to attempt a distinction between morphogenetically relict features and dynamic forms. But, while some processes are confined to a particular range of climates (e.g. aeolian processes), others take place with varying intensity or frequency across a wide range of conditions (e.g. landslides, formation of stone-lines). In many cases the 'stratigraphic position' or landscape context can offer relative ages for landforms and deposits, and complex weathering and pedological sequences can be determined from stone-lines, datable artifacts and soil micromorphology (Bigarella, *et al.*, 1965; Kadomura & Hori, 1990).

Widespread *fans* and *glacis* are present as major landscape elements in the humid tropics, and have been documented from Brazil, West Africa and from West Malaysia. In Brazil, Bigarella and Andrade (1965) recorded three pediments, each

TABLE 8.3   Common terms applied to piedmont slopes

Generic terms:	(1) *Piedmont*: any extensive landsurface extending from a neighbouring upland
	(2) *Terrace*: any depositional surface of minimal slope bordering a valley or drainage channel
	(3) *Fan*: an accumulation of channel sediments and flow deposits marking the exit of an erosional catchment on to the piedmont
Specific terms:	(1) *Rock pediment*: a gently inclined surface, originally undissected, exhibiting exposed rock or only a thin veneer of regolith or sediment
	(2) *Detrital pediment*: a gently inclined surface of accumulation or erosion within unconsolidated sediments, normally found beyond the distal limit of a rocky pediment
	(3) *Glacis d'érosion*: an originally undissected surface of erosion cut across weak rocks, unconsolidated sediments or saprolite
	(4) *Glacis d'accumulation*: a constructional surface built of unconsolidated sediments
	(5) *Slope pediment*: a type of glacis (3), normally cut within the saprolite and associated with a low scarp and the accumulation of coarse surface debris forming a stone-line
	(6) *Rampa*: a ramp-like slope, partly erosional and partly constructional, often associated with hillslope hollows and colluvium in rampa complexes.
	(7) *Pedisediment*: sediment accumulated on or beyond the distal end of a rock pediment; usually a sandy colluvium
Proposed usage:	(1) *Pediment*, a term reserved for features resembling rock pediments, as in (1) above
	(2) *Glacis*, a term to describe forms encompassed by (3), (5) and (6) above, therefore covering surfaces cut across unconsolidated materials; the term 'glacis' is used here as the equivalent of 'slope pediment'.

associated with coarse detrital material, and these were attributed to more than one phase of Pleistocene aridity (Figures 8.15 and 8.16). Although this early work lacks firm dates (revisions by Bigarella and Andrade-Lima (1982) are used here), it advanced a coherent model for the development of Brazilian relief. After the formation of a summit plain (Pd$_1$) (possibly Oligocene), a further pediplain was formed during the Upper Miocene (Pd$_2$), cut below the former level and carrying residual inselberg relief. The detritus from this erosional episode supplied sediment that now forms the Barreiras Formation on the margins of the Atlantic shield. The last extensive pedimentation (Pd$_1$) was thought to date from the Upper Pliocene–Lower Pleistocene, and cut back selectively, supplying sediments to subsiding basins (e.g. the Guabirotuba Formation in the Curitiba Basin). According to Bigarella and Andrade (1965) this surface is warped down towards the Atlantic coastline and has been dissected into the multi-convex 'half orange' topography that is very typical of the humid areas of eastern Brazil.

Into this pediment, which he re-labelled P$_3$ (=Pd$_1$), are cut two further pediments (P$_2$,P$_1$), each with its correlative deposits (Figure 8.16). Significantly, Bigarella and Andrade (1965) attributed each phase of pediment extension to prolonged aridity during the Pleistocene, while the dissection took place during interglacial intervals of humidity. The initial correlations were updated in 1982 (Bigarella & Andrade-Lima, 1982) and these studies (Bigarella & Ab'Saber, 1964; Bigarella & Mousinho, 1966; Bigarella & Becker, 1975) remain pivotal contributions to the study of the geomorphology of the Brazilian Quaternary.

However, Klammer (1981, 1982) has argued that fossil pediments and talus fans in both the Pantanal Mato Grosso and in southeastern Brazil, between the coast near Rio de Janeiro and Rio Paraguay date from the pre-glacial Neogene and record a single major arid period which was succeeded by fluctuating humidity during the Pleistocene. Klammer (1981) held that the humidity of the Atlantic coast of Brazil is dependent on incursions of cold air from Antarctica, and that,

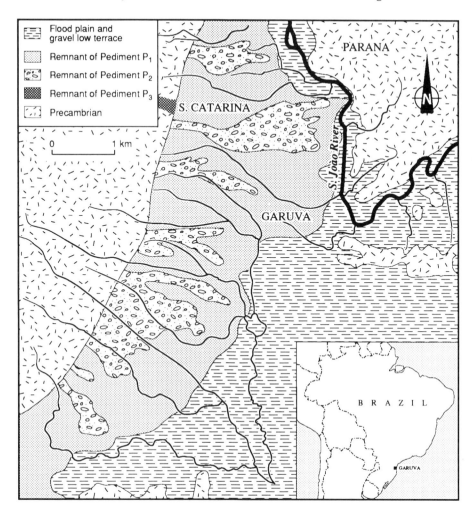

FIGURE 8.15   Distribution of pediments and correlative deposits near Garuva, at the Paraná–Santa Catarina border, Brazil. The schematic profile of the area is shown in Figure 8.16. The Precambrian area is made up of crystalline rocks and is mountainous; the pediments are composed mainly of detrital material, with rocky pediments only in the western part near the mountains (after Bigarella & Andrade, 1965; reproduced by permission)

prior to the extension of the Ice Cap, this area would have experienced prolonged aridity.

There is now evidence to support a combination of both sets of ideas. Periods of aridity within the humid tropics probably occurred repeatedly during the late Cenozoic (Chapter 9), and most probably embracing the Middle Oligocene (31–29 Ma), Middle Miocene (15–14 Ma), the Miocene–Pliocene transition (6–5 Ma) and the Pliocene-Pleistocene threshold (2.6–2.4 Ma). The importance of these

episodes is commonly underestimated but they are likely to have been responsible for many of the major, relict landforms in the humid tropics.

Similar phenomena are widespread in West Africa, and Michel (1973) has also described three glacis from the basins of the Gambia and Senegal rivers:

(1) the *high glacis* is present as fragments carrying weathered duricrust blocks on low interfluves and is probably mid-Tertiary;

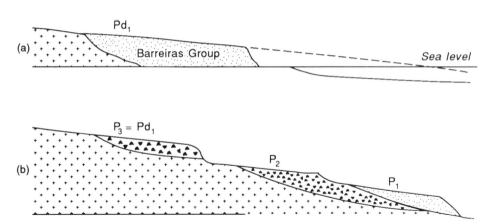

FIGURE 8.16   Schematic long-profiles through the detrital and erosional pediments in the Garuva area, Brazil (see Figure 8.15) (after Bigarella & Andrade, 1965; reproduced by permission)

(2) the *middle glacis* is strongly duricrusted with ferricrete, forming piedmont benches around many residual hill masses (mostly of mafic rocks) and is considered Pliocene or early Pleistocene; while

(3) the *low glacis*, according to Michel (1973), is currently undergoing ferrallitic pedogenesis and is developed within present-day river valleys.

Slope pediments from Nigeria and Brazil described by Rohdenburg (1969/1970, 1982) and from West Malaysia by De Dapper (1989) belong to this group of forms and may be considered as glacis. De Dapper and Debaveye (1986; Debaveye *et al.*, 1986) established a date of *c.* 30 000 BP from volcanic tephra for one such glacis, correlated with an alluvial terrace.

Similar features can also be found in the present forest zone of Sierra Leone (Thomas, 1980, and Figure 8.17), and, as in the case of the fans and pediments described by Klammer (1981, 1982) from Brazil, the two forms are closely associated. Glacis are prominent in the landscape of Koidu, Sierra Leone (Figures 8.18 and 8.19), and glacis around the coast and interior valleys of the Freetown Peninsula contain coarse, fanglomerates, now heavily duricrusted by iron percolating from the surrounding basic igneous rocks (Figure 4.15). Similar duricrusted platforms are found around the Kasewe Hills and other interior massifs. They must

mark prolonged periods of aridity interspersed with more humid climates encouraging migration of Fe^{2+} (Figure 8.17). They appear to correspond with the middle glacis of Michel (1973) and predate the glacial Quaternary (hematite in the ferricrete of one example has reversed polarity, and therefore is older than 0.73 Ma).

These older features of prolonged aridity (Goudie, 1983a; Klammer, 1978, 1981) are widely found and possibly correspond with the greatest extensions of the Kalahari sand sheet (De Dapper, 1988), though these and other dune occurrences such as those of the Llanos of Venezuela (Schubert, 1988) were re-mobilised in the late Pleistocene.

Colluvial deposits and stratigraphy

Superposed layers of colluvium characterise some lower hillslopes and also valley fills, where alluvial and colluvial processes may have alternated frequently. Buried soils are often associated with these sequences (Botha *et al.*, 1990). Stratigraphic principles applied to soils and sediments are well documented (Butler, 1959, 1967; Ruhe, 1960; Mabbutt & Scott, 1966; Morrison, 1978; Finkl, 1980, 1984; Birkeland, 1984), and have frequently been used in establishing Quaternary sequences, mainly in non-tropical areas. In many cases, however, buried soil horizons are not easily distinguished or the organic-rich A horizon may have been removed

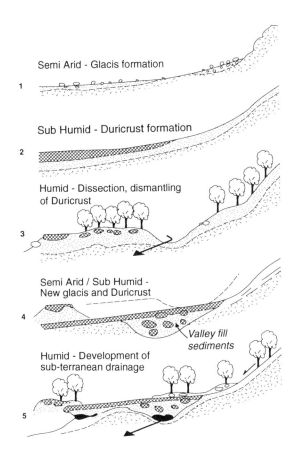

Semi Arid - Glacis formation

Sub Humid - Duricrust formation

Humid - Dissection, dismantling of Duricrust

Semi Arid / Sub Humid - New glacis and Duricrust

Valley fill sediments

Humid - Development of sub-terranean drainage

FIGURE 8.17 Schematic reconstruction of Quaternary hillslope-glacis development in Sierra Leone. An upper glacis is being dismantled and a middle glacis is strongly ferricreted and with subterranean drainage channels. The extensive middle glacis may be largely early Quaternary in origin. A lower glacis is not shown but extends locally towards present drainage

by erosion, making dating very difficult if not impossible.

Colluvium in the forested tropics of southeastern Brazil

Large areas of South America and Africa, and significant portions of other tropical landmasses are underlain by extensive plateaus, frequently dissected into a multi-convex, *demi orange* (Portuguese *meias laranjas*) topography. On these landsurfaces,

complex mosaics of surface deposits are found. To illustrate these important phenomena, some specific examples will be used.

Studies from Brazil have revealed widespread mass movement, and colluviation contributing to what have been described as *rampa complexes*. These comprise amphitheatre-like hollows and associated glacis, which may be both colluvial and alluvial in origin. The deposits are complex in detail and a tentative chronostratigraphy has been advanced (De Meis & De Moura, 1984; De Moura *et al.*, 1989). These features have been described from three sites on the faulted basement migmatites in the hinterland of Rio de Janeiro: in the middle Rio Doce valley (De Meis & Monteiro, 1979), the Bananal area in the Rio Paraiba basin (De Meis & De Moura, 1984; De Moura *et al.*, 1989), and on the high plateau of the Campos de Jordao, São Paulo State (Modenesi, 1983, 1988) (Figure 8.18). These locations vary in altitude from < 500 m to > 2000 m asl, and each displays a characteristically steep, multi-convex topography incised 50–100 m below an older planate surface. Hollows have often been formed by deep-seated slump-slides in thick saprolite, though some have been extended by slope retreat into the regolith, producing *glacis d'érosion* or erosional *rampas* (De Meis & Monteiro, 1979). The landscape of hollows and rampas is one of dominantly concave elements cut into the multi-convex relief, the distal ends of the rampas merging and interfingering with the alluvial deposits of the valley floors (Figures 8.11, 8.12 and 8.19). In places, gently convex depositional surfaces have been termed '*lombas*' (Modenesi, 1988). On the Campos de Jordao, Modenesi (1988) described three generations of deep-seated slump-slides and recognised that these could arise from prolonged rains within a humid climate (De Ploey & Cruz, 1979), but preferred to interpret them as having occurred during the wet–dry transitional climate of the Middle and Late Pleistocene. This view was based mainly on the association of the slides with coarse terrace deposits. These events were considered to have pre-dated colluvial phases of the Holocene which led to the rampa complexes. Shroder (1976) has described similar features on the Nyika Plateau of Malawi.

252

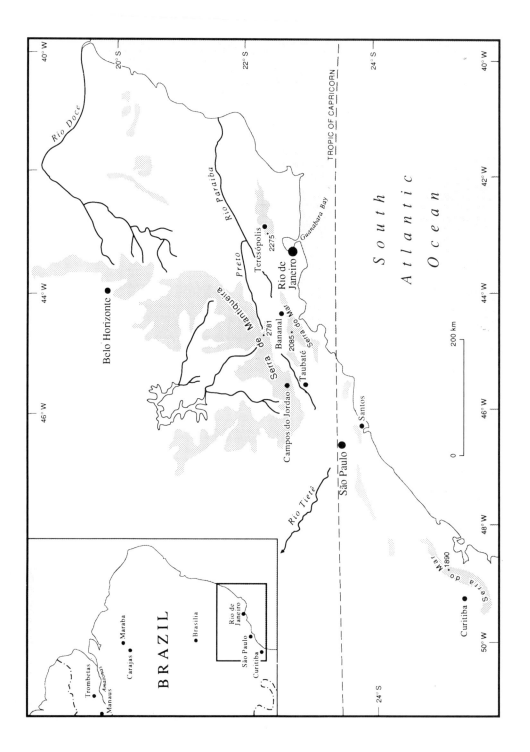

FIGURE 8.18   Location map of areas in southeastern Brazil mentioned in the text

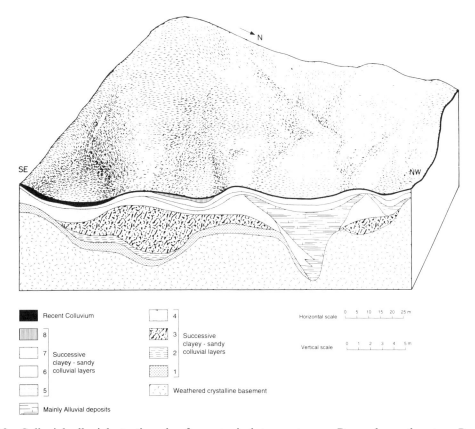

FIGURE 8.19 Colluvial–alluvial stratigraphy for a typical transect near Bananal, southeastern Brazil. Local stratigraphic terms have been omitted for clarity, but polyphase hillslope development can be correlated across a wide area (derived from De Moura *et al.*, 1989, De Moura & Mello, 1991)

Many of the valley floors are thickly alluviated, giving the effect of a 'drowned' relief, and in the Rio Doce valley (20 °S), east of Belo Horizonte, there are also a number of lakes and some dry lake beds. According to Pflug (1969) these are due to excessive sedimentation in the main river channel, raising its bed by as much as 20 m, and effectively damming many tributary valleys. This phenomenon was explained in terms of high rates of erosion on the mountain fronts rising 750 m above the valley, during a semi-arid period coinciding with the last glacial advances of the Pleistocene. One radiocarbon date of *c.* 14 000 BP was obtained from the aggraded sediments. The tributary valleys draining the smaller catchments within the dissected piedmont could not aggrade their channel levels at

the same rate and therefore became impounded as lakes.

The complex stratigraphies revealed in cuttings and boreholes from these different sites have contributed to a regional picture of repeated episodes of enhanced areal erosion correlated with downslope colluvial fan formation (De Meis & Monteiro, 1979). Most of these deposits have been attributed to the Holocene and are thought to be post-10 000 BP, but there are very few dates available. To produce these features, the present-day humid climates would need to have been drier, perhaps supporting a *cerrado* or savanna woodland over an extended period. This view obviously requires quite important climatic swings during the Middle Holocene, following the re-establishment of

FIGURE 8.20   *Glacis* or pediment forms cut across saprolite and less resistant rock, probably during periods of aridity in the Pliocene and Pleistocene, Koidu, Sierra Leone

the rainforest, probably by 9000 BP. Evidence for several dry periods during the Holocene, each lasting at least 500 years, has come from other tropical areas (see Chapter 7).

Account must also be taken of the late Tertiary and Pleistocene tectonic movements in this area. The impact of these movements on stream thalwegs and sedimentation, and on the generation of landslides may have been decisive, and they appear to have been continued alongside the climatic oscillations of the later Quaternary. All these sites lie along the Serra do Mar rift system which runs parallel with the coast of Brazil from near Rio de Janeiro through São Paulo. Several sedimentary basins (graben) are contained within this system and both normal and transcurrent faulting have been

recorded (Hasui & Sadowski, 1976; Hasui & Ponçano, 1978; Riccomini *et al.*, 1989).

Furthermore, normal faults have been shown to affect the stone-line and the late Quaternary colluvium in the Rosende Basin (Riccomini *et al.*, 1989), indicating continued crustal movement at least until the late Pleistocene, and probably continuing until the present day. It is, therefore, important to recognise that both climate changes and tectonic activity have occurred. With care, their separate signals may be understood, but with largely undated sediments, this remains difficult. Furthermore, the recurrence of landslide-forming events in eastern Brazil is such as to warn that some features may be a part of the contemporary climatic system.

FIGURE 8.21   Detail of the Koidu landscape, showing exposed tors, glacis surface and incision of the present drainage (mining camp, Yengema)

An adequate summary of the late Quaternary chronology for southeastern Brazil is not yet available but Riccomini *et al.* (1989) have compiled a section for the southwest Taubaté Basin, near São Paulo. This indicates at least two major cycles of sedimentation, with two stone-lines or palaeo-pavements dated to *c.* 52 000 BP and *c.* 18 000 BP, both dates having significance elsewhere (Schubert, 1988; Thorp *et al.*, 1990). The second cycle of colluviation is considered to be entirely Holocene which corresponds with the work already cited.

Studies of colluvium in West Africa

Another humid tropical area that has revealed widespread colluvial deposits is southern Cameroon (Hori, 1977, 1986; Tamura, 1982). But, although considerable thicknesses of fine sediment were observed (3–14 m), a less complex stratigraphy was described. Tamura (1982) described infilled hillslope hollows that appear to be similar to those in California and Oregon in the USA (see Chapter 5; also Dietrich & Dunne, 1978; Reneau & Dietrich, 1987). As in many humid tropical landscapes, there appears to be only one migratory layer above the residual soil. This layer usually has a fine upper horizon and a distinct gravel layer or *stone-line* marking the junction with the residual soil. Such layers may be only 1–2 m thick and largely attributable to *bioturbation* which involves no lateral transfers on gentle slopes.

The landscapes examined by Hori (1977) in the Yaoundé area were of the *demi orange* type (Kadomura, 1977), but the complexities of the

FIGURE 8.22   Alluvial fans near Chipata in eastern Zambia (900 mm year^{-1}); probably formed during late glacial
dry climates

sections were attributed mainly to horizons in the weathering profiles (Figure 8.23). In most sections, the superficial materials thicken downslope where, *slope pedimentation* may have occurred, as a result of the retreat of a low scarp, usually in duricrusted saprolite (Figure 8.25).

This process was described by Moss (1965; also Figure 8.24) from southwest Nigeria and also by Rohdenburg (1969/1970) from southeast Nigeria. More recently, studies from peninsular Malaysia have confirmed these findings (Debaveye & De Dapper, 1987; De Dapper, 1989; Embrechts & De Dapper, 1987). These authors tend to use the term 'pedisediment' to describe the colluvial layer (Figure 8.26). The wider significance of these observations was indicated by Moss (1965) who suggested that some more or less structureless sandy deposits which had been given stratigraphic names

on the basis that they were conventional alluvial sediments, were probably colluvium formed by local transfer of saprolitic material during scarp retreat. The *acid sands* of southern Nigeria and the *terre de barre* of Benin (former Dahomey) were cited as examples.

ARID ZONE DEPOSITS AND COLLUVIUM IN THE SAVANNA TROPICS

It has taken some time for the importance of colluvial sediments to become accepted for the humid tropical zone, but in the savanna and semi-arid areas such deposits have long been widely known, and their importance has been stressed by Watson *et al.* (1983) and by Dardis *et al.*, 1988. If large areas of the humid tropics became open thorn woodlands or savanna, then much of the present

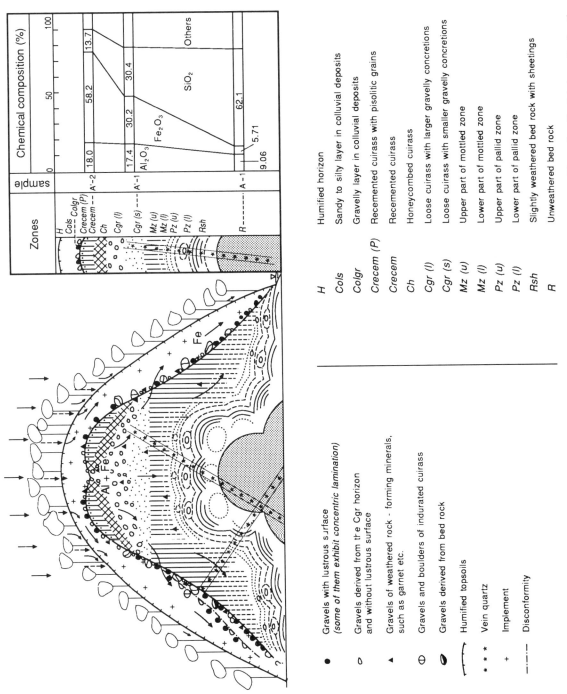

FIGURE 8.23   A model showing the relationship between demi-orange topography, weathering profile and colluvial deposits for the Yaoundé area of Cameroon (after Hori, 1977)

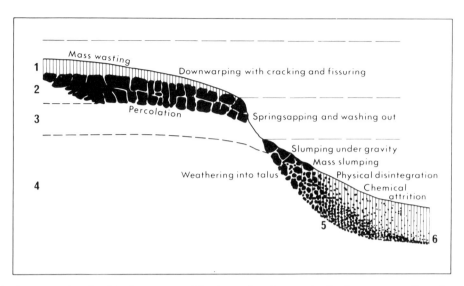

FIGURE 8.24   Processes involved in the retreat of ferricrete breakaways in the forest zone of southwestern Nigeria (after Moss, 1965; reproduced by permission of Blackwell Scientific Publications). (1) Surface soil mantle; (2) hard duricrust (ferricrete); (3) altered and weathered sandstone; (4) ferruginous sandstone with mudstone bands; (5) weathering talus deposits; (6) non-mottled ferrallitic sandy clay on lower pediment

savanna zone must have become steppe-like in appearance during the glacial maxima. There are several reasons for thinking that the strongly seasonal climates became very dry.

The existence of sand dunes

Many areas of Africa, Australia and northern South America have revealed actual dune sands beneath present-day woodlands. In South America the Llanos of Venezuela are notable, while the palaeodunes of the present Sudanian zone of West Africa are well documented (Grove, 1958; Grove & Warren, 1968). In the southern hemisphere, similar features are found in Western Australia, and in the Kalahari of southern Africa (Lancaster, 1981). The so-called Kalahari Sands may be an older pre-Pleistocene cover, but they were probably re-activated in many areas during the Pleistocene. Satellite images of western Zambia and parts of Angola testify to these events. A small but significant dune field is also found in the São Francisco Valley of eastern Brazil (Ab'Saber, 1991) some 500 km south of the present-day northeastern dry zone. The age of the dunes is uncertain but they

appear young and probably relate to the LGM. Dunes also occur in the Pantanal depression in the Mato Grosso, where present-day rainfall exceeds 1200 mm. Here they are associated with large fans or alluvial cones, fronting the pediments along the foot of the Palaeozoic escarpment. Klammer (1982) described this relief as 'that of a desert come under the influence of a humid climate', but he did not relate this change to glacial–interglacial oscillations. As in other publications, Klammer (1981, 1982) thought that most of the arid forms in Brazil were pre-glacial but Tricart (1985) considered these forms to be late Pleistocene.

The occurrence of pans

The fans and dunes of the Pantanal are also associated with shallow clay or salt pans. This association is also found on the central African plateau, in southeastern Angola and western Zambia. The pans may have formed during the drying of the climates, when the drainage became disorganised, and maybe deflation follows. Many of these small pans are still visible on satellite images, although drainage channels now

FIGURE 8.25 Colluvial rubble derived from a duricrusted weathering profile in the forest zone of Sierra Leone (see Figure 8.22)

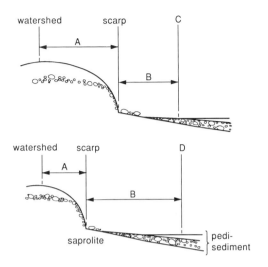

Deposition of the pedisediment with retreating scarp. The stone-layer is deposited at the foot of the scarp. The fine material is :
(A) transported by sheet wash and gully erosion ;
(B) transported over the stone-layer without important deposition or take-up of material ;
(C) deposited after transportation over a *short* distance over the stone-layer (basis of the hill wash cover) ;
(D) deposited after transportation over a *long* distance over the stone-layer and the hill wash cover (top) and hence after a relative increase of the amount of fine grains and an improvement of the sorting with regard to situation (C).

FIGURE 8.26 Model for the deposition of pedisediment with retreating scarp. The stone layer is deposited at the foot of the scarp, while fine sediment is deposited at the distal end of the slope pediment or glacis (after Embrechts & De Dapper, 1987; reproduced by permission of the Royal Academy of Overseas Sciences and Geo-Eco-Trop)

traverse them (present rainfall 600–800 mm year^{-1}). Some of the pans may have formed subsequent to the dunes and during a period of quite strong slope wash. Comparable conditions are found extensively in the Australian semi-arid zone.

In the dry zone of southeast Ghana, which today experiences a low rainfall and savanna conditions (*c*. 800 mm), Brückner in 1955 described ventifacts in the ubiquitous stone-lines and claimed that arid conditions prevailed in the Quaternary. This area is not far from the now-forested mudflows of the middle Birim valley discussed above, and this claim now needs to be taken very seriously. What appears to be a mudflow deposit on a low-angled slope in the Accra Plains of southeast Ghana is illustrated in Figure 8.28.

Alluvial fans

Fans are not specifically related to dry conditions (Nilsen & Moore, 1984; Kesel, 1985), but tend to accumulate at the base of mountain fronts or other abrupt slopes. In many instances they record long-term spasmodic uplift of faulted blocks. Nonetheless, fans will form in specific locations as a consequence of changes to the local relief (as in formerly glaciated terrain) or to the factors controlling stream energy and sediment load. The latter usually result from important environmental changes.

The influence of differential weathering on relief forms in the tropics is one reason for the widespread occurrence of a sharp *piedmont angle* around

FIGURE 8.27   A polygonal block of weathered phyllite, coated with coarse bleached sand, excavated from beneath possible mudflow sediments in a Birim Valley tributary, in the forest zone of Ghana (~50 km from Lake Bosumtwi)

residual hills on the cratonised landscapes of the Gondwana continents. Most such hills carry little residual saprolite, so that water flows, though often ephemeral, carry little sediment. However, around more complex plateaus and groups of hills escarpments have developed, sometimes in response to powerful uplift. A combination of relief, tectonism and climate change appear to have produced some important fan accumulations in these situations.

Very large fans have formed in the Pantanal, Brazil. The largest, currently dissected by the Rio Taquari, covers 50 000 km², and falls from 135 m to 95 m asl over a distance of more than 250 km. Klammer (1982) considered this pre-glacial but

Ab'Saber (1991) is of the opinion that it was active during the 23 000–13 000 BP dry period of the last glacial. Actual dates have yet to be obtained. Large *talus fans* were also described by Klammer (1981) from the Rio Paraguay valley near Urucum. Fans, in places associated with groundwater laterites, are also prominent features in the landscape of the Chipata area of eastern Zambia where rainfalls are *c.* 900 mm year^{-1} (Figure 8.22). It is also apparent that duricrusted pediments bordering the mountains of the Freetown Peninsula in Sierra Leone (<2500 mm year^{-1}), are comprised of a coarse iron-cemented breccia, and were probably at one time debris or talus fans (Figure 4.15) (Thomas, 1983).

FIGURE 8.28   Possible mudflow deposits on the Accra Plains, southern Ghana (present-day rainfall ~700 mm year^{-1})

The Jos Plateau of Nigeria, which although 12 °N has a present-day rainfall in excess of 1200 mm and was formerly forested, illustrates fan formation, both internally between residual granite massifs and peripherally along the escarpment zone. Accumulations of probable *mudflows* form a major part of the plateau landscape (Hill & Rackham, 1978, and Figure 8.29), and R. Zeese (1991) has dated material from the base of one fan sequence to 18 000 BP. The combination of stripped granite hillslopes and colluvial/alluvial deposits in the valleys is most striking, and although earlier phases of alluviation have complex relationships with volcanic activity, the late Pleistocene sediments are certainly climatically determined. The deposits are poorly stratified and attain thicknesses of 5–15 m (exceptionally 20 m). The surface slopes are gentle, 1–7° (Plate 8-I and Figure 8.30). It can also be noted that these deposits have been subject to rapid

erosion and deep gullying, which although apparent in 1952 (Grove, 1952), advanced rapidly during the following 20 years, during which many gullies were cut back by some 300 m (Jones, 1975; and Chapter 5. Figures 5.9 and 5.10).

The escarpments of the Jos Plateau are also marked by large *fans* which are dissected by the present lineal drainage, and by an abrupt *piedmont angle* between steep rocky hillslopes and surrounding pediments or glacis. The total combination of landforms and deposits characteristic of the Plateau and its margins is a powerful demonstration of the combined and sequential effects of pre- or early Quaternary deep weathering, and later Quaternary stripping and sedimentation in response to climatic oscillation.

The colluvium of the non-forested tropics is thus widespread and often deep. It tends to form low-angled surfaces of accumulation abutting steep

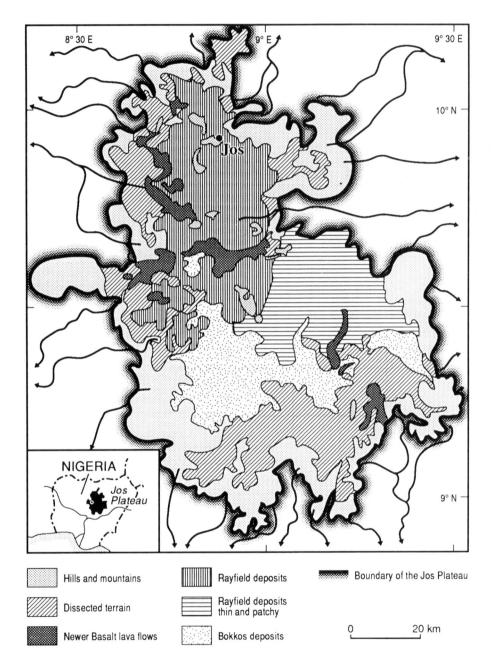

FIGURE 8.29   Aspects of the geomorphology of the Jos Plateau, northern Nigeria, showing distribution of colluvial–alluvial deposits (Rayfield and Bokkos deposits), associated with stripping of saprolite from Younger Granite ring complexes (after Hill and Rackham, 1978; reproduced by permission of Gebrüder Borntraeger)

FIGURE 8.30 Saprolite truncated by a stone-line and covered by fine colluvium, marking phases in the development of relief on the Jos Plateau, Nigeria (see also Plate 8-I)

hillslopes denuded of all but coarse regolith. The processes involved must have included sheetflow and rill flow, but also generated mudflows and debris flows. Subsequent weathering and pedogenesis in these deposits have been rapid, tending to homogenise the upper 3 m. Evidence is also available to show that lateral transfer of colluvium has occurred on the low-angled slopes and within the shallow valleys of the central African plateaus. This sediment often shows some sorting and probably indicates the predominance of surface water flows. A consideration of this topic is offered in conjunction with a discussion of dambo valleys.

STONE-LINES

No discussion of superficial deposits in the tropics and sub-tropics would be complete without an attempt to tackle the topic of stone-lines. Taken literally, 'stone-lines' are thin stringers of resistant, little-weathered clasts, oriented downslope and found within many soil/weathering profiles at a depth of 0.5–2 m below the groundsurface. But the reality is often more complex and gravel layers may be more than 1 m thick in places and merge upslope into residual interfluve gravels, while elsewhere there may be superposed stone-lines within sections 4–5 m deep (Ojanuga & Wirth, 1977). Some stone-lines are mainly unrolled quartz or quartzite clasts, but others contain rolled pebbles, clasts of other resistant minerals or rocks, pieces of ferricrete and iron oxide nodules.

So diverse are these accumulations that it might be better to adopt different terms to describe what may be deposits having very different origins and significance in the landscape. Stone-lines *sensu stricto* might be regarded as rather simple, usually thin, stone layers in soil profiles lacking evidences of truncation and unconformity. These are likely to be the products of continuous pedogenetic and/or morphogenetic processes. Stone layers forming compact accumulations of coarse clasts are possibly best described as *palaeopavements* (after Bigarella & Andrade, 1965) or pediment gravels. Other stone layers are clearly degraded alluvial terraces (channel gravels), or they may be a result of colluvial processes; a few form talus fans, and it is probably unfortunate that the topic has been so often tackled as though stone-lines were a single phenomenon.

Theories and processes of stone-line origin

The characteristics of many stone-lines favour theories of origin involving the residual accumulation of the stony material as a 'lag' deposit which is later re-covered, usually by loamy sand. To explain such a sequence, environmental change involving a period of intense sheetwash is often invoked (De Ploey, 1964; Vogt, 1966). However, residual accumulation can occur as weathering penetrates the subjacent rocks, and coarse clasts which remain unweathered become concentrated into a recognisable layer. Three continuous mechanisms, requiring no environmental changes, have been advanced to account for this phenomenon. These are: *faunalturbation* (bioturbation by soil fauna, otherwise called 'pedoturbation'), *dynamic lowering* of the landsurface, and *soil creep*; all three of which may act simultaneously or dominate different parts of a landscape.

As weathering attacks the subjacent rock, and material is exported in solution and as fines by surface wash or by infiltrating groundwater, the saprolite undergoes collapse, while simultaneously bioturbation (pedoturbation) may renew the topsoil. No environmental change is required to explain this process which, with some variation, has been described by Aleva (1983, 1991), McFarlane and Pollard (1989), and Thomas and Thorp (1985). On sloping ground, stone-lines can develop from quartz veins (or other resistant bedrock fragments) by soil

creep (Chapter 5) and downslope movement (Aleva, 1983; Moeyersons, 1989b). However, De Ploey and Poesen (1989) argued for the importance of pluvial processes such as sheet wash and rill erosion in the formation of stone-lines, bringing about the exposure and re-covering of coarse clasts by alternating removal and deposition of fine sediment. Other examples clearly have alluvial origins, being the degraded remnants of river terraces subjected to colluviation downslope. The role of colluvial processes in the burial of stone-lines and in the formation of multiple sequences has been stressed by Wells *et al.* (1990) from studies in Madagascar.

Thus the palaeoclimatic significance of stone-lines is not straightforward. None the less, Ab'Saber (1982) has concluded from observations in Brazilian Amazonia that a majority of occurrences, 'represent a single widespread period of climatic aridity'. Bigarella and Andrade (1965) described a palaeo-pavement carrying coarse stony detritus over pediment landforms in eastern Brazil, and Bigarella *et al.* (1965) provided detailed descriptions of stone-line complexes associated with alternating surface wash and mass movement. These authors attributed the palaeopavement to widespread aridity during the last glacial maximum. Ruhe (1956, 1959) and De Ploey (1964) were equally emphatic on this point from their studies in Congo (Zaire). Carefully dated stone-line stratigraphies with palaeolithic implements have also been documented by Kadomura and Hori (1990) from southern Cameroon and western Kenya, and they concluded that such deposits record intense 'glacial' aridity in the humid tropics (20 000–13 000 BP).

The action of termites and ants

Some pedologists and geomorphologists have argued that most stone-lines arise from faunalturbation, mainly by termites, on otherwise stable landsurfaces (Vogt, 1966; Williams, 1978; De Dapper, 1978; Soyer, 1989; Johnson & Watson-Stegner, 1987, 1990; Johnson, 1993) (Figure 8.31). This argument refers to the common '3-Tier Soil' in which a homogenised and fine grain-sized (<2 mm) layer forms the mineral topsoil (M horizon) and this is underlain

FIGURE 8.31   A thin stone-line beneath a gently sloping landsurface in the southern Nigerian forest zone. The upper, fine deposits may be partly due to termite activity

by a coarse stony layer (S), beneath which is the saprolite (W), as in Figure 8.31.

The role of soil biota in homogenising surface soils has been known for more than 100 years and was discussed by Charles Darwin (1881). In tropical areas ants are sometimes important, but the role of termites has received far greater emphasis. Watson (1974) has pointed to some of the important features of termite mound building. The large, fungus growing, *Macrotermes* produce large mounds as high as 5–10 m and up to 30 m in diameter. These mounds are constructed by the termites from material brought up from depths of more than 3 m (termite penetration goes to much greater depths) and the following features have been observed:

(1) mounds are constructed from material brought from below the stone-line;

(2) termite galleries are observed below the stone-line;

(3) stone-lines are depressed into bowl-shaped formations below termite mounds;

(4) there is a close mineralogical correlation with the underlying rock in varied terrain;

(5) the presence of a soil (M) mantle on hilltops, underlain by a stone-line precludes a colluvial origin on those sites.

The termitaria also have a higher moisture content and higher base status than the surrounding soil. This is partly attributed to the accumulation of vegetal matter by the termites, and partly by penetration of the termites to bring material from the deeper saprolite into the M horizon.

A simple model for faunalturbation has been advanced by Moeyersons (1989b) and this is shown

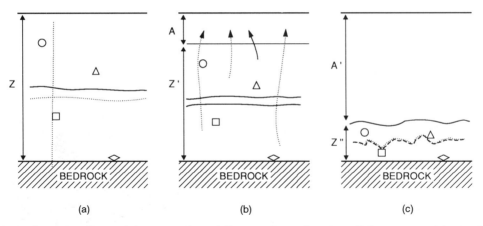

FIGURE 8.32   Possible effects of the upworking of fines to the surface by soil fauna: (a) original soil with depth
Z; (b) collapse of small cavities and galleries created by removal of particles reduces soil thickness by A to Z'; (c)
continued reduction of soil surface leads to a concentration of coarse elements at the base (after Moeyersons, 1989b;
reproduced by permission of the Royal Academy of Overseas Sciences and Geo-Eco-Trop)

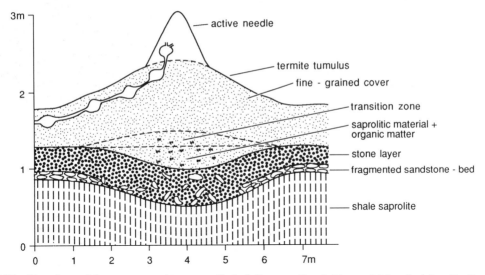

FIGURE 8.33   Termite activity on a stone layer near Lubok Perong, Perak Terap, Malaysia (after De Dapper, 1989;
reproduced by permission of the Royal Academy of Overseas Sciences and Geo-Eco-Trop)

in Figure 8.32. It indicates how an initially random distribution of coarse, heavy clasts becomes transformed into a layered soil as the fine material is brought to the surface and larger stones settle to the base of the disturbed layer (Figure 8.32). However, De Dapper (1989) takes a contrary view, suggesting that in fact termites operate *on* the pre-exisiting stone layer, bringing saprolite up through it into the M horizon, and possibly occupying old surface channels and depressions (Figure 8.33 and Plate 8-II). This process can either contribute to the stone layer or dismantle it.

If termites are responsible for transferring material from the weathered rock or saprolite to the surface, then the rate at which this takes place is of considerable significance in geomorphology and to the renewal of surface soil eroded by runoff and throughflow. In many tropical areas there is a

FIGURE 8.34   Large termitary (*Macrotermes natalensis*) some 6 m high and 10 m in diameter, in the Muhulu dry forest zone of Shaba, Zaire

combination of large termitaria with densities of 2–5 Ha^{-1} and small termitaria perhaps exceeding 1000 Ha^{-1}. The large mounds which can reach 10 m in height and 30 m in diameter, are more usually around 5 m high and 10–12 m in diameter (Aloni, 1975, and Figure 8.34). The species involved are termed *Macrotermes* (e.g. *natalensis*, *facilger* or *goliath*). According to Watson (1974) many very large mounds may be as much as 700 years old, though the evidence for this is slender. The smaller mounds are built by many species (e.g. *Microtermes*, *Tumulitermes hastilis*) and, according to Williams (1968), 50% of the mounds at a site in the Brock's Creek area of northern Australia had been built in the previous 25 years, while 42% had been abandoned (65% of the larger mounds were defunct in this area).

Once abandoned, the mounds become subject to rapid erosion and the fine material brought up by the termites to build the mounds is spread out over the intervening ground. It was thought that the larger mounds might take 10 years to erode in this way (Williams, 1968). By measuring the thickness of the upper mixed (M) soil layer and the density and volume of the termite mounds, and by deducing the likely age of the soil M horizon, usually on the basis of contained artifacts (hand axes have commonly been found), the rate of formation of the topsoil can be estimated within broad limits. Table 8.4 summarises some typical figures obtained for this process.

On the basis of the data given in Table 8.4, homogenised surface soil layers of 3–5 m thickness appear to have minimum ages of 6000–10 000

TABLE 8.4   Some rates of topsoil formation by faunalturbation

Location	Rate cm ka^{-1}	Rate mm year^{-1}	Source of data
Central Africa	5–50	0.05–0.5	De Heinzelin (1955)
Southwest Nigeria	2.5	0.025	Nye (1955)
Congo/Zaire	10–20	0.1–0.2	De Ploey (1964)
Northern Australia	3	0.03	Williams (1968)
Australia	40	0.4	Lee & Wood (1971)
Tropics	10^0–10^2	10^{-2}–10^0	Tamura (1991)

years, but these values do not account for all pedoturbation processes and ignore rates of surface erosion. M. A. J. Williams (1968, 1978) considered the landsurface dynamics at Brock's Creek, on a supposed Cenozoic planation surface remnant. The termite mounds are built from material in the underlying saprolite, and according to M. A. J. Williams (1978) this was being truncated at a rate of 0.068 mm year^{-1} (6.8 cm ka^{-1}). The mounds contain around 20% clay, but the M horizon gave figures nearer to 10%. The clay was being lost by eluviation, to be deposited at the base of the S layer, where it comprised 20% of the total clay on summit sites, and also by slope wash.

According to Williams (1968) the time taken to produce a thin M horizon of 0.15 m underlain by a stone-line 0.3 m thick, at Brock's Creek, was 12 000–17 700 years. Much thicker M horizons should by this reasoning be older, but rates of accumulation vary, and Tamura (1991) recognised a range of rates, all of which are higher than those suggested above. If much of the fine M horizon was removed during the period of late Quaternary desiccation, lasting from 20 000 BP to 12 500 BP, then this could place a limit on the topsoil age.

Dynamic lowering of the landsurface

Very few rates of surface erosion under natural vegetation exceed the figures for topsoil renewal (*cf.* Tables 5.6 and 8.4). What this tells us is that the dynamics of the landsurface, where vegetation cover is not disturbed, permits the overall lowering of the soil profile during prolonged denudation. This point has been emphasised by M. A. J. Williams (1978), who pointed out that primitive *Mastotermes* had been active since the Eocene, and that the lowering

of the entire profile by 100 m could have been accomplished during the Quaternary (at a rate of 5 cm ka^{-1} over 2 Ma). This reasoning is also the basis of several important constructs in geomorphology and in soil science, including *dynamic etching* or *etchplanation* (Büdel, 1982; Thomas, 1989 a,b); the development of *triple planation surfaces* (Aleva, 1983) and the composite ideas of Johnson (1993) on *dynamic denudation*. These ideas go beyond the specific action of termites and are discussed further in Chapter 9.

The accumulation of residual (autochthonous) interfluve gravels and of angular (unrolled), stream valley, quartz gravels can be thus attributed to a *dynamic etching* process (Thomas & Thorp, 1985; Thomas, 1989a), whereby rock weathering, chemical denudation, and the gradual removal of fine-grained materials (mostly clays) effect the down-wasting of the landscape under conditions of low energy surface erosion, under forest (Figure 11.6). The sinking of heavier clasts through the clay soil was proposed by Laporte (1962; reported by Wells *et al.*, 1990), but this process is not thought to be effective. The low energy conditions described here also favour bioturbation and the renewal of fine topsoil. This appears comparable with the interpretation of stone-lines in Malawi and Zimbabwe by McFarlane and Pollard (1989), who considered them to have formed at a dissolution front separating the saprolite from the superficial cover (Figure 8.35). Shallow trenching of interfluve surfaces in Koidu, Sierra Leone, revealed a complex lag gravel within 0.5 m of the ground surface, in places resting directly on a bedrock surface with a strong iron patina, and exhibiting clusters of heavy clasts in rill-like depressions.

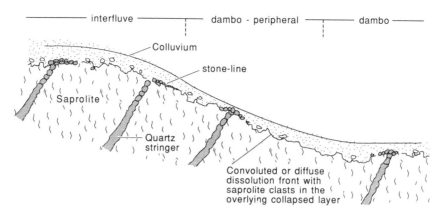

FIGURE 8.35 Diagrammatic representation of the interface between saprolite and surficial materials across a dambo catena in Malawi (after McFarlane & Pollard, 1989; reproduced by permission of the Royal Academy of Overseas Sciences and Geo-Eco-Trop)

FIGURE 8.36 Heterogeneous surface gravels on a planate interfluve in Koidu, Sierra Leone (see Figure 8.40). Larger fragments are mainly ferricrete duricrust

The complex lithology of these gravels (Teeuw, 1989) indicates their derivation from a variety of local rock types, while their packing and variations in thickness suggest transport by localised water flows under an open vegetation canopy. The occurrence of rounded quartz rising to >20% in some zones is an indication of former river activity at the interfluve level, while the presence of Fe sesquioxide nodules and some actual duricrust fragments, adds to the conviction that some complex surface gravel layers can record a prolonged period of sub-aerial landscape formation (Figure 8.36) (Thomas *et al.*, 1985). A comparison of these stone-line gravels with surface gravels formed in the semi-arid zone ('gibber plains' in Australia) illustrates the probable condition of many of these slopes at the LGM (Figure 8.37).

The action of soil creep and colluvial processes

Thin stone-lines can be traced within soil profiles of connecting valley-side slopes which may be regarded as conveyors for coarse clasts moving downslope. But these features may have more than one origin. In Malawi, McFarlane and Pollard (1989) demonstrated both the vertical collapse of the profile and the downslope creep component leading to thin stone-lines derived from single quartz veins. Studies of soil creep processes by Moeyersons (1989a,b) in Rwanda (see also Chapter 6)

FIGURE 8.37   Glacis surface truncating ancient weathering profiles with silcrete cappings in the arid zone of New South Wales, Australia. The surface gravels ('gibber plain') are made up of silcrete fragments. Glacis in the humid tropics may have developed in similar conditions, but with vein quartz clasts and ferricrete fragments in dominance

led him to conclude that accelerated creep, and some sliding, took place along discontinuities between soil horizons developing at depths of up to 0.5 m on slopes over 25°. Throughflow along such discontinuities causes loss of fine material, and according to Moeyersons (1989b) coarse clasts may converge towards this zone (Figure 8.38). In Puerto Rico, Simon *et al.* (1990) recorded shallow slope failures at similar depths, and found that the horizon discontinuity resulted from clay illuviation down profile. Soil creep does not, however, appear as a likely (dominant) process across shallow slopes (<5°), where stone-lines are equally common.

The complex stone-lines in Madagascar were considered by Wells *et al.* (1990) to be derived from hillslope lag and valley gravels, buried by colluviation; taking place within a continuous process of landscape development involving 'autocyclic' or complex response mechanisms. The

process was seen as an integral part of the lateritisation and denudation of the landscape. The authors found widespread redistribution of laterite, and they speculated as to whether any of the material was formed in place. However, the area is an elevated (1200–2000 m asl) grassland, dissected into a multi-convex relief (100–600 m) with steep slopes. The proximity to a major Neogene volcanic centre may also be significant.

The impact of climate change

Many stone-lines appear to date to arid phases of the Quaternary. Fairbridge and Finkl (1984) referred to them as 'winnowed' mudflow deposits, but more commonly they are regarded as lag gravels remaining after sheetflow events which, significantly, also led to the truncation of soil profiles, sometimes to the extent that the stone-lines

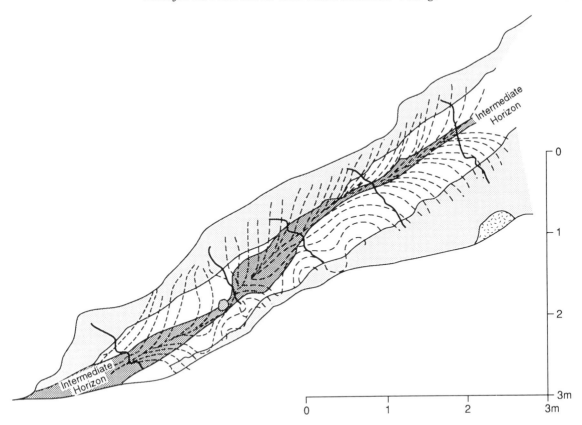

FIGURE 8.38   Creep-line pattern in the wall of a soil (Young) pit at Rwaza Hill, Rwanda. Convergence of creep lines to an intermediate, crushed horizon measured over 3 years (1981–1984). Tracer (broken) lines are exaggerated ×10 scale of pit wall (after Moeyersons, 1989b; reproduced by permission of the Royal Academy of Overseas Sciences and Geo-Eco-Trop)

rest directly on eroded saprolite. In the Curitiba basin in southeast Brazil, the stone-lines form a compacted stone layer or *palaeopavement* (Bigarella & Andrade, 1965) that truncates the older saprolite (Plate 8-III); this situation has been more recently demonstrated in Madagascar (Wells *et al.*, 1990). The topsoil that has formed above the stone-line is quite different in colour and texture from the saprolite below and may have resulted partly from mass movement.

Stone-lines may occur in complex sequences, under considerable thicknesses of colluvium, as seen in southern Ghana (Figure 8.39) and in the valley-head colluvium observed in Koidu, Sierra Leone (Figure 8.40), and few contain datable material. In southern Nigeria the stone-lines have been described by Fölster (1969, and Fölster *et al.*, 1971b) and by

Burke and Durotoye (1971) all of whom considered these to be 'pediment' gravels dating from Pleistocene dry periods. In many sections it can be seen that the stone-lines have a steeper slope than the landsurface which makes it very unlikely they could result primarily from creep, termite activity, or downwasting (Figure 8.41).

Many stone-lines probably relate to the aridity of the LGM, but multiple stone-lines will be more complex. A point that is seldom addressed is the provenance of the stones that formed the palaeo-pavements during the dry climates of the Quaternary. There appear to be two main possibilities: that they were derived by concentration and packing of pre-existing stone-lines *sensu stricto*, formed by the continuous dynamic processes outlined above; or that they were concentrated following the

FIGURE 8.39   A thick and complex stone-line undulating below the present landsurface and buried by colluvium, in a forest zone, near Kumasi in southern Ghana

incision of the landscape and the formation of slope pediments in the manner described by Rohdenburg (1969/1970) and De Dapper (1989). In the model derived by the latter author both of these mechanisms are implied (Figure 8.26).

An outline of the processes that may be involved is given in Table 8.5. Using this table, it is possible to postulate combinations of causal processes that either imply environmental change or could operate under more or less constant formative conditions. Thus A/1 + A/2 + B/1 + B/2 denotes a combination of surface and near-surface processes common throughout the humid tropics, and a single process system may be capable of producing the profile with (M), (S), and (W) horizons. Such combinations constitute the processes of *dynamic etching* (Thomas, 1989a,b), whereby surface gravels and

stone-lines may be generated by steady-state conditions. It can also be suggested that A/4 + B/4, although probably denoting a change towards more humid conditions, could record the short-term effects of storms leading to strong sheetflood, followed by rapidly declining overland flow. However, this event-based analysis would not act cumulatively, since the surface fines would be washed out by the succeeding storm(s).

More experienced students of geomorphology will already have noted that the processes listed under A and B are not necessarily exclusive, or sequential, and such phenomena as surface and near-surface gravels are usually the cumulative results of complex sequences of processes over extended timescales (10^4–10^5 years). As such they have commonly experienced many shifts of climate,

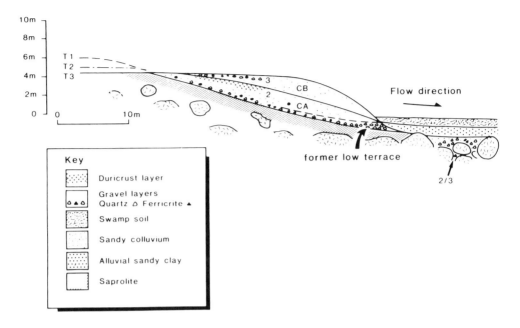

FIGURE 8.40   Colluvial deposits accumulated in valley heads in Koidu, Sierra Leone. Each phase of lowering of the interfluve has been associated with the deposition of sediment in the valley head, so that time period T1–T2 = CA, and T2–T3 = CB. These phases mark periods of strong sheet flow and the absence of forest. CA is probably late Pleistocene; CB is probably Holocene (after Thomas, 1988; reproduced by permission of the Royal Scottish Geographical Society)

vegetation cover and soil fauna. Thus while they may denote environmental change, they may not imply specific directions or sequences of change, though the general hypothesis that many are 'lag' gravels reflecting periods of increased energy in surface processes stands. It is probable that a majority of stone-line complexes evolve through several stages:

(1) residual accumulation resulting from dissolution and removal of weatherable and finer material under humid conditions;
(2) redistribution and concentration of gravel by surface water flows and related colluvial activity;
(3) modification and burial by bioturbation, soil creep, tree throw, and possibly by anthropogenic activity.

A model for their development was developed by De Dapper (1989), building on earlier work by Rohdenburg (1969/1970) and Fölster (1969). This postulates the probability of landscape instability due to aridification, leading to the retreat of low scarps in the saprolite-covered slopes; the fine materials being washed into the valley floors, while the vein quartz, duricrust fragments and other clasts were left as a lag deposit which later became buried by fine soil material. A similar model integrating events on interfluves, valley sides and small stream channels was developed for west Africa (Thomas, 1989a, and Figure 8.42).

CHANNELLESS VALLEYS (DAMBOS)

Many headwater valleys in areas of low relief are channelless and in humid areas may contain swamps. These are known as *dambos* in central Africa, and as *vleis* in southern Africa. Other local names are also common, and they are similar to the *baixas* of Amazonia, the *bolis* of Sierra Leone (both in humid tropical regions), and also to the *fadama* of northern Nigeria. Although the bolis are found in forested areas, typical dambo forms

FIGURE 8.41   A stone-line in the southern Nigerian forest zone. Note that the buried stone-line dips more steeply than the present landsurface and is covered by colluvium, and that a surface gravel caps the present soil, probably due to anthropogenic activity

are associated with highly seasonal, savanna climates. Such 'saucer-shaped valleys' were called 'Flachmuldentäler' by Louis (1964), and 'Spülmulden' or *wash depressions* by Büdel (1957, 1965, 1982). In the type areas of Zambia, Ackermann (1936, p. 150) stated, 'the (dambo) form is characterised by a shallow linear depression (washed basin floor) at the upper end of a drainage system without a marked stream channel. The vegetation consists of treeless grassland and seepage plants.'

Raunet (1985) placed these forms into a framework of tropical valley bottoms typical of deeply kaolinised, ferrallitic basement landscapes throughout Africa, and this is useful because it widens the scope of the discussion. The east African *dambo* according to Raunet (1985), is found in

areas with rainfalls of 900–1400 mm year^{-1}. In Africa, different valley forms were associated with the Sahelo–Sudanian zone (500–1000 mm year^{-1}), the Sudano–Guinean zone (1000–1400 mm year^{-1}), and the humid tropical (forest) zone (>1400 mm year^{-1}). The account indicates the presence of flat, alluviated valley floors in *demi orange* landscapes with their deep saprolitic mantles, a feature associated with steep slopes and amphitheatres caused by mass movement in eastern Brazil where dambo-like forms have also been described (Coltrinari & Nogueira, 1989) (above). These humid tropical valleys were contrasted with the forms of seasonal and semi-arid landscapes where ferricrete duricrusts (cuirasses) form cap rocks on butte-type summits from which long rectilinear glacis lead via a sandy marginal zone into valley floors with a clay

TABLE 8.5 Processes associated with the formation of stone-lines

Processes	Vegetation	Climate
A. PROCESSES LEADING TO ACCUMULATION/CONCENTRATION OF GRAVEL		
A/1 Bedrock weathering and down-wasting by loss of solutes and fines, residual accumulation of large clasts	Mainly humid to sub-humid forest and woodland climates	Humid
A/2 Upward transfer of fines by soil fauna, with settling of heavy clasts (combines with B/1)	Forest and woodland vegetation	Humid and sub-humid
A/3 Mass transfer by soil creep concentrating quartz from veins (can combine with B/2)	Under forested slopes	Humid
A/4 Loss of fines by sheetwash and rill action leaving lag gravels	Open canopy woodland or scrub	Seasonal, perhaps Semi-arid
A/5 Loss of fines via macropores and horizon discontinuities	Wide variety of forest and savanna vegetation	Humid or sub-humid
A/6 Retreat of low scarps (often in ferricrete) leaving gravel lag	Mainly open woodland and scrub	Semi-arid
A/7 Gully formation with deposition of channel gravels	Open or disturbed vegetation	Semi-arid
B. PROCESSES ASSOCIATED WITH REGENERATION OF TOPSOIL COVER		
B/1 Action of soil fauna, especially termites (combines with A/2)	Undisturbed woodland or forest	Humid
B/2 Mass transfer of topsoil, creep, colluviation, slides (creep combines with A/3)	Humid climates, mainly forested	Humid to sub-humid
B/3 Flooding and settlement of fine sediment	Seasonal and humid climates with active floodplains	Humid (seasonal monsoon)
B/4 Weak sheetflow and settlement of fines	Under most woodland canopies	Humid or sub-humid

fill. The *bolis* of Sierra Leone do not quite fit this scheme; they are found in both sedimentary and crystalline terrains, and in the latter are associated with etched valley floors and exposure of corestones in valleys and on slopes.

It may be helpful to think in terms of shallow *degradational* features etched into the plateau surfaces of central Africa and elsewhere on the one hand, and *aggradational* (cut-and-fill) dambos associated with areas of slightly higher relief on the other (Mäckel, 1974). However, these categories overlap and some dambos on planate landscapes have cut-and-fill morphologies (Figure 8.43).

Dambos as 'zero order' basins

Although dambos are associated with areas of low relief, channelless extensions of present-day channel systems are well known from areas of high relief and from many non-tropical environments. These *zero order basins,* as they are sometimes called, frequently drain from *hillslope hollows,* and they

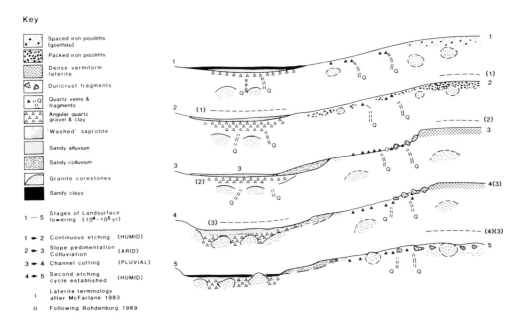

Key

FIGURE 8.42   Schematic representation of slope forms and deposits resulting from one Quaternary humid–arid–humid climatic cycle, based on studies for low relief areas of Koidu, Sierra Leone. 1→2 indicates slow degradation of the surface with an accumulation of Fe_2O_3 pisoliths under humid conditions; 2→3→4 indicates shallow slope pedimentation with colluviation on lower slopes under dry conditions; 4→5 indicates a return to humid conditions (after Thomas, 1989a; reproduced by permission of Gebrüder Borntraeger)

have been the subject of detailed work, especially in the Pacific coastlands of the USA (Dietrich *et al.*, 1987). The essence of all these occurrences is their origins as concavities in the landscape, attracting convergent flow that in turn imports colluvial sediment to the hollow floor. Because surface and subsurface flows of water continue to take place, the hollow is kept moist and is subject to a rapid rise in pore water pressures and water levels after prolonged rain. This can give rise to liquefaction (Tsukamoto & Minematsu, 1987), and on steep relief landslides are preferentially generated in these landforms and deposits (Reneau & Dietrich, 1987). There is field evidence to suggest that hillslope hollows are common in tropical landscapes and that a distinction between hollows and dambos may be arbitrary, and based mainly on the very low gradients in the latter.

Like many hollows, dambos frequently contain evidence of Quaternary environmental changes. Some are even linked to pans and may have histories

closely linked to these features. Many display buried channels with clay infills, sometimes over ferricrete benches (Clark, 1974, and Figure 8.40).

Mäckel (1985) recognised a characteristic zonation of vegetation and soils in the dambos of the central Zambian plateau. The surrounding landscape is typically *Miombo* woodland characterised by *Brachystegia*, *Julbernadia*, and *Isoberlinia* tree species. This woodland terminates abruptly and gives way to a marginal grassland (*Loudetia simplex*) or 'upper wash zone' with up to 50% exposure of the sandy soils. A 'lower wash zone' of fluctuating water-table and iron enrichment is typified by a dense mat of *Andropogon* and *Hyparrhenia* grasses, and leads into the central 'seepage zone' with associated plants such as *Hyparrhenia bracteata*. This type of zonation, associated with a marked soil catena, involves a transition from deep residual soils on interfluves, via a sandy fringe, where soils are mottled at depth, to *vertisols* dominated by smectite clays in the valley floor.

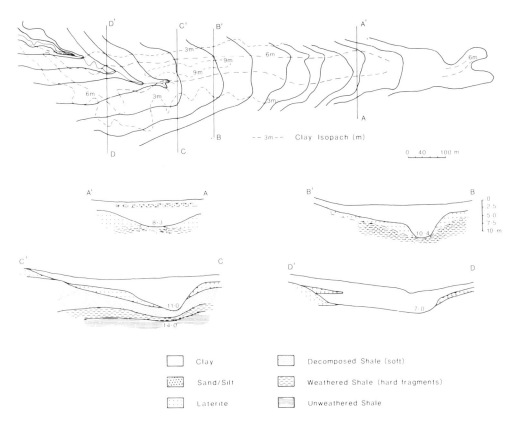

FIGURE 8.43   Four cross-sections through a dambo valley at Kankamo, Zambia, showing the presence of a buried channel form and infill of sand, silt and clay deposits (redrawn after Clark, 1974; reproduced by permission of Chapman & Hall)

Processes of dambo formation

Much interest centres on the dynamics of these drainage systems. Their hydrology was investigated by Balek and Perry (1973) and also Balek (1977), who concluded that they were fed mainly by water falling on the area of the dambo itself, rather than by groundwater seepage. This is confirmed by the observation that dambos contribute little to downstream base flows in the dry season. In most of these valleys a clay fill is found in the valley floor. According to Mäckel (1974, 1985) and Pullan (1979), the dynamics of the typical dambo lead to the washing out of fines from the marginal zone into the centre, with the fines continually being brought to the surface by termites. The clays of the central, seepage zone are generally smectites and may be sodic. Mobile ions leached from

surrounding slopes are therefore retained in the dambo floor, contributing to the neoformation of 2:1 clays (McFarlane & Whitlow, 1990).

Cut-and-fill processes

The flat-bottomed valleys often conceal former, buried drainage channels beneath alluvial fills of sand and clay, and such examples are therefore both aggradational and fluvial in origin. Mäckel (1985, p. 11) commented that dambo development, 'was preceded by active phases of incision and accumulation through a river. With decreasing river activity more fine material was washed into the depression until the present-day level was reached.'

In Malawi, Meadows (1985) found evidence for cut-and-fill phases triggered by climatic changes in dambos on the Nyika Plateau. Channel cutting commenced prior to 12 000 BP, with a fill phase starting shortly after 12 000 BP and continuing until shortly after 5000 BP. He quoted Bond (1967) who, in a survey of dambo forms in southern Africa, concluded that below a critical rainfall of ~ 900 mm year^{-1} aggradation gave way to channel cutting. This suggests that, in seasonal climates too dry for evergreen forests, a *rise* in rainfall appears able to trigger widespread erosion and sedimentation. In Nyika, Malawi (Meadows, 1983), and Inyanga, Zimbabwe (Tomlinson, 1974), the onset of this sedimentation was around 11 700 BP. This date is similar to those obtained from Koidu and Birim for basal sediments in some larger streams, and both may reflect the rapid changes in climate and vegetation cover at the termination of the last glacial period.

It is often difficult to place period(s) of active slope wash and colluviation accurately into a time frame. In wetter areas, the gradual onset of dry climates would have led to a slow increase in sediment transfer around 20 500 BP, and slope wash would have continued throughout the dry glacial period (thick slope deposits on the Jos Plateau appear to have accumulated after *c.* 18 000 BP). The rates of sediment transfer would, however, have increased rapidly as climates became wetter after *c.* 13 000 BP, when higher rainfalls possibly also led to widespread gullying and to channel cutting with deposition of coarse sediments in the river channels. In the broad, shallow dambos of the Lilongwe Plain in Malawi, Meadows (1985) found abundant evidence of sedimentation in the dambo margins and floors, but he was unable to attribute these to any chronology of climate change, due to the many local variations in stratigraphy and lack of datable material. These instances are not far removed from the origins of colluvial layers in more steeply dissected terrain, and the process sequence matches that associated with stone-line formation as previously discussed.

Etching processes

There is also strong evidence for etching and chemical denudation into the bedrock beneath such valleys (Thomas & Thorp, 1985; McFarlane, in Boast, 1990), and recent extensions of the valley network also appear to have taken place by chemical sapping processes (Figures 5.10–5.12, 5.14 and 5.15). McFarlane and Whitlow (1990) have even suggested that, '*dambos* are not "fossil" fluvial systems, nor is their clay infill alluvial', and they attributed their basic features to differential leaching (or 'etching'), and to the precipitation of ions transferred from interfluves, as reconstituted, smectite clays. In this scheme such dambos would be degradational, and a function of groundsurface lowering by weathering and chemical denudation. This argument is close to that of Büdel (1982), but Büdel acknowledged the role of surface sediment transport into the depressions, and only in situations of extreme planation would lateral transfer of fine sediment virtually cease.

The chemical component of denudation is undoubtedly important in these landscapes but it does not operate in isolation from gravitational and wash processes. The presence of pipes tends to confirm this view, since these conduits experience turbulent flow and transport sediment as well as solutes (Ziemer & Albright, 1987). In California Dengler *et al.* (1987) have demonstrated that colluvium can be as thick in the gentlest of swales as it is in steep hillslope hollows, and this relationship may well be found in the dambo landscape. There is no logical reason why sediment should not be stored within valleys that are fundamentally degradational, and this appears to be the case in the bolis of Sierra Leone and in many dambos of central Africa. It is, however, a situation in which rapid excavation of the sedimentary fill can result from unwise use of the valley floor (Whitlow, 1988a,b).

The wider landscape context of dambo formation

In the context of dissected, rocky landscapes, Giardino and Mäckel (1985) considered the exposure of convex rock surfaces (*dwalas, bornhardts*) and the alluviation of valley floors as concurrent and related processes, possibly triggered by climatic change. This study illustrates the need to place typical dambo forms into a much wider framework. It was shown in northern Nigeria that beneath the valleys of some seasonal streams, saprolite is preserved (Thomas, 1966a). This denotes the ineffectiveness of fluvial erosion, and the reason for this must surely be a combination of infrequent surface flows and a high rate of sediment transfer into the valley floor.

There are also intriguing pointers to the role of chemical processes in the dambo complex, because upslope soils, often beyond the dambo margin and beneath the *miombo* woodland, may resemble the *white sands* discussed below. These sands are today found on dry, upper slopes and even on low interfluves. To bring them into the reach of a hydromorphic soil forming regime might require valley floors to stand higher than they do today, but it must be borne in mind that podzolisation can take place on broad flat plateau surfaces, where impeded drainage occurs (Lucas, *et al.*, 1987; see Figures 2.14 and 2.15). Some instances of elevated white sands in Zambia are, however, of obvious fluvial origin, being associated with rounded pebble gravels. These cases must be remnants of former valley landscapes as well as markers of climatic change (Fairbridge & Finkl, 1984).

Some conclusions about dambo geomorphology

Dambos are but a special category of channelless extension of the drainage network, and there are many parallels between the forms described from central Africa and those found in other parts of the tropics and beyond. Some conclusions from this discussion are:

(1) channelless extensions of the drainage network constitute *zero order basins* and may be floored by combinations of bedrock, saprolite, colluvium and alluvium;

(2) some such features may be *seepage zones* for groundwater rather than 'fossil' extensions of fluvial channel systems;

(3) the formation of saucer-shaped depressions and open valleys may be a consequence of *chemical etching processes* and may be associated with sapping and the extension of the dambo margins;

(4) in steeper relief, headwater valleys may begin as colluvium-filled *hollows* that transform downslope into stream valleys;

(5) most channelless valleys have a *colluvial/ alluvial fill* which varies in depth from < 1 m to > 10 m and may provide evidence for Quaternary climate change;

(6) the transfer of fines towards the centre of the valley arises from a combination of termite activity and surface flows;

(7) leachates and kaolinite combine to form *smectite clays* in the valley-floor sediments; deep cracking and formation of pipes and other macropores in these promotes subsurface water flows.

Opposition between sapping (or etching) processes and sedimentation in dambos is misplaced. The bolis described from the crystalline rocks of Sierra Leone (Thomas & Thorp, 1985) display both characteristics, and the sediment appears to be flushed out of the system by periodic major floods. There is also a necessary link between the downwasting of valley floors and interfluves, and this link brings together the formation of headwater valleys and stone-lines (see Figure 11.6).

PODZOLISED WHITE SANDS

Podzolised *white sand* deposits have been noted from many different areas. They fall into two main groups: those that occur as *sand plains* on ancient cratonic surfaces (Mulcahy, 1964; Fairbridge & Finkl, 1984), and those that have developed in sandy sediments. However in both cases the loss of clay from the formation usually needs to be explained. In Guyana, Bleackley and Kahn (1963) concluded, 'that the white sand areas of the Berbice Formation were not originally laid down white as they now appear . . . ; the sands have been bleached by the removal of iron as organo-metallic compounds both downwards and laterally. This removal of iron has been accompanied by the loss of clay.'

In neighbouring Surinam, Heyligers (1963) considered that the white sands had originated as a superficial blanket by colluvial redistribution of older fluvial sediments under increased rainfall, and that they have since become bleached by leaching of iron and other constituents. As with studies of stone-lines and dambos, white sand formations have also led to apparently opposing theories, proposing chemical leaching and mechanical transfer as alternatives. It is also clear that all three phenomena are closely linked to the manner of landscape evolution in low relief areas.

White sands as hydromorphic podzols

White sands are described by Duchaufour (1982) as *tropical hydromorphic podzols*, and they are characteristic of the sandy coastal plains of very humid equatorial regions. They occur in response to special site conditions that contrast markedly with the ferrallitisation of well drained areas. According to Duchaufour (1982, p. 332), the conditions for podzolisation are, '(i) the presence of a permanent high water-table; and (ii) a parent material impoverished in weatherable minerals and with a sandy texture'. Many Tertiary sediments and Quaternary alluvial deposits satisfy these requirements. Andriesse (1968, 1970), from a study of Sarawak soils formed under such conditions, concluded that leaching of lateritic (ferrallitic) soils could lead to podzolisation as the soil becomes progressively more acid, and these ideas have been taken up by later workers (Lucas *et al.*, 1987, 1988) and explored in Chapter 2. Young (1976, p. 173) strongly contested this view, and argued that two separate processes of soil formation, leading respectively to either ferrallitic or hydromorphic podzolic soils occur in the perhumid tropics. The problem here is twofold: first, some ferrallitic soils show signs of further evolution (Chauvel, 1977; Buurman & Subagjo, 1978, Lucas *et al.*, 1987, 1988) and leaching of Fe^{2+} and Al^{3+}, and second, there is a close juxtaposition of ferrallitic soils (oxisols/ultisols) and podzols (spodosols/histosols) across soil toposequences (catenas) (Bravard & Righi, 1990). As discussed in Chapter 2, these observations can be construed to support an hypothesis for the degradation of ferrallitic mantles, by the loss of Fe^{2+} and Al^{3+} along with SiO_2, leading to the breakdown of kaolinitic clays and the formation of a progressively more sandy residue (Figures 2.14

and 2.15). In this section, other aspects of white sand phenomena will be explored.

Beneath the white sand is a thick Bh horizon which locally becomes a *humicrete*. This Bh horizon often develops at a level where fine silty sands occur in the sediments, but it is not always possible to say whether this reflects the original stratigraphy or whether it is a result of subsequent translocation of fines in the profile. Such thick Bh horizons have been described from North Carolina (Daniels *et al.*, 1975; Holzhey *et al.*, 1975). The upper, structureless white sand (Spodic) horizon has led to some conjecture concerning a possible marine origin for these sands, where they occur in coastal plain environments. There is, however, no evidence from the grain morphology to support this view, except where the sands occur as part of an unambiguous raised beach deposit.

White sands, lithology and relief

The lithology of the rock substrate and the shape of the relief forms can be decisive factors in the formation of white sands, and Brabant (1987) has shown the correspondence of podzols in central Kalimantan (Indonesian Borneo) with the distribution of broad flat interfluves and glacis developed on Tertiary and Quaternary sandy sediments, a relationship seen clearly in western Kalimantan (Thorp *et al.*, 1990) and in Amazonia (Lucas *et al.*, 1987, 1988).

However, the development of sand plains on ancient cratonic surfaces, in Guyana (Bleackley & Kahn, 1963) and Surinam (Heyligers, 1963), for example, implies the operation of these processes on crystalline feldspathic rocks, where hydromorphic conditions can develop widely on uplifted surfaces of low relief, frequently containing dambo forms. According to Fairbridge and Finkl (1984, p. 64), 'large areas of the tropics became covered by sands that had been progressively leached of their soluble components during repeated wet phases of the Pleistocene, but distributed largely by wind action during the arid phases'. The reactivation of ancient sand sheets such as the Kalahari Sands during each arid phase, and the subsequent leaching during pluvial periods must have recurred many times; the latest movements of the Kalahari Sand probably being late Pleistocene (De Ploey, 1965).

Ab'Saber (1982) noted that different modes of genesis attach to contrasting types of white sand formation in the Brazilian Amazon. Some result from pedogenesis on intra-formational sandy lenses in (Tertiary) bedrock, a feature seen also in Kalimantan. But the most important occurrences are fluvial sands which Ab'Saber (1982, p. 49) considered 'were deposited during a savanna environment'. The evidence for this conclusion includes old channel fills containing fossil dunes, and, near Rio Caeté, sands extend beneath present-day riverine plains dominated by clays and organic sediments, a feature similar to coastal Kalimantan. Ab'Saber attributed different occurrences to dates ranging from early Quaternary to Holocene for sands beneath small tributary valleys.

White sands in coastal plain alluvium

In other areas, white sands typify coastal plain alluvium, as in the Guyanas (Heyligers, 1963; Lucas *et al.*, 1987) and in Kalimantan (Sieffermann, 1988; Thorp *et al.*, 1990). Here aeolian transfer may often be ruled out, but rapid deposition of already leached quartz sands, derived from upslope saprolites by sudden floods in open savanna environments, probably contributed the bulk of the sediment which has subsequently become podzolised under humid tropical conditions. The sediments were part of an alluvial plain extending across areas of former sea bed exposed during Pleistocene low sea levels.

Chauvel (1977), Lucas *et al.* (1987, 1988) and Bravard and Righi (1990) all emphasised pedological transformations on a stable landsurface. In this connection, Sieffermann (1988) has emphasised the important role of the peat swamps in leading to loss of iron and aluminium in a process he described as *deferrification*. This appears to be similar to the concept of *ferrolysis* put forward by Brinkman in 1970 and discussed in Chapter 2. Large areas of Amazonia and the Guyana coastal plains, together with those of Kalimantan thus exhibit sediments containing sequences of pure sands intercalated with kaolinitic and gibbsitic clays. Their particular features appear

to reflect the influence of the humid tropical climate within past or present hydromorphic pedogenic regimes (Bravard & Righi, 1990).

Interestingly, Sieffermann (1988) has pointed to the geological history of the Amazon Basin which, according to his interpretation, could have developed into a vast lake as a result of the rise of the Andean chains at the close of the Miocene, cutting off the former exit towards the Pacific. Widespread peat swamps would have bordered the lake, leading to the transformation of underlying ferrallitic soils to podzols. In Indonesia the submergence of the Sunda shelf also provides an environment for the development of coastal peat swamps, but there is little evidence for former extensions of the peats to higher levels. However, a more detailed approach to the evolution of the basinal forelands of the rising Andean chains has led to the suggestion that alluvial sedimentation and widespread flooding of the *Várzea* have been largely controlled by basin subsidence (Räsänen *et al.*, 1987), and that fluvial reworking has produced the observed mosaic of *várzea* and *tierra firme* environments.

The formation of white sand horizons

The processes associated with the formation of white sands can be summarised:

(1) further alteration of saprolite mantles: leaching of ferrallitic saprolite by 'lessivage' and hydrolysis of kaolins under hydromorphic conditions, often swamps, in a per-humid climate, to form a sandy podzolic residue;
(2) weathering and podzolisation of sandstones;
(3) podzolisation of unconsolidated sandy deposits formed by:

   (a) deposition of sand splays, fans and braided alluvial deposits by seasonal sand-bed rivers in a semi-arid climate;
   (b) marine deposition and coastal dune formation;
   (c) colluvial (sheetflood) deposition around hills and valley heads under open canopy woodland or scrub in a semi-arid climate;
   (d) aeolian deposition by redistribution of fluvial and residual sands in an arid climate.

The unconsolidated, sandy sediments which provide the parent materials for many white sand profiles appear to require a major shift of climate, from humid to dry, leading to rapid deposition, and then from dry back to humid, to induce hydromorphic conditions leading to podzolisation. Over basement rocks, it is not clear whether an arid phase of a aeolian redistribution of the saprolitic mantle has commonly occurred as suggested by Fairbridge and Finkl (1984). But the drying out of climates repeatedly during the Quaternary has been shown to have led to widespread surface wash, infilling valleys and accumulating sandy sediment on piedmont slopes. There is thus the possibility that some white sands encountered on cratonic plateau surfaces may have had a sedimentary origin, while elsewhere the transformation from a saprolite may have taken place *in situ*.

The intervention of swamp conditions may have been important. These occur quite widely on the Gondwana continents, as a result of the low river gradients, in places affected by basinal structures causing subsidence and back tilting of river courses. Slight incision of drainage or a fall in the water-table due to other causes, would be sufficient to expose these formerly hydromorphic environments as dry surface sands. Thus the sandy fringe of the dambo valleys in central Africa, and their analogues on other continents, may mark former higher levels of the poorly drained, swampy floors of these features.

It emerges from this discussion that the apparently distinct deposits, labelled 'stone-lines' and 'white sands' may in reality be closely connected. Such connections can also be more widely drawn, since both are also part of the tropical weathering system and relate to saprolites and to laterites (*sensu lato*). Equally, they are part of the denudational or morphogenetic system operating across the landscape and must be related both to landforming processes (erosion and sedimentation) and to landscape history. Attempts to explain either phenomenon entirely by pedogenic or by geomorphological processes and reasoning are therefore bound to fail. It is also worth noting that many writers regard the laterite–bauxite transition as marking the extremity of tropical weathering

(e.g. Bardossy & Aleva, 1990). But a wider view might regard the podzolised white sand formations as the ultimate result of chemical denudation for, while complete dissolution of quartz is theoretically possible, it is probably rare.

Some concluding remarks

It has been shown here that the continuous operation of steady processes, including the progression of weathering, the lowering of the ground surface and the formation of topsoils by soil fauna, are capable of producing complex profiles containing leached horizons, stone-lines on interfluves, clay complexes in valley floors, and also ferricretes.

None the less, few if any aspects of geomorpholgy can be pursued without reference to environmental changes on different timescales. Individual deposits which are sometimes assumed to mark periods of different or changing environments persisting over 10^3–10^4 years, may mark events of much shorter duration. Some such events could have arisen within the present-day climatic regime. It is therefore necessary to seek corroboration for reasoning that invokes widespread major oscillations of climate and perturbation of vegetation cover.

The routes to this corroboration are mainly via dated chronologies of change and palynological interpretation of palaeoenvironments. This evidence has been reviewed in Chapter 7. There remain in the literature many undated sequences of deposits, some of which have been cited here. The work on colluvium-filled hollows in the Pacific northwest of the USA (see also Chapter 6) has revealed recurrence intervals for debris flows that place these features within the range of environmental disturbances due to rainstorms within the present-day climate. On the other hand, basal dates for the colluvium have indicated a climatic signal which suggests much greater activity in the past, during the Pleistocene–Holocene climatic transition.

This work emphasises the importance of viewing the changes in landscape as part of a *continuum*. Events may have been more frequent or of higher magnitude in past environments, but there were no sudden breaks in the continuity of pedogenesis or morphogenesis that can be attributed to periods of

10^3 years. Where such breaks are evident from unconformities within deposits, they relate to the occurrence of short-term events within these longer periods. The level at which we try to resolve puzzles about landform development largely dictates the nature of our conclusions. In this chapter that level has been that of the longer periods of Quaternary climatic change (10^3–10^4 years). There is strong evidence from the landscape to show that this level is of very great significance to our understanding of surface deposits in particular, but also some other features of the landscape (see also Roberts & Barker, 1993). The reader is, however, urged to think back to the earlier chapters on processes and the occurrence of events on the shorter timescales.

In the subsequent chapters much longer term evolutionary timescales, commonly 10^6–10^7 years will become the focus of attention, for these also are vitally important to our understanding of landform patterns and characteristics.

PART IV

THE EVOLUTION OF TROPICAL LANDSCAPES

9

Etching Processes and Landform Development

THE CONCEPT OF
THE ETCHED PLAIN

The concept of the 'etched plain' was first advanced by Wayland (1933; also in Adams, 1975, pp. 365, 367), who stated simply that,

> absence of any marked relief, a flat gradient and a seasonal climate lead to vertical rather than horizontal movements of groundwaters and the consequent rotting of all but chemically resistant rocks, such as certain quartzites, to a depth of tens of feet. This zone of rotted rock or saprolite, is largely removed by denudation if and when land elevation supervenes, and the process may be repeated again and again as the country rises slowly and discontinuously. Thus the surface (or large parts of it whose areas are determined by drainage basins) is lowered and kept at or near base level by superficial removal against elevation from below.

Such surfaces of marked flatness were described by Wayland as 'etched plains' which he thought indicative of slow, continuous upward crustal movement. However, he also recognised that such etched plains were individualised and separated by marked topographic breaks which he attributed to periods of rapid uplift. According to Wayland (1933), while peneplanation is a product of denudation of mountainous country over periods of $r \times 10^7$ years, the etched plain could be developed from an original peneplain in a much shorter time which he did not seek to define.

It is significant that Wayland attributed the concept to Willis (1936), who embodied the ideas in his celebrated monograph on the *East African Plateaus and Rift Valleys*. Willis was concerned with the origins of the Tanganyika Plateau which he continued to describe as a peneplain, thus distracting attention from his comments on etching. However, these were pertinent and considered the interaction of stream 'corrasion' and 'rock decay' in producing the relief. Willis (1936, p. 135) stated that, 'it is an etched peneplain, locally incised and even planed by streams and widely mantled with residual and transported soils'.

Willis was also interested in rock decay as an agent of planation independent of altitude, and he related the processes to the formation of granite inselbergs as high as 1000 m, as a result of many cycles of saprolite formation and removal to a general depth of ~ 10 m. He described the rock floor of the Tanganyika Plateau as an 'etched surface'.

This early association of etched plains with peneplains, together with the remoteness of the field areas and, in the case of Wayland (1933), obscurity of the publication, greatly reduced the impact of these ideas in Europe and North America, where the controversy between the ideas of Davis and Penck was dominant, at least in the anglophone literature. Yet this concept of etching did not arise in a vacuum; many of the early geologists testified to the importance of deep chemical weathering, and some, like Branner (1896) in Brazil, and Falconer (1911, 1912) in Nigeria, had developed concepts of weathering and stripping of saprolite to account for inselberg formation (Chapters 1 and 10).

During the ensuing decades, however, etchplains became at most a footnote to debates about planation by other mechanisms. Thornbury (1954,

p. 193) for example, commented, 'it is conceivable that the process designated by Wayland as etching might contribute to local differential lowering of a peneplain surface . . . but it is difficult to visualise this process operating widely enough to produce etchplains of regional extent.' Bishop (1966, p. 149), referring to the peneplains of Uganda, discarded the use of Wayland's terminology because, 'there is no stripping to expose considerable areas of the basal surface of weathering'. The restriction of the term etchplain to describe extensive areas of exposed rock is found also in the work of Mabbutt (1961) on Western Australia and was later called for by Ollier (1969).

Nevertheless, it is possible to trace the development of ideas concerning deep weathering and stripping from continued research in east Africa during a period of more than 30 years (Wayland, 1933; Willis, 1936; Handley, 1952; Pallister, 1956; Ollier, 1959, 1960; Harpum, 1963; De Swardt & Trendall, 1969; Doornkamp, 1968; McFarlane, 1971). It is significant that Linton's (1955) seminal paper on 'The Problem of Tors' drew heavily on the interpretation of tor formation in Tanzania (Handley, 1952) so that there was a direct connection between early work on deep weathering and stripping in east Africa and the formulation of a workable hypothesis for the evolution of similar rock forms in high latitudes.

Büdel and the concept of *doppelten Einebnungsflächen*

One of the most important inputs to the central question of weathering and landform development in the tropics came from Büdel in 1957. His concept of *doppelten Einebnungsflächen* or *double planation surfaces* (the wash surface and the basal weathering surface) has had a major impact on geomorphological thinking about the tropics in particular, but also in a palaeogeomorphological context concerning higher latitude landscapes (see also Kiewietdejonge, 1984b). More recently, during the translation of his work on *Climatic Geomorphology* (Büdel, 1982), it was decided to describe inselberg studded plains or *Rumpfflächen* as 'etchplains' (see Translators' Foreword and p. 36 in Büdel, 1982).

This has extended the use of the concept into questions unforeseen by Wayland (1933) and Willis (1936).

Conditions for etchplanation according to Büdel

In Büdel's (1957, 1982) view, the formation of etchplains is associated particularly with the seasonal tropics (rainy season 6–9 months) and with tectonically quiet areas, and his type area was situated in the Tamilnad Plains of southern India (Büdel, 1965). Büdel's work was never clearly focused on the forested (perhumid) tropics, and in his book Büdel (1977, 1982) admitted that observations from these areas had been few (only 15 pages were devoted to the inner tropical zone as compared with 65 to the peritropical zone). Where etchplains are found in the inner tropics, Büdel (1982) is inclined to attribute them to past periods of seasonal climate. Discussion of this point will follow later.

In the seasonally wet tropics, four circumstances were considered indicative of etchplanation:

(1) soil thicknesses of 3–30 m;
(2) homogenisation of the soil profile indicating a constant, dynamic evolution;
(3) a sharp transition from weathered to fresh rock over wide-jointed, quartziferous rocks, indicating intense weathering;
(4) the widespread existence of a deeper zone of decomposition 100–150 m (perhaps 200 m) thick.

The last of these is marked by the formation of *grus* or partially decomposed rock to great depths, particularly where rocks are close jointed. The sharp transition to fresh rock occurs mainly where the basal weathering surface approaches the land-surface, in-between the zones of deeper weathering.

Over most of the plain, erosion operates only on the upper soil surface, so that the basal and wash surfaces have different functions; saprolite development at the basal surface of weathering, and soil erosion at the wash surface. Intensive weathering is active all year at the basal surface which remains moist throughout the dry season. The intensity of weathering is promoted by the high CO_2 content of

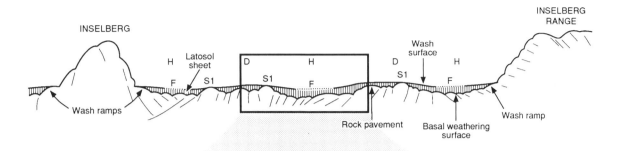

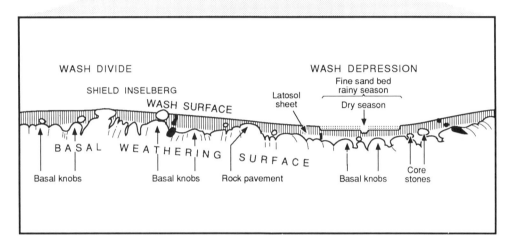

FIGURE 9.1   Typical N–S cross-section through the Tamilnad Plain (east side of Deccan) as an example of an active etchplain. H: Wash depressions; D: wash divides; S1: shield inselbergs; F: fine sand in the rainy season riverbed. Detail of the wash divide and the wash depression is shown in the expanded box (after Büdel, 1977, 1982; reproduced by permission of Gebrüder Borntraeger and Princeton University Press)

the soil atmosphere (20–40 times the levels in mid-latitudes, according to Büdel) and the occurrence of organic acids.

The relief of the basal surface is typified by basal knobs (*Grundhöcker*) 1–3 m high, but larger rock masses form both taller hills and more extensive rock pavements. Progressive lowering of the weathering surface may isolate detached *cores* as weathering penetrates via inclined joints, while larger *shield inselbergs* are also formed. The intervening plains are typically drained via *wash depressions*, i.e. shallow saucer-shaped valleys lacking fluvial channels which are described elsewhere as 'dambos' (Chapters 8 and 11). According to Büdel (1957, 1982), it is this characteristic

etchplain–inselberg relief that forms the prime evidence for an hypothesis of gradual landsurface reduction or *etchplanation*.

In seasonal climates, the dry season desiccation leads to cracking of the clay-rich soils which are also riddled with root channels and tunnels made by soil fauna. With the onset of the rains, water is initially absorbed via these macropores and also into smaller pores created by bacterial action. However, water is soon adsorbed onto the clay particles which then repel each other and go into suspension. These and loose, fine sand brought to the surface by termites are readily washed from the surface by rainy season rivulets that flow across the entire slope without forming channels with fluvial features. These wash

TABLE 9.1   Terms proposed for the description of etched plains

Willis (1936) East Africa	Büdel (1982) Southern India	Mabbutt (1961) Western Australia	Thomas (1965a,b, 1969) Nigeria	Ollier (1960) Australia	Finkl & Churchward (1973) Finkl (1979) Western Australia
Etched peneplain		Weathered landsurface	Lateritised etchplain	Deeply weathered	Incipient etchplain
			Dissected etchplain		Partial etchplain
			Partially stripped etchplain	Partial etchplain	Semi-stripped etchplain
	Grundhöcker relief (etchplain)	Stripped landsurface (etchplain)	Dominantly stripped etchplain	Etchplain	Stripped etchplain
			Incised etchsurface	Complex etchplain	
			Pedimented etchplain		
	Etchplain escarpment				
			Aggraded etchplain		Buried etchplain

(Reproduced by permission of Gebrüder Borntraeger from Thomas, 1989a)

processes serve to create and preserve the etchplain surface as a low gradient plain. In Tamilnad this rises only 200 km inland over a distance of 100 km (equivalent to a slope of 0.2% or 0.11°). *Wash divides* and *wash depressions* form gentle undulations across the plain, providing open-ended erosional concavities (Figure 9.1). These have the following characteristics:

(1) they are always gently saucer-shaped;
(2) there is no perennial river; only rainy season rivulets;
(3) floors are sandy and unincised;
(4) there are no undercut or slip-off slopes or terraces;
(5) shield inselbergs may break the surface at any point;
(6) they are active plains, not ancient planation remnants.

Bremer (1971) has emphasised the similarity between wash depressions and larger tropical rivers in many of these characteristics, but over-generalisation in this area of tropical geomorphology has obscured many important aspects of stream activity which were explored above (Chapter 5).

It is important to stress here that the essential conditions for etchplanation are the preservation of a thick, moisture-retentive 'latosol' (or saprolite) cover, separating the double planation surfaces. Former ideas concerning the restriction of the term 'etchplain' to areas of exposed basal weathering surface (*Grundhöckerrelief*) therefore rob the term and its related concepts of their essential meaning.

Further studies of weathered landscapes

Ollier's (1959, 1960) work on tropical pedology and on the formation of inselbergs in Uganda advanced the development of these ideas, and in the decade following Büdel's (1957) original paper, a number of more specific attempts to apply the concepts of double planation and the formation of etchplains were published. Significantly, many of these came from Australia, where Mabbutt's (1961) study of 'A Stripped Landsurface in Western Australia', described the results of saprolite stripping as an etchplain, and this was followed (Mabbutt, 1962, 1965) by his important statements on 'The Weathered Landsurface of Central Australia'. The terms 'weathered landsurface' (for the duricrusted Old Plateau of central Australia) and 'stripped landsurface' were preferred, but their connections to the concept of the etchplain were also made clear (see Table 9.1), and later extended (Mabbutt, 1988).

A thorough consideration of these concepts was undertaken by Bakker and Levelt (1964) in the context of climatic change in both the tropics and in western Europe since the late Cretaceous. They denied the possibility of a monogenetic explanation for both tropical and temperate latitudes, and equated pediments with etchplains, and altiplanation with the lowering of interfluves by alternate deep weathering and stripping of saprolite.

Extensions of the etchplain concept

The present author sought to extend the etchplain concept to a range of land surface types in Nigeria and Sierra Leone (Thomas, 1965a, 1968, 1974a, 1989a,b), to encompass deeply weathered terrains, stripped landsurfaces, and (schematically) intermediate situations (Figure 9.2). Some of the terms proposed to describe this range of landsurfaces are set out in Table 9.1. A revised scheme, developed from these ideas is proposed in Table 9.2.

This typology was challenged by Ollier (1969), who argued that the term 'etchplain' be restricted to stripped landsurfaces; other instances were to be called deeply weathered plains or partial etchplains. The objections to this restriction are twofold. First, it is not clear that Wayland (1933) intended to restrict his term to conditions of complete stripping; he was more concerned with the repeated sequence of weathering and stripping through geologic time

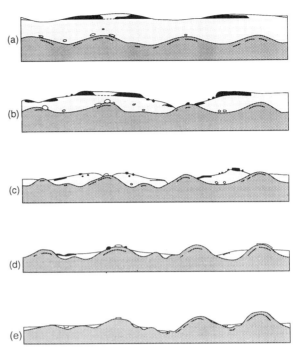

FIGURE 9.2 Some common types of etchplain and etchsurface over crystalline rocks as derived from a deeply weathered landsurface (see Table 9.2): (a) mantled etchplain (palaeo-etchplain); (b) partly stripped etchplain (dissected, duricrusted etchplain, >50% mantle); (c) partly stripped etchplain (isolated duricrusts, 10–50% mantle remaining); (d) stripped etchplain or etchsurface (regosols, <10% mantle); (e) complex etchsurface (incised, pedimented, type 4A/B) (after Thomas, 1965a; reproduced by permission of the Macmillan Press Ltd)

(it is also the case that Wayland referred to weathering of 'tens' of feet and not just 'ten' feet as indicated by Ollier, 1969, p. 209). Second, very few landscapes outside of arid and former glacial zones are completely stripped, and the type areas in Uganda would hardly conform with this restriction (Bishop, 1966). Yet the processes of chemical weathering and the formation of etched rock surfaces beneath variable depths of saprolite are most evident in the humid tropics, and it is this active or *dynamic etchplanation* that provides the justification for describing particular landscapes as etchplains (Büdel, 1982; Thomas & Thorp, 1985).

Application and modification of the expanded etchplain theory have come from Nigeria (Thomas,

TABLE 9.2   Revised scheme for the description of etched landsurfaces

1	Mantled etchplains	(A) Relict or palaeo-etchplains (old weathered/duricrusted landsurfaces) (B) Active etchplains (dynamic lowering, active etching)
2	Partly stripped etchplains	(A) >50% of mantle remaining (possibly with extensive duricrusts) (B) <50%, >10% of mantle remaining (isolated duricrusts, except on footslopes)
3	Stripped etchplains (and etchsurfaces)	<10% of mantle remaining (thin regolith/regosol, boulder slopes)
4	Complex etchplains	(A) Incised etchsurface (valleys cut along fracture lines) (B) Pedimented etchplain (local pediment formation) (C) Re-weathered etchsurface[a] (re-formation of saprolites)
5	Buried etchplains	(with extensive depositional cover)
6	Exhumed etchplains and etchsurfaces	(from beneath sediments or lavas)

[a]Probably most active etchplains of type 1(A) will inherit landforms and deposits from prior weathering/stripping episodes.

1965a; and Figures 9.3 and 9.4); Western Australia (Finkl & Churchward, 1973, 1976; Finkl, 1979; Fairbridge & Finkl, 1980; and Figure 9.5); Guyana and Surinam (Eden, 1971; Kroonenberg & Melitz, 1983), and India (Demangeot, 1975; Pardhasaradhi, 1976), where Büdel (1965) had led the way with an application of his concept of *doppelten Einebnungsflächen* to the Tamilnad plains. It is important to note, however, that subsequent work in the Tamilnad and neighbouring areas of southern India (Demangeot, 1975), while confirming Büdel's (1957) concept of double planation surfaces, provided evidence of important climatic changes, producing what Demangeot called *pédiplaines polygéniques* as distinct from a specific form of monogenetic planation in the manner of Büdel (1965). Mabbutt (1988, p. 465) has also returned to the interpretation of the duricrusted Old Plateau of central Australia and indicated that, 'the mode of evolution postulated for the Old Plateau, of progressive lowering preceded, accompanied, facilitated and to some extent limited by antecedent chemical weathering, provides a more rational justification for regarding it as an etchplain in its own right.'

The convergence of these concepts with the findings of recent research in the Appalachian region of the USA on the one hand (Chapter 2, pp. 48–51), and with ideas concerning laterite formation on the other (Chapter 4, pp. 98–100), is striking and leads inevitably to the conclusion that in the past many mistakes have been made, concerning the apparently static nature of planate landsurfaces, and the association of deep weathering and the formation of laterite (ferricrete) horizons with these conditions of advanced planation.

In reality, large parts of the older continental surfaces have a complex history of incremental lowering and redistribution of surface deposits. Shallow dissection into saprolitic mantles has taken place by the formation of *slope pediments* (Rohdenburg, 1970; and Chapter 8) or *glacis*, while during periods of landsurface stability, the landscape can be lowered by *dynamic etchplanation*.

However, comparison of depths of erosion into kimberlitic diatremes of Cretaceous age or older (Hawthorne, 1973), which intrude some Precambrian cratons, demonstrates that whereas some, as in Koidu, Sierra Leone, have been deeply eroded into the feeder dyke zone (>1000 m depth), others show negligible lowering of the landsurface during these very long time periods. This is evidenced by the preservation of surface sediments associated with the initial volcanic event. It is therefore necessary to limit conditions of dynamic

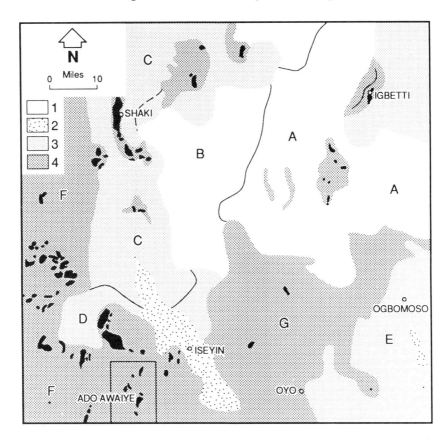

FIGURE 9.3   Etchplains in a part of southwestern Nigeria. (1) Mantled etchplains: A, with sandy saprolites over acid gneiss and quartzite; B, with deep red-brown loams and clays over biotite gneiss and amphibolite. (3) Partly stripped etchplains: C, grey-brown coarse sands with frequent outcrops; D, sandy loam soils, with few outcrops; E, shallow soils over quartzites, but few quartzite outcrops. (4) Stripped etchplains and etchsurfaces: F, grey-brown coarse sands, with many outcrops often of domed inselbergs (shown black); G, similar to F but with small low outcrops. The insert is of the area in Figure 9.4. (after Thomas, 1969; reproduced by permission of Methuen & Co)

lowering to landscapes where integrated drainage systems are able to lower the landsurface, thus maintaining a sufficient hydraulic gradient to promote weathering penetration and also erosional energy at least capable of removing fine-grained sediments as well as solutes from the surface. Some landscapes in semi-arid Australia have altered little over vast spans of time as noted by Ollier *et al.* (1988a,b). Büdel (1982) also stressed the durability of etchplains once they had been formed close to base level.

When and where vertical (but not lateral) stream erosion effectively ceases, the final stages of planation become possible. The progressive flattening of interfluves has been attributed by Pavich (1986, 1989) to the volume loss in the conversion of saprolite to soil (as high as 70% ?). In the seasonal tropics Chauvel *et al.* (1987) and Millot (1980, 1982) have also claimed that the collapse of argillo-ferric structures towards sandy podzolic residues involves a volume reduction of 30%. Similarly, iron crusts in many tropical landscapes, not only induce an apparent surface of planation (Trendall, 1962), but also appear to undergo degradation and dissolution (Nahon, 1986), though this is partly achieved by mechanical attrition.

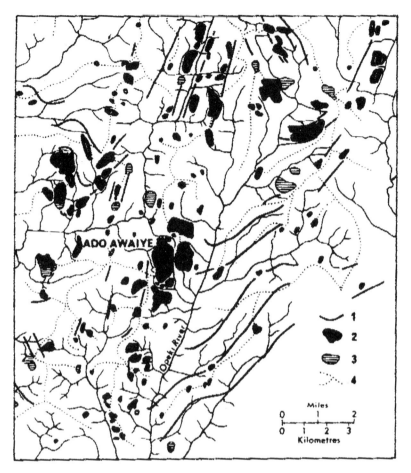

FIGURE 9.4 Landforms of a stripped etchsurface around Ado Awaiye, southwestern Nigeria. (1) Major structural lineaments as mapped from air photos; (2) complex, domed inselbergs mainly on migmatitic gneiss; (3) inselbergs with substantial soil cover; (4) drainage divides. In this area only 7% of the landsurface is outcropping rock, but residual soils tend to be thin and sandy, and outcrops occur on 25% of recorded catenas. There are very few ferricretes (after Thomas, 1969; reproduced by permission of Methuen & Co)

In highly seasonal climates of low rainfall (<700 mm), surface wash will redistribute the sandy residuum, further flattening the landscape, while stream channels continue to achieve some wider planation. As Twidale (1983) has pointed out, this condition of flatness is unconnected to the specific occurrence of pediments or to the formation of pediplains. Opinion remains divided (if it is offered at all) concerning the mechanisms for final planation, but it is interesting to note that lateral planation by rivers has received some renewed support (Osborn & Du Toit, 1991).

Bremer's principle of divergent weathering

Bremer (1971, 1981, 1985b) has drawn attention to the role of the saprolite cover in promoting relief development, by creating conditions for the continuous lowering of the landscape under low slope, low energy conditions of denudation. Adjacent hillslopes, because they become stripped of their soil cover, are effectively arid and thus immune to large-scale weathering attack. In this way, plains (etchplains) are lowered, while hills and escarpments can increase in relief. This was seen by

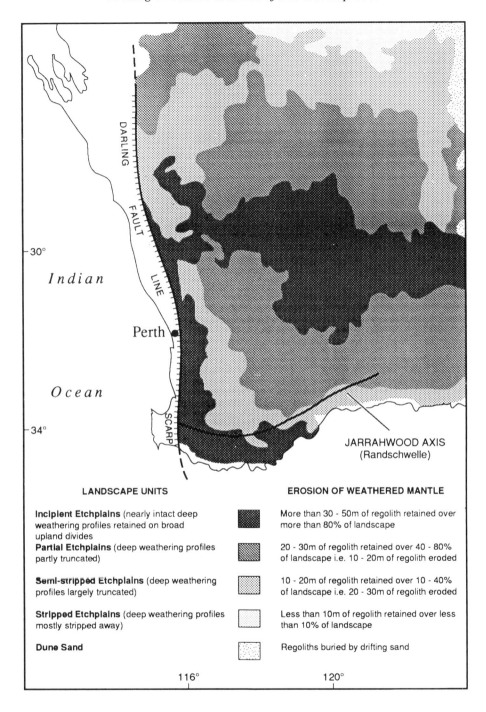

FIGURE 9.5   Etchplains of southwestern Australia (Yilgarn Block). This map uses the nomenclature introduced by Finkl and Churchward (1973; see Table 9.1). An ancient 'primary' cratonic landscape has been dissected during the late Cenozoic to form distinctive landscapes reflecting regional patterns of stripping (after Fairbridge & Finkl, 1980; reproduced by permission of the University of Chicago Press)

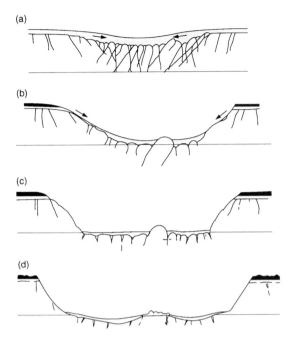

FIGURE 9.6   Outline of intramontane plain evolution by divergent weathering. (a) Stronger jointing leads to inward water flow and a shallow depression forming under thick soil. (b) Continued lowering of the depression floor with increased water supply from hillslopes; duricrusts harden due to water-table fall. (c) Accentuation of the piedmont angle with stripping of hillslopes and the accumulation of thick soils in the plain; duricrusts thicken. (d) Change to the arid-cold climate with mechanical attack on hillslopes and fragmentation of duricrusts. (Mainly after Bremer, 1965; reproduced by permission of Gebrüder Borntraeger and Princeton University Press from Büdel, 1977, 1982)

Bremer (1971) as a reversal of the balance of erosion in arid or cold climates, where frost shattering can cause rapid hillslope erosion.

In this way intramontane basins are formed in the seasonal tropics by the process of *divergent* weathering (Figure 9.6), and can persist when the climate changes to colder or drier conditions. However, Bremer herself noted (1971) that the 'relief-forming soil' could be retained on slopes of 20°, and in fact slopes on many rock formations can retain saprolite to much higher angles, even above 35°, in seasonal as well as perhumid, forest climates.

It was established in Chapter 6 that landslides were particularly effective at slope angles of 28–38°. This

implies that soil can re-form on such slopes, and according to some authors such a cycle of soil formation and removal by mass movement is characteristic of steep hillslopes in the humid tropics, and must lead to the retreat of the hillslope where moisture availability for weathering is available in the saprolite across the entire hillslope. This is in contrast to seasonal climates where the weathering attack may be concentrated at the base of the slope.

Etch processes on escarpments

The role of etch processes in scarp retreat becomes clearer when footpoint weathering is considered. Büdel (1982) referred to 'subcutaneous basal sapping' (Figure 9.7) as a form of etching process that attacks the base of hillslopes and this is the same process already described from the work of Ruxton (1958). Büdel (1982) also argued for the formation of *etchplain escarpments* (Figure 9.8) and for the formation of *stepped landsurfaces* which were also described by Wahrhaftig (1965) from the front ranges of the Rocky Mountains in the USA. Such stepped landsurfaces ('etchplain stairways') are similar to the *Rumpftreppen* of Penck (1924). The formation of these planation steps on the flanks of crustal uplifts remains a matter of important and active research (Bremer, 1985a; Klein, 1985; Summerfield, 1985; Thomas & Summerfield, 1987), and is linked closely both to the concept of pediplanation (King, 1950a, 1955, 1962; Pugh, 1955; Ollier, 1985), and to the issue of differential etching (Warhaftig, 1965).

According to Büdel (1982), etchplain escarpments will form on the flanks of crustal uplifts, and around intramontane basins that have been lowered by etchplanation ('divergent weathering'). Although these may be dissected by deep V-shaped valleys, often following structural weaknesses exploited by prior weathering, a major process is basal sapping by subcutaneous weathering (Ruxton, 1958; Mabbutt, 1966, Büdel, 1982). This opens out re-entrants at the base of the scarp that can form 'etchplain passes' into interior basins.

The question of lateral expansion of etchplains, when accompanied by parallel lowering of the

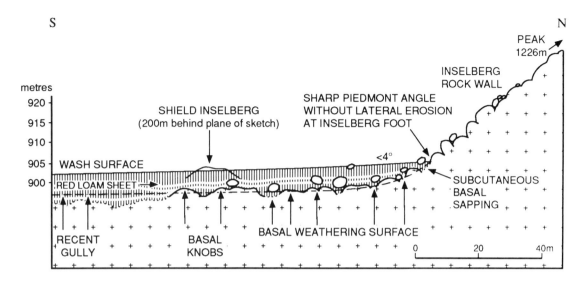

FIGURE 9.7 Accentuation of an inselberg foot by subcutaneous basal sapping (based on Kolar, east of Bangalore, southern India). Loss of soil from the steepened hillslope leads to a concentration of weathering at the piedmont angle and beneath the surrounding plain, possibly leading to the formation of a marginal depression. (Reproduced by permission of Gebrüder Borntraeger and Princeton University Press from Büdel, 1977, 1982)

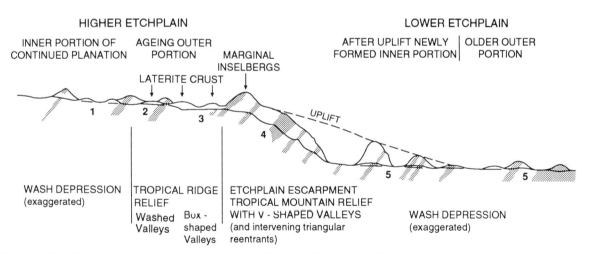

FIGURE 9.8 Creation of an etchplain escarpment in an uplifted area. Continued planation occurs on the lower etchplain; the uplifted (higher) etchplain develops ferricrete crusts and becomes dissected. 1 and 5, wash depressions; 2, a wash valley; 3 a box-shaped valley; 4, a V-shaped mountain valley. (Reproduced by permission of Gebrüder Borntraeger and Princeton University Press from Büdel, 1977, 1982)

double surfaces of planation, was also addressed by Büdel (1982), and his views are pertinent to wider debates in geomorphology. He observed that the gradient of the etchplain surface increased to ~4° across a 100–300 m piedmont ('pediment') near rocky slopes, including individual inselbergs. The 'latosol' blanket is, none the less, preserved at thicknesses of 5–6 m across this piedmont, and where it is retained against the hillslope itself, it enforces slope retreat by basal sapping. Since the upper slope may be of bare rock, this leads to the steepening of the slope and, finally, to the loss of its soil cover. At this stage the slope or inselberg will be subject to only very slow erosional processes and slope retreat effectively ceases. This argument was also applied to the evolution of inselbergs (Chapter 10).

Preferential weathering at the slope foot can also lead to subsequent evacuation of the weathered material as the plains are lowered and the piedmont becomes dissected. This leads to the formation of a circumferential 'moat' or scarpfoot depression (*Bergfuss-Neiderungen*) which has been described by several other writers (Clayton, 1956; Pugh, 1956, 1966; Mabbutt, 1965; Twidale, 1967). Sometimes these depressions connect to radial drainage systems carrying water away from the inselbergs.

ETCHPLAINS IN DIFFERENT ENVIRONMENTS

Büdel (1982) first described the occurrence of etchplains in the humid tropics, but thought they may persist there, following climatic change from savanna conditions of formation. He also gave much detail regarding the survival of etchplains in the extra-tropical regions, especially in central Europe, where he thought they dated from the late Mesozoic–early Paleogene period of near-tropical climates (the 'tropicoid palaeo-earth').

Etchplains have now been described from humid tropical regions (Thomas, 1965a; Kroonenburg & Melitz, 1983; Thomas & Thorp, 1985), arid and semi-arid areas (Mabbutt, 1961, 1988; Finkl & Churchward, 1973), and from temperate and cold environments (Büdel, 1982; Lidmar-Bergström, 1982; Hall, 1986, 1988, 1991; Söderman, 1985; Kaitanen, 1985). It is also claimed that etchplanation has occurred repeatedly throughout the Phanerozoic record, alternating with stripping and pediment formation (Fairbridge & Finkl, 1980; Lidmar-Bergström, 1982). The different varieties of etched plain must therefore be considered widespread and etching processes central to regional landform description and interpretation.

Weathered landscapes and etched plains in high latitudes

With his publication on 'The Problem of Tors', Linton (1955) launched a new debate about the contribution of Tertiary weathering to the formation of upland landscapes in Britain. In this he drew freely on the work of Scrivenor (1931) in Malaysia and Handley (1952) in present-day Tanzania. That debate has now widened and converged with much stronger traditions in this field represented in Europe (see Thomas, 1978). The palaeoclimatological argument, embracing ideas from the tropics, including etchplanation, was clearly stated by Bakker and Levelt (1964) in a paper that still deserves to be widely read. The importance of weathering was also incorporated into the work of Fitzpatrick (1963) and Godard (1965), on the geomorphology of Scotland, and by Gjessing (1967) in a study of Norway's *Paleic* surface. A recognition that pre-glacial conditions may have contributed to the course of glacial erosion (Bakker, 1965), and that many landforms had survived glaciation unscathed (Sugden, 1968, 1976; Kaitanen, 1969), aroused greater interest in this subject (Gellert, 1970; Montford, 1970; Dury, 1971; Feininger, 1971; Touraine, 1972; Klein, 1974).

As new evidence for palaeotropical or near-tropical conditions in high latitudes became more widely available (Nilson & Kerr, 1978; McGowran, 1979; Isaac, 1983; Hall, 1985, 1986, 1987), so interest in the pre-glacial modelling of high-latitude relief increased. Whilst the etchplanation model is not universally applied, the contribution of palaeo-weathering covers to the formation of relief is now very widely accepted, and special issues of *Fennia* (**163**(2), 1985), *Geografiska Annaler* (**70A**, 1988) and *Zeitschrift für Geomorphologie, N.F.* (Supplementband, **72**, 1989), together with other recent works, have provided a broad spectrum of ideas on this topic, spanning the Atlantic provinces of North America (Bouchard, 1989; Wang & Ross, 1989), upland Britain (Battiau-Queney, 1984; Le Coeur, 1989; Hall, 1986, 1987, 1988, 1991) and Fennoscandia (Kaitenan, 1985; Söderman, 1985; Lidmar-Bergström, 1989; Peulvast, 1989).

It is no longer feasible to include a proper review of this work in a study of the tropics, but two fundamental points can be made.

(1) Stripping of palaeo-weathering mantles is held responsible for many pre-glacial landforms at different scales from regional etchsurfaces to minor etch forms.
(2) Although clay-rich weathering mantles may have been Mesozoic or Palaeogene in origin, a widespread *gruss*, or sandy weathering mantle was probably formed during post-Miocene time, under temperate conditions (Bakker & Levelt, 1964; Hall *et al.*, 1989).

This sandy weathering was recognised by Bakker (1967) as important and requiring explanation. It is a partial answer to those who claim that tropical climates are not required to form ferrallitic mantles, and that, whether because of the dryness or coolness of climate, or for the lack of time, no widespread ferrallites (gibbsitic residues, laterites) have formed during the later Tertiary in northern Europe. Indeed, the earlier Tertiary saprolites were themselves clayey grusses rather than ferrallitic tropical residues, and in Northern Ireland they have been likened to *Terra Rossa* soils (Smith & McAlister, 1987). Three tentative conclusions concerning these observation can be offered:

(1) that high latitude passive marginal landscapes in Europe and North America have experienced several periods of deep weathering under varying climatic conditions and for different lengths of time, from the Mesozoic onwards;
(2) that etching processes are associated with the initial breakdown of the rock fabric to produce a saprolite, whether a gruss or more thoroughly altered mantle;
(3) that both the deep weathering and the stripping to reveal etched bedrock forms can, therefore, occur under different climatic regimes.

There remain some difficulties over terminology, but in Tables 9.1 and 9.2 a proposed typology has been compiled from published studies, and it is hoped that this will be sufficiently flexible to allow for most occurrences where saprolite formation and removal is encountered. When exposed rock surfaces become subject to direct chemical denudation, as opposed to saprolitic (isovolumetric) weathering, it is also possible to distinguish *solutional etching* from *saprolitic etching*.

ETCHPLAINS, PENEPLAINS AND PEDIPLAINS

It is necessary to consider the concept of the etchplain alongside those of pediplain and peneplain, all three of which were given equal coverage in the collected papers edited by Adams (1975). It is clear that, while the translators of Büdel's (1977) work wished to emphasise the link between double planation and etchplanation, many French geomorphologists have adopted the principles of etchplanation but retained the term 'pediplain' (Demangeot, 1975; Millot, 1980, 1983), or proposed alternative designations ('plaine de corrosion'—Demangeot, 1978). In the volume on *Theories of Landform Development* (Melhorn & Flemal, 1975) there is, however, no reference to the work of Büdel nor to other German or French geomorphologists with the exception of Penck (1924), and Ahnert (1970). Some protagonists of etchplanation such as Fairbridge and Finkl (1980, p. 82) have argued for the term peneplain to be be retained, 'stripped of . . . genetic overtones . . . for describing . . . complex polygenetic surfaces of low relief'. Twidale (1983) also referred to extensive areas of almost flat terrain as peneplains containing pediments of limited extent, and reflecting primarily structure. This tendency to use the term peneplain generically for any surface of extensive planation removes it from its original and many subsequent theoretical connotations. The term thus becomes useful for description but retains no power of explanation.

Both Klein (1974, 1985) and Touraine (1972) considered tectonism and the influence of marine transgression to be decisive in planation, but also acknowledged the importance of chemical weathering. Touraine (1972) concluded that the great geological episodes of planation have been associated with: (1) an absence of large detrital material, (2) the presence of residual ferruginous clays, and (3) consecutive marine transgression. The importance of the residual mantle was also stressed by Dresch (1978), Demangeot (1975), and Millot (1980, 1983) amongst French researchers. Birot *et al.* (1974) also considered that the extensive planations of the northeastern Transvaal were due mainly to downwearing and not to backwearing and pediplanation.

The role and relations of pediment landforms

Many of these conclusions point to the need to separate the discussion of *pediments* as individual landform units from the possible formation of *pediplains* by the backwearing of hillslopes and coalescence of pediments (Chapter 11). The notion of the pediplain, developed by lateral wasting of hillslopes (King, 1957, 1962) is seldom seen today as a universal process (Hack, 1960; Thomas, 1974a; Twidale, 1983), but it remains pertinent to the possible retreat of major escarpments (Ollier, 1985), and where there are sub-horizontal rock formations or a cap rock (Oberlander, 1989).

Some definitions of different pediment and other piedmont landforms were given in Chapter 8 (Table 8.2), where the importance of climate change in their evolution over time was emphasised. The question as to whether, in the longer timescales of landform evolution, the lateral extension of pediments is more effective than the lowering of the pediment surface will be discussed in Chapter 11. It is clear from many studies in the semi-arid zone that lowering of pediment surfaces by a combination of etching and surface wash takes place, either simultaneously (Dohrenwend *et al.*, 1987), or by alternate mantling and stripping (Mabbutt, 1966; Oberlander, 1974, 1989; Busche, 1972, 1976). It was also envisaged by Twidale (1981b), though he emphasised the exposure of bedrock steps on the flanks of inselbergs as indicators of past pediment encroachment. However, all of these investigations have been in the semi-arid zone, and their applicability to the humid tropics has seldom been tested.

Büdel (1982), however, was clear that hillslope retreat could accompany the formation of etchplains, but was limited by the lowering of the plains and dissection of the piedmont itself. Hack (1960) clearly separated pediment evolution from scarp retreat, and saw pediments commonly, 'fixed in space by geological controls' or indicating, 'a condition of disequilibrium between two zones', a view in part supported by Mabbutt (1978), Twidale (1981a) and by the present author (Thomas, 1974a, 1989a,b). Oberlander (1989), however, does not subscribe to this view. His contention has been that parallel rectilinear slope retreat depends critically upon the presence and retention of a saprolite mantle across the slope, and proceeds in a sub-humid savanna environment. There is no implied dependence on the former existence of a deep, antecedent weathering profile.

Links between etchplanation and pediplanation

A close link between pediplains and etchplains is explicit in the work of Bakker and Levelt (1964), but is discussed in different terms by Fairbridge and Finkl (1980), who considered the processes to act sequentially within a *cratonic regime* (Chapter 11), involving up to 10^9 years of biostasie (Erhart, 1955) and deep weathering (etchplanation), alternating with shorter (10^5–10^7 years) periods of rhexistasie and saprolite stripping (pediplanation). This view was developed particularly with reference to the conditions of present-day aridity and semi-aridity in western and central Australia.

The recession of low scarps and the formation of pediments within deep saprolite covers has been described by many writers including Pallister (1956), Thomas (1965a), and Rohdenburg (1969/1970), who developed the concept of slope pedimentation (Chapter 8) to describe the removal of successive, shallow layers of soil and saprolite by the backwasting of low scarps. The results of this process of stripping are also embodied in the work of Mabbutt (1961, 1962) in arid and semi-arid Australia, and of both Doornkamp (1968, 1970) and Ollier (1977, 1981) concerning the interpretation of landsurfaces in Uganda. More recently, Rohdenburg (1983) extended his analysis to claim that landscapes of prolonged crustal stability will become *panplains* in the sense of Crickmay (1933,

1974), developed by extensive lateral planation by streams, and associated with (but not caused by) very deep weathering.

Slope forms developed as pediments and the formation of peneplains of low internal relief are, therefore, both compatible with an expanded theory of etching and etchplanation. However, there is an important difference between studies concerned principally with the description of a sequence of periods of deep weathering followed by stripping, possibly repeated through geologic time (Linton, 1955; Fairbridge & Finkl, 1980; Finkl, 1981; Lidmar-Bergström, 1982; Elvhage & Lidmar-Bergström, 1987), and the provision of a convincing account of the processes involved and their geomorphological consequences. Some principles for the formation of etchplains have been adduced above, but many questions remain that require further study.

A study of the Koidu etchplain, Sierra Leone

A detailed study was made of the landforms and surface deposits in an area of Sierra Leone, referred to as the Koidu Basin or etchplain (Thomas & Thorp, 1985), a part of the uplifted West African Craton (Thomas, 1980; Summerfield, 1985; Figures 8.1, 8.2 and 11.3). This is an area of low relief (<100 m), developed across Basement Complex gneiss and granite at an altitude of 350–450 m asl. Supracrustal rocks form ranges and smaller hills capped by ferricrete, and stand high above the basement plains. The climate is tropical monsoonal with a mean annual temperature of 27 °C, and rainfall exceeding 2000 mm year^{-1}. These conditions probably position the area within Büdel's peritropical zone of extreme planation, although the area was until very recently rainforest and is not close to sea level.

Beneath the valley floors of small headwater streams, weathering has penetrated the subjacent rock to variable depths, and solutes, clays and silts are removed from the system by low-energy drainage systems forming channelless swamps known locally as 'bolis' (Figures 5.11–5.13). This process allows the accumulation of coarser and heavier mineral fragments as a lag gravel. As the

sediment accumulation in the valley floors deepens, so the proportion of runoff taking place as subsurface flow will increase. The coarse sediment protects the rock floor from abrasion, and promotes the open drainage system of the swamp floor. These factors will reinforce the weathering of the bedrock, and a gradual process of ground lowering takes place by removal of fines and solutes from both interfluves and the valley floor (Figure 11.6). The bolis drain as much as 80% of some land areas but also join to form stream valleys with permanent channels (third order and above) and the sediments of these larger streams are often deposited amongst accumulations of massive corestones (Figure 8.3).

Below the small channelless valleys, the bedrock relief shows the effects of differential etching, and corestones and tors break the swamp surfaces (Figure 5.11). The overlying gravels are mainly unrolled vein quartz liberated during the chemical decay of the rocks. Although this mechanism of dynamic etchplanation may appear self-limiting, as quartz gravels accumulate on the landscape, pulses of increased stream energy appear to have evacuated the residual gravels in response to bioclimatic oscillations during the later Quaternary, allowing the processes to continue. Dissolution of quartz also takes place. Quartz grains not only become etched, but are also subject to disintegration along incipient stress-release fractures. In hydromorphic environments this process leads to the formation of quartz shards and similar angular products of disintegration. In redox environments where Fe^{2+} is able to penetrate the cavities, oxidation to Fe_2O_3 in the form of goethite can lead to the formation of an iron oxide skeleton from which silica may eventually become leached. The penetration of the iron oxide has been described as *plasma infusion* and the resulting grains as *runiquartz* (Eswaran *et al.*, 1975).

Coarse gravels are also found on low interfluves and include both weathered runiquartz and ferricrete fragments (Figures 8.36 and 8.38). Some are actual surface gravels, but in undisturbed soils they are usually found beneath a biogenic layer, as a stone-line. These stone-lines are seen in many sections to be continuous from the interfluve towards the lower slopes and swamps. In part they must form by the lowering of the complete slope, but it has been shown (Chapter 8) that stone-lines often mark surface lag deposits and may have experienced some lateral transfer of materials downslope. In the swamp environment of the valley floor, the iron pisoliths and other fragments become slowly reduced and removed in solution (Sivarajasingham, 1968). This can be seen in terms of the leaching and disappearance of iron oxides from nodules within the gravels of the swamps (Teeuw, 1986; and Plate 9-I), and is corroborated by reports of high iron content in the swamp marginal soils (1700 mg litre^{-1} according to IITA, 1975).

The formation of the ferricrete appears to follow closely McFarlane's (1983a) views of the accumulation of iron oxide pisoliths during land-surface lowering, eventually forming a lateritic duricrust (Chapter 4, Figures 4.9 and 4.10). Over felsic rocks, such duricrusts are thin and discontinuous, and they are readily fragmented by shallow dissection and erosion of slopes which lead also to a falling water-table and continued weathering penetration beneath the interfluves. As in the valley floors, corestones emerge on interfluves and slopes, as fine materials are removed (Figure 8.21). The entire landscape, with the exception of some resistant rock masses and thickly duricrusted outcrops of supracrustal rocks, is therefore being progressively lowered by the normal processes of sediment transfer, but over these mechanically resistant, high grade metamorphic and igneous rocks this can only take place because of the efficacy of the weathering processes at all levels in the landscape. It is therefore possible to regard this area as an active etchplain, and a continuous etching model is illustrated for activity in headwater valley/interfluve systems (Figure 11.6).

There are at least two limiting factors in the applicability of this model over long time periods. First, accumulation of coarse clasts will armour the landscape and inhibit further lowering of both interfluves and valley floors. Second, analysis of stream sediments and colluvial deposits from the same area (Chapter 8) reveals a strikingly episodic behaviour in landforming processes, largely determined by late Quaternary environmental changes. The model must, therefore, be modified to a concept of episodic etchplanation through time (Figures 8.35 and 8.42).

It is also readily seen that beneath the floodplains of larger rivers, where substantial thicknesses of sediment may survive for 10^4 years, there is also time for effective etching to occur, and it seems likely that although rivers often flow over bedrock, effective downcutting occurs as the rivers migrate laterally on to bedrock weathered beneath the floodplain sediments. Such deeper weathering adjacent to river channels has been shown to occur (Chapter 3). The preparation of saprolite and its removal by surface denudation are therefore essentially connected in the periodicity of environmental change and its relationship to rates of saprolite production. The Quaternary oscillations of 20 000 BP, 40 000 BP and 100 000 BP (Jouzel *et al.*, 1987) must play an essential role in this equation, as does the storage time of sediments in a floodplain. A model of floodplain development is required which will link fluvial processes to the progress of etching beneath floodplain sediments within the 10^4–10^5 year timescales commonly recorded. A tentative approach to such a model is shown in Figure 8.42 which can be compared with the findings of Avenard (1973) on the Ivory Coast.

In comparison with the Tamilnad etchplain described by Büdel (1965), the Koidu etchplain exhibits several differences:

(1) it is not a true plain but exhibits a varied, if low, relief;
(2) not all valleys are saucer-shaped; many are etched into the landsurface by 15–30 m and terminate in steep, spring-sapped valley heads which are sometimes plugged with colluvium;
(3) valleys with linear stream channels are floored with sediments recording episodes of sedimentation and erosion;
(4) most rivers have previously excavated saprolite to expose the fresh bedrock; larger rivers have bedrock channels today;
(5) 'low terrace' deposits are widespread, bordering most drainageways, and higher ferruginised terraces occur;
(6) inclined planar surfaces comprise more than 25% of the land surface, resembling glacis and possibly containing 'high terrace' remnants; many are dissected by the present drainage (Figure 9.9).

FIGURE 9.9   Dissected *glacis* slope below a duricrusted summit on schists on Koidu, Sierra Leone

The conclusion must be that this is an eroding landsurface, undergoing groundsurface lowering by chemical weathering, solutional loss and erosion of surficial fine sediment. Climatic and vegetational changes of the Quaternary have led to pulses of energy being expended on the surface slopes and in valley floors and river channels, probably at times of rapid environmental change (Knox, 1972). There is also an important subdivision within the area, between more planate and more eroded drainage basins. The landforms of the area have been analysed by Teeuw (1991b). The two areas define partly stripped and stripped etchplains or etchsurfaces as described in Table 9.3.

These differences are attributable in part to the influence of granite bedrock which is more abundant in the Yengema area, but also to a greater degree of dissection and stripping of the Yengema area via the Gbobora–Bandafayi river system. Increased erosional energy is marked by a decrease in the percentage of valley swamps and a rise in the amount of bare rock, and one could argue that the dissection has been sufficient to interrupt the etching regime.

ETCHING AND THE DIFFERENTIATION OF RELIEF

The application of etchplanation to the formation of landsurfaces of extreme planation has tended to be the primary emphasis given to the theory, perhaps because the explanation of extensive plains has always been elusive, and the continued operation of weathering with the gradual removal of the fine-grained or solutional products, provides a logical mechanism for the destruction of relief.

TABLE 9.3   Dissection of an etchplan: landforms from two drainage
basins in Koidu

Landform	Landform frequency (%)	
	Yengema area (Bandafayi basin)	Kania area (Meya-Moinde basin)
Bare rock	17.0	2.1
Residual hills	10.6	10.4
Glacis slopes	38.6	13.3
Near-planate interfluves	10.5	40.9
Low terraces	12.2	8.3
Valley swamps	12.1	25.0

(After Teeuw, 1991; reproduced by permission of Gebrüder Borntraeger)

However, the basal weathering surface (weathering front) which may become exposed as the *Grundhöckerrelief*, is generally quite irregular, and most instances are better described as etchsurfaces rather than as etchplains (Thomas, 1974a). This irregularity is usually contained within some 30 m of internal relief, but it is associated also with large isolated hills or *inselbergs*, which rise abruptly, perhaps by as much as 600–1000 m above the etched plain. Although some writers (Ollier, 1988b) claim that depths of weathering may have attained 600 m at the end of the Mesozoic, from which these inselbergs might have been excavated by erosion, most of these larger hills must have involved more than one period weathering and stripping of regolith from surrounding plains, unless they have arisen by other mechanisms. This discussion is taken up in a specific consideration of inselbergs in Chapter 10.

Arguments about inselbergs can be extended to the much wider issue of relief differentiation as a function of geochemically controlled weathering processes, operating on uplifted segments of the earth's crust. This constitutes not only a logical extension of the etching concepts of Wayland, but also offers a link to tectonic processes (Hack, 1975; Thomas & Summerfield, 1987).

Relief differentiation during the last 100 Ma

The last 100 Ma since the late Mesozoic have witnessed the formation of new coastlines along the passive margins of drifting continents, and models to describe the geomorphic response to these events are briefly introduced in Chapter 11. The differentiation of relief appears to be a necessary corollary of this non-orogenic uplift and tilting of the crust. The principles of groundsurface lowering and the differential erosion of relief forms according to the concept of *dynamic equilibrium* in the Appalachian Highlands following the ideas of Hack (1960, 1975, 1979) are relevant here, as are the related studies of Pavich (1985), who analysed a long-term 'balanced system of saprolite development, erosion and associated uplift' (p. 316) for the Appalachian Piedmont. Moreover, Hack (1979, p. B15) linked his ideas of landform development to formation and stripping of a deep saprolite mantle: 'under conditions of slower erosion in the crystalline rock areas . . . a residue of saprolite would collect at the surface. When the residual mantle became generally thick, differences in relief were largely eliminated over broad areas.' Hack (1979) commented also that the present average thickness of saprolite in the Piedmont and Blue Ridge is about 18 m and may exceed 90 m. Partial removal of the mantle clearly has led to the differentiation of the relief.

Comparisons between Africa and South America

Studies from two tropical cratons on either side of the Atlantic have come to very similar conclusions. In Surinam, Kroonenberg and Melitz (1983, p. 397) concluded that lithological controls, operating primarily through the agencies of deep chemical weathering, were 'the most powerful means of vertical differentiation of relief' (Figure 11.11). In Sierra Leone, the present author considered that evidence points to 'an ever increasing differential erosion since the early Mesozoic' (Thomas, 1980). What is also interesting in all the accounts of etching processes, from Falconer (1911), Wayland (1933) and Willis (1936) onwards, is the requirement for a repetition of events leading to formation and removal of the saprolite mantle and the emergence of massive inselbergs and other landforms into higher and higher relief. This progressive increase in relief development has been remarked by Bremer (1981) on the basis of studies in Australia, Brazil, Sri Lanka and West Africa.

Kroonenberg and Melitz (1983) have emphasised the importance of alternating humid and arid climatic conditions, leading to periods of etching and stripping of the saprolite mantle. Garner (1974) proposed the incision of rivers during humid phases with lateral planation intervening during arid periods, with the effect of increasing relief.

Hack (1979) laid great stress on *differential uplift* accompanied by small-scale movements along pre-existing faults in the Appalachians. The role of faulting in generating relief on oldlands remains uncertain, but it appears to have made a major contribution to the formation of relief in Venezuelan Guiana (Demangeot, 1985), Brazil, Sierra Leone (McFarlane *et al.*, 1981; Thomas & Summerfield, 1987), and in southern Africa (Summerfield, 1991a,b).

In the Koidu area of Sierra Leone, discussed above, there is a relief difference of 350 m between the widely duricrusted supracrustal rocks (Nimini Hills and Sula Mountains) and the Koidu etchplain developed over the basement gneisses. This appears to be largely attributable to the protective role of the duricrusts, and is a relationship that is common elsewhere. As the Koidu etchplain has been lowered, so deeper groundwater circulation beneath the duricrusted schists took place, leading to the formation of weathering profiles more than 130 m thick (Gaskin, 1975). There is evidence from this area for two periods of more rapid water-table fall, resulting in complex weathering profiles (Gaskin, 1975, and Figure 9.10). A similar situation appears to occur within the bauxite profiles on the Kasila Series gneisses, nearer the coast at Mokanje, and it is tempting to draw the conclusion that they are related to pulses of uplift. Each pulse would lead to the lowering of the weathered gneiss landsurface; to the downward progression of the basal weathering surface beneath the duricrusted hills, and to an increase in the height of rocky inselberg massifs on the neighbouring porphyroblastic granites. It is probable that some differential movement and faulting of the land-surface has also been involved, particularly in respect of the elevated granite massifs of the Loma (1945 m) and Tingi (1709 m) mountains (Figures 8.1, 8.2 and 11.3). This trend towards a more differentiated landsurface appears to have been continuous since the late Mesozoic (Figure 9.11).

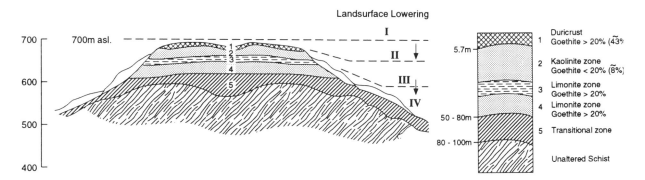

FIGURE 9.10 Schematic cross-section of the Sula Mountains, Sierra Leone. Successive stages of profile development in the schists (1–5) (after Gaskin, 1975) are considered responses to lowering of the relief on adjacent basement gneiss and granite terrane (I–IV); superficial landslide debris is left unshaded (after Thomas, 1989b; reproduced by permission of Gebrüder Borntraeger)

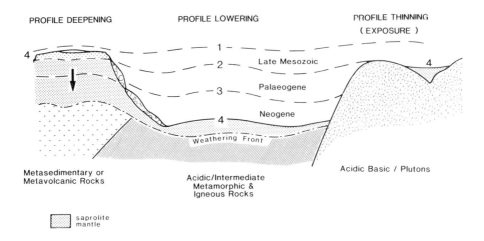

FIGURE 9.11   Schematic model to show geochemically controlled differential etching and denudation on an uplifted cratonic landsurface during the last 100 Ma. (Reproduced by permission of Gebrüder Borntraeger from Thomas, 1989b)

The problem is to reconcile this model with the claim for both regional and local planation levels in this area (Dixey, 1922; Hall, 1969). These claims rest on altitudinal relationships of the major topographic levels observed in this part of West Africa (Dixey, 1922; Hall, 1969; Michel, 1973; McFarlane *et al.*, 1981), and discussed elsewhere (Thomas, 1980). Two salient points emerge. First, only two widespread surfaces have received consistent recognition: the 'Main Plateau Surface' which is undoubtedly internally complex (and probably relates to the 'African' surface elsewhere), and the 'Coastal Plain Surface' which is developed mainly across the softer rock formations below 100 m asl, and may be partly depositional. Second, the postulated age for the last regional planation across the Basement Complex rocks is given as no later than the Palaeocene/Eocene transition.

This leaves the probability that domal uplift accompanying the opening of the Atlantic Ocean has led to tilting and dissection of an ancient Palaeogene surface. This dissection may have been episodic, in response to pulses of uplift, but has led to progressive differentiation of relief. According to the theory of ferricrete formation by groundsurface lowering and accumulation of Fe_2O_3 pisoliths advanced by McFarlane (1983a), the present-day duricrusted summits of the schist belt hills (Nimini Hills, Sula Mountains) will record a long period of downwasting, and need not relate to some ancient regional base level of erosion. Similarly, according to the theories of dynamic etchplanation elaborated by Büdel (1982) and for the Koidu basin (above), progressive lowering of the granite and gneiss terrain could have produced the Koidu etchplain during the ensuing Cenozoic Era. In effect, this is a specific application of Bremer's principle of divergent weathering (Figure 9.6), and also derives from models of domal uplifts along the rifted margins of Africa and America (Le Bas, 1971; Summerfield, 1985, 1991a).

Rates of denudation in cratonic regimes

This style of landform development appears to be a cratonic regime recognisably different from that proposed by Fairbridge and Finkl (1980) for Western Australia. The role of pediments within the West African cratonic regime appears limited to the formation of local glacis which truncate the weathering profiles and some local rock formations. Michel (1973) recognised three phases of glacis formation in neighbouring Senegal, and these can be detected in Sierra Leone (Thomas, 1980).

Comparisons between West Africa and Australia reveal two major differences. First, although relict duricrusts and landsurfaces dating from the Mesozoic or early Cenozoic occur in both areas, the subsequent development of the Australian terrain has been described mainly in terms of the stripping and redistribution of the ancient weathered mantle (Mabbutt, 1965; Finkl & Churchward, 1973; Butt, 1981). In contrast, the *continued development of weathering profiles* is evident from humid areas such as West Africa and also Guyana. Second, very low rates of denudation were determined for the Yilgarn Block in Western Australia, used as the type area for cratonic regimes (Fairbridge & Finkl, 1980). Finkl (1982) quoted 1 mm year[4] for cratonic interiors and 1 mm year[3] for cratonic margins.

In Sierra Leone, the Koidu etchplain appears to have been lowered relative to the duricrusted schist belt hills by a minimum of 400 m in *c*.65 Ma; a rate of 6 mm year[3], or six times that estimated for the Western Australian cratonic margin. But if the duricrusted surfaces themselves also record an earlier period of landscape lowering, and the depth of erosion into local kimberlite pipes (dated at 92 Ma) is taken to be 1000–1500 m after the Hawthorne model (1973), then this figure could be much higher, possibly 10.8–16.3 mm year[3]—an order of magnitude higher than Western Australia. This difference could be explained by a combination of crustal tilting and higher weathering rates in the humid tropical environment. A comparable rate of 8.4 mm year[3], was calculated for southern Kenya during the Tertiary, when the climate was wetter than today (Dunne *et al.*, 1978). This rose to 29 mm year[3] during the Quaternary, when vegetation cover was thinned by aridity of climate.

By comparison, Hack (1979) calculated a rate of lowering during the Cenozoic for the eastern USA of 40 mm year[3], and Velbel (1985) has recently argued for the penetration of the weathering front into fresh rock at a rate of 38 mm year[3] in the southern Appalachians. Pavich (1985), however, considered a denudation rate for the Cenozoic of nearer 25 mm year[3]. These rates clearly make possible a steady-state lowering of the saprolite, in a humid, warm temperate environment at a rate of $2.5–3.8 \times 10^1$ such as that experienced in Western Australia. In equatorial northwest Kalimantan, mid-Tertiary (*c*. 30 Ma) granitoid intrusives at Gunung Ibu must have been unroofed from beneath 1–1.5 km of country rock, indicating a possible rate of denudation as high as 50 mm year[3] in a landscape of widespread deep weathering and evident active etchplanation.

If such rates of denudation and saprolite formation are considered realistic over periods of 10^7 years, within which it is recognised that periods of profile deepening and differentiation will have alternated with profile thinning and stripping, then an etch theory of landform development becomes necessary, not only to explain certain forms of tropical plain, but also to account for landscapes of differentiated relief, in a wide range of tropical and non-tropical and post Archaean landscapes. Even in the Mojave Desert, Dohrenwend *et al.* (1987) have estimated the lowering of both the weathering front and pediment surfaces at an average rate of 20 mm year[3] over a period of 4×10^6 year within basaltic lavas.

TOWARDS AN INTEGRATED THEORY OF ETCHING AND ETCHPLANATION

The argument advanced in this chapter is designed to develop etching concepts to account for landscapes of both extreme planation and those of marked relief, and to acknowledge the influences of both tectonics and climate in development and elimination of relief. It is suggested that the concept of etchplanation provides an organising principle within which otherwise disparate observations can be brought together. The progress of chemical weathering is favoured by warmth of climate; high, continuous precipitation; high organic productivity, and good drainage (Curtis, 1976). But, although the humid tropics remain the most favoured environment for the formation of deep, clay-rich saprolites, the development of sandy weathering types (gruss, arênes) is widespread in the cooler climates (Bakker, 1967; Dixon & Young, 1981; Hall, 1985). There is therefore no reason to limit etching concepts or theory to the tropics and sub-tropics; only the influence of glaciation has selectively removed most traces of weathering mantles from substantial parts of higher latitude landscapes. Along the humid eastern margins of both North America and Australia there is both continuity and transition of conditions from cold terrains towards tropical conditions, and there is growing evidence for the efficacy of chemical etching processes in the formation of the relief throughout these zones, under present as well as past environmental conditions.

Very deep penetration of the weathering systems tends, however, to be associated with jointed or fissile rocks in elevated positions, and also along shatter zones which may occupy lower sites. The preservation of these deep saprolites results from either buttressing by resistant rocks such as quartzite intercalated with biotite schists, or protection by duricrusts. The duricrust itself, however, commonly marks the concentration of oxides of silica, iron or aluminium in more humid climates or carbonates of calcium and magnesium in semi-arid environments. For such concentrations to accumulate the ions in question must be kept within the profile once mobilised. Such features (other than residual bauxites) therefore tend to mark previous conditions of impeded drainage, and possibly highly seasonal or aperiodic fluctuations in soil moisture status. They are associated with strong differentiation of the weathering profile, but not necessarily with very deep saprolite mantles. If the McFarlane (1983a) model for iron accumulation is adopted, extensive duricrusted surfaces may record periods of groundsurface and profile lowering, prior to the progressive deepening of the profile beneath the protective crust.

A distinction can therefore be made between profile differentiation and profile deepening (see pp. 84–85), although both take place beneath groundsurfaces experiencing minimal erosion. In many instances, deepening appears to have occurred following the differentiation of the upper horizons and as a result of increasing local relief. Beneath landsurfaces of low relief (remote from escarpments or dissecting valleys), the geochemical cycling redistributes the mobile ions with minimal removal. Thus sodic soils can form within depressions on high plateaus in central Africa, where rainfall may be 800–1000 mm year^{-1}. In a comparable way surface detritus can be activated, but will be mainly redistributed to form varieties of sand plains, clay

pans and other superficial deposits (Fairbridge & Finkl, 1984).

Although the water-table does not place any finite limit upon the depths of rock decay, there must be a deep circulation of water if the weathering systems are to penetrate the bedrock deeply. It is therefore necessary to be sceptical concerning the formation of very deep weathering profiles beneath base-levelled plains, particularly if they remain close to sea level. These seem to be the sites par excellence for the mobilisation and accumulation of iron, and the breakdown of clays (Brinkman *et al.*, 1973). When such plains are subject to warped uplift, selective dissection of the landsurface brings about a lowering of the water-table and the conditions may become ideal for the formation of duricrusts and for the progressive deepening of the weathering profiles beneath them, while profile lowering possibly in dynamic equilibrium takes place over susceptible rocks elsewhere in the landscape.

At this stage, the type and rate of uplift are clearly important in determining the future behaviour of weathering systems and profiles. Faulted and monoclinally warped continental margins are likely to lead to the formation of escarpments, creating a disjunction between landform development in the hinterland which may experience basin subsidence and sedimentation, and along the coastal margin which may become deeply dissected. Saprolite mantles must then evolve independently beneath upper and lower landsurfaces in the two-storey landscape of Ruxton and Berry (1961), and as shown by Büdel (1982; and Figure 9.8). Thus, along the Great Dividing Range of eastern Australia deeply etched granite embayments have been lowered against the escarpment. Although Dixon and Young, (1981) have argued for hydrothermal alteration in the Bega Batholith of New South Wales, Australia, this was quickly challenged by Ollier (1983b), and there appears to be no cogent reason to invoke this factor as decisive. Above the escarpment, the Monaro forms an extensive etchplain; the Bega Basin below is associated with a dissected, multi-convex landscape exhibiting occasional boulder outcrops and tors. Weathering profiles persist across the scarp itself which might be described as an etchplain escarpment after Büdel (1982; and Figure 9.8).

TABLE 9.4 Behaviour of weathering profiles through geological time

A Following crustal uplift

(1) Profile deepening	—Minimal surface erosion (a) beneath dissected plateau surfaces (b) beneath faulted blocks (c) beneath protective duricrusts
(2) Profile lowering	—Weathering column lowered in dynamic equilibrium; saprolite replaced by weathering as the groundsurface is lowered (no duricrusts)
(3) Profile thinning	—Weathering profile becomes shallow over massive crystalline rocks
(4) Profile truncation	—Stripping by accelerated surface erosion (commonly slope pedimentation) (a) resulting from crustal tilting and regional erosion (b) resulting from climatic crises (usually aridity)
(5) Profile renewal	—Re-weathering of partly stripped etchplains may occur during the uplift of stepped relief

B Under quasi-static conditions

(6) Profile collapse	—Breakdown of clays (*ferrolysis*) with progressive dissolution of saprolite under hydromorphic conditions

C Following crustal subsidence

(7) Profile burial	—(a) Burial under fluvial sediment and organic peats (b) burial beneath terrigenous and marine sediments

Where crustal tilting has occurred, differential etching has been advocated to explain the landforms of the Atlantic margins of the West African Craton. This argument can be extended to younger and tectonically more active areas, as in New Zealand, where the Separation Point Granite has been sharply tilted ($\sim 70\,\mathrm{m\,km^{-1}}$) and uplifted to $\sim 1000\,\mathrm{m}$ in the west. A summit plateau forms an etchsurface and its tilted flank has become subject to selective deep arenisation along structurally controlled, arcuate belts (Thomas, 1974b).

There are too few data available to postulate any general relationship between rate of uplift and response of weathering systems, but the typical cratonic regimes which appear to have long-term denudation rates of $k \times 10^{-1}\,\mathrm{mm\,ka^{-1}}$, contain highly differentiated weathering profiles at advanced (ferrallitic–allitic) stages of alteration. Where denudation rates are an order of magnitude higher, as suggested for the Appalachians, it is probable that high surface erosion rates will maintain weathering profiles at equilibrium in a state of partial alteration. However, this will be affected by climate and rock type. Models of repeated differential etching and stripping over different geological time periods have been proposed for basement landscapes in Surinam (Kroonenberg & Melitz, 1983 and Figure 11.11), the Appalachian Highlands and Piedmont (Hack, 1979; Pavich, 1985), the Buchan area of northeast Scotland (Hall, 1986), and southern Sweden (Lidmar-Bergström, 1982). There is therefore a convergence of views regarding the development of relief on uplifted basement terrains, involving repeated etching and saprolite formation followed by partial stripping, and then by further differential weathering and relief development. Bremer (1971) described divergent weathering in a similar context to account for intramontane plains, as discussed above.

The behaviour of weathering profiles

During the process of relief differentiation, weathering profiles may evolve in different ways, as discussed in Chapter 3. These trends are usually related to crustal states and have been summarised according to the context of uplift, quasi-stability or subsidence in Table 9.4.

Profile lowering in dynamic equilibrium over the long term may characterise active etchplains, when due allowance is made for short-term episodic changes (Schumm, 1975; Thomas & Thorp, 1985).

Profile deepening occurs beneath the duricrusted etchplains which may undergo dissection but experience only localised landsurface lowering. Over the long-term evolution of landforms (Ollier, 1979) the behaviour of weathering profiles may undergo a sequence of changes due to feedback mechanisms within the controlling systems. Profile lowering leading to the accumulation of iron pisoliths and the formation of a duricrust (McFarlane, 1983a) will then afford the profile protection during subsequent dissection. A sequence of profile lowering → profile differentiation → profile deepening → profile truncation → profile renewal, may take place over 10^7 years. Bioclimatic crises will induce medium-term fluctuations in the course of events, retarding the advance of the weathering front and increasing rates of surface erosion, and this leads to the truncation of profiles, where they are not protected by duricrusts.

The role of sedimentary covers

The sedimentary covers which form many coastal plains may have an important geomorphological role in relation to the basement topography. In southern Nigeria, weathering beneath the Cretaceous sandstones has been recorded (Jeje, 1972) and Savigear (1960) suggested that much of the basement terrain in this area could have been exhumed from beneath this cover. More recently,

Pavich and Obermeier (1985) have documented up to 15 m of saprolite formation beneath the sediments of the Atlantic coastal plain near Washington, while Blanck (1978) recorded a pre-Cretaceous laterite near New York. Kaye (1967) previously also called attention to deep kaolinisation of bedrock in Massachussetts. In Natal and at the Cape in South Africa, where the marginal uplift has resulted in the exhumation of basement relief from beneath the Table Mountain Sandstones, there is evidence that saprolite formation took place beneath the sandstones, but possibly after uplift and during their removal by scarp retreat (see Thomas, 1978; and Figure 10.14). There appears to be no reason to doubt that basement weathering beneath sedimentary formations is very widespread.

However, some caution is required; in the case of the Sunda Shelf between Borneo and Sumatra (Figure 8.7), an old weathered landsurface containing significant local relief and surmounted by former inselbergs appears to have foundered, and Neogene sedimentation has subsequently buried the old saprolite mantle (Aleva, 1973; Batchelor, 1979a,b). In this way downwarped etchplains may have their weathering profiles preserved for very long periods of geological time, and subsequently they may become exhumed and reactivated under sub-aerial conditions. This sequence has probably occurred repeatedly through geological time.

10

Erosional Terrain and Residual Hills

TERRAIN UNITS IN TROPICAL LANDSCAPES

Uncertainty and controversy has surrounded most attempts to identify distinctively tropical landforms or terrain parameters. *Inselbergs*, long associated with the geomorphology of the tropical shields, now appear more characteristic of deeply eroded plutons than of tropical environments *per se*. Their presence in deserts (Ollier, 1978; Selby, 1977b), as well as in many other environments (Thomas, 1978; Büdel, 1982) has made arguments concerning their origins turn on issues such as the operation of universal processes of slope development (King, 1957; Ahnert, 1982), control by rock mechanics (King, 1975, Selby, 1977a,b), and the need for long-term evolution spanning important environmental changes over periods of 10^6–10^8 years. Yet the development of these hills still has a place within a geomorphology of the tropics as will be shown below.

Quantitative parameters of slope and drainage also fail to provide adequate numerical data on which to base a tropical geomorphology, yet this study has already explored many aspects of both hillslopes and stream valleys that pertain particularly (but not exclusively) to warm climates.

A more promising approach to this question was introduced in an earlier study (Thomas, 1974a), which emphasised the importance of the vertical zonation of the weathering profile: its duricrusts, mottled, pallid and other zones of alteration, and also the etched rock forms, all of which can become exposed as the profile is dissected and thinned by erosion (Figure 3.1). This approach will be further developed here, utilising suggestions made by Butt (1987), and the discussion of transported colluvium discussed in Chapter 8. A list of landform facets and materials characteristic of the tropics is shown in Table 10.1. No such scheme can ever be definitive, but it may help to focus attention on common facets of tropical landscape that are absent or unfamiliar in studies in other environments.

Slope and drainage characteristics

No unique combination of slope values or drainage densities can readily be construed as typical of tropical or sub-tropical areas, and the fact that large regions within the tropics are developed on ancient cratonic terranes has probably confused interpretations based solely on climatic criteria. Nevertheless, low drainage densities have been recorded from basement areas. In Nigeria, low densities of 0.5–1.5 km km^{-2} appear characteristic of forested and wetter savanna areas (Wigwe, 1966; Jeje, 1970), while in Uganda, Doornkamp and King (1971) found that 54.7% of third-order basins exhibited densities of 1.25–2.5 km km^{-2}. These low figures are not representative of tropical landscapes, but are found over a limited range of rather planate landscapes, notably in Africa. They almost certainly reflect the presence of deep, weathered mantles and the high infiltration capacities of ferrallitic soils and regoliths (Nye, 1954, and Chapter 5).

With increasing relief and slope, regoliths become thinner and runoff also increases, leading to higher

TABLE 10.1   Common landform facets and materials in the tropics

A   Residual materials
  I Interfluve and plateau residuum
    (1) Residual soil on weathering profile
    (2) Duricrust of weathering profile (tabular summits and plateaus), includes alucrete, ferricrete, silcrete, calcrete cappings
    (3) Duricrust rubble (convex summits; slopes around (2))
 II Hillslope residuum
    (4) Duricrust cliffs (cliffs around (1) and (2); see also BI).
    (5) Upper zones of weathering profile (usually upper slopes)
    (6) Lower zones of weathering profile (saprolite + corestones on steep and lower slopes)
    (7) Boulder-controlled slopes; with tors as residual hills (convex summits and steep slopes >20°)
    (8) Soil and thin regolith on actively weathering rock in juvenile or rejuvenated terrain
III Bedrock surfaces
    (9) Angular, jointed rock, e.g. steep hillslopes
    (10) Smooth, bare rock surfaces, e.g. domed inselbergs and rock pavements
    (11) Truncated rock surfaces, e.g. piedmont footslopes

B   Transported materials
  I Interfluve and plateau deposits
    (1) Duricrusts containing transported clasts (tabular summits), e.g. ferricrete, silcrete cappings (often impregnating sediments)
    (2) Gravel layers (exposed or as stone-lines) comprised of: duricrust fragments, rolled and unrolled quartz
    (3) Sand cover — either aeolian sand sheets, including dunes or 'white sands' of eluvial, alluvial and colluvial origins
 II Hillslope deposits
    (4) Colluvium — often in hillslope hollows or former gullies
    (5) Landslide debris, including:
       — boulder accumulations from rock falls or slides
       — earthflow debris from slumps and translational slides
       — debrisflow and mudflow sediments, often channellised
III Lower hillslope and piedmont zone deposits
    (6) Coarse talus and debris fan deposits, including:
       — coarse, clast-supported debris and talus fans
       — debris flows similar to (5) above
       — earthflows associated with rotational slides
       — channel sediments associated with debris fans
    (7) Colluvium ('colluvium–alluvium'), including
       — mudflow deposits, sometimes as gully-fills
       — sheetflood sediments, laminated sands and silts
       — shallow, quasi-alluvial fan and channel sediments
       — unstratified colluvium (process obscure)
    (8) Cemented deposits, including:
       — ferricrete benches in colluvial and alluvial deposits
       — silcreted sands and gravels
       — calcrete layers and benches
 IV Valley floor deposits
    (9) Alluvial deposits, including:
       — alluvial fan facies, frequently fine-grained
       — sheetflood, sandy sediments ('pedisediment')
       — deep channel fills; flood sediments, fining upwards
       — shallow braided valley sediments
       — clay fills in dambo valleys; clay pans
       — floodplain facies from meandering streams
    (10) Lacustrine deposits, often associated with alluvial plains
    (11) Aeolian deposits, including dunes and sandplains, see also (3) above
    (12) Cemented deposits, as in (8) above

TABLE 10.1 *(continued)*

C *Erosional slope units*
   (1) Rock faces subject to spalling and rockfalls:
       —dominated by near-vertical jointing
       —dominated by curvilinear sheeting fractures
   (2) Weathering-limited rock slopes with thin regolith:
       —steep hillslopes with a migratory debris layer
       —rocky pediments
   (3) Bare rock surfaces subject to sheet wash
   (4) Landslide scars due to shear failure, usually crescentic headwall slopes (amphitheatres) in saprolite:
       —rotational slump scars with well-defined shear surfaces
       —shear surfaces at gullyheads or marking debris flows
   (5) Erosion gullies cutting weathering profile zones:
       —cut by debris avalanches and flows
       —cut by channel flow in streams
       —resulting from collapse and washout of tunnel gullies

Developed from Thomas (1974a) and Butt (1987) with modifications and extensions

drainage densities; and it is possible that these increase more rapidly than in temperate areas, as a function of rainfall amounts and intensities. Eyles (1971) found that densities rose to 12.5 km km^{-2} in West Malaysia while Morgan (1976) was able to draw morphogenetic regions for this area on the basis of drainage density variations (<8– >11 km km^{-2}). An early study by Peltier (1962) demonstrated a much steeper rise in drainage density with increasing mean slope in tropical areas than in any other (non-glaciated) climatic zone.

Some scepticism about the significance of these findings may be necessary, however; first, because the present patterns of stream channels are seldom fully adjusted to the prevailing bioclimatic environment, and second, because the quality of data inputs to some of these analyses must be questioned. The widespread occurrence of shallow, channelless valleys (dambos), for example, is an indication of the first factor, and is bound to create problems of definition (for stream valleys), leading to the second objection. It was demonstrated above (Chapter 8) that dambos, which often occupy broad, flat-floored and alluviated valley floors, can conceal buried stream channels. The broad valley floor is subject to non-channellised flooding after heavy and prolonged rain, while the base flow is transmitted down valley through the channel-fill deposits. The recognition and mapping of such drainageways poses operational difficulties when using aerial photographs or conventional maps.

Some relief and slope studies have emphasised contrasts between hilly relief, plains relief with isolated hills, and plains relief with only minor development of intermediate foothills (Eyles, 1969, 1971; Speight, 1967, 1971), but low convex hills (the *meias laranjas*) are widely encountered in other humid tropical regions such as parts of Brazil and the island of Borneo.

The grouping of slopes on Bougainville Island into hillslopes of 32° ± 10° and alluvial plains with mean slopes of 0.5° (Speight, 1971) is striking, but this is a tectonically active island arc subject to rapid denudation and coastal deposition. Nevertheless, a sharp *piedmont angle* at the junction of hillslope and plain has been widely recognised and is striking in many sub-humid to semi-arid areas (probably because it is not obscured by vegetation). This feature has been attributed to a number of different factors, not all of which are climatically driven (Cooke *et al.*, 1993), including: (1) lateral planation by streams; (2) faulting; (3) lithological and structural contrasts; (4) basal sapping of hillslopes by intense weathering in the foothill zone accompanied by concentrated subsurface and surface erosion; (5) contrasts in debris size between hillslope and plain; and (6) differences in processes operating on the hillslope and the piedmont plain. Certain of these factors can only operate once the hillslope–plain junction has come into being, but may act to sharpen the angle over time.

Weathering and regional landform systems

More useful, but less easily quantified, are characteristics of tropical terrain arising from the presence of the weathered mantle. The landform facets attributable to these residual materials are listed in Table 10.1, but such individual facets or elements of the landform are usually related in repeated patterns as *land systems* (Christian & Stewart, 1968; Mabbutt, 1968) developed by the erosion and truncation of weathered landsurfaces, forming a major group of tropical landscapes here described as *etchplains* or *etchsurfaces* (see Chapter 9).

This concept has provided a number of regional terrain classifications. One of the first areas mapped on the basis of the stripping of a weathered landsurface was described by Mabbutt (1961) from Western Australia, and this author subsequently grouped the land systems of the Alice Springs area of central Australia (Mabbutt, 1962) into:

(1) the erosional weathered landsurface, representing more or less intact remnants of a polygenetic landsurface, together with its deep weathering profiles, of late Mesozoic to mid-Cenozoic age;
(2) the partially dissected erosional weathered landsurface, where partial removal of the old surface and its weathered mantle has occurred;
(3) erosional surfaces formed below the weathered landsurface, where only younger forms survive;
(4) depositional surfaces.

The progressive stripping of old weathered landsurfaces is also the basis for defining categories of etchplain (Thomas, 1965b, 1969, 1974a; Table 9.1 and Figure 9.2; Finkl & Churchward, 1973; Finkl, 1979). A similar interpretation (though using different terminology) has been applied to the major geomorphic divisions of Uganda, where the Buganda Surface has been progressively stripped to form extensive etchplains (Ollier, 1960, 1981). Doornkamp (1968) also reported on this area and used descriptive terms to describe five principal terrain types: (1) residual hills standing above the highest landsurface; (2) the upland landscape carrying many duricrusted hills; (3) inselberg areas on granite rocks, where the upland surface has been destroyed

by dissection into deep weathering profiles; (4) the lowland landscape covered by a thinner mantle and less well developed laterite; and (5) infill or aggradational landscapes which in this area have resulted from drainage reversal within the Lake Victoria basin.

The erosion of deeply weathered landsurfaces (mantled etchplains) produces characteristic soil catenas and surface materials, both residual and colluvial (Finkl & Churchward, 1976; Churchward & Gunn, 1983). The frequent occurrence of ferricrete, appearing as duricrust in dissected landscapes, has led to more general statements of the patterns and occurrences of duricrusts in soil catenas (Ollier, 1976 and Figure 3.22). The classification of soil catenas into a few recurrent categories indicates how closely the soil patterns relate to the weathering profile. Thus Moss (1968) proposed the following categories:

(1) *Catenas with hard laterite* (ferricrete):
    (a) with ferricrete as an upper slope or summit feature with (i) massive duricrust or (ii) detrital ferricrete;
    (b) with duricrust as a lower slope feature.
(2) *Catenas with rock outcrops:*
    (a) associated with extensive areas of flat land with (i) or without (ii) extensive pre-weathering;
    (b) in mountainous areas.
(3) *Catenas without rock outcrops or hard laterite:* found widely over different rock types and subdivided accordingly.
(4) *Soil associations on relatively flat land* (i.e. not forming toposequences).

Regolith terrain mapping

A more recent but similar approach to terrain description and mapping involves the concept of *regolith terrain mapping* which has been described by Chan (1988) and developed by the Bureau of Mineral Resources in Australia (Chan *et al.*, 1986; Pain *et al.*, 1991). Its aim is to assist mineral exploration, and the objective is to systematise terrain description for areas underlain by deep and extensive regolith mantles. A *regolith terrain unit* is conceived as an area of terrain associated with a distinct regolith profile and its description emphasises its geological and geomorphological relationships.

A method for defining deeply weathered rock units was proposed by Senior and Mabbutt (1979), who described six parameters for the definition of a deeply weathered profile: (1) reference section (analogous to a type section); (2) lithology (physical and chemical properties of weathered and parent rocks); (3) thickness, including thicknesses of constituent horizons; (4) age; (5) distribution; and (6) relationships with parent rocks.

Similar ideas using the *land systems survey* approach have long been applied to engineering problems (Dowling, 1968; Newill & Dowling, 1968). The patterns, units and correlated physical properties for a duricrusted landscape are shown in Figures 10.1, 10.2 and 10.3). Many engineering classifications depend on the development of an adequate *terrain model*, and two further examples from South Africa (Brink & Partridge, 1967; Brink *et al.*, 1970; Cooke & Dornkamp, 1990) and from Hong Kong (Hansen, 1984) (Figures 10.4 and 10.5) serve to emphasise that not all tropical landscapes can be fitted into a few restricted categories. They may be as diverse as landscapes in other parts of the world.

RESIDUAL HILLS IN TROPICAL LANDSCAPES

Erosional forms in the landscape have been the traditional emphasis of geomorphology. In the 'cycle of erosion' as proposed by W. M. Davis (1899a), and in older texts dealing with this question, a succession of landscapes defined by the erosional forms of interfluves and residual hills were the focus of attention. In this study, much greater emphasis has been placed upon the formation of the weathering profile or mantle and on the processes which modify the saprolite by surface runoff and mass movement. The erosional landforms that have been discussed have largely been specific forms, resulting from these processes acting on non-cohesive materials.

This may appear strange to many students of geomorphology, and aberrant to some. However, the erosional residual, whether of fresh rock or saprolite, is generally the product of a long history, and is what remains when weathering and erosion have produced many of the other features discussed in this study. Slopes are often considered to be either *transport limited*, in which case they arise from erosion of the weathered mantle, or *weathering limited*, in which case they depend on active processes of rock decay. But there are also rock slopes that are modelled within the mantle by etching of the rock surface, and which are later exposed by removal of the regolith. Such slopes cannot easily be attributed to either of the two classes, but are *weathering derived* slopes, the forms of which do not depend on surface erosional processes at all. Many forms of rocky inselbergs appear to belong to this class, wholly or in part. These landforms will be considered separately, following an enquiry into landforms on saprolites with or without a duricrust capping.

EROSIONAL FORMS IN THE RESIDUAL MANTLE

Saprolite hills

Hills modelled in saprolite, unprotected by ferricrete or other duricrust, often form weathered compartments within granitoid rocks or gneiss. They have been referred to extensively in this study as multiconvex landscapes (known as *demi oranges* in French and as *meias laranjas* in Portuguese). The thick, permeable saprolite readily forms convex summits under the probable influence of creep processes, but the steep flanking slopes then become subject to landsliding and to rapid erosion by gullies, when cleared of protective vegetation. These have been detailed above from work in Brazil, and they have also been studied in Madagascar (Wells, *et al.*, 1991).

The unprotected saprolite seldom survives to form isolated residual hills; the 'demi oranges' are usually closely grouped, separated by relatively

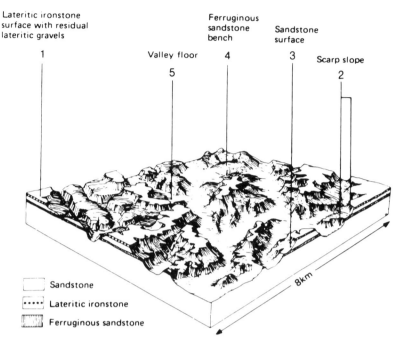

FIGURE 10.1 Block diagram and general description for the Doto Land System, Nigeria (after Newill & Dowling, 1968; reproduced by permission of Chapman & Hall)

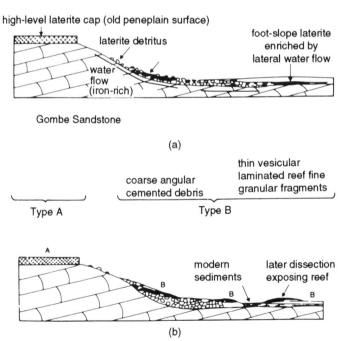

FIGURE 10.2 Relationship of surface materials to slope facets in the Doto Land System. (a) Formation of footslope laterites and their relation to high level duricrusts. (b) Distribution of laterites (ferricretes) in the present landscape (after Newill & Dowling, 1968; reproduced by permission of Chapman & Hall)

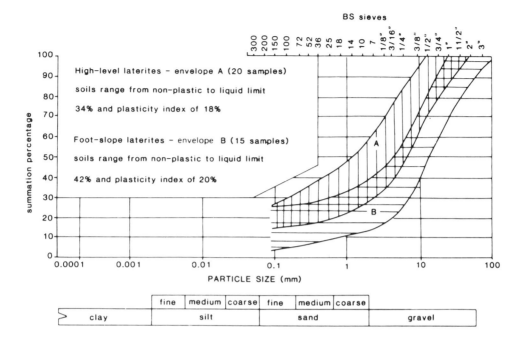

FIGURE 10.3 Particle size distribution and plasticity data for high level and footslope ferricretes from the Fika–Nafada area, Nigeria (after Newill & Dowling, 1968; reproduced by permission of Chapman & Hall)

narrow, and sometimes alluviated valleys. The flanking slopes of many of these hills can be steep and remain quasi-stable at angles of 20–25°. The detailed forms of these landscapes are discussed above in Chapter 8.

Duricrusted residual hills

Many duricrusts in elevated positions prove very durable in the erosional landscape. Ferricretes in particular develop a continuous, highly crystalline carapace under favourable conditions, and this may withstand further weathering and the effects of surface water flows for indefinite periods. The vulnerability of ferricrete to erosion arises principally from the susceptibility of the underlying saprolite to removal, thus undermining the surface crust which can undergo cambering, fracturing and fragmentation. This leads to marginal collapse of

the duricrust and to the formation of a residual gravel comprising the duricrust fragments. These can be of boulder size, but often break down subsequently into angular cobble-sized pieces and pisoliths. This process also affects silcretes and, in Australia, the resultant gravel-littered surfaces are known as *gibber plains*. Alucretes (or bauxites) may be less resistant if poor in iron, and tend to be more earthy; the aluminium sesquioxides are less crystalline and more amorphous. Their appearance as hill cappings in the tropical landscape is often attributable to the presence of a surface ferricrete layer.

Duricrust cappings are often described as quasi-horizontal, and this observation then led to the idea that they had formed on landsurfaces of minimal relief. These ideas need substantial modification, and this has been considered above (Chapter 4). Many duricrusts have cemented both hillslope debris- and alluvial-fan gravels, creating inclined

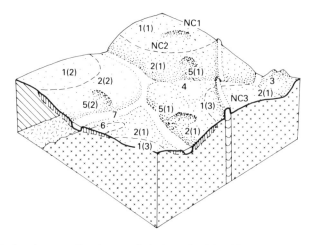

Engineering properties of some soils in the Kyalami Land System

Unit	Variant	Form	Soils and hydrology	Statistic	Liquid limit	Plasticity index	Linear shrinkage
1	2	Hillcrest, 0–2°, width 450– 1800 m	Residual sandy clay with collapsing grain structure on granite (0–23 m). Ferralitic soil. Above groundwater influence except at depth.	X̄ S V	44.1 11.0 0.25	22.6 10.8 0.48	9.78 4.27 0.44
2	1	Convex side slope 2–12°, width 450–910 m	Hillwash of silty sand derived from granite (0.3–0.9 m) on granite schist or basic metamorphic rocks. Occasionally saturated.	X̄ S V	17.3 8.98 0.52	4.9 3.70 0.76	2.04 1.04 0.51
7		Alluvial flood– plain 1–3°, width 45–230 m	Expansive alluvial clays and sands (3–6 m) on granite schists and basic metamorphic rocks. Mineral hydromorphic soil. High watertable (data for black alluvium)	X̄ S V	58.5 14.7 0.25	37.1 13.7 0.37	13.7 10.3 0.75

x̄ mean
s standard deviation
v coefficient of variation

FIGURE 10.4 Schematic block diagram of the Kyalami area, near Johannesburg, South Africa. Land units are labelled and sample engineering properties of the soils are indicated. (Reproduced by permission of Oxford University Press from Cooke & Doornkamp, 1990, after Brink *et al.*, 1970)

surfaces resembling pediments. Many laterites become indurated only when valley incision takes place, and eventually they appear as interfluve cappings by a process of relief inversion. After each period of incision, valley-floor widening creates new sites for the influx of Fe^{2+}, possibly resulting in a succession of duricrusted benches, often incorporating both duricrust rubble from upslope and former channel gravels (effectively these are river terraces).

A check-list of duricrusted landforms is offered in Table 10.2.

The landforms created by these duricrust cappings can be deceptively simple, with quasi-horizontal sheets capping residual hills of tabular morphology, sometimes surrounded by one or more lower benches of duricrusted debris. But in reality the situation is often more complex. Hills that are duricrusted around their margins may lack a continuous lateritic formation over much of their area. In other cases there may be a marked inclination on the tabular summit, leading to different details of morphology and process at each

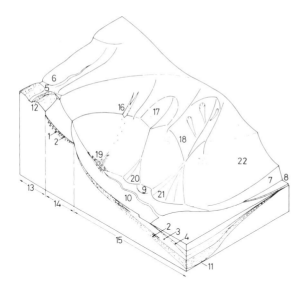

FIGURE 10.5 Evolutionary model of Hong Kong's terrain. Materials: 1, old colluvium; 2, young colluvium; 3, alluvium; 4, marine deposits. Landforms: 5, upland valley; 6, deeply weathered hills; 7, coastal cliffs; 8, wave-cut platform; 9, alluvial terrace; 10, floodplain; 11, submarine buried valley; 12, ridgecrest gully erosion; 13, relict landforms on upland; 14, older landform assemblage; 15, younger landform assemblage; 16, stream incision into superficial deposits; 17, incision widened to small valley; 18, deep valley between spurs, subject to instability; 19, boulders in stream channel; 20, small colluvial fan; 21, large colluvial fan; 22, coastal slope with thin soils (after Hansen, 1984; reproduced by permission of Gebrüder Borntraeger)

or silcrete (less well developed in alucrete and calcrete). These are aptly named, since the duricrust often literally breaks away, falling as blocks that become progressively broken into smaller fragments downslope.

This process comes about because of the undermining of the ferricrete cap rock by subsurface water flows. The ferricrete itself, once indurated, is often impermeable and water will run off the surface during rain. However, it is also usually well fractured due to subsidence or cambering, and may be penetrated by macropores due to root channels. The continuity of the sheet is usually also poor when viewed over a large area. Water is thus able to penetrate to the sub-ferricrete layers which are often enriched with clay eluviated from the duricrust, and may be very poorly permeable. Because water penetrates from the surface via fissures and macropores, subsurface flows become concentrated and often erode tunnels or caves (Figure 10.8). Some of these have small diameters and are little more than pipes, but others form caves large enough for large mammals or people to enter. Water flows in these tunnels become turbulent and significant quantities of sediment from the parent material of the ferricrete (saprolite or sediments) can be removed. This leads to a loss of support for the duricrust sheet above, which may subside or fragment. Deep duricrust-walled gullies can form at the margins of the plateau-like sheets.

The retreat of duricrust breakaways, to form landscapes patterned by tabular hills surrounded by steep slopes, is recorded from many areas of the tropics, and some aspects of their morphology are illustrated by Figures 3.12, 8.22, and 10.1. In savanna and semi-arid climates, the duricrust fragments are strewn on the slope pediment, becoming comminuted by attrition as they become affected by intense rainfall and surface water flows. In more humid areas similar forms can be traced, but they often comprise features attributable to slumping of the duricrust and underlying saprolite, with the accumulation of thick colluvium on the lower slope. This colluvium may contain quite large duricrust fragments in a non-mottled sandy clay (Moss, 1965 and Figures 8.22 and 8.25).

end. The intensity of processes leading to the dismantling of duricrusted surfaces will vary with climate, but their effectiveness will also be influenced by the thickness of the ferricrete and by other factors.

Where they become dissected into compartments, the tabular hills form a variant of the multi-convex landscapes previously described, with the tendency for summit flattening greatly exaggerated by the cap rock. This type of relief is common around the shores of western Lake Victoria in Uganda (Figure 10.7).

Processes of duricrust fragmentation

The margins of duricrusted hills usually show sharp *breakaways* or low cliffs formed in the ferricrete

TABLE 10.2   Some erosional forms associated with duricrusts

A   Interfluve or summit forms:
   (1) tabular hills with a duricrust 'cap rock', usually very gently inclined (<0.5°), but sometimes visibly sloping (1–2°);
   (2) cambered or undulating surfaces underlain by ferricrete, often over Fe-rich rocks, in humid tropical areas;
   (3) interfluves carrying a dense cover of duricrust fragments (possibly >1 m day) and/or iron nodules, sufficient to inhibit surface runoff;
   (4) large depressions due to subsidence or collapse of duricrust, and often infilled with sediment or peat; maybe flooded.

B   Hillslope forms:
   (5) duricrust 'breakaways' forming near-vertical margins to small butte or mesa forms; often with visible fragmentation;
   (6) re-cemented landslide debris, forming irregular, armoured hillslopes, particularly over Fe-rich rocks in humid areas;
   (7) re-cemented talus or debris-fan fragments on slopes of 10–16°, possibly marking a climate change from dry to wet;
   (8) ferricrete benches, possibly marking former levels of valley floors.

C   Piedmont forms:
   (9) cemented fan sediments from both debris and alluvial flows, inclined at slopes up to 7° (continuous with (7) above);
   (10) cemented river terrace sands and gravels;
   (11) hillfoot (groundwater) laterite formed in various parent materials and possibly concealed by or currently forming in superficial sediment.

(Some typical combinations of these forms are shown in Figures 3.12, 3.13, 3.22, 4.15, 10.1–10.4, 10.6 and 11.8.)

Duricrusted benches or terraces

The incision of drainage into a weathered land-surface can lead to the formation of new valley floors into which fluxes of Fe^{2+} can arrive from upslope as well as upstream. Any abandoned, higher level, laterite or plinthite will become indurated as the water-table falls. When this process is repeated, flights of benches or terraces preserved by ferricrete duricrusts along their streamward margins can be formed. This type of morphology was described for the River Niger terraces in the Kankan region (Pelisier & Rujeri, 1981, and Figure 10.9), but is widely encountered.

Earlier studies (De Swardt, 1964) emphasised the common occurrence of a single bench, leading to a division of laterites into 'Older' and 'Younger' formations. But this approach is very schematic and is useful mainly for illustrative purposes (Ollier, 1976). More complex patterns of duricrust formation result from cyclic sedimentation and pedogenesis adjacent to the flanks of crystalline uplands, and in these circumstances differential tilting and subsequent erosion can produce apparently confusing sets of forms (Sombroek, 1971; also in Goudie, 1973).

In Uganda, McFarlane (1976) has elucidated the complex character of duricrusted surfaces related to the so-called 'African' or 'Buganda' landsurface, and has made it clear that correlation of duricrusted residual hills as remnants of major planation surfaces cannot be justified on either theoretical or morphological grounds.

Duricrusts on mafic rocks

It is a common experience in the humid tropics to find that relatively fissile rocks rich in iron stand high in the landscape, capped by thick ferricrete duricrusts. This is the case with quartz diorites, amphibolites and many metasedimentary greenschist belts. Over greenschist facies in central Sierra Leone, elevated linear ridges or plateaus have been

FIGURE 10.6   Ferricrete duricrust breakaway on the Jos Plateau, Nigeria. Large blocks camber and then break off due to removal of saprolite from beneath the duricrust by throughflow

formed during differential etching. These are heavily duricrusted and associated with very deep weathering, re-cementation of marginal landslides, and internal depressions such as Lake Sonfon, which is thought to be a result of saprolite collapse and eluviation beneath the duricrust (Figures 9.7 and 9.10). This feature is 60 m below the summit surface and several square kilometres in area, but most depressions in duricrusted terrain are much smaller, perhaps 3–30 m across and 3–4 m deep, as described from Uganda (Pallister, 1956) and many other areas. It is interesting that, despite all the evidence for destructive forces operating on duricrusted landsurfaces, they remain relatively high in relief.

By contrast, some mafic plutons (gabbros, norites, peridotites) and metavolcanic rocks form elevated, rocky relief, by virtue of their relative resistance to weathering. In these instances Fe^{2+} is released by active weathering of the steep hillslopes and a thin (<10 m) mantle of debris is formed. The flow of water carrying Fe^{2+} in complexed form across the slope can be such as to provide ample iron for the cementation of surrounding footslopes. Again, this is very evident around the Freetown Peninsula and other mafic hill complexes in Sierra Leone.

EROSIONAL FORMS IN ROCK—INSELBERGS

The tropical landscape is certainly not made up predominantly of inselberg landscapes. Neither are

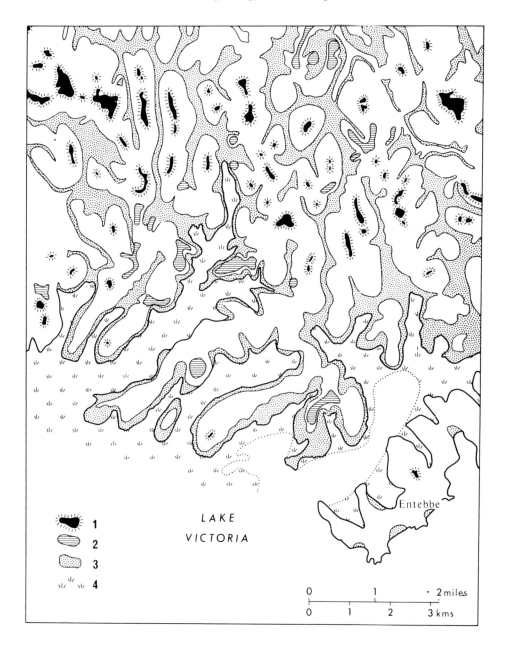

LAKE

VICTORIA

Entebbe

1

2

3

4

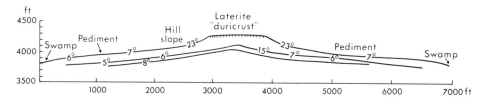

FIGURE 10.8   A subterranean tunnel below ferricrete carries substantial runoff after rainfall

rock pediments ubiquitous landforms. Indeed it can be argued that pediments characterise semi-arid areas in both hot and cold environments, and that inselbergs are forms determined largely by rock type and structure. But while these contentions may in part be true, they also obscure the importance of both types of landform to the study of geomorphology in the tropics.

Confusion arises (as so often) in part from questions of definition, but also because the longevity of these landforms can cause them to occur in climatic environments different from those of their formation. Pediments for example are probably not desert landforms, but products of seasonal tropical and sub-tropical environments (Chapter 9). Inselbergs, on the other hand, are quite probably landforms favoured by a change of climate from prolonged warm and wet conditions towards more seasonal regimes (from biostasie to rhexistasie in effect).

As a consequence of these formational factors, both types of landform are commonly encountered within the tropics and sub-tropics. Furthermore, tropical karst relief, which is included in this section, has many parallels with granite relief, although the dissolution of calcium carbonate contrasts with the formation of saprolites in silicate rocks in important respects.

FIGURE 10.7   *(Opposite)* Landforms associated with the dissection of a ferricrete duricrust in Uganda. 1, Nearly flat summits capped by duricrust (possible remnants of 'African' surface?); 2, intermediate benches; 3, alluvial silts and lacustrine beds; 4, papyrus swamp (after Pallister, 1956)

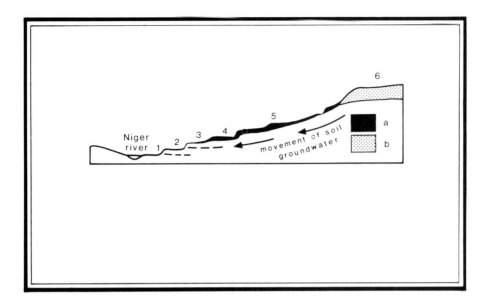

FIGURE 10.9   Multiple laterite crusts formed as benches on River Niger terraces in the Kankan region. 1, Floodplain; 2–4, succession terrace levels at 3, 7 and 25 m; 5, piedmont; 6, Tertiary plain with ferallitic weathering crust; a, laterite horizon; b, ferrallitic crust. (Reproduced by permission of Balkema from Glazovskaya, 1984, mainly after Maignien, 1958)

Categories of inselberg

The term *inselberg* means 'island hill' (or mountain) in German, and these forms are characterised by their separation from surrounding terrain and, frequently, by their independence of the regional drainage network. Some occur within escarpment zones, but many more stand out from encircling plains as more or less isolated hills or groups of hills. Thus inselbergs have been described as *zonal* (inselbergs of position) or *azonal* (inselbergs of durability) according to their location with respect to major hillslopes (Birot, 1958; Büdel, 1982). They can vary in size from less than 100 m in both height and extent, to very large and mountainous massifs approaching 1000 m in height and more than 1 km in diameter.

Attempts have been made to categorise inselberg forms (Young, 1972; Thomas, 1978; Gerrard, 1988):

(1) tabular hills (buttes, mesas) and rock towers, usually found in horizontally bedded sediments;
(2) conical hills, often reflecting uniform, boulder-controlled slopes, in fissile and jointed rocks;
(3) castellated hills (boulder inselbergs, koppies or tors), developed in well-jointed but rooted rocks, most frequently granodiorites and granites;
(4) domed hills (bornhardts) of massive rock, commonly granite or migmatite.

Limestone karst hills can be matched in part to these categories, but their special features will be recognised in a separate discussion. The account will mainly concern granite inselbergs (and inselbergs of other silicate rocks), because the dominant argument can develop directly from the previous discussion of etching and the existence of double surfaces of levelling in tropical landscapes.

However, the other categories should not be overlooked, since many of these appear to develop mainly by slope retreat coupled to slope decline on soil mantled slopes. Since these subjects of pediment formation and hillslope retreat are discussed in Chapters 9 and 11, the matter will not be further pursued here.

FIGURE 10.10   Bedrock-rooted granite tor on the Jos Plateau of Nigeria

GRANITE INSELBERGS

The origins of bedrock landforms have not so far been emphasised in this study, but they should take their place alongside concerns about saprolites, laterite, and transported deposits. Much emphasis has been given to the origins of inselbergs in earlier studies, including more than 40 pages in a previous book by the writer (Thomas, 1974a), while Twidale (1982) has provided an important survey of granite landforms. Outline discussions are also given in many standard texts (e.g. Ollier, 1969; Chorley *et al.*, 1984; Gerrard, 1988), and the author has previously offered a review of the study of inselbergs (Thomas, 1978).

The main argument for the association of inselbergs with etchplains has already been outlined (Chapter 9). The stripping of saprolite to expose fresh rock as detached cores, bedrock-rooted *boulder inselbergs* or 'tors', rock pavements and small *domed inselbergs*, is well established through comparative studies from many regions underlain by granitoid bedrock.

Boulder inselbergs (tors)

Boulder inselbergs take different forms, but share a characteristic morphology, defined by intersecting fracture sets. The individual blocks so defined remain *in situ* and are usually visibly rooted in bedrock (Figure 10.10). Although some blocks may have split under tension, most have been fashioned by weathering into spheroidal or 'woolsack' forms of widely varying size. The formation of these cores within the saprolite, by the weathering of joint blocks, can be verified from a very large number of field sections and landscapes, where the boulders are seen as *corestones*, wholly or partially embedded in the regolith (Figure 2.2 and Plate 3-III). It is a short logical step to propose that these cores become exposed at the surface by stripping of the saprolite,

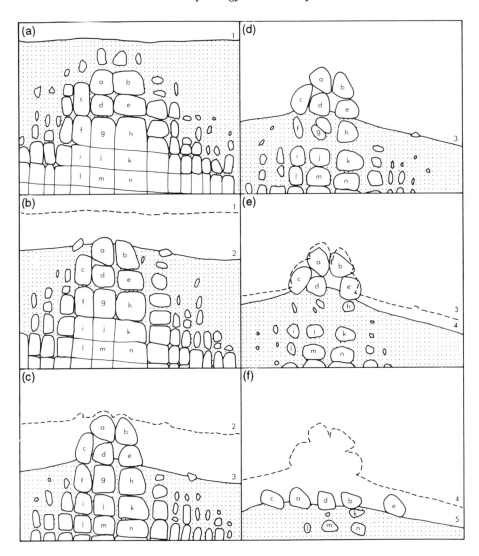

FIGURE 10.11   Schematic model for the development and decay of boulder inselbergs (tors) in jointed granite (after Linton, 1955 and Thomas, 1965b; reproduced by permission of Gebrüder Borntraeger). (a) Uneven development of the saprolite in response to jointing patterns; (b) lowering of the groundsurface by erosion; (c) continued lowering of the surface exposes the boulder group; (d) foundations of the tor become progressively weathered in the saprolite zone; (e and f) further groundsurface lowering leads to the collapse of the boulder group into scattered corestones

and this mode of formation is almost universally accepted for the well rounded examples found in warm climates. Other views may prevail when discussing the rock forms of cold climates (Thomas, 1976).

Many early writers observed this relationship, in a wide variety of tropical and non-tropical environments, including Cornwall, England (Jones, 1859), Brazil (Branner, 1896), Canada (Chalmers, 1898), Nigeria (Falconer, 1911, 1912) and New Zealand (Cotton, 1917). This hypothesis was elaborated by Handley (1952) for parts of Tanzania (then known as Tanganyika), and his conclusions were drawn upon by Linton (1955) for his

FIGURE 10.12   Deep weathering adjacent to a degraded tor along the Snowy Mountains Highway, New South Wales, Australia. The collapse of the boulders is probably due to weathering also beneath the tor group which are now no more than core boulders (Figure 10.11)

discussion of the Dartmoor Tors of southwest England. Despite its local usage in southwest England, the word 'tor' is frequently used to describe these boulder outcrops, while the word 'kopje' (or koppie) is often applied to these forms in southern Africa.

These boulder inselbergs or tors have a world-wide distribution in granitoid rocks, and also in some schists and arkosic sandstones. Although tors are in many cases markers of environmental changes, they cannot be always be used as indicators of a tropical environment (present or past). Linton's (1955) study was seen as controversial because it postulated Tertiary pre-weathering under near-tropical conditions which had not, at that time, been widely acknowledged for the British islands. The removal of the saprolite by periglacial processes (gelifluction) posed few problems, since the deposits

were evident in the field, but good evidence for Tertiary weathering remained hard to find. However, by proposing a 'two stage hypothesis' for tor formation, Linton (1955) formally attached a climatic change hypothesis to the formation of tors, and this has found wide application in other environments.

If boulder inselbergs are not dependent on the tropical environment for their formation, they are none the less favoured by the rapidity and thoroughness of decay that gives rise to a sharp interface between the sound rock and saprolite (Figure 2.2). The exposure of core boulders and bedrock-rooted tors can occur at all levels in the landscape; on summits, spurs, side slopes and even in valley floors as a result of etching and surface water flows (Figure 5.12). However, without the intervention of environmental change or dissection due to uplift, in rainforest environments their exposure would probably be limited to sites of relatively rapid downwasting.

To some extent the widespread multi-convex or 'demi orange' relief of weathered compartments, found in particularly humid areas of the tropics and sub-tropics (Figure 3.2), can be seen as an alternative form of dissected granite relief, where either rock susceptibility or environmental rates of weathering promote the complete decay of core boulders before exposure by erosion. A few boulders and tors can be seen in some areas of half orange relief.

The argument for environmental change is dramatically illustrated by the landforms and deposits of the Jos Plateau, Nigeria (11°N). Here basinal landscapes on the Jurassic, Younger Granites (Thorp, 1967a,b), are commonly surrounded by boulder and tor fields, or by boulder-strewn ramparts of granite rock, while the intervening ground is deeply alluviated (Hill & Rackham, 1978), and large alluvial fans have developed around the plateau escarpments. Rapid stripping of the saprolite has exposed boulders of all sizes, including some very large blocks up to 20 m in diameter, and also some small hemispherical domes.

Once exposed, the constituent blocks of boulder inselbergs dry rapidly after rain and remain subject to only slow weathering under lichen patches or beneath shaded overhangs, where moisture is retained in rock contact. By contrast, the concealed foundations of the tor are kept moist within the soil and continue to decay. In suitably jointed granites, this process will completely undermine the tor which tends to collapse or settle as denudation removes the fine clastic constituents of the saprolite from between the core boulders. This observation allows a sequence of boulder inselberg formation and destruction to be built on earlier ideas (Thomas, 1965b; Figures 10.11 and 10.12). As the landsurface is lowered, so cores will emerge on to the slopes according to the fluctuating balance between rates of weathering and surface denudation. Continuous lowering of the whole landscape may be envisaged, so long as the depth of erosion does not go below the weathering front over large areas. When this

FIGURE 10.13   Boulder inselberg or tor on Precambrian granite in southern Ghana. A rocky pediment extends from the foot of the tor towards the camera

occurs, corestones and boulder inselbergs are replaced by unweathered rock ledges, steps and larger domes.

Notwithstanding these ideas, boulder inselbergs may become isolated by lateral retreat with extension of pediments across more closely jointed and weathered rock. In other words the removal of saprolite and breakdown of corestones takes place slowly by lateral movement of water as well as by dissolution at depth (Figure 10.13).

Domed inselbergs (bornhardts)

Domed inselbergs can be the most striking of landforms, and are often described as 'bornhardts' (Willis, 1936) after the name of the German

FIGURE 10.14 Core boulders formed within arched sheets of granite at the foot of the Jos Plateau escarpment, Nigeria

geologist who first described them from east Africa (Bornhardt, 1900). The domed form can be seen in excavations (Boye & Fritsch, 1973), and low domed rock surfaces can be seen emerging from surrounding saprolite, or encircled by boulders. Arched basal rock surfaces are also seen in many road and rail cuttings (Plate 2-I). The link between such arched structures and pressure release joints (sheeting), and the influence of these on both boulder and domed inselbergs is everywhere apparent (Figure 10.14). This also demonstrates that not all tors are underlain by well-jointed rocks, and those with massive rock foundations will not undergo the sequence of decay referred to above.

Troughs and basins of deep weathering, surrounded by broadly domal bedrock highs have also been recorded from deep drilling (Thomas,

1966a,b; and Chapter 3, Figure 3.16). There is, therefore, little reason to object to an hypothesis of saprolite stripping and bedrock exposure for the origins of at least some domed inselbergs. These landforms can be quite small or very large, and their relief above the surrounding landsurface can vary from almost nil to many hundreds of metres. The gently domed outcrops were termed 'shield inselbergs' by Büdel (1982), and have been called 'whalebacks', 'ruwares' and 'dwalas' by others.

The great height of many inselbergs, which can stand 300–600 m above local piedmont surfaces (Figure 10.15), clearly requires further explanation, when known depths of weathering in comparable rocks are usually less than 100 m. Such high relief can, however, be attributed to repeated deep weathering and stripping through geological time

FIGURE 10.15   Massive granite domes or bornhardts formed in Precambrian granites at Idanre, southern Nigeria

(Falconer, 1912; Willis, 1936; Büdel, 1957; Twidale, 1964, 1978; Thomas, 1965b, 1978), and many large bornhardts are probably of great geological antiquity—a fact that also ensures that they will have experienced many changes and fluctuations of climate. A simplified model for this type of development is offered in Figure 10.16.

Domed inselbergs frequently carry remnants of former weathering covers on their surfaces, and possess many small-scale etch forms which are discussed below. The steep marginal slopes are subject to landsliding and stripping of the bedrock surface (Figures 6.5 and 10.17), while at the same time vegetation mats, and possibly soil development, may re-cover some formerly stripped rock surfaces. The possibility that some domes emerge as incision takes place along major fracture zones between fresh rock compartments, and only ever

develop a thin saprolite cover, which is then subject to periodic catastrophic stripping by shallow landslides, offers an alternative mode of development that does not, of itself, require widespread, regional deep weathering.

Nevertheless, the existence of these hills is important to the central position of etch processes in the differentiation as well as the elimination of relief forms. Very high, domed inselbergs (bornhardts) are difficult to explain by mechanisms other than the lowering of the surrounding terrain, and many, such as the Idanre Hills in southern Nigeria, are remote from escarpments or are much higher in altitude than older landsurfaces in reasonable proximity. The sequence of such lowering may of course have been by means of the layer by layer removal of weathered and broken rock by laterally migrating low scarps. In Uganda

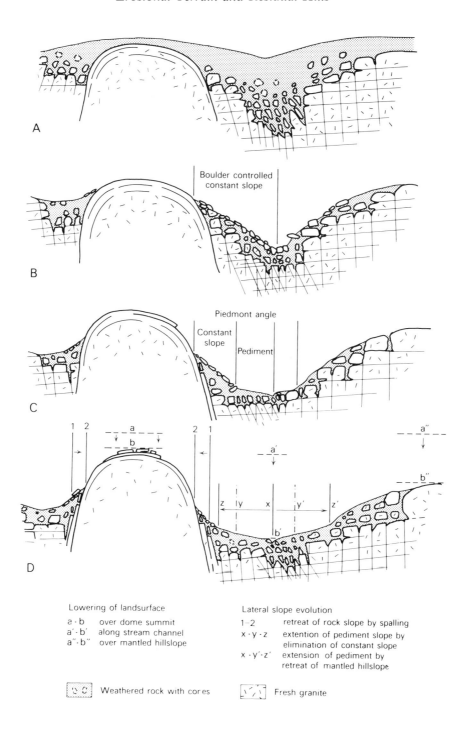

FIGURE 10.16   A sequence of bornhardt (domed inselberg) development by continuous weathering of mantled rock slopes during dissection. (Reproduced by permission of Gebrüder Borntraeger from Thomas, 1978)

FIGURE 10.17 Stripping of the thin regolith from a steep dome face in granite–gneiss, near Tongo in Sierra Leone

(Doornkamp, 1968) and in Zimbabwe (Twidale, 1988), claims have been made that inselbergs have been exposed by the dismantling of the early Cenozoic, African Surface, and similar relationships with the weathered landsurface of Australia were noted by Mabbutt (1961, 1966).

However, it is not possible to ignore alternative hypotheses for inselberg development, and these landforms have been attributed to hillslope retreat in sound rock (King, 1962, 1966, 1975; Selby, 1977a,b).

In some instances, disagreement occurs because forms of different kinds are being compared; in others it may be because a single hypothesis is sought for a group of forms that can arise in different ways. A detailed analysis of inselbergs in

the Machakos area of Kenya (seasonal sub-humid climate, rainfall 600–1300 mm year^{-1}) led Ahnert (1982) to conclude that the 'double planation' process is not at work today. He showed from morphometric studies that there is a highly significant relationship in the area between the total height of the inselberg-plus-pediment and the width of the pediment, and he reasoned that the convex forms of the inselbergs were subject to slope retreat and to summit lowering, finally becoming tors. He described rock-cut pediments with few signs of deep weathering.

However, the inselbergs in this region are varied in morphology and soil cover, and few are massive bornhardt domes. It is certainly true that boulder-controlled slopes undergo retreat, and in this process massive rock kernels may become isolated as domes or tors, as the larger hill masses become reduced in the manner described by Ahnert (1982).

Perhaps one of the more interesting, initial questions concerns the locational factors for inselberg formation.

Factors in inselberg location

It is possible to assert that granite or gneiss inselbergs may be independent of the regional drainage network, and this argues against a general origin by the retreat of steep valley walls, for such a pattern of evolution should produce a (zonal) distribution related to the drainage/interfluve networks. This observation is supported by Brook (1978) from studies in South Africa and by Twidale (1980) from work in southeast Australia. It is a feature also of many West African landscapes (Thomas, 1974a, 1980). Groups of inselbergs formed from granite plutons will produce their own radial/rectilinear drainage networks and the larger individual hills may feed stream heads, but these do not determine the regional drainage patterns.

It is also clear that domed inselbergs develop from their initial exposure as individual hills and do not evolve from larger rock masses by slope retreat. Thus bornhardts are often closely spaced, and separated by deep clefts. They occur as individual features along fault scarps, as near Rio de Janeiro, and elsewhere along the Serra do Mar in eastern Brazil, and often define the outcrop of

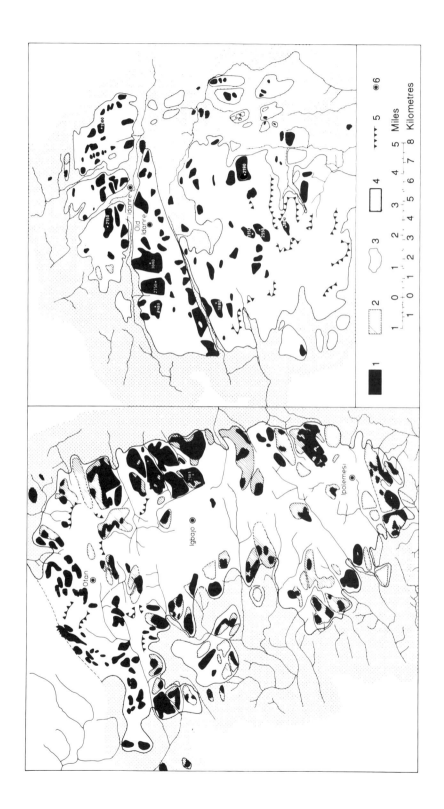

FIGURE 10.18   Distribution of domed inselbergs in two Older Granite massifs in southern Nigeria (forest zone). 1, Domed inselbergs and other convex rock outcrops; 2, convex summrits with soil cover; 3, areas of dissected terrain; 4, Basement Complex rocks, mostly weathered; 5, major breaks of slope; 6, towns. The Igbajo Plateau (left) retains a weathered upper surface that is deeply weathered, and domes are emerging today around the periphery. The Idanre Hills (right) are heavily dissected into massive domes (Figure 10.15) and hilly terrain. Mapped from 1:40 000 air photographs (after Thomas, 1974a; reproduced by permission of the Macmillan Press Ltd)

deeply eroded granite intrusions, as in the Idanre and Igbajo Hills of southern Nigeria (Figure 10.18). It should, therefore, be anticipated that they must be located by geological factors such as: (1) rock type (mineralogy, rock fabric); (2) structure (fracture patterns and density); or (3) tectonic factors (horst features, compression zones). There is some evidence to support each of these groups of factors, but there are also apparent anomalies, suggesting that erosional forces can override these controls, given time. Each factor group will be briefly considered, and pointers towards a composite theory for inselberg formation will then be offered.

Lithological factors

Lithological factors (rock type) clearly influence the occurrence of inselbergs. They are mainly restricted to crystalline rocks of felsic to intermediate composition or to massive sandstones and limestones, in which they form tropical karst hills. King (1975) clearly stated that bornhardts arise in rocks with a low porosity, consequent upon metasomatic infusion of a host rock, usually granite or gneiss. Brook (1978), reviewing evidence from Transvaal, South Africa, concluded that, 'inselbergs are preferentially developed in rock masses that have undergone potash metasomatism or in younger plutonic bodies which may also be potash rich.' In Sierra Leone, Marmo (1956) found similar relationships, associating bornhardts with porphyroblastic granites, in which microcline porphyroblasts had developed by potash enrichment of granitised sediments. Brook (1978) took the view that this relationship was widespread and included most of the Cambrian Older Granite suite in West Africa, and similar settings in India and Brazil. The reasons for this relationship include the sealing of fissures by the metasomatic process and intergranular silicification. Brook (1978) also suggested that many such rocks are more massive because of dominance by primary-α quartz which is less fissured than primary-β quartz. Birot (1958) also referred to the absence of quartz fissuration as an important factor in the formation of the 'suger-loaf' (Portuguese, Pão Açucar) domes around Rio de Janeiro.

Certain famous inselbergs, such as Ayer's Rock in central Australia, are formed in silica-cemented arkose (Bremer, 1965), while sandstone towers in the Bungle Bungle of east Kimberly in northwest Australia, have been interpreted as results of silica solution (Young, 1987a,b, 1988). Such an explanation demands long periods of development, since SiO_2 dissolves only very slowly in rainwater. The similarity of the sandstone towers to *tower karst* formed in limestone rocks within tropical regions is striking, and there are many parallels between all these categories of residual hill.

However, despite these important correspondences between rocks and relief, the actual margins of individual inselbergs may not follow outcrop boundaries, and other factors must be invoked to explain this. It has also proved difficult to demonstrate any petrographic differences between rocks forming boulder inselbergs or tors and those giving rise to bornhardts, although Pye *et al.* (1984) found indications in the Matopos Hills, Zimbabwe, that bornhardts may possess a greater frequency of microcline phenocrysts.

Structural factors

Structural factors undoubtedly influence inselberg form and position, and these range in scale from zones where inselbergs occur within plutons, or along structural 'highs' (Twidale, 1964), to the influence of fractures on inselberg plan form and profile (Twidale, 1964, 1980, 1982; Moeyersons, 1977; Thomas, 1967, 1974a). Structural lineaments in metamorphic terrain are equally influential. The detailed control of fracture planes over inselberg forms is frequently very striking. In many plutons, there are several fracture sets that have arisen during cooling or from subsequent tectonism, while exposure of the granites induces pressure release fractures (sheeting), that may follow the outer form of the inselberg hill. This topic will not be explored further here (see Twidale, 1982; Gerrard, 1988).

Tectonic controls

Tectonic controls may also be involved, and some mountainous, inselberg areas have been attributed to the direct influence of faulting. Barbier (1967) proposed this origin for Pic Parana in southeastern Brazil, while McFarlane *et al.* (1981) postulated a *graben* structure running between the Loma Mountains and Tingi Hills in Sierra Leone which they thought were *horsts*. Since these hills rise well above adjacent ferricreted surfaces (early Cenozoic) on supracrustal rocks and can be seen on satellite imagery to be delineated by major fractures, this hypothesis is appealing (Thomas & Summerfield, 1987; and Figures 8.2 and 11.4).

Tectonically induced stress fields are also held to influence the location of inselberg domes and to determine fracture patterns and densities by Twidale (1964, 1982). Compressed compartments may be relatively free of internal fractures and, according to Twidale (1982), lateral stresses due to tectonic forces can greatly exceed those leading to pressure release from unloading. These stresses may be sufficient to induce some upward bulging as well as sheeting fractures, on massive rock compartments forming bornhardt domes. No adequate explanation for the distribution of such stressed rock compartments has been adduced, but there is some possibility that massive rock compartments will be encountered at different depths as well as different locations within a major rock mass (Thomas, 1965b; Twidale & Bourne, 1975). If this is correct then the position of an inselberg must be viewed in three-dimensional space, and therefore also in a time frame that allows for deep-seated rock masses to emerge at different locations as the landscape is lowered.

A further factor may be the need for deep erosion into ancient plutons, where compressive stresses have been stored. This seems to be the situation for the Cambrian Older Granite suite of West Africa, in which major bornhardts have formed, in contrast to the boulder inselbergs that typify the high-level, Jurassic Younger Granites of the Jos Plateau.

In the search for single explanations it can sometimes be forgotten how complex the combination of geological factors can be. What Büdel (1968, 1982) called 'petrovariance' as a factor in landform development, is a combination of many subtle differences in rock mineralogy and fabric, textural and structural features and tectonic

(endogenetic) forces. For this reason claims for or against geological control over inselberg location and morphology should be treated with some caution. Nevertheless, the view of Büdel (1982) that inselberg margins will retreat from outcrop boundaries by slope retreat and steepening until loss of saprolite cover takes place, should be held in mind when other aspects of inselberg formation are discussed. Twidale (1980) holds that inselberg margins are usually fracture controlled and this has been the present author's finding in many instances.

Inselbergs in multicycle landscapes

It has been commonly observed that inselberg landscapes (if not all inselbergs) are usually associated with landscapes where surfaces of planation are preserved at higher levels. Twidale (1980) has stated that two good reasons for this situation are:

(1) deep, differential weathering would have had time to take place beneath the ancient surface, and
(2) the presence of two landsurfaces separated by several hundred metres of relief implies deep erosion.

This situation has been noted in Uganda (Doornkamp, 1968; Ollier, 1981), Zimbabwe (Whitlow, 1980) and Natal, South Africa (Thomas, 1978; Twidale, 1980), in all of which locations it has been suggested that the inselbergs have been formed as a result of deep erosion into the African Surface of erosion (Figure 10.19). However, again caution is necessary, because, as in Sierra Leone and Nigeria, some inselberg groups rise to heights greater than nearby planation levels, and the domes form the highest points in the regional landscape.

The striking domes that give the tropical city of Rio de Janeiro (23° S, rainfall 1500 to >2000 mm year^{-1}) its scenic setting arise in a cratonic terrain broken by major faults, associated with the opening of the South Atlantic. For several hundreds of kilometres the coastal plains are backed by the Serra do Mar, a mountainous escarpment up to 1500 m high, but broken by a number of step faults inland

FIGURE 10.19   Granite domes in the Valley of the Thousand Hills, Natal, South Africa. These bornhardts are exposed after removal of the Table Mountain Sandstone seen in the distance. Foreground slopes are in weathered granite, near Botha's Hill (see Figure 10.20)

from Guanabara Bay at Rio. The terrain here is deeply weathered at all levels, from the coastal lowlands to the inland plateau (the 'Sul Americana Surface'; see Table 11.1). The treads of the step-faulted terrain are dissected into typical half-orange relief (*meias laranjas*) within which the compartments are often completely altered. The bornhardts mostly occur along the steps, and are associated with metasomatic rocks. Erosion between the domes is associated with deep weathering and frequent landsliding (Plate 10-I).

A similar but less humid setting occurs in the Valley of the Thousand Hills, Natal, South Africa, where the African Surface developed on Table Mountain Sandstone has been deeply dissected along the warped and faulted margins of the continent. The domes here appear to be initiated

by deep weathering which is evident near the rim of the dissected plateau (Thomas, 1978), and as an escarpment capped by the sandstone has retreated (Figures 10.20 and 10.21). But, as in the case of Rio de Janeiro, dissection has since proceeded well below the likely depths of antecedent decay beneath the ancient plateau, and the inselberg development requires continual renewal of the saprolite with repeated episodes of dissection and stripping (Figure 10.16).

Given that bornhardts are very durable forms, they may be very ancient, and could have developed over a long period of general downwearing of the landscape. The fact that few large inselbergs appear on the Mesozoic–Cenozoic landsurfaces, such as the so-called African Surface, is probably testament to the persistent very low rates of denudation on the

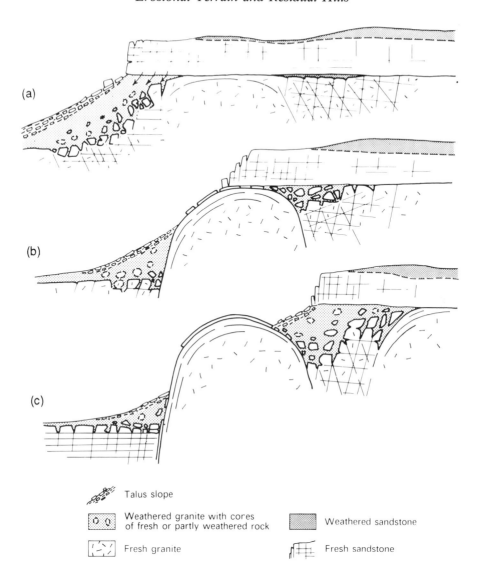

Talus slope

Weathered granite with cores
of fresh or partly weathered rock

Weathered sandstone

Fresh granite

Fresh sandstone

FIGURE 10.20 A model of bornhardt development by the removal of a sedimentary cover rock, based on sites in the Valley of a Thousand Hills, Natal, South Africa. Deep weathering penetration is thought to occur towards the plateau edge, beneath the Table Mountain Sandstone, and is continued beneath the talus slopes (Figures 10.19 and 10.21). (Reproduced by permission of Gebrüder Borntraeger from Thomas, 1978)

former Gondwanaland supercontinent prior to fragmentation and continental drift.

Inselbergs and the tropical environment

To the extent that inselbergs appear strongly influenced by geology and by rock mechanics (King,

1975), they are not specifically tropical landforms. The granite domes in the Rocky Mountains (Cunningham, 1969, 1971) and Sierra Nevada (Yosemite Valley) are justly famous and exhibit all the classical features of bornhardts (massive domes, sheeting joints), except an association with tropical plains. Similarly, a whole range of granite domes

FIGURE 10.21   Deep weathering in granite below the sandstone scarp near the site of Figure 10.19, and associated with a long talus slope

and monoliths can be seen in the core of the Guadalajara Mountains of central Spain, in Corsica and many other parts of the world. The most massive bornhardts are exceptional in that the structural determinants are so strong that they almost certainly override influences from different environments. In any case, many are not true inselbergs, but occur in the compressed and deeply eroded cores of mountain chains.

There remain good reasons, however, for discussing inselberg development in the context of tropical geomorphology. Principal amongst these is the way in which contrasts in rock susceptibility to weathering are highlighted by the arguments, and the importance of the basal weathering surface in terms of erosional mobility and landscape reduction. There are also reasons to believe that

some mid to high latitude occurrences of inselbergs may have their origins in palaeotropical environments of the early Cenozoic.

In the summary that follows, the principal reasons for advocating a connection between etchplain formation and inselberg development are outlined (adapted from Thomas, 1974a,b):

(1) The extent, depth and patterns of weathering on crystalline rocks in the tropics provide circumstantial support.
(2) The details of weathering profiles reveal patterns of basins, troughs, domes and ridges with a relief of up to 50 m, while the basal weathering surface may plunge steeply around many outcrops.
(3) In inselberg-forming rocks, the basal weathering

FIGURE 10.22   Partially weathered, sheeting layers exposed by the stripping of saprolite from a dome surface in the central Nigerian savanna zone

surface is commonly sharply defined on upper dome surfaces and around boulders; it therefore forms a major break in erosional mobility of earth materials.

(4) Gently domed surfaces (shield inselbergs) and core boulders can be seen as both surface landforms, and within weathering profile exposures.

(5) Large accumulations of corestones are found in valleys between inselbergs and on lower slopes around massive domes; it is logical to conclude that these have settled by gravity as saprolite has been removed.

(6) Surface forms and materials (e.g. ferricrete blocks, etched and rotted rock slabs, patches of saprolite) on exposed rock surfaces often suggest former burial and subsurface etching.

Freise (1938, quoted by Bakker & Levelt, 1964), advocated a cycle of formation and removal of a soil and vegetation cover on these hills. He started with bare gneiss domes and ridges which became colonised first by lichens and then by savanna plants, to be followed by the immigration of trees and the development of rainforest with strong weathering. Crusting of the soil was then accompanied by degradation of the forest and soil erosion, leading to grassy savannas and ultimately to the re-emergence of the inselberg domes and ridges. This hypothesis implies the operation of possible feedback mechanisms leading to a switch from soil development towards soil loss and rock exposure. It is not at all clear, however, why this should take place in the absence of external forcing

FIGURE 10.23   Polygonal (or 'mural') weathering on the surface of exposed granite corestone

mechanisms such as climate change. This will be considered below.

Small-scale surface forms on inselbergs

A number of distinctive, small-scale features on inselbergs has attracted attention, and most of these can be attributed to weathering processes, though whether they result from sub-aerial, or sub-soil (etch) processes, or indeed a sequence involving both environments, in some cases remains unclear (Thomas, 1978).

Ferricrete blocks rest on some dome summits and a reasonable inference would be that they were once located in the upper zone of a former weathering profile, most of which has now been stripped to expose the dome summit. However, this might not have applied where footslope laterites are concerned, and evidence of transformation of

weathered granite into laterite on glacis surfaces in the Tingi Hills has been described (Thomas, 1978). Nevertheless, loose blocks of vesicular ferricrete seem likely to be remnants of past profiles.

Etched rock slabs are also widely found (Thomas, 1967; Twidale, 1987) resting on unweathered rock surfaces (Figure 10.22). They may therefore have been partially weathered as separated rock sheets beneath a former saprolite, and then exposed to rapid wetting and drying cycles in the sub-aerial environment. Polygonal weathering, sometimes known as 'mural weathering' is also seen on many exposed boulders. This takes the form of superficial cracking (to a depth of 5–10 cm) in an outer rim of stained and etched rock, and suggests again that the exposure of the boulder subjected an outer, slightly altered shell, to a sub-aerial environment which has led to solutional loss and to shrinkage (Figure 10.23). Sometimes granite cores become

FIGURE 10.24   Weathering pans on a large granite dome at Dombashawa, Zimbabwe

case hardened in the sub-aerial environment. This process involves the precipitation of oxides of iron and manganese within the partially weathered rock fabric, effectively sealing the rock and reducing its porosity.

Weathering pits and grooves (karren) diversify many rock surfaces in both temperate and tropical climates (Thomas, 1976). Sometimes the two forms interconnect at the edges of boulders. These forms diversify bare rock surfaces and may have their origins in small-scale depressions at the weathering front. Once exposed they retain water after rain and become deepened by further weathering, often assisted by lichens and sometimes higher plants (Figure 10.24). The *karren* or *rillen* features resemble the karst forms from which the terms are derived, and form channels a few centimetres deep on steeper slopes.

Tafoni which are weathering hollows that develop on near-vertical rock faces have led to quite a lot of comment and speculation. Frequently these are aligned along major fractures or other rock structures (Figure 10.25), but in some cases there is no obvious geological cause for their location. Twidale and Bourne (1975) have argued that more or less horizonatal lines of tafoni on Ayer's Rock, central Australia indicate former levels of subsurface weathering. They also argued that these forms mark the extent of lateral attack or backwearing on massive bornhardt slopes during each phase of stable base level. On the other hand, salt weathering has been held responsible for the formation of tafoni in arid (and shoreline) environments (Wellman & Wilson, 1965; Evans, 1971). The retention of water, and possibly also the flow of water along major rock discontinuities, must be considered a possible reason for the development of these features. The moisture encourages the growth of algae, bacteria, lichens and also higher plants at such locations. Evidence

FIGURE 10.25   Small tafoni forming along the vertical or incurved face of a gneiss inselberg in southern Nigeria

for the sub-soil formation of tafoni is therefore tenuous.

Flared slopes could be considered a major form associated with some inselbergs, but will be included here. Twidale (1962, 1980, 1982, 1987, 1990) has laid great emphasis on the importance of steepened margins and concave, flared slopes around certain bornhardts in Australia. He has argued that the upper shoulders of these features mark former levels of the weathered mantle, and that the 'flares' are part of the former weathering front. His view of the excavated floor of the Yarwondutta reservoir in South Australia is persuasive of this view, and similar photographs of the celebrated 'Wave Rock' in Western Australia are impressive (Twidale, 1990). However, it has to be said that instances of this kind are relatively rare, though basal concavity of a lesser kind is common (Thomas, 1967).

Twidale (1987) saw the merging of concave slopes with lines of tafoni as further confirmation of the former relationship with a higher saprolite level. The problem with many such reconstructions is the likelihood that structural factors have dictated some elements of the morphology. Large-scale *sheeting*, truncated downslope, is often the locus for tafoni and larger caverns. In fact, incurved slopes enclosing significant caves are as common as concave flares, and these may help corroborate the subsurface origins for both forms. Where caverns occur in granite and gneiss, there is nearly always a master joint or other discontinuity.

The problem here is whether they represent an aspect of sub-soil weathering similar to the formation of boulders, but on a larger scale, or whether they are in effect karst forms dependent on the free passage of water. Examples of water conduits in granites may be rare, but examples are found. An overhang cave at Dombashawa, not far from Harare in Zimbabwe, contains a shaft that penetrates the granite roof for perhaps 20 m, and is large enough for a man to stand upright in its lower end. It is difficult to see how such conduits could form within a sub-soil environment, and it is perhaps more likely that they have developed within the exposed bornhardt dome. As with the sandstone towers of the Bungle Bungle in Australia (Young, 1987a,b), very long time periods would be required to form such features.

Despite the ambiguities surrounding some of the current surface forms of rocky inselbergs, there is sufficient observational evidence to associate both domed and boulder inselbergs with deep weathering, and it is a simple logical step to suggest that they emerge as rock landforms when the saprolite mantle is removed by erosion. This process of stripping, and the exposure of the domes is described as *dégagement* by some French writers (Rougerie, 1955). This can be seen as one aspect of the process of formation of *stripped etchplains* as described above and by Büdel (1982).

Climate change and inselberg development

Inselbergs are durable forms and the larger hills must have experienced many changes to the external

FIGURE 10.26   Large granite domes rising from rainforest in central Sierra Leone. It is not clear whether these are now being colonised by forest once more after stripping during the glacial dry climates

environment. Since these changes were instrumental elsewhere in the landscape, in altering the balance between weathering and erosion, leading to a variety of slope forms and distinctive sediments, it must be anticipated that they will have been important in inselberg development.

The most obvious manner in which climate change could have affected inselberg formation is to alter the balance between weathering penetration and surface erosion rates. Under rainforest, slopes remain covered with vegetation and a thin soil layer develops on all but the steepest slopes ($>45°$?). This is seen well displayed in exposures in granite inselbergs in Johore, West Malaysia (Figure 3.23). In the study of landsliding (Chapter 6) under humid tropical conditions it was shown that repeated sliding and re-vegetation of slopes with re-formation of saprolite can occur.

However, both domed and boulder types of inselberg forms are common in the humid tropics as well as being striking features of the savannas (Figure 10.26). Early associations of inselberg landscapes with savannas (Cotton, 1942) depended on limited observations which came mainly from the more open grasslands of Africa and India. In Africa, however, major inselberg landscapes have been described from Uganda (0° Latitude), Tanzania (5–10°S), West Africa (6–>10°N), Zimbabwe (15–23°S), and South Africa (30°S) for example. In fact their distribution appears not to be controlled by climatic zone.

The weathering associated with these inselberg landscapes may have commenced in the Mesozoic and continued through the Cenozoic, while tectonic events leading to deep dissection may have occurred in both Miocene and Pliocene times. The Oligocene to mid-Miocene was a period of climatic deterioration (aridification) in tropical latitudes and serious aridity may have intervened towards the end of the Pliocene (see Chapter 11). The example of the Kalahari Sands which extend across 25 ° of latitude must give pause for thought about late Cenozoic climatic aridity in Africa. The Quaternary shifts of climate have been discussed in Chapter 7.

It is probable therefore that, during periods when the relief of the Gondwana continents was increasing, during the last 100 Ma, major episodes of cooler, drier climates swept across the tropics and sub-tropics. Surface erosion rates would have increased as the wetter zones became drier and covered with more open vegetation communities. This could have been the trigger for the dismantling of the old deeply weathered landsurfaces. As streams incised between massive rock compartments, increased surface erosion rates would have led to the emergence of dome summits and of boulder fields from the corestone horizons of granitic weathering profiles. Once bornhardt summits had lost their saprolite mantle they became both dry for most of the year and subject to immediate powerful surface water flows during rain. Weathering processes became localised in joint cavities and other irregularities of the inselberg surfaces which retained debris, while any vegetation mats that developed were liable to rapid removal by rain and runoff.

The progressive development of inselberg relief on basement terrain as a consequence of repeated environmental changes is graphically illustrated by the work of Kroonenberg and Melitz (1983, Figure 10.11) from their studies in Surinam, and also by Garner (1974) from studies in southern Venezuela. In this latter case, however, lateral planation by scarp retreat is envisaged, particularly as it affects the Roraima Sandstones (Briceño & Schubert, 1990; Schubert & Hubert, 1990). Over long timescales these ideas match those of Fairbridge and Finkl (1980) concerning the *cratonic regime* of landform development. This concept invokes the concepts of *biostasie*, a condition of ecological and geomorphic stability associated with deep weathering penetration and landsurface stability (Erhart, 1955), and *rhexistasie*, when delicate equilibria become

broken and sediment is progressively transferred from uplands into sedimentary basins. Fairbridge and Finkl (1980) considered periods of 10^9 years during which biostasie, combined with etchplanation, prevailed, interrupted by shorter (10^5–10^7 years) periods of rhexistasie, stripping and pediplanation. Such changes were not solely a result of climatic change, but also due to tectonic movements and rejuvenation of landscapes.

The question of *timescales* for these developments remains unresolved in this account, but care must be taken to distinguish long-term changes in the landscape ($> 10^5$ years) from relatively short-term fluctuations in the climatic environment (10^3–10^4 years). Many inselbergs and inselberg landscapes are ancient forms, standing high above encircling plains that carry ferricretes and silcretes of early Cenozoic ages. But equally, other domes have been exposed as a result of recent deep dissection of mountain ranges, as in the western Cordillera of North America. Twidale (1980) cited examples from Australia of granite domes at least 60 Ma in age, and Savigear (1960) considered it likely that some inselbergs in Nigeria had been exhumed from beneath Cretaceous sediments. Willis (1936) considered that some east African bornhardts may be of Palaeozoic age. However, direct evidence of the ages of these hills is generally unavailable, and associations with surrounding landsurfaces and their deposits depend of the presumption that there has been no faulting or other differential movement between them. This has not always been the case.

It is probable that we should see the long-term history of the larger bornhardt domes and inselberg ranges in terms of prolonged groundsurface lowering (but possibly involving shallow pedimentation cycles), and sometimes deep erosion, into ancient plutons and other migmatised basement rocks. This will involve the thinning of the saprolite over the resistant dome tops as they are encountered within the rock mass (Figure 9.10 and Table 9.2).

The rock surfaces could become exposed without invoking an environmental crisis, particularly in seasonal, savanna climates, solely because of the resistance of massive rock compartments to weathering and soil formation. However, it is more likely that they will emerge during periods of

FIGURE 10.27 A widely stripped surface (etchsurface or *grundkockerrielief*) within a semi-enclosed depression on granite, Kagoro Hills, Jos Plateau, Nigeria

accelerated surface erosion. Across the dissected margins of continents, this acceleration may have taken place along eroding escarpment fronts, generated by faulting or warping of the basement. But climate change is likely to have been concurrent with these events. With the older forms we can only guess at these events, and there is a possibility that rock domes may have been uncovered and then re-covered by vegetation and soil more than once.

The Quaternary environmental changes came late in the development of many of these forms, but would have increased the likelihood of exposure and extension of rock surfaces during periods of rapid climatic transition, especially from dry to wetter conditions. Many inselberg landscapes involve both tors and domes across a wide size range (Thomas, 1976), and many of these forms are less durable in the landscape. In many parts of the tropics granite

landscapes form great dome and boulder fields that are remarkable for their lack of soil cover. Examples include the Matopos Hills of Zimbabwe (rainfall ~ 600 mm year^{-1}), several Younger Granite massifs on the Jos Plateau of Nigeria (Figure 10.27) (rainfall ~ 1200 mm year^{-1}), and parts of the Koidu basin and its surrounding hills in Sierra Leone (rainfall ~ 2000 mm year^{-1}). Other examples occur in sub-tropical and temperate latitudes such as the Monaro plateau in New South Wales, Australia. These landscapes appear to have undergone relatively recent, comprehensive stripping of saprolite. In the case of the Koidu basin, discussed above, the present rainfall supports a monsoonal forest cover, and these forms have been revealed by recent deforestation. Because this area is situated within 50–80 km of the forest/savanna boundary in present climatic conditions, it is likely that it has

FIGURE 10.28   Karst tower underlain by granite, near Kuala Lumpur, West Malaysia. Note the alluviation and excavations of gravel for tin mining in the middle and foreground

experienced repeated severe desiccation of climate during the Quaternary. Only this proposition can explain the prominent pediment or glacis forms, the widespread exposure of rock, and the ubiquitous and complex surface gravels and stone-lines, that occur in this area (Thomas & Thorp, 1985; Teeuw, 1989, 1991a; see Chapter 9).

KARST RESIDUALS (KARST-INSELBERGS)

Similarities between granite relief and limestone karst have been recognised by a number of writers (e.g. Twidale, 1990), yet the two topics have been researched quite separately by groups of people who seldom meet to exchange ideas. As a consequence there are two separate literatures, and many of the works on subterranean karst are of a specialised nature and will not be reviewed here. Valuable summaries of karst geomorphology have been

offered by Sweeting (1972), Jennings (1985) and by Ford and Williams (1989).

In Chapter 2 a brief introduction to limestone weathering acknowledged that the rate of dissolution of limestone in the tropics may not always be higher than in the temperate zone. Temperature differences affect the system in opposing ways, and water availability is a crucial factor. However, the soil CO_2 levels and the contribution of microbiological processes are likely to be decisive in weighting the balance in favour of more rapid humid tropical karstification. Comparative measurements of denudation rates are few and unsatisfactory, and views on the subject remain in conflict. Nevertheless, there is some agreement that karstic landforms in the tropics *do* differ from those in cooler climates. Thus Jennings (1985, p. 201), considered the 'botanic hothouse' of the tropical forests an extreme environment for karstification and stated: 'it is in the tropical humid climates, with or without a dry season, that the greatest variety of karst and the most striking expressions of it have been found.'

Ford and Williams (1989, pp. 115–116) have made similar observations but concluded that: 'the contrasts in karst landform styles in humid tropical and temperate zones appear not to be attributable to radically different solutional denudation rates. Therefore it follows that contrasts in the spatial distribution of erosion must be the explanatory factor—unless our comparative denudation estimates are seriously in error.'

It is of course this contrast in the spatial distribution of denudation in the tropics that is so evident on many silicate rock terrains and is responsible for inselberg landscapes. Some limestone hills have been called *Karstinselberge* (Klaer, 1957), and the similarities between inselberg development on silicate and carbonate rocks are intriguing (Twidale, 1987). The karst landform styles of the humid tropics that are considered distinct would not be agreed by everyone, and differences due to structure have been emphasised in Papua New Guinea (Löffler, 1978). Nevertheless, the striking *tower karsts* (Figure 10.28) are most frequently attributed to warm–wet conditions, and in China both climate and structure are considered important (Lu, 1987; Yuan, 1987).

The range of karst landforms

In order to elaborate on this topic, some consideration of karstic landforms in general is required, and for a number of reasons the surface morphology will be emphasised here. Rather few cave systems have been explored in the humid tropics and conclusions about their features would be very preliminary. The *length* and *depth* of cave systems do not appear to be strongly influenced by warmth of climate, but the existence of some caves with very large *volume* in the humid tropics may be significant. Collapse dolines (*dolines avens*) in New Britain (rainfall 5700–12 000 mm year^{-1}), for instance, measure up to 350 m deep and 300 m wide. This is, however, a specialised topic, better explored via the relevant literature.

A good starting point for discussion of karst topography is the simple scheme outlined by Williams (1972) from studies in Papua New Guinea. He recognised three major karst landscapes:

(1) *doline karst* characterised by depressions on an otherwise intact landsurface, but not necessarily a plain;
(2) *polygonal karst* in which the prior surface has been consumed, and karst depressions are surrounded by residual hills on polygonal divides; and
(3) *plains karst with residual hills* which often take the form of 'towers' protruding above a planate surface.

This scheme emphasises the context within which individual features are found; others have paid specific attention to the types of karst residuals. Balazs (1973), for example, classified karst hills, using type areas from tropical environments. He used a ratio of diameter to altitude (*d:a*).

(1) *Doline karst* or 'Tual' type from Ketjil Island, Indonesia. This has *d:a* values of >8; hills are 10–50 m high, with up to 50 hills km^{-2}. This does not quite fit with P. W. Williams' requirement for an intact surface between depressions.
(2) *Cone or cockpit karst* (German: *kegelkarst*) or 'Sewu' type from Gunung Sewu in Java, Indonesia. This has *d:a* values of 3–8; hills are 30–120 m high, with 15–30 hills km^{-2}.
(3) *Tower karst* (German: *turmkarst*) or 'Yangzhou' type from Guangxi, China. This has *d:a* values of 1–5; hills are 100–300 m high, with 5–10 hills km^{-2}.

Lehmann (1936), who also studied in Java, sought to place karst landforms into an evolutionary sequence, in which an uplifted plain would gradually transform into a series of domed hills separated by solutional depressions or 'cockpits'. This is the type of polygonal karst relief analysed by Williams (1972). It has been suggested that tower karst is a later development, but much opinion now believes tower karst to be a separate development.

Jennings (1985) also pointed out that in Papua New Guinea, karst development did not begin on an erosional plain.

Cockpit karst

Cockpit karst is not unique to the tropics, but examples containing fewer but larger cockpit depressions and larger hills on the polygonal divides mostly come from the tropics. A variant, called *cone karst*, with prominent mamillated hills was observed by Williams (1978) to be confined to elevations below 1500 m in Papua New Guinea. These areas experience a tropical climate with mean annual temperatures $>20\,°C$. Again, the comparison with polyconvex relief types on weathered granites in humid tropical environments is striking. Polygonal karst is thought to be favoured where limestones are well fissured.

Research suggests that some 70% of all autogenic solution takes place in a superficial zone within 10 m of the surface. This implies that surface lowering is likely to be more important (volumetrically) than solutional activity in conduits. This *epikarst* solution leads to rapid widening of fissures in the superficial zone, while permeability at depth remains poor, with water flow focused on infrequent connecting pipes to deep aquifers or cavern systems. This process of *doline* development leads to polygonal karst when the depressions become contiguous. Where there is a superficial cover, *suffosion dolines* can be formed as seepage through the cover deposits concentrates at depth into sub-soil karren. In studies of Jamaican karst depressions, Day (1979) observed greater rates of solution beneath talus and on inclined bedrock surfaces within the depression floor, contributing to the maintenance of the depressions. Crowther (1984) established a positive relationship between weathering potential and soil depth in peninsular Malaysia, and noted that recharge areas are associated with deep soils in depressions between karst hills. He concluded that the high volumes of water, combined with high weathering rates produced a 'mechanism of self-enhanced depression formation' which he compared with the findings of Jakucs (1977, reported by Crowther, 1984). These processes appear to contribute to the formation of steep, often enclosed depressions, between high karst hills (*tower karst*).

Tower karst

Tower karst appears most frequently in tropical and sub-tropical wet environments. Conditions for the development of tower karst were outlined by Verstappen (1960): (1) low primary porosity, (2) great compaction of the limestone, (3) widely spaced planes of weakness, (4) great depth of karst development, and (5) subjacent development of intervening depressions beneath superficial deposits. These conditions appear at variance with those favouring cockpit karst, but Williams (1985; and Figure 10.29) has advanced a model for the development of tower karst from cockpit karst, as the phreatic zone is approached, and lateral corrosion and slope retreat become favoured by the hydrological regime. Also striking in the present context, however, are the similarities with the favourable factors giving rise to domed inselbergs (or bornhardts) in silicate rocks. The concentration of weathering and waterflow in the hillfoot zones of both types of terrain appears critical.

According to Williams (1987, p. 441), tower karst may develop via a cockpit karst phase as above, or direct, in response to especially favourable factors. He also divided tower karst into four different types:

(1) residual hills protruding from a planed carbonate surface, veneered by alluvium;
(2) residual hills emerging from carbonate inliers in a planed surface cut mainly across non-carbonate rock;
(3) carbonate hills protruding through an aggraded surface of clastic sediments that buries the underlying karst topography; and
(4) isolated carbonate towers rising from steeply sloping pedestal bases of various lithologies.

The famous tower karsts of south China (Guilin) have been widely divided into two types: the 'peak cluster' (*fengcon*) and the 'peak forest' (*fenglin*) (Yuan, 1987). In the former, groups of rock towers

(a) DOLINES INCISE VERTICALLY THROUGH THE VADOSE ZONE

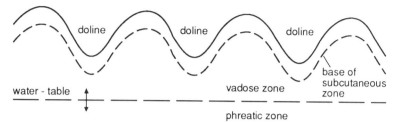

(b) INCISING DEPRESSIONS REACH THE EPIPHREATIC ZONE

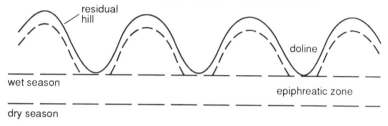

(c) HILLS REDUCED BY PARALLEL SLOPE RETREAT AND UNDERCUTTING, DEPRESSION FLOORS COALESCE, CORROSION PLAIN DEVELOPS

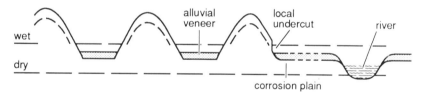

FIGURE 10.29   Development of residual hills ('towers') from cockpit karst as the phreatic zone is approached (after Williams, 1987; reproduced by permission)

protrude from a common rocky basement, which may contain closed depressions between the peaks; in the latter the peaks are separated by a flat stony surface, sometimes covered by a thin veneer of sediments. A third landscape of isolated peaks within plains is also recognised. Attempts to place these types into an evolutionary sequence can be faulted, and Yuan (1987) has observed that 'peak forests' occur close to non-limestone regions and receive allogenic river flows (Figure 10.28). They are often floored by intercalated, non-carbonate rocks supporting surface drainage, or occur where

soluble rock is protected by an impermeable overburden. By contrast, the 'peak cluster' is developed mainly by karst solutional processes and possesses little surface drainage.

Some features of tower karst include:

(1) vertical and overhanging lower slopes and footcaves,
(2) swampy alluvial plains around the towers,
(3) polje type depressions within the towers,
(4) deep karst corridors dividing the towers,
(5) horizontal tunnel caves through the towers,

(6) abandoned high-level caves containing stalagmites, and

(7) the presence of marginal plains and depressions.

The intervening plains have been described as
corrosion plains and are thought by Ford and
Williams (1989, p. 433) to 'develop by solutional
removal of irregularities down to a surface
controlled by the watertable'. The plains become
the focus for inflows of water and clastic sediments
and may form swamp environments with high
acidity. Incoming rainfall adds aggressive water to
the system and is thought to promote lateral
extension of the plains by attacking the marginal
slopes of the towers, a form of basal sapping. There
is evidence from existing hillfoot notches and some
abandoned caves, of lateral planation by the
(possibly allogenic) rivers of the plains. Ford and
Williams (1989) considered these plains to resemble
pediplains but in reality they are quintessentially
etchplains.

McDonald (1976a,b, 1979) studied karst towers
in Sulawesi, Indonesia, and in Belize, where he has
examined hillfoot caves and depressions. He found
that exposed hillfoot caves contained stalagmites
and concluded there was no current solution of the
cave roof. The solutional undermining had in his
opinion come about under sub-alluvial conditions,
or during initial flooding after heavy rainfall (see
also Jennings, 1976). There is in these landforms
much evidence for long-term lowering of the
surrounding plains, in the form of abandoned caves
containing rolled pebbles, and speleothems which
can be dated (Figure 10.30). In Guilin, China, there
is evidence for continuous development of karst
towers since pre-Pliocene times, with a rate of
lowering of the plains on average of 23 mm ka^{-1}
(Williams, 1987; and Figure 10.31).

A further factor that must enhance the resistance
of the towers to denudation is *case hardening* which
is the development of calcareous weathering crusts
of low porosity (only 10% that of natural
limestone). These crusts appear to result from the
degassing of carbonated water which leads to
precipitation of calcite in solution cavities, and also
to a wet recrystallisation process within the
limestone itself. The indurated layer can develop to

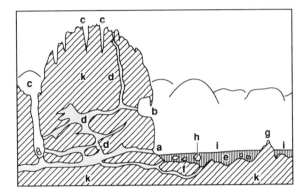

FIGURE 10.30 Characteristics of tropical karst hills
(towers) based on Sierra do los Organos, Cuba. Features
shown include: a, active hillfoot cave exit; b, abandoned
caves with speleothems; c, deep karstified fissures; d,
vadose cave systems, related to previous levels of plains;
e, alluviated valley floor; f, sub-alluvial phreatic cave
systems; g, limestone knobs: h, detached limestone
blocks; i, flat surface of intervening plains, frequently
flooded; k, limestone tower. (Modified after Lehman,
1954; from Sweeting, 1968)

1 m thickness in 10 ka (Ford & Williams, 1989). The
process also affects other exposed limestone
surfaces in warm climates.

The evolution of tower karst landscapes involves
an understanding of the geological history, and
research is hampered by the lack of correlated
sediments. Evidence from Guilin in south China,
for example (Yuan, 1987), indicates that the
Devonian limestone was previously covered by late
Cretaceous Red Beds to depths of several hundreds
of metres. This implies that much of the spectacular
tower karst of this region has developed as the Red
Beds have been stripped by erosion during the
Cenozoic. Yuan (1987, p. 1122) concluded that, 'it
is possible that the varieties of tower karst . . . are
all different forms developed on carbonate rocks
at different times, and under different geological
and hydrological settings.'

Karst geomorphologists also recognise both
palaeokarst and *relict karst*. In the former case the
karst-containing rocks have been isolated from
the karst-forming system, for example by burial.
In the case of relict karst, the karst system remains
integral with the relief, but has become inactive due

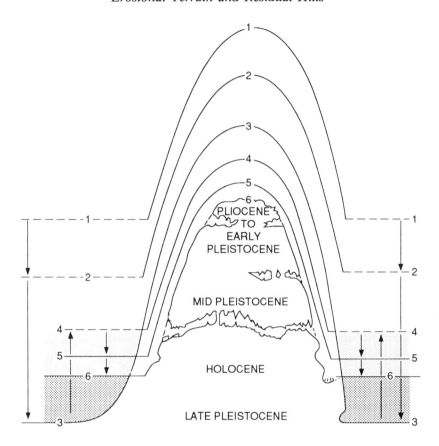

FIGURE 10.31   Schematic model for tower karst development near Guillin. Stage 1 is inherited from previous cockpit karst (Figure 10.29). Simultaneous lowering of the peak and the basal corrosion plain results in lowering of the tower form without a change in its geometry (1–3). Floodplain aggradation buries the foot of the tower to level 4, followed by stripping of the surface and lowering of the peak in stages 4–6. Stillstands in this process are marked by swamp notches and water-table caves. (Reproduced by permission from Williams, 1987)

to climatic change or some other factor. In southern Africa, Marker (1987) has drawn attention to relict karst which was probably developed during the wetter climates of the Tertiary, with some reactivation during wetter periods of the Quaternary.

Parallels between karst relief and granite relief

It is clear, even from this brief survey, that the understanding of karst geomorphology should be brought more firmly into the wider thinking about landforms on silicate rocks. The reasons why this has not occurred are all circumstantial rather than scientific. Speleology has many dedicated

adherents, while to some other geomorphologists the subject appears arcane. As in many other topics, process studies have tended to concentrate on karst waters in temperate limestone areas, with a strong emphasis on the influence of P_{CO_2} relationships, and it is only the striking tower karst landscapes of the humid tropics and sub-tropics that have brought other processes into focus. The fact that limestone dissolution is both complete yet reversible appears to set karst processes apart from the incongruent, saprolitic weathering of the silicate minerals. Yet the controls over the progress of both systems are similar.

The chemical *reactions* in both cases are advanced by continued water inputs and favoured by warmer temperatures. They are also mediated by microbial productivity and the formation of soil acids that donate protons to the weathering systems. As a consequence of these factors, warm and wet conditions favour etch processes, whether they are congruent or incongruent in their operation. If higher temperatures *per se* are less important than water availability, they have great significance in terms of biological productivity.

The parallels between karst relief and granite relief are striking, and it is only the different terminologies that hinder their joint consideration. In terms of the role of etch processes, perhaps the most important observation is to note that limestone dissolution can lead both to highly accidented relief (tower karst) and to almost flat corrosion plains, often in close juxtaposition in humid tropical regions. White (1984) has maintained that the intensity of tropical rainfall will strip some areas of soil and vegetation cover, leading to differential rates of dissolution. The heavy rainfall also ensures that undersaturated solution kinetics will maintain the aggressiveness of water in the system, leading to pronounced differential solution and dramatic landforms.

These factors operate with similar impact on massive silicate rocks, leading to inselberg landscapes. There remain important differences, however, and these mainly attach to the roles of surface and subterranean drainage. Although suffosion forms develop over silicate rocks, beneath the saprolite cover they are not as widespread, nor are they as well developed as on limestones. For this reason granite basins are different from polygonal karst depressions, being linked by surface drainage channels in most cases. But in more basic silicate rocks sink holes are found and there is a trend towards karstic development. Even mamillated hill landscapes on limestone strongly resemble the 'demi-orange' landscapes on weathered granite. Such comments invite the reaction that landscape forms experience both convergence and equifinality, but some thought about the operation of the formative processes in each case should convince sceptics that there is much common ground between granite and limestone geomorphology. The arguments most commonly adduced to account for the major, and many minor, landforms in both rock types also represent applications of etching theory. It can also be seen that this theory operates in three different contexts: (1) where deep pre-weathering and saprolite formation has taken place; (2) where shallow weathering is accompanied by surface lowering (or alternates with it over short time periods); and, (3) where total (congruent) bedrock dissolution occurs in both carbonate and quartz-rich rocks.

11

Long-Term Landform Evolution

INTRODUCTION

The aim of this section of the book is to discuss the nature of debates about landform evolution through time periods of as long as 10^8 years. Some very ancient, Proterozoic (700 Ma) landsurfaces have been identified from Australia (Ollier *et al.*, 1988a, Gale, 1992), but over most of the more humid parts of the tropics, and where Cenozoic tectonic disturbance has taken place, the last 100–140 Ma probably encompasses most of what we see in the landscape today (except of course for the rocks themselves and some exhumed landsurfaces). Indeed, very little of the landscape that derives from the Mesozoic Era has survived unaltered, and even where great plateaus form the markers, their surface forms and deposits contain the imprints of many changes of environment. It will assist the following discussion if a brief summary of some of the changes to the position of the continental plates and to the earth's climates over this longer period is offered.

Plate tectonic movement

Two hundred million years ago the universal landmass of Pangea may have been virtually intact, but the volcanic episode of the Karroo period (Triassic, 220–180 Ma) in Africa presaged the modern era of plate tectonic motion and continental drift. By the mid-Jurassic (160 Ma), the North Atlantic had begun to open between northwestern Africa and eastern North America, but Gondwanaland probably remained intact until the late Jurassic (135 Ma). Thereafter, India and Antarctica + Australia became separated from Africa, with Australia centred around Latitude 55°S, and India moving northwards from 30–40°S. The Equator then passed through northern Amazonia as now and traversed Africa obliquely from around 10°N in the west to 25°N in the east.

By the mid-Cretaceous (100 Ma) the North and South Atlantic Oceans became joined and a proto-Indian Ocean spread northwards to the Tethys. These events gradually brought into being the present coastlines of Africa, South America and India, with Australia joined to Antarctica. This must have brought in train major changes to the course of denudation on the continents, as a consequence of faulting and warping of the marginal zones. Changes to ocean and atmosphere circulation must also have taken place as the new oceans asserted their influence over global climates, and it is in the late Cretaceous that humid tropical conditions in Africa and South America became established.

In the early Tertiary, continental and coastal sedimentary basins were filling with lateritic facies from the dismantling of the Cretaceous weathering profiles, building the coastal plains of the Guianas and West Africa, for example. India appears to have moved northwards across the Equator during Palaeocene and early Eocene times, compressing sediments in the Tethys and eventually forming the great Himalayan mountain chains in the Miocene. During the same period, Australia moved across the 30°S parallel, with its northern area coming into the tropical latitudes for the first time. Northward

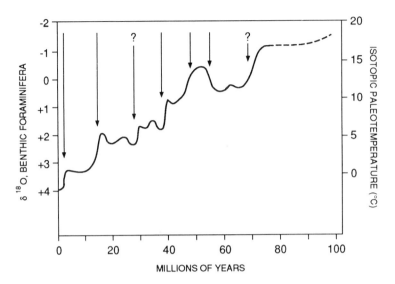

FIGURE 11.1 $\delta^{18}O$ record of deep-water temperatures for the last 100 million years. Ice volume changes complicate the estimation over part of the period. The arrows indicate times of relatively abrupt transitions. (Reproduced by permission of John Wiley & Sons Inc. from Crowley & North, 1991)

movement of the new arrangement of continents continued through the later Cenozoic, amounting perhaps to 10° of latitude during the last 40 Ma.

One hundred million years of changing environments

The mid-Cretaceous thermal optimum lasted at least from 120 to 90 Ma, and during this time there were ice-free polar seas (Douglas & Woodruff, 1981, and Figure 11.1). Sea levels were high and marine transgressions were widespread across landscapes of low relief. Shallow seas covered large areas of northern Africa and South America, but the altitude of the pre-drift Gondwanaland possibly rose to 2000 m asl in southern Africa (Partridge & Maud, 1987), also encompassing southeastern Brazil and possibly Antarctica. Although most of the present-day tropics experienced palaeolatitudes between 30 °S and 60 °S during this period, climates were warmer in these latitudes than they are today. Antarctic sea surface temperatures were probably 10–20 °C warmer. This high latitude warmth was probably combined with high seasonality (Crowley & North, 1991).

Slight cooling towards the end of the Mesozoic is considered likely, but the event that caused widespread extinctions around 66 Ma may not have had a long-term effect on climate (Crowley & North, 1991). Whether this was an asteroid impact or possibly associated with the vulcanism marked by the Deccan Trap basalts (Duncan & Pyle, 1988), which has been associated with the passage of the Indian plate over a mantle plume, remains uncertain. Notwithstanding this event(s), tropical conditions extended to 45° Latitude into the early Eocene (55–50 Ma) (Frakes, 1979), and Antarctic SSTs remained 10 °C warmer than today (Shackleton & Kennett, 1975). But significant cooling of climate set in after 50 Ma, coinciding with the opening of the Norwegian–Greenland Sea and with major volcanic events throughout the North Atlantic region. These 'Thulian' basalts may have been equal in volume to the Deccan basalts (Roberts et al., 1984) and there were further major extinctions around 40–38 Ma. The first signs of Antarctic ice may date from this time and ice-rafted debris has been dated to 34 Ma along with a major change in benthic $\delta^{18}O$. Rapid cooling over a period of 100 ka took place, and sea level may have fallen by 50 m. The ice cover expanded during the Oligocene (31–29 Ma) and SSTs may have declined in the tropics (Barrett *et al.*, 1987). However there was some amelioration of climate in the early Miocene (20–15 Ma). Ruddiman and Kutzbach (1989) have argued that the late Cenozoic climate changes may have been reinforced by plateau uplift in southern Asia and western America.

Between 15 and 14 Ma there appears to have been a major increase in ice volume and a fall in deep-water temperatures by 4–5 °C (Shackleton & Kennett, 1975). This mid-Miocene change in climate was accompanied by the spread of arid conditions in many parts of the tropics. This has been claimed for many parts of Australia and also in sub-tropical Chile (Chapter 3). A further advance of Antarctic ice may have occurred at 11 Ma, when sea levels fell by 50–100 m (Haq *et al.*, 1987) and fluctuations in climate occurred around the Miocene–Pliocene boundary, with the ice front repeatedly advancing and shrinking. By 3.2 Ma northern polar ice was forming (Icelandic tills from 3.15 Ma), and an abrupt change in climate took place between 2.6 and 2.4 Ma at the Pliocene–Pleistocene boundary, with incursions of ice-rafted debris in the Arctic (see Crowley & North, 1991, for a summary).

It is almost certain that cooling took place in the tropics over the last 5 Ma, and there is evidence for tropical aridity. Aridity in Africa increased and the Saharan flora was established around 5–6 Ma (Van Zinderen Bakker & Mercer, 1986), and climates in northwest Africa and in the Siwalik Hills of India climates became drier still by 3.2 Ma (Gaur & Chopra, 1984; Stein & Sarnthein, 1984), and an increase in grasses in Ethiopia between 2.5 and 2.3 Ma is thought to denote cooler and drier climates (Bonnefille, 1983). Cooling of climate at Bogota, Colombia *c.* 2.4 Ma is reported by Hooghiemstra (1984, reported by Crowley & North, 1991).

There is therefore much evidence to suggest that some tropical climates experienced repeated phases of aridity, in the mid-Miocene and within the Pliocene, as well as during the Pleistocene. The extent of these changes is difficult to establish, and of course the continental plates were migrating towards their present positions throughout this long

period of time. The establishment of modern atmospheric circulation patterns probably postdates not only the fragmentation of Gondwanaland and the building of the Alpine mountain chains world-wide, but also the (consequent ?) build-up of, first the Antarctic ice-cap (Oligocene–Miocene), and subsequently the Arctic Ice (Pliocene–Pleistocene).

Some geomorphologists would argue that the era of deep weathering and lateritisation took place before these later events led to the modern disposition of landmasses and climates, and that such profiles are inherited from the warm, humid late Mesozoic and early Cenozoic eras (much earlier in some instances). For some summital and highly indurated, lateritic duricrusts this argument remains powerful, but the formation of saprolite has continued throughout the long intervening period (> 50 Ma), and has been stimulated by the break-up of Gondwanaland, and by important tectonic events during the Neogene, both of which have led to uplift, dissection and increased hydraulic gradients.

As the mid-latitude high pressure cells evolved to produce continental aridity, so those landmasses that moved into these positions became drier, and this appears to have been the experience particularly of Australia. The anomalous position of Australia becomes very clear when the antiquity of its landforms and the very slow rates of denudation calculated for such areas as Kimberley and the Yilgarn blocks of Western Australia (Ollier *et al.*, 1988a,b; Wilford, 1991; Gale, 1992) are considered. This topic will be discussed further.

The pattern of climate oscillation in the Quaternary is better known (Chapter 7), but there remains a serious question concerning the proportion of this period that experienced interglacial warmth and humidity, with estimates ranging from 50% down to as little as 10%. When all these inferences about past climates are assembled, it is very clear that few assumptions can be made about the distribution and duration of humid tropical environments after the Eocene, yet it would be unwise to jump to the conclusion that warm moist climates have not always been present. Few of the long-range data come from lowland tropical land sites and many problems have yet to be solved concerning the evolution of tropical climates over the last 100 million years.

Debates about planation

Planation has been a central theme in classical geomorphology from the time of Davis (1899a,b). It is not a topic confined to tropical or sub-tropical environments, but many of the arguments are well focused by discussion of the landform evolution on the fragmented former supercontinent of Gondwanaland.

This aspect of geomorphology received considerable emphasis during the decades prior to the 1960s, but as attention turned to measurable processes on the one hand, and to the datable Quaternary on the other, interest in long-term, evolutionary geomorphology waned. At least, this has been the situation, until recently, in the anglophone worlds of Europe and North America, where process studies have received almost exclusive attention. In Australia, however, there has been continued interest in evolutionary geomorphology, represented by the work of Ollier (1979, 1982, 1985, 1991), Wellman (1979), Pain (1985), Twidale (1976, 1981b, 1983), and Young (1981, 1983). Finkl (1979, 1982) and Fairbridge and Finkl (1980) have also contributed important new ideas on the development of cratonic landscapes. A modest revival of a wider interest in long-term landform evolution is now evident, and this arises particularly from the dating of lavas on land, the accumulation of results from sea-floor drilling, and the application of isotopic techniques to the dating of regoliths (Chivas, 1983; Bird & Chivas, 1988a,b; Bird *et al.*, 1989).

The theory of continent-wide pediplanation, as developed by Lester King (1950a,b, 1953, 1957, 1962, 1976, 1983, and Table 11.1), has continued to influence thinking in this field. King's focus was particularly on the landscapes of southern Africa, and his ideas have stimulated a continuing debate in that region. This debate has led to alternative views (De Swardt & Bennett, 1974; Le Roux, 1991); searching critiques of the tectonic history (Gilchrist & Summerfield, 1990, 1991; Summerfield, 1985, 1991a; De Swardt & Trendall, 1969), and to a comprehensive revision of King's own scheme for southern Africa (Partridge & Maud, 1987, 1988 and Table 11.2).

A notable feature of King's (1976) analysis is the recognition that little of the Gondwana and Post-Gondwana surfaces survive, and that it is only the *African* planation cycle that is largely perfected as a surface of low relief, deep weathering and lateritisation. Following this planation, subsequent 'cycles' produced rolling and dissected terrains with local planations over less resistant rocks. Partridge and Maud (1987) confirm this view and recognise the African surface as polygenetic. The complexity of this surface has been documented by McFarlane (1971, 1976).

In South America, Bigarella and Mousinho (1966) developed their own nomenclature for pediplains and their correlative sediments, but Briceño and Schubert (1990) have recently reviewed ideas about the relief of the Guyana Shield in southeastern Venezuela, in the light of King's (1956) proposals for Brazil, and similar schemes advanced for Guyana (McConnell, 1968), and Surinam (Krook, 1979; Zonneveld, 1982). They claim that two Mesozoic surfaces can be recognised on the table-mountain remnants of Precambrian quartzites (Roraima Group). Aleva (1984) has related the composition of bauxite and laterite crusts to cyclic planation surfaces on the Guiana shield and is of the opinion that the main aluminous laterite (bauxite) level is to be found on the Sul Americano surface. The ideas of King therefore underlie much of what has been written on the topic of continental planation, and Melhorn and Edgar (1975) have supported arguments for world-wide correlative erosion surfaces by evidence from studies of tectonics.

Renewed interest in such questions has been stimulated in recent years by the development of new dating techniques, in particular the application of stable isotope dating (mainly $\delta^{18}O$) to determine the age of the Australian regolith (Bird & Chivas, 1988a,b). In Australia, this is thought possible because of the steady northward progression of the continent across the climatic zones since the Permean, allowing calculated palaeotemperatures to be attached to specific periods in climatic

TABLE 11.1 World-wide pediplanation cycles according to L. C. King: King's (1962) scheme for southern Africa (in italics) with South American (Brazilian) equivalents

Cycle	Age
VI The 'Youngest' Cycle (*Congo*—in Brazil known as 'Paraguaçu')	Quaternary—deep valley dissection
V The 'Widespread' Landscape (*Post-African 2*—in Brazil known as 'Late Velhas')	Pliocene—extensive in drainage basins
IV The 'Rolling' Landsurface (*Post-African 1*—in Brazil known as 'Early Velhas')	Miocene—incomplete planation
III The 'Moorland' Planation (*African*—in Brazil known as Sul-Americano)	Late Cretaceous to mid-Miocene
II The 'Kretacic' Planation (*Post-Gondwana*)	Early to mid-Cretaceous
I The 'Gondwana' Planation (*Gondwana*)	Jurassic

After King (1962, 1976), with additions from Aleva (1984).

TABLE 11.2 Summary of the principal geomorphic events in southern Africa since the Mesozoic according to Partridge and Maud (1987)

Age	Events
Holocene to late Pliocene	Climatic oscillations; sea level changes
	Post-African II cycle of major valley incision; local planation (Low Veld)
Late Pliocene (2.5 Ma)	*Major Uplift* by 900 m in eastern marginal areas
Early mid-Miocene to late Pliocene	*Post African I cycle* of erosion; imperfect planation
End of early Miocene (*c.* 18 Ma)	*Moderate Uplift* by 100–300 m; slight westward tilting & monclinal warping
Late Jurassic/early Cretaceous to end of early Miocene	*African Cycle* of erosion (polycyclic); advanced planation of subcontinent; deep weathering
Late Jurassic/early Cretaceous	*Break up of Gondwanaland* through rift faulting; initiation of Great Escarpment

Simplified after Partridge & Maud (1987).

evolution. It is also interesting that the antiquity of much of the Australian landsurface, together with its comparative aridity and low elevation, have made it difficult to find correlative surfaces below the complex 'weathered landsurface' (Mabbutt, 1965).

Much of the more recent work has turned from issues to do with the mode of landform evolution to address questions of dating and the age of landsurfaces, based on studies of the weathered mantle, the tectonic history and rates of landform change. In respect of this last topic, radioisotope dating of lavas in eastern Australia has led to some surprising results concerning the age of the Eastern Highlands of Australia and their escarpments and valleys, certain of which have apparently changed little since the mid-Cenozoic (Gale, 1992). The use of apatite fission track analysis as a tool for evaluating the thermal history of rocks at temperatures below 125 °C has also led to some conclusions of value in evolutionary geomorphology (Browne *et al.*, 1990).

The need to address these questions of long-term landform evolution is focused by the relief of 'Gondwana' continents, where extensive elevated plains and dissected plateaus call for explanation, and where the weathered mantle frequently provides evidence for development through long periods of geological time. But the issues raised by studies of Archaean or Proterozoic cratons are equally relevant to studies of many Palaeozoic terrains and apply also to certain Mesozoic and Cenozoic provinces, wherever tectonic relief has been substantially modified by prolonged weathering and erosion and cratonisation is advanced.

During periods of 10^6–10^8 years, relief is not modified by the operation of a constant set of processes and does not continue independently of tectonic movement. A striking instance of these problems can be cited from studies of the supergene alteration of a porphyry-copper deposit in the Atacama desert of northern Chile (Alpers & Brimhall, 1988; 1989), where mid-Miocene desiccation of the climate 'switched off' the weathering systems and led to the preservation of very deep, ancient profiles of alteration, a situation probably also mirrored in the development of the 'weathered landsurface' of central Australia (Mabbutt, 1965, 1988).

The importance of these questions for the understanding of mineral deposits is considerable, and both prospecting and extraction can benefit from workable models of landscape inheritance and evolution. Many of the bauxitic ores (Chapter 4), including nickeliferous laterites, illustrate important conclusions about landscape denudation. Such studies should demonstrate for sceptics the value of long-term evolutionary studies in geomorphology.

A field geomorphologist working with exploration geologists will always be asked questions such as: 'how old is this landsurface?' or 'how much erosion has there been since the Cretaceous?' These questions are often difficult to answer, but they bring the geomorphologist face to face with professional geologists who needs to have the answers. The lack of interest in such questions, and perhaps lack of competence to attempt the answers, is a weakness in the training of geomorphologists, and possibly part of the 'disenchantment' regretted by Baker and Twidale (1991).

FACTORS AFFECTING LONG-TERM LANDFORM EVOLUTION

A short check-list of factors that influence the evolution of relief over long time spans must include:

(1) the plate-tectonic framework, inter-plate rifting, and the varying nature of plate boundaries;
(2) thermal uplift and the changing location of domal uplifts and basin subsidence;
(3) isostatic adjustments due to loading and unloading of the lithosphere;
(4) tectonism involving faulting or flexuring of the landsurface, including the influence of neotectonics;
(5) world-wide sea level changes and their causes;
(6) the formation of escarpments by tectonic, structural and lithological influences;
(7) the evolution of world climates and changes in ocean/atmosphere circulation;
(8) continental drift and the climatic implications of palaeolatitudes;
(9) the history of rock weathering, mantle formation and removal (stripping);
(10) mechanisms for the development or decay of relief forms;
(11) the timing of formative events and rates of landform adjustment.

All of these factors are internally complex and this study cannot attempt to analyse each in detail. Nevertheless, it is important to identify the contributory factors appropriate to the study of long-term landform development (Schumm & Lichty, 1965; Brunsden, 1980; Brunsden & Jium-Chuan, 1991). Ollier (1981) in his study of *Tectonics and Landforms* addressed many of these questions and reproached geomorphologists for failing to accept tectonic ideas and the 'new geology'. In the discussion that follows, some key issues in the long-term evolution of tropical (and non-tropical) terrains will be addressed.

Geodynamic and tectonic models

Much of the reasoning employed in this section will relate to the most recent period of plate-tectonic movement, involving the break-up of Pangea, and although this began in the Triassic (200 Ma), the opening of the Atlantic, Indian and Pacific Oceans and the strong northward movement of the continental plates probably commenced no more than 100 Ma ago.

This last 100 Ma since the late Mesozoic have witnessed the formation of new coastlines along the passive margins of drifting continents, but models to describe the geomorphic response to these events remain tentative. Fairbridge and Finkl (1980) considered the concept developed by Wise (1972, 1974) of an increasing continental *freeboard* (viewed as relief above sea level) through Phanerozoic time, which they related particularly to poleward motion of the continents and to a decreasing global spin rate.

Associated with this phenomenon has been the strong northward movement of India and Australia, the former moving at the rate of 16 cm year^{-1}. In addition to the implications of this motion for collision tectonics in the Himalayan region, it has been associated with the positioning of continental plates over mantle plumes, leading to major volcanic episodes, of which the Deccan flood basalts are a striking example (Roberts *et al.*, 1984). This movement has also caused the continental plates to traverse several (present-day) climatic zones. It seems almost certain, however, that these zones were not in existence during the Mesozoic and Palaeogene, and therefore the implications for

weathering and landform development are not straightforward (Taylor *et al.*, 1992).

Summerfield (1985) on the other hand, has emphasised the importance of domal uplifts along passive continental margins, particularly in Africa, following Holmes (1965), Le Bas (1971), Dingle (1982) and others. These are thought to be associated with the rifting that brought about the fragmentation of Gondwanaland. Bremer (1985a) has also drawn attention to the phenomenon, which has been described as *'Randschwellen'* by Jessen (1943), demonstrating that it has a world-wide occurrence. Such uplifts are thought to be thermally driven by mantle convection, and they are frequently accompanied by rifting. An increased rate of denudation will take place on the flanks of the domes, which are progressively steepened by further isostatic uplift, due to denudational unloading inland and subsidence resulting from sediment loading offshore. In effect, the margin of the continent undergoes rotation, and this effect has immediate and lasting impact on the course of denudation. The intervention of faulted scarps will also increase erosion rates abruptly along a narrow zone. Summerfield (Thomas & Summerfield, 1987) has advanced six major tectonic factors controlling the long-term morphological evolution of passive margins (Figure 11.2):

(1) thermally driven uplift U_T
(2) isostatic uplift due to denudational unloading inland U_I
(3) thermally driven subsidence S_T
(4) isostatic subsidence due to sediment loading offshore S_I
(5) rotation of continental margin resulting from U_I and S_I
(6) escarpment 'retreat' (?)

It has proved difficult to demonstrate strict correspondence of sedimentary sequences and their unconformities offshore, with the development of planation surfaces inland, despite numerous attempts to do just this (King, 1983; Summerfield, 1985). Nevertheless, it has been a corollary of continental drift as a consequence of plate tectonic motion that the dismemberment of Gondwanaland

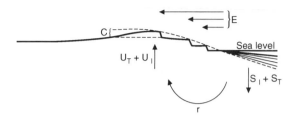

FIGURE 11.2   Six major tectonic factors controlling the long-term morphological evolution of passive margins. See text for main explanation of uplift (U) and subsidence (S) terms. C represents the constructional relief resulting from volcanic and intrusive rocks; rotation of the margin is designated r. (Reproduced by permission from Thomas & Summerfield, 1987)

has been accompanied by the dissection of the marginal swells and uplifts of the trailing edges of the Americas, Africa, India, and to a less extent Australia, and possibly Antarctica. According to viewpoint, this has led either to world-wide correlative planation following the concepts of Penck (1924) and Meyerhoff (1975) or King (1962, 1976; Melhorn & Edgar, 1975; Ollier, 1985), or to complex developments of cyclic and acyclic forms that defy easy categorisation (Hack, 1979; Thomas, 1980; Klein, 1985; Pavich, 1985; Summerfield, 1985).

Summerfield (1985, p.27) summarised the most likely tectonic sequence:

> models of continental rifting and passive margin evolution predict an initial zone of thermal uplift along a nascent margin following continental rupture. This is subsequently replaced by a coastal uplift induced by flexure and rotation of the margin associated with post-rifting thermal subsidence, sediment loading off-shore and denudational loading inland.

He also concluded concerning Africa that:

(1) there are significant discrepancies between proposed landscape cycles and the stratigraphic record of the continental margin;
(2) such discrepancies may be due to errors in dating and chronology and to the complex response of the crust;

(3) generation of landscape cycles due to base level changes is likely to be confined to the vicinity of marginal upwarps;
(4) mechanisms of epeirogeny would not lead to extensive uniform uplift.

Ollier (1985) referred to uplift mechanisms for plateau landscapes and escarpments stated by McGetchin *et al.* (1980). It is clear that initiation of uplift is commonly associated with thermal expansion over mantle plumes (or *hot spots*), but may also occur where a continental plate overrides (by subduction) hot mid-oceanic ridges or spreading centres. Other factors include shear heating between the lithosphere and asthenosphere, expansion due to partial melting or to hydration reactions (serpentonisation). The positioning of parts of the continental plates over mantle plumes resulted in highly differentiated relief development on and between the resulting domes (Le Bas, 1971; Summerfield, 1985, 1991a,b), and also led to major volcanic episodes such as the lava emissions represented by the Deccan Traps of peninsular India (c. 63 Ma), and similar outpourings in eastern Brazil, or the volcanoes and basalts of Kenya.

In all of these cases the uplift is restricted in extent by the formative processes, and in Africa, for example, a series of basins and swells (arches, domes) can be described and mapped in some detail (Holmes, 1965; Le Bas, 1971; Burke & Dewey, 1973, Dingle, 1982). The swells tend to be close to the continental margin, where they are usually faulted, and are commonly associated with vulcanism. They are therefore taken as the precursors of continental splitting. This is made clear in Africa by comparing the series of swells along the great Rift Valley System with the sheared, half-swells along the Atlantic margin or Africa–South America junction (Ollier, 1981). By comparison, the vast interior basins of Africa occur on a different spatial scale (10^3 km as against 10^2 km in diameter), and must be matched to the overall marginal uplift of the southern and eastern part of the continental margins. Gilchrist and Summerfield (1990, 1991) have noted that lithospheric thickness increases markedly from the continental rim, where it may be only 5–10 km to cratonic interiors where it can range from 50–100 km. This will greatly affect the compensation function and may be associated with magmatic underplating.

Gilchrist and Summerfield (1990, 1991) have also distinguished between mature (>60 Ma) margins and much younger, rifted margins. The mature passive margins classify broadly into: (1) low-lying margins, as in southern Australia and west Africa, and (2) steeply rising margins found in southwest Africa, eastern Australia and eastern Brazil. The mature, steeply rising margins rise inland over 50–300 km to a marginal upwarp 300–900 m high and then decline in altitude towards the continental interior. Facing these coastal margins are major escarpments such as the Drakensberg in South Africa, the Serra do Mar in Brazil and the Great Dividing Range in eastern Australia.

Because these are set back from the present coastlines, and also because in other areas planation surfaces were considered to form a 'staircase' of progressively older and higher treads inland, simple tectonic models for the formation of such flights of surfaces have been sought. One such model depends on an interpretation of isostatic adjustment that may be mistaken, namely that there is some threshold which must be exceeded by scarp retreat into an old continent, before isostatic readjustment to the reduced load following denudation can occur. This view was advanced by King (1955), and Pugh (1955), who suggested that a new plain some 400 km wide would need to form by scarp retreat, before a further rise in the crust could take place. This view offered a simple mechanism by which to explain the flights of erosion surfaces. This idea has not always been followed, but it is specifically refuted by Gilchrist and Summerfield (1990, 1991), who examined the basis for isostatic recovery and concluded that time-bound models of crustal behaviour applied only to shorter timescales (10^4–10^5 years), and that continuous crustal displacement takes place over the timescales of planation (10^6–10^7 years). They considered that *continuous flexure* due to denudation inland, consequent upon scarp retreat, and sediment loading offshore, has led, around the margins of southern Africa, to a significant marginal upwarp

with an amplitude of several hundred metres. In effect, this has arisen following continental rifting in the late Mesozoic, as the new scarps retreated inland from the coast. This view appears to conflict with the concept of relative rapid uplift in the Pliocene, advocated by Partridge and Maud (1987).

The complexities of the tectonic framework are therefore such as to warn against reconstructions of landform history based on models dependent on global synchroneity, or uniformity of uplift. Ollier (1981) has pointed out that swells and uplifts did not occur synchronously around Australia and therefore could not have induced continent-wide cycles of planation. There are also wide variations in sedimentation rates offshore (Summerfield, 1985, Table 1) which argues for differential rates of denudation on land. Summerfield (1991a, p. 560) has concluded that,

> although erosion surface sequences are ubiquitous on the Gondwana continents many, if not all, appear to be subject to direct structural control, or to indirect control through the creation of local base levels. Consequently, the vertical separation of such erosion surfaces cannot be taken as unambiguous evidence of discrete episodes of surface uplift initiating relative falls in base levels.

The tectonic arguments, therefore, require a fresh view to be taken of the common models for the development of world-wide correlative planation surfaces.

Extensive planation and the ages of landsurfaces

Evolutionary geomorphology must make assumptions about the relative and absolute ages of landforms, but defining landscape 'age' presents serious problems to the geomorphologist.

First, to ascribe a single 'age' to an extensive surface of planation can be misleading, since different parts of the surface may have formed at different times. If parallel slope retreat is adopted as the evolutionary model, then surfaces should decrease in age with distance from coastlines and major river valleys, as the wave of backwearing has moved away from these features. In this model the surface at the foot of an escarpment will be near-contemporary. There is, however, little evidence to confirm this relationship.

Second, landsurfaces remain dynamic, after major planations have been 'completed'. If down-wearing or the concept of *dynamic equilibrium* is accepted, then the actual surface is in a constant state of change. There is, therefore, a difference in concept between the date of first development of an extensive planation and the age of the present groundsurface.

We must conclude, therefore, that extensive surfaces of planation are complex, probably diachronous in age, and exhibit strong lithologic and structural controls, multi-layer regoliths and transported mantles (Boulvert, 1985). Frequently they are described as 'polygenetic' or 'polycyclic', and they must be considered as a mosaic of forms and deposits with ages more recent than the gross form of the plateau landsurfaces on which they may be found. In order to avoid these pitfalls, surfaces of widespread planation are usually given 'ages' bracketed by formative events, usually uplift at the start and sedimentation at the end of a 'cycle'.

The too-ready acceptance of pediplanation as a near-universal model for landscape development has led to the rather simplistic view that a wave of rapid denudation associated with the retreating escarpment is succeeded by the formation of a pediplain, which then experiences gradual elimination of relief perhaps to become an *ultiplain* (Twidale, 1983), on which most processes are inactive. Such ultiplains, with erosional origins, are rare on a global scale and, although planation remains a central theme in evolutionary geomorphology, the maintenance of equilibrium and the formation of new relief should be equally important concerns.

Planation surfaces in Africa and South America

It was a part of the reconstruction by King (1962, and Table 11.1) that it should be possible to recognise the pre-continental drift landsurface of the Gondwanaland continent as a surface of very extensive planation, forming the highest (non-volcanic/constructional) remnants of planation. As such it should be pre-Cretaceous in age. A number of independent studies have, however, failed to identify these ancient remnants in Africa, except perhaps where recently exhumed from beneath a sedimentary cover.

Partridge and Maud (1987; 1988, and Table 11.2), in their analysis and appraisal of evidence for a chronology of landscape evolution for southern Africa, stated (1988, p.11) that they, 'cannot find any conclusive evidence for the preservation of any subaerial erosion surface dating from before the breakup of Gondwanaland'. The thickness of offshore Cretaceous sediments points to rapid denudation of the continent following the opening of the Indian Ocean, and they indicate that at least 300 m of material has been lost from the Lesotho Highlands. Similar limitations upon the ages of supposedly ancient landsurfaces apply to areas such as the Jos Plateau in Nigeria (Thomas, 1978, 1980), where Jurassic (160 Ma) Younger Granites have been exposed by deep erosion in areas where King (1962) had originally mapped his 'Gondwana' surface.

Partridge and Maud (1988) stated clearly that they, 'find the widespread fundamental erosion surface of the subcontinent to be the "African" surface of King'. This they found associated with a deeply kaolinised, weathered mantle and with a duricrust capping (laterite in the east; silcrete in the west). In their view, 'the African cycle was initiated by the breakup of Gondwanaland and persisted until its termination, in the early Miocene, by marginal uplift and tilting of the subcontinent.'

This reasoning places the formation of this planation surface between the late Cretaceous and the end of the Oligocene (a period of 50–60 Ma). Equally important to their reconstruction is the recognition (Partridge and Maud, 1987, p.189) that, 'of fundamental importance to an understanding of all subsequent landscape evolution within the subcontinent is an appreciation of the fact that, throughout this and later cycles, erosion proceeded simultaneously at different levels above and below the Great Escarpment.' This view differs from King's later work (King, 1962), but was reflected in his earlier writings (King, 1949, 1951). Few studies of erosion surfaces on the Gondwana continents have understood or applied this argument. In any case it depends crucially on the timing of the uplift leading to scarp formation.

In South America, King (1956) recognised the 'Sul-Americana' surface as equivalent to the African surface, and further representatives of this planation have been suggested for Venezuela (Briceño & Schubert, 1990), Guyana (McConnell, 1968; Aleva, 1984) and Surinam (Krook, 1979; Zonneveld, 1982). The 'age' of this surface is considered to be early Tertiary and terminated by deposition of Oligocene–Miocene sediments in Surinam (Zonneveld, 1982, 1993). However, the view that these landscapes comprise a series of stepped erosion surfaces of successively younger age towards the present drainage lines, is not universally upheld by other writers. Petit (1985) has compared different schemes for the identification and dating of the planation surfaces in this region, and also alluded to the problems of altimetric dating in an area of lithologic and tectonic contrasts. Demangeot (1985) emphasised the importance of block faulting in southern Venezuela, stating that tectonic rejuvenation is probable. In Surinam, Kroonenberg and Melitz (1983) challenged the interpretation of relief based on the work of King (1956) and Zonneveld (1982), and stressed the close correspondence between relief units and lithological discontinuities, interpreting the appearance of many different altitudinal levels in terms of differential weathering (etching). More recently, Zonneveld (1993) has reviewed these interpretations. Eden (1971) also differed from McConnell (1968) in his interpretation of relief in Guyana, emphasising the importance of the saprolite mantle and differential weathering.

These views are presented in outline to stress two points. First, that many geomorphologists continue to claim the widespread occurrence of stepped planation surfaces in the tropics. Second, that alternative explanations for some of the features have been advanced. The fundamental differences of view can be distilled into a conflict between the proponents of backwearing and pedimentation which must eventually plane all rocks to local, and subsequently continental base levels, and those who consider the tectonic, structural and lithologic controls on denudation to be decisive in producing contemporary patterns of elevation.

It is common, however, for both interpretations to ignore the tectonic settings and histories of the landscapes described. A point stressed by Partridge and Maud (1987) is that, 'a significant part of the controversy . . . in relation to the continent-wide correlation of erosion surfaces can be ascribed to the failure of recent workers to appreciate the vertical separation of landsurfaces of the same age which was dictated by the high initial elevation of the continent (of Africa) prior to rifting.'

Because Africa was situated at the centre of Gondwanaland, it has been estimated that its central parts had an altitude of ~2000 m (Partridge & Maud, 1987, 1988) at the time of break-up. This position may have favoured the formation of the major escarpment around the southeastern rim of the continent at that time, and contributed to the present high altitude of the Lesotho Highlands (>3000 m). But if the post-Gondwana wave of erosion has not penetrated to the interior of the continent, except along some major river valleys, the recoil of this feature must have been very slow.

It is equally important to realise that the high elevation of the southern landmasses did not extend into the northern hemisphere in the Cretaceous or early Cenozoic. Much of northwestern Africa was low and the late Cretaceous (100 Ma) sea extended from the Tethys (present Mediterranean) to the Gulf of Guinea (Atlantic). The Amazon Basin represented a submerged structural trough. In the southern hemisphere, Australia, still joined to Antarctica, was largely a lowland continent. The uplift of the Eastern Highlands of Australia probably dates from the Palaeocene (50–60 Ma), though agreement on the details has not been reached.

Planation in Australia

In Australia comparable debates have pursued a different course. An extensive 'weathered landsurface' was recognised in western and central Australia, especially by Mabbutt (1961, 1965, 1967), and this has received widespread attention. However, it was not conceived as a cyclic erosion surface but 'as an entire landscape which was subjected to weathering' (Mabbutt, 1965, p.109). The denudational history of this landscape has proved difficult to unravel, probably because of its polycyclic character.

Interpretations of the eastern continental margin and the Great Dividing Range have also proved contentious (Wellman, 1979; Wellman & McDougall, 1974; Jones & Veevers, 1982, 1983; Ollier, 1982, 1988b; Taylor, 1983; Pain, 1985), illustrating the difficulty of determining the timing, duration and magnitude of uplift along continental margins. It is perhaps this problem, amongst many, that has delayed the resolution of other debates about the generation and retreat of escarpments.

The dating of Australian landsurfaces and regoliths has been pursued along four different lines of enquiry.

(1) K–Ar dating of lavas overlying landsurfaces and specific forms. These can only give us a *minimum* age for the surface on which the lavas have solidified.

(2) Palaeomagnetic dating of iron accumulations in weathering profiles. This relies on the accuracy of reconstruction for Australia's continental drift history, and also on assumptions about the timing of formation of iron minerals in the saprolite which may be variable (Idnurm & Schmidt, 1978; Schmidt & Ollier, 1988).

(3) Oxygen isotope ($\delta^{18}O$) dating of the regolith, based on the the northward drift of Australia since rupture with Antarctica at 95 Ma. This depends on the $\delta^{18}O$ of meteoric waters which would have become progressively heavier with drift into lower, warmer latitudes since the Cretaceous (Chivas, 1983; Bird & Chivas, 1988b).

(4) Stratigraphic relationships, unconformities and the recognition of exhumed landsurfaces remain important to the determination of limiting ages (Ollier *et al.*, 1988a,b; Wilford, 1991).

By plotting $\delta^{18}O$ values for Australian regolith samples against both present latitude and ages assigned on the basis of geological evidence

(overlying sediments or lavas), Bird and Chivas (1988b) showed that many regoliths are probably much older than formerly thought, some being before the late Mesozoic. In New England, Schmidt and Ollier (1988) derived palaeomagnetic evidence of late Cretaceous to early Tertiary weathering on the upland plateau surfaces which form part of the Eastern Highlands of Australia, and similar conclusions have been reached for southern New South Wales by Pillans (1977). Ollier (1982) and Pain (1985) have summarised much of the evidence from Australia concerning the antiquity of the continental landsurfaces, and doubts about the age and relationships of the uplifts giving rise to the Great Divide and the escarpment which lies to the east of it.

Mabbutt (1988), meanwhile, has returned to the interpretation of the continental landsurface and its margins in Western Australia. With respect to the Wiluna–Meekatharra area, a part of the Yilgarn Block, he referred to the ideas of King (1950b), who sought to identify the Gondwana surface and an 'Australian' surface in this area. Mabbutt (1988) concluded that the separate recognition of different planation surfaces in this region cannot be verified, and that the landsurface is a polycyclic feature reduced in height only little from the pre-Permean unconformity and landscape. The work of Mabbutt on the formation of weathered and stripped landsurfaces was considered above in the context of etching and etchplanation, but in the present context, the important element of his recent thinking is that many landscapes of partial or extreme planation are not the products of a single major cycle of denudation, but are the results of progressive lowering and planation through long periods of geological time, during which deep weathering has preceded phases of stripping, and both have left complex superficial deposits on the present landsurface.

If this view is accepted then no great wave of backwearing, involving the retreat of a major escarpment has occurred in Western Australia as a consequence of the break-up of Gondwanaland. On the other hand, the formation and role of the escarpment which forms the 'plateau edge' in Australia remains contentious, and, although its status as an erosional feature has been confirmed by Pain (1985), its meaning in terms of continent-wide planation is unclear.

The great geological 'age' of some landsurfaces in Australia has been emphasised by Ollier *et al.* (1988a), who has identified in the Kimberley Plateau, remnants of a Proterozoic planation surface, striated by a Precambrian glaciation and carrying tillites. There is no trace in this area of submersion and deposition of a sedimentary cover during the last 700 Ma. There are also Cambrian fluvial sediments found on remnants of the Ashburton Surface south of Tennant Creek (20–21°S) (Stewart *et al.*, 1986; Gale, 1992). Similarly, the Yilgarn Block in southwestern Australia still preserves the remnant drainage system once flowing across the area from Antarctica, and now seen as a series of aligned salt lakes. According to Fairbridge and Finkl (1980), the Yilgarn landsurface has not been effectively lowered by denudation over 1.5×10^9 years, and it has been little affected by dissection during emergence of 250–300 m since the Eocene.

Exhumed landsurfaces of great age are also found, and the well-preserved Ordovician glaciated pavement in the central Sahara represents a good example. These have been revealed following the stripping of Silurian rocks, so that the modern period of exposure to sub-aerial modification is not known (Beuf *et al.*, 1971; Rapp, 1975).

Major erosional events on the Gondwana continents

Despite the recognition of some very old landscape elements, especially in Australia, the following observations appear to reflect our present understanding of the evolutionary events that shaped the continents which derive from the former Gondwanaland supercontinent.

(1) Evidence for the *widespread* survival of a Gondwana (pre-continental drift) landscape is lacking.
(2) Claims for a Cretaceous-early Miocene *polycyclic* planation are substantiated from a number of areas and this possibly includes the African, Sul-Americano, and Australian surfaces.
(3) This planation took place at *different elevations* above and below the Great Escarpment formed by the rifting of Gondwanaland, and did not eliminate this feature.
(4) Miocene uplift (*c.*18 Ma), followed by aridification of climate (in Australia and Chile at *c.*15 Ma) terminated this phase.
(5) Partial planation (e.g. Post-African 1 cycle) may have occurred during the mid-Miocene to late Pliocene.
(6) Late Pliocene uplift is claimed for southern Africa (possibly 900 m in the southeast), but is not universally supported; such an event did not take place in Australia.
(7) Deep incision ensued, with only localised planation during the Pleistocene.

The concept of a 'cratonic regime'

The distinctive nature of what might be called 'Gondwana' geomorphology, with its emphasis on long timescales of evolution, widespread planation, and deep weathering mantles, has led to the development of models to match what Fairbridge and Finkl (1980) have called the *cratonic regime*. Other terms such as the 'epeirogenic realm' (Garner, 1974) or reference to passive margin landscapes, convey a similar message. This regime has both a geological rhythm and a gemorphological evolution.

The essential geological characteristics comprise the foundation of crystalline metamorphic basement rocks, traversed by greenstone belts and intruded by granites. Superposed on this cratonic platform are thin transgressive wedges of sediment with widely separated ages; early to middle Ordovician, early Carboniferous, and late Cretaceous. The steady evolution of the relief has been interrupted by short-term (10^4–10^5 years) 'climatic accidents' or crises; by epeirogenic motions bringing into being domes and basins; and by the megatectonic events leading to rifting and continental drift, and causing more abrupt upwarps and downwarps, accompanied by faulting.

As a consequence of these events the cratonic regime was seen by Fairbridge and Finkl (1980) to alternate between high and low relief states over 10^7–10^8 year periods. The so-called 'normal' state has been one of low relief and high sea level (*thalassocratic*), associated with humid climates and prolonged deep weathering under a forest cover. These periods were interrupted by increasing relief and low sea levels, accompanied by changes of climate (*epeirocratic*). The periods of warm–humid morphogenesis correspond to the ecological concept of *biostasie* (Erhart, 1955). Weathering penetrated deeply (possibly down to 600 m), saprolites became ferrallitic, and very little sediment was transported from the system. Rivers carried mainly solute loads with significant quantities of Ca^{2+} and Si^{2+} being transported to the oceans (*cationic denudation*). Ocean basin sediments from these periods reflect this regime and contain important quantitites of chert and limestone.

Disruption of this system due to the onset of the epeirocratic regime and the increasingly seasonal nature of climates ushered in a state of *rhexistasie* (Erhart, 1955), whereby surface erosion rates became increased and clastic sediments were transferred to neighbouring basins. These deposits were dominated by ferrallitic clays (red beds) and by quartz sands, reflecting the stripping of the former saprolite mantle (Millot, 1970, 1982, 1983). Fairbridge and Finkl (1980) thought these episodes could be regarded as 'catastrophic' in terms of the long timescales involved, and that they lasted 10^4–10^5 years (Finkl, 1981, 1982).

These authors offered a further scenario of three climatic states, typical of the Quaternary:

(1) perennial warm–humid (the rainy tropics), associated with biostasie and typified by the Brazilian Amazon today;
(2) seasonal wet–dry tropics (monsoon tropics), favouring vertical (up–down) transfers of ions in solution, leading to the formation of ferricretes and duricrusts;
(3) semi-arid environments (experiencing hyper-arid to semi-arid fluctuations), representing only 10% of each glacial cycle of 100 ka.

Since glaciations occupy only 10% of the geological record, extreme conditions favouring stripping of saprolites and lowering of the cratonic landscape were thought to have been brief and infrequent in terms of these longer timescales (10^6–10^7 years). The lowering has been mainly achieved by shallow pedimentation cycles, stripping the saprolite mantle to expose the weathering front (Mabbutt, 1961; Thomas, 1965b; Rohdenburg, 1969/1970). In relatively stable areas, these events brought about only local transfers of material, leading to polygenetic surfaces that carry complex surface deposits. This alternation according to Fairbridge and Finkl (1980) is between *etchplanation* and *pediplanation*.

These views are very close to those of Millot (1970, 1982, 1983) who also employed the concepts of biostasie and rhexistasie to explain the continental deposits that accumulated in sedimentary basins in Africa and Europe during the later Tertiary. Millot (1970) saw the need for prolonged *lessivage* of the stable continental surfaces existing in the late Mesozoic and early Cenozoic eras, in order to explain the red beds, limestones and clastic sequences observed in section. Ollier (1988b) too has used the same general model in his appraisal of the evolution of the Australian landscape, which was warm and wet during the Mesozoic, but became progressively more arid from the Oligocene onwards. He paid particular attention to the residual deposits left behind by this evolution on the Yilgarn Block, including residual red beds corresponding with greenstone outcrops and sandplains resulting from the breakdown of kaolinite in the saprolite (podzolisation; see Chapter 2). The old drainage systems in this area are plugged by Eocene clays derived by erosion of the saprolite mantle.

An important conclusion from these studies is that the cratonic landscapes which straddle the tropics illustrate an 'evolutionary geomorphology' (Ollier, 1979); a sequence of events influenced by the prior history of weathering and planation, uplift and dissection, and not simply a repetition of cycles or episodes. This emphasises the importance of *inheritance* in the landscapes of today (Twidale, 1976; Twidale *et al.*, 1974). These principles also extend to other landscapes, beyond the confines of the Archaean cratons of Gondwanaland, and apply with greater or lesser force to Palaeozoic and even Mesozoic terrains, including areas such as Kalimantan (Indonesian Borneo) (Barber, 1985), and also the Hercynian massifs of Europe and America.

The formation and role of escarpments

It is clear that the origins and functional relationships of escarpments are central to many debates about landform evolution, yet the topic remains controversial; the ideas of King (1962) being far from universally accepted (Hack, 1960; De Swardt & Bennett, 1974; Crickmay, 1975; Summerfield, 1985). No single morphotectonic model has so far proved universally applicable, though differences between the domal uplifts of West Africa and the plateau uplift described by Ollier (1985) for Australia and southern Africa, may depend largely on the incidence of major faults. The division into *rifted, sheared* and *sunken* margins proposed by Dingle (1982), and the stages of rifting discussed by Veevers (1981) are equally significant contributions to this question.

Many striking escarpments are formed in layered sedimentary or volcanic formations (the Drakensberg of southern Africa; the Voltaian Scarp in Ghana; the Western Ghats of India), and some are also strongly faulted as along the Atlantic coast of eastern Brazil (the Serra do Mar) and in Venezuela and Guyana. Others arise from differential denudation of adjacent rocks as in the case of the south-facing granitic escarpments of the Jos Plateau in Nigeria. The recession of escarpments by parallel slope retreat in the manner advocated by King (1962) remains a favoured model for landscape development (Ollier, 1982, 1985; Pain, 1985), but, as has been discussed, there are many unresolved questions surrounding this issue (Summerfield, 1985).

In eastern Australia the timing and number of uplifts (Pain, 1985) and the cyclicity of erosional/depositional events remain uncertain (Jones & Veevers, 1982), and some parts of the escarpment have been generated by differential denudation, as in the case of significant embayments on weathered granites (Dixon & Young, 1981).

However, the idea of 'The Great Escarpment' (Ollier, 1985) remains central to the evolutionary geomorphology of the southern continents. Both Ollier (1985) and Pain (1985) contrasted the landscapes above and below the Great Escarpment of eastern Australia in terms of morphology and rates of denudation. In view of the conclusion from the previous section, that only one subcontinental surface has been reliably identified, and this at levels both above and below the Drakensberg in South Africa, the argument for scarp retreat needs closer scrutiny.

Even if the notion that 'great escarpments' are erosional (Ollier, 1985; Pain, 1985; Partridge & Maud, 1987) is accepted in principle, there is little evidence to support rates of retreat capable of eliminating high relief on crystalline basement terrain (cratons), within the time periods so far elapsed since their formation. The Drakensberg, the Western Ghats and the Great Escarpment of eastern Australia all lie within 50–150 km of the coast. The strongly faulted Serra do Mar of Brazil approaches the coast itself in several places.

The argument here is not so much that escarpments do not slowly recoil from their structurally defined, initial positions, but rather that retreat is slow and strongly influenced by lithological and other factors. In the evolutionary geomorphology of the largely tropical, southern continents, the notion of repeated 'cycles' of effective planation is therefore inappropriate. There has instead been a sequence of tectonic and erosional events which have achieved partial planations, but have led, during the last 100 Ma, to an episodic increase in relief and diversification of terrain (Thomas, 1989b; Twidale, 1991).

Landscapes composed of layered sedimentary or volcanic rocks may evolve in a manner influenced by a 'cap rock'. This refers to any situation where a resistant formation overlays conformable strata, and where undermining of the cap rock can induce scarp retreat. It is not always necessary for the cap rock to be self-evidently more resistant to denudation than the underlying strata, though this will speed up the process. This principle can be applied equally to the stripping of saprolite below duricrust cappings, and to the removal of

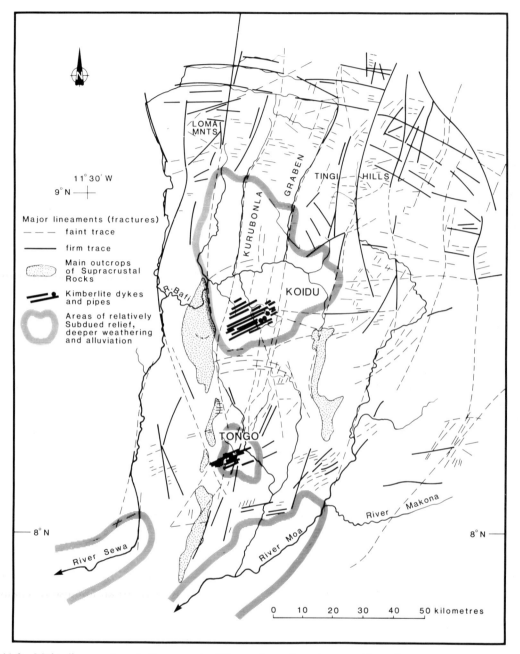

FIGURE 11.3 Major lineaments and other aspects of the geology of northeastern Sierra Leone (as shown in Figures 8.1 and 8.2). The presence of the kimberlite dykes and their significance to groundsurface lowering are discussed in the text

sedimentary strata from a crystalline basement disconformity.

That such events occur in long-term landform evolution can be verified in the field. In central Sierra Leone the roots of Cretaceous (92 Ma) kimberlite diatremes are exposed in the Koidu basin diamond field (Figures 8.1, 8.2 and 11.3). Sedimentary xenoliths from these exposures have been analysed by Hubbard (1983). They resemble rocks of the Saionya Scarp Series (Ordovician) which outcrops along a pronounced escarpment near the northern border with Guinea (Figure 8.1). According to Hawthorne's (1973) model of a kimberlite pipe, the present exposure of the dyke zone in Koidu should be 1000–1500 m below the landsurface of the time of formation. The highest local relief, on strongly duricrusted, supracrustal metamorphic schists is only 350–400 m above the granitoid basement. It is likely, therefore that a sedimentary pile has been removed from this area, and that the Saionya scarp represents the contemporary expression of that event. It stands about 300 m high.

Two conclusions are drawn from this example. First, it would appear that the sedimentary rocks have been removed by one or more periods of scarp retreat, to expose the basement surface. Second, that the subsequent evolution of the landforms has produced a further 350 m of relief that strongly reflects the lithological boundaries in the Archaean and Proterozoic basement rocks (Figures 8.1 and 11.3). This simplified account is open to the objection that it ignores the apparent recognition of several planation surfaces by other authors (Dixey, 1922; Hall, 1969; McFarlane *et al.*, 1981). Aspects of this problem have been discussed elsewhere (Thomas, 1980; Thomas & Thorp, 1985), and it is clear that the elements of the debate are similar in both West Africa and in Guyana and Venezuela (Kroonenberg & Melitz, 1983; Briceño & Schubert, 1990). In the Gran Sabana of southeastern Venezuela, the tabular relief of the 'Tepuis' or table mountains, is almost self-evidently destroyed by scarp retreat, but the evolution of the basement landscape is subject to different arguments which are discussed below. It is, however, the position of the writer that many instances of erosional surfaces are dependent on lithological variation and structural patterns in the basement.

MODELS OF RELIEF REDUCTION

Further discussion of such examples will be unproductive, unless knowledge of the possible processes or mechanisms of relief reduction in

broadly tropical conditions is better focused. The foregoing account has concentrated upon the problems of recognition and dating of planation surfaces. This section will discuss the processes at work. Models of landscape evolution that are widely used, but which require further examination for their application to tropical landscapes, include the following:

(1) the elimination of relief by downwearing (vertical lowering) of interfluves in the *Davisian cycle of erosion* (Davis, 1899a,b);
(2) the destruction of higher surfaces by *parallel slope retreat* (Penck, 1924; King, 1953, 1957);
(3) the application of Büdel's (1957, 1982) concept of 'Doppelten Einebnungsflächen' or *double surfaces of levelling*, also involving patterns of saprolite stripping;
(4) the formation and maintenance of a *dynamic equilibrium* on slopes typifying different lithologies (Hack, 1960, 1975, 1979).

Several other concepts are important to the discussion of long-term landform evolution. First, Crickmay (1975) emphasised the spatial contrasts in rates of relief reduction in the landscape, citing the survival of stream gravels on isolated plateaus as the denial of the Davisian model of downwearing, and argued for the importance of lateral stream erosion as a mechanism of localised planation within an *hypothesis of unequal activity*.

Second, Brunsden and Thornes (1979) introduced the concept of *landscape sensitivity* to changes in the operation of controlling processes. Brunsden (1980, p. 18) sought to embody this concept of landscape sensitivity in three ideas of time: (1) 'the time taken to react to an impulse of change (*lag time*)'; (2) 'the time taken to attain the characteristic state (*relaxation time*)', and (3) 'the time over which that state persists (*characteristic form time*)'. In this trio the relaxation time is critical, because landscape forms can only reach stability if the interval between form-changing events is greater. In most cases, we do not possess the information to define these concepts in numerical terms, but in more sensitive areas, usually where slopes are steeper or on floodplains, unsteady behaviour and the occurrence

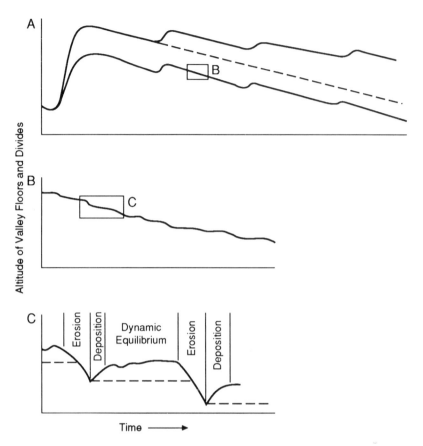

FIGURE 11.4   Modified concept of the geomorphic cycle. (A) Erosion cycle, as envisioned by Davis (dashed line), following uplift and as affected by isostatic adjustment to denudation; (B) a portion of the valley floor in (A) above, showing the episodic nature of the decrease in valley floor altitude; (C) a portion of the valley floor in (B) above, showing periods of instability separated by longer periods of dynamic equilibrium. (Reproduced by permission of Chapman & Hall from Schumm, 1975)

of transient landforms and deposits will be common.

These ideas throw light upon the problems of landscape inheritance and the possibilities of persistent dynamic equilibrium being widely established within the landscape. The concept of *episodic erosion*, as explained by Schumm (1975), offers a way of incorporating short-term periods of instability within the concept of a cycle of erosion. Schumm also emphasised the importance of thresholds within the geomorphic system and the way in which the passing of such thresholds could trigger a *complex response* in the modelling of the landscape, especially by surface water flows.

The building of models of landscape evolution that embody both the functional (process) and the historical (evolutionary) approaches to geomorphology has proved elusive, though the way was charted by Schumm and Lichty in 1965 and extended by Schumm (1975, 1977). A plea for a more integrated approach to the subject was made by Starkel (1987), who emphasised the persistence of forms and deposits of widely differing ages in the landscapes of the Carpathians, and went on to identify the loci of stability and rapid change in the landscape.

The models developed from the application of these concepts are not necessarily mutually exclusive, when viewed on the scale of a complex landscape comprising many different slopes and lithologies, and during different timescales of enquiry for 10^2–10^8 years. However, for a particular slope, during a defined period, it may be necessary to determine which outcome the operative processes will produce. This is another way of saying that not all slopes in a landscape need behave in the same way, and a single slope may change its behaviour through time. Over the time periods involved in true planation (10^7 years), there may be a series of stages through which a landscape will pass (Figure 11.4). It is also clear that the formation of true escarpment faces will be dependent on the generation of sufficient relief along a downwarp or fault zone.

Vertical lowering of the landsurface

There can be little dispute over the lowering of sharp interfluves, where opposing slopes draining to adjacent valleys, meet along a well-defined ridge, because any loss of material from the slope must reduce the height of the interfluve, though local relief may remain constant if the river channel is also lowered. What is more disputed is the potential for generalised lowering of the landsurface. This question depends crucially on the role and behaviour of the saprolite layer, and much important research on this topic has come from the warm temperate areas of the southeastern Piedmont and Appalachian areas of the USA. It is likely that these principles will apply widely to all saprolitic landscapes, so the ideas will be explored in more detail.

Since the time of W. M. Davis and D. W. Johnson (1931) the Appalachian region of the USA has been one of the testing grounds for geomorphic theory in the humid temperate zone. The area ranges from 35°N to 40°N latitude and cannot therefore be regarded as tropical today. However, the processes operating in the non-glaciated Piedmont landscapes will be similar to those active in warmer climates, even if they proceed more slowly. Also, of course, the time span of landform evolution may include periods of warmer climate in the early Cenozoic.

Modern work in this area derives much of its inspiration from the seminal paper by Hack (1960) which offered a new, non-cyclic, view of denudation in humid temperate areas, and which was later elaborated by detailed studies in the Appalachian region (Hack, 1965, 1975, 1979, 1980, 1982). The recording of process data, however, springs from innovative studies of geochemical mass balances in small watersheds by Cleaves *et al.* (1970, 1974), which were subsequently continued into further discussions of landscape evolution (Cleaves & Costa, 1979; Cleaves, 1983; Costa & Cleaves, 1984). Geochemical research has been continued by Velbel (1985, 1986), while Pavich (1985, 1986, 1989) has provided a fresh perspective on saprolite formation, by introducing the concept of 'regolith residence time' in the saprolite-covered landscape. The findings of these authors tend to be convergent, but their interpretations differ.

The recognition that saprolite is widespread and often deep in the Appalachian Piedmont is common to the work of all these authors, and its formation, along with concepts of chemical denudation are essential to the development of new ideas in landform evolution for this area. It is also found that, in a gently undulating forested landscape, the saprolite thickens beneath the interfluves (Cleaves & Costa, 1979; Pavich, 1989, and Figures 3.19 and 3.20), and this is a key observation with respect to landscape evolution. In the tropics this relationship was previously established for sites in Nigeria by Thomas (1966a, and Figures 3.18 and 3.21) and in Columbia by Feininger (1971, and Table 3.5), and it has been substantiated by many different observations.

Geological evidence in the Appalachian area indicates substantial uplift and tilting of the landsurface since the mid or late Cretaceous, and a substantial depth of erosion has been accomplished during the past 100 Ma, to account for present relationships amongst the landforms and coastal plain deposits. Hack (1979) calculated that the projection of the unconformity associated with the Fall-line Peneplain indicates that surface lowering has been at least 700 m, 50 km inland (a rate of 7 m Ma^{-1} over 100 Ma).

Pavich (1989) has observed that the classical convex divides, arising from soil creep processes under forest, as described by Gilbert (1909), appear flattened in the Piedmont landscape of low relief and gentle slopes ($< 8°$). This flattening is attributed to three main processes of groundsurface lowering:

(1) volume reduction during the compaction of saprolite into soil;
(2) loss of mass in dissolved solids draining the B soil horizon;
(3) loss of mass from the soil surface by erosion of clay.

These observations converge with those derived from studies of podzolised white sand terrains in the tropics, where lowering and flattening of interfluves are implied (Chauvel, 1977; Schnütgen & Bremer, 1985; Lucas *et al.*, 1987, 1988 and Figures 2.14 and 2.15).

These processes combine to achieve the continual lowering of the so-called 'peneplain' surface, and can also result in relief inversion, leading to river gravels perched on residual hills. If Ollier and Galloway (1990) are correct and ferricrete has formed mainly within valleys, then the appearance of duricrust-capped hills across tropical landscapes is a vivid demonstration of the principle of relief inversion. The effect is also reproduced by lava flows which formerly filled valley floors. It is difficult to see how such features could arise by parallel slope retreat, though they may subsequently become modified and destroyed by this means.

The study of base-flow stream geochemistry led Pavich (1986, 1989) to calculate a minimum rate of saprolite production of $4 \, m \, Ma^{-1}$. But, other indices of denudation increase this figure perhaps fivefold. Profile mineralogy indicates a total loss of mass during chemical alteration of saprolite to clay of as much as 75%, as indicated by ZrO and clay fraction increases. Pavich *et al.* (1985) also took cores from the saprolite and measured the ^{10}Be content. ^{10}Be is formed in the atmosphere by cosmic ray spallation reactions and then washed into the soil by rain, accumulating in the soil and saprolite over periods of 10^4–10^6 year. This provided a new rate of production and removal of

saprolite of $20 \, m \, Ma^{-1}$. Calculation of the loss of soil by surface erosion produced a figure of $5 \, m \, Ma^{-1}$, and this led Pavich (1986) to conclude that the other processes of soil reduction must accomplish the volume reduction. A lower figure of around 30% loss of volume during the breakdown of clays to form sandy mantles was advanced by Chauvel (1977).

The figure of $20 \, m \, Ma^{-1}$ for saprolite production and removal, over 70 Ma would bring about a relief reduction of around 1400 m. The importance of this work, in the present context, is its agreement with much field evidence for surface lowering by similar processes in the tropics. It implies a downward movement of both the weathering front and the soil/saprolite boundary with a net upward mass flux. The continuity of these processes is attested by the relatively constant thickness of the *soil* profiles on undulating upland surfaces, where they are less developed than on level terraces with an age of less than 1 Ma (Markewich *et al.*, 1987).

There remains some uncertainty about these figures, however, since Velbel (1985) derived a rate for the penetration of the weathering front in the southern Blue Ridge area of $38 \, m \, Ma^{-1}$ which he claimed was close to the 'average' denudation rate for the Appalachians. Cleaves (1989) also considered tilting of the basement to have been more effective in inducing erosion than faulting, and that the selectivity of later erosion points to the operation of Crickmay's (1975) hypothesis of unequal activity, rather than a simple application of Hack's (1960, 1975) concept of dynamic equilibrium.

Ahnert (1970) had come to similar conclusions in his study of relief evolution in mid-latitude drainage basins, in which he stated (p.247) that, 'for long periods of denudational surface lowering the soil thickness at any locality remains approximately constant, since the processes of weathering and of removal tend towards a state of dynamic equilibrium'. However, in the case of Appalachia, Cleaves (1989) has pointed out that, 'geochemical modelling suggests that two to four times the volume of saprolite may have been generated since late Miocene time than is currently present on the uplands.' This leads to the view, expressed frequently in the context of other areas such as Australia (Mabbutt, 1961, 1966; Ollier, 1969, 1988b), that there has been progressive stripping of the regolith over the period since the last widespread planation was achieved. This view will be discussed further below.

Many of these ideas had been foreshadowed by Nikiforoff (1959) in his 'reappraisal of the soil', and are also found in the work of Büdel (1957, 1970, 1982), in his concept of 'doppelten Einebnunsflächen' or double surfaces of levelling which he applied to the seasonal tropics and which is discussed in Chapter 9. Other versions of these ideas are to be found in the work of Aleva (1983, 1991), Johnson (1993); Kroonenberg and Melitz (1983), McFarlane (1976, 1983a), Millot (1982, 1983), Thomas (1980, 1989a,b) and Thomas and Thorpe (1985).

Evidence of groundsurface lowering in tropical landscapes

Some specific mechanisms of groundsurface lowering by combined chemical and physical processes can be identified in certain tropical environments. These include:

(1) the accumulation of Fe, Ni and other heavy minerals in weathering profiles;
(2) the accumulation of gravel layers on interfluves and upper slopes;
(3) non-abrasive lowering of valley floors by etching of bedrock and loss of fines and solutes;
(4) the ferrolysis (breakdown) of clays in hydromorphic environments with formation of white sands;
(5) the dissolution of rocks and the formation of karst hollows in crystalline terrain.

In addition, the bedrock control over topographic detail in erosional terrain can be construed as evidence for differential downwearing as against lateral planation.

Heavy mineral accumulation and ferricrete

In Chapter 3, the ideas of McFarlane (1976, 1983a) were presented as one model for the addition of Fe to the developing laterite profile. This model (Figures 4.9 and 4.10) requires the wholesale lowering of the weathering profile, as solutes and fines are removed from the surface and upper horizons of the soil, and weathering penetration advances the weathering front into the fresh rock at depth. In this way, Fe segregations form and become indurated as pisoliths in the vadose zone. In conditions of low erosional energy, these heavy concretions collect in the upper profile, as fines and solutes are removed downslope and down profile. Further Fe is added from below as the weathering front advances and this continues until a continuous skeleton of Fe_2O_3 creates a massive plinthite or duricrust.

Mineral exploration has commonly found that concentration of minerals in the weathering profile by downward movement of both the weathering front and the groundsurface is common (see Chapter 14). This is essentially the same process as ferricrete formation, and leads to the accumulation of nickel for example in some laterites. According to Esson and Surcan dos Santos (1978), 10.7 m of nickeliferous saprolite over a serpentonite at Liberdade, Brazil, represents the residue of as much as 240 m of rock weathering. The exact figure may be questionable, but the principle is important. Löffler (1978) made a similar calculation based on element loss during the weathering of dunite in Papua New Guinea, and arrived at a figure of 150 m of surface lowering to produce an observed 6 m of saprolite, in an area with many small basins and sink holes. Both these studies give a ratio of 20–25:1.

Aleva (1991) has recorded the accumulation of heavy minerals such as cassiterite (sg 7.0), monazite (sg 5.1) and zircon (sg 4.6) at the base of sandy weathering products derived from Permo-Triassic sandstones on Belitung Island, Indonesia (Figure 11.5). He claimed that this took place under gravity following elutriation of clay from the system. There is of course some movement away from interfluves towards adjacent valleys by mass movement which is recorded by stone-lines (Thomas & Thorp, 1985) and by hydromorphic dispersion, recorded by geochemical anomalies (Govett, 1987). The existence of *Gossans* (weathered rock profiles

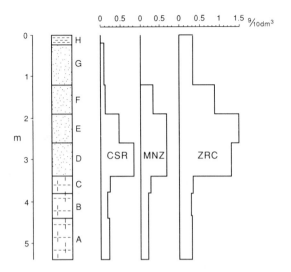

FIGURE 11.5 Section through the upper 5.5 m of the weathering mantle over a granitic basement (island of Belitung, Indonesia). Samples A, B and C represent normal kaolinitic *in situ* weathering products of granitic rocks; whitish, and retaining rock textures. Samples D, E, F and G are mainly coarse sand; a mixture of granitic weathering material without relict textures and rounded quartz grains from weathered sedimentary rock and its accessory zircons. H is humic topsoil. The three histograms show amounts of heavy minerals—cassiterite (CSR), monazite (MNZ) and zircon (ZRC)—present at different levels quantified as grammes, from a 10 m³ sample (after Aleva, 1991; reproduced by permission of Kluwer Academic Publishers)

containing suphides) also requires an hypothesis for the concentration of Fe_2O_3 in the upper profile (Chapter 4). This could occur by the McFarlane (1983a) model, but Ollier and Galloway (1990) favour lateral movement of Fe^{2+} into valley floors where it becomes indurated with time and subsequently forms ridge tops by relief inversion. However, if the iron enrichment directly overlies the sulphide ore body, this latter explanation is unlikely to apply.

Surface gravels and stone-lines

It is interesting that Hack (1960) showed the effect of chert nodules concentrated as a surface *armour* on particular limestones, and thereby inducing steeper slopes, in his concept of dynamic equilibrium.

On undulating, often multi-convex, but sometimes on more planate groundsurfaces in the tropics, colluvial layers containing gravel concentrations are commonly found. These *stone-lines* have been discussed in Chapter 8.

One hypothesis proposes the dynamic lowering of the landsurface and accumulation of resistant, mainly quartz, clasts, as weathering penetrates the subjacent rock, freeing the vein quartz from the now-decayed rock fabric. This appears to be a more or less continuous process but poses certain problems over long timescales, because of the steady accumulation of coarse clasts in the S horizon. In effect, the stone layer must be removed, to permit dynamic lowering of the landsurface to continue. Three possible mechanisms exist for this to take place:

(1) the stony material is removed by more energetic surface erosion, which would require the prior removal of the M horizon (topsoil);
(2) it is subject to soil creep processes, taking it into adjacent drainage basins;
(3) it is destroyed by *in situ* weathering processes such as *plasma infusion*.

There is some field evidence to support all three of these mechanisms. Removal by erosion probably requires the formation of low scarps and slope pediments as described in Chapter 8. Although soil creep processes are conventionally regarded as more rapid close to the surface, it was demonstrated above (Chapter 6) that, at least on steeper slopes, shearing and accelerated creep processes may take place at depth. The clasts in stone-line gravels are undoubtedly subject to aggressive weathering and there is evidence both of silica dissolution and fabric collapse, and of re-cementation and duricrust formation which effectively halts the process of surface lowering.

Sandplains

White sands have been discussed in Chapter 8. They appear to arise from the *lessivage* of saprolites formerly containing kaolinite clay which breaks down to alumina and silica in solution in

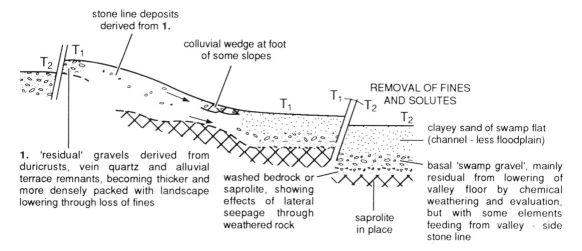

FIGURE 11.6 Continuous etching model for small valleys and interfluves, based on sections in the Koidu etchplain, Sierra Leone. The model proposes low energy lowering of interfluves and valley floors by chemical etching and removal of solutes and fines, between times T_1 and T_2. Accumulation of residual clasts as interfluve and channel gravels and as stone-lines may limit the continuation of this process, but high energy pulses have periodically evacuated coarse clasts through the system (adapted from Thomas & Thorp, 1985)

hydromorphic environments. This is the process of *ferrolysis* (Brinkman, 1970) and it probably occurs in planate landscapes that are seasonally flooded. The loss of clay must give rise to loss of volume, though this has seldom been quantified. Chauvel *et al.* (1987) and Millot (1982, 1983), however, have claimed that the breakdown of the argillo-ferric structure into a sandy material is accompanied by a 30% loss of volume and contributes to the lowering and flattening of plateau surfaces, a view also supported by Ollier (1988a), and by Schnütgen and Bremer (1985).

Etching of valley floors

A similar result comes about in small, shallow headwater streams, where bedrock is etched rather than abraded (Figure 5.13), and unrolled quartz, tourmaline and other resistant minerals accumulate in the 'channel' floor. This is a process seen in the 'bolis' of Sierra Leone, where channel flow is virtually non-existent (Thomas & Thorp, 1985). In this instance, it is hypothesised that the interfluve and valley floor become lowered together (Figure 11.6) in a process described as *dynamic etching* (see Chapter 9). A similar approach has been taken to

the interpretation of shallow valleys known as *dambos* (Chapter 8) by McFarlane and Pollard (1989).

Karst and pseudo-karst landforms

This is a diverse grouping and not all karstic landforms denote landscape lowering. It is important to make certain distinctions at this juncture, between true limestone (carbonate) karst and sandstone (silicate karst), both of which can evolve through direct dissolution of rock minerals, and 'pseudo-karst' features that commonly develop below a saprolite or ferricrete mantle. Some authors use the term 'pseudo-karst' for any non-carbonate solution features, but this fails to make the distinction drawn above, and Busche and Erbe (1987), Pouyllau (1985), and Young (1988) have clearly demonstrated that silicate solution gives rise to true karstic forms (see also Jennings, 1983; Mainguet, 1972).

In the present context of landscape lowering, the formation of closed depressions or basins is clearly relevant and includes various types of *sink, doline* and *polje* landform. In ferrallitic terrain, enclosed depressions are common on basic to ultrabasic rocks

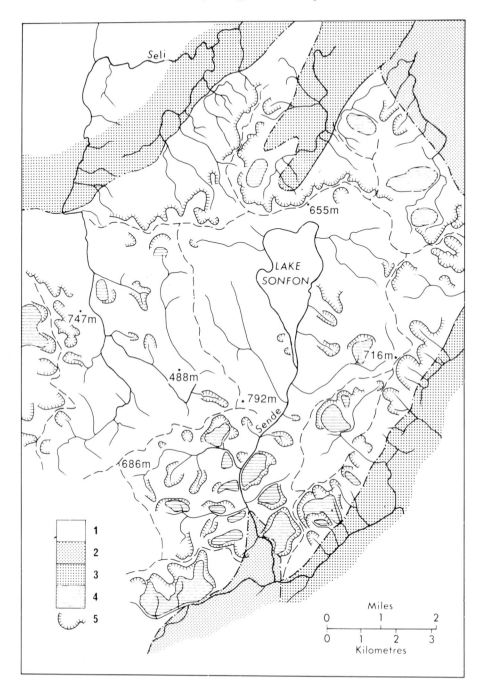

FIGURE 11.7   Geomorphic features on the lateritised landsurface of the Sula Mountains, Sierra Leone. 1, Duricrusted surface over metamorphic rocks (amphibolite, talc and chlorite schists); 2, granite basement; 3, gneiss and migmatite; 4, earthflows, commonly ferricreted; 5, landslide (slump) scars. Lake Sonfon occupies a depression 60 m deep in the plateau surface and may be due to loss of material beneath the duricrust surface. (Reproduced by permission of the Macmillan Press Ltd from Thomas, 1974a)

FIGURE 11.8   Lake Sonfon, an enclosed depression in the Sula Mountains, Sierra Leone (see Figure 11.7). It is postulated that this may be due to loss of volume in saprolite beneath a duricrust

subject to high rates of dissolution. Löffler (1978) described such depressions in a summit surface on dunite in the Bowutu Mountains of Papua New Guinea, including one quite large lake (Lake Tryst), approximately 6 × 1.5 km in size, that is probably a karst depression of the polje type, exhibiting strong structural control.

In duricrusted terrain, karst or pseudo-karst depressions are common; often they are tens or at most hundreds of metres in diameter, generally shallow and bordered by ferricrete bluffs. However, larger enclosed depressions are found, as at Lake Sonfon in the Sula Mountains of Sierra Leone (Thomas, 1974a, Figures 11.7 and 11.8), where it is possible that the lake represents an internal, solutional, collapse of the underlying saprolite. Pseudo-karst collapse features due to the mechanical removal of debris from beneath duricrust cappings have been recorded from the Kasewe Hills in Sierra Leone (Bowden, 1980), and

from the Sturt Plateau in Australia (McFarlane & Twidale, 1987), for example. They do, however, appear to be more or less discrete features, characteristic of nearly flat terrain. They contribute to the fragmentation of duricrusted surfaces and are due to the presence of substantial subterranean cavities and stream channels. The processes at work may not be dominantly solutional, but due to the mechanical work of subterranean waters (*suffosion*) (see the discussion by Ollier in McFarlane & Twidale, 1987). This view is supported by field observation of sub-laterite caves in Sierra Leone, where poorly cohesive and poorly sorted, coarse valley-fill sediments are selectively eroded beneath the groundsurface (see also Figure 8.17). The widespread survival of duricrusted surfaces at higher altitudes (relative to non-duricrusted surfaces) is, however, circumstantial evidence for the persistence of these deposits, and of their protective role in landscape formation.

FIGURE 11.9   Silicate karst hollows on crest slopes, over gneiss in southeastern Brazil

Enclosed depressions on ridge tops and plateau surfaces have also been recorded in the absence of a duricrust capping, frequently on siliceous, as well as carbonate rocks. They appear to arise from dissolution of the subjacent rock, and range in size from a few metres to more than 1 km in diameter. Some may be drained by shallow ephemeral channels, but others remain enclosed, suggesting that the products of weathering must be exported largely in solution via pipes and other macropores and through rock fractures at depth. Examples can be cited from multi-convex, saprolite-covered interfluves in southeast Brazil (Filizola & Boulet, 1993 and Figures 11.9 and 11.10), and they have been recorded on different terrains in Colombia (Feininger, 1969; Khobzi, 1972), Papua New Guinea (Löffler, 1978), Venezuela (White *et al.*, 1966) and Uganda (Brosch & Gerson, 1975).

In Uganda, these forms are common in meta-sedimentary rocks, on the summits of the Upland or African Surface (Doornkamp, 1968; McFarlane, 1976), under a sub-humid (650–950 mm year^{-1}) climate, close to the Equator. Brosch and Gerson (1975) measured 100 of these features with diameters from 50 to 350 m. Most (80%) were less than 150 m across, and situated on crests 200–500 m broad (slopes <7°). Some had a compound form and cross-sections indicate the presence of clay and organic horizons underlain by stone-lines. They were considered to have formed by solutional processes resulting from percolation of rainwater beneath gentle slopes.

The role of these features in more generalised landscape lowering is difficult to evaluate. They assist in the destruction of landsurfaces, but generally contribute to localised rather than

FIGURE 11.10   As in Figure 11.9; some of these hollows are linked downslope and resemble small dolines

generalised downwasting. Some feed into finger-tip stream heads and are thus breached and incorporated into the slope hydrology–stream system of surface denudation.

Small basinal rock forms are widespread in planate granitoid landscapes and probably reflect the shape of the weathering front. They become exposed by stripping, possibly due to wind erosion, as on the Monaro of New South Wales (Thomas, 1974b), or by narrow surface channel exits (Figure 5.11); also by subterranean water flow. Comparable shallow basins, really weathering pits and pans, are also found on sandstones. These are all etch forms which are strongly influenced by fractures and other structures in the rocks. When subterranean weathering patterns in granite (Thomas, 1966a, Figure 3.16) are compared with small, sub-aerial

basins in the same rock (Thorp, 1967a,b), it is clear that these respond to structure in a negative sense in the same manner as granite inselbergs reflect positive structural and mineralogical controls.

When it is realised that these solutional features on crystalline rocks are found equally on interfluves and in valley floors, their significance to landscape reduction by lowering is strongly suggested.

Bedrock control over topography

There are disagreements in the literature concerning the extent to which topographic features are controlled in detail by lithology or structure, but many examples occur where this is so. There is sometimes an attempt to use small-scale geological maps to indicate lack of any bedrock change at

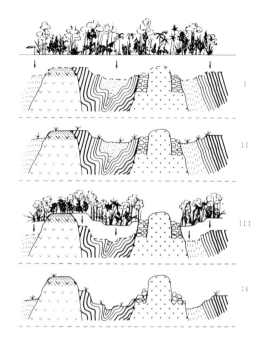

FIGURE 11.11   Schematic model of the development of
summit levels by differential etching and stripping. Based
on conditions in Surinam and commencing in the early
Tertiary (after Kroonenberg & Melitz, 1983; reproduced
by permission of Kluwer Academic Publishers)

topographic boundaries. But, in the field, changes
in bedrock composition and fabric can be quite
subtle and they may not be recognised in hand
specimens alone; frequently they occur at scales
which would prohibit the mapping of boundaries
on available maps.

Some geological boundaries are marked by major
escarpments or hillslopes, rising more than 300 m
from adjacent lower ground. The south-facing
scarps of the Jos Plateau, thus, often closely
correspond with the petrographic boundaries of the
Older Granites. Similarly, the Younger Granite ring
complexes which intrude the basement in this area,
provide striking examples of rock control over
relief, and particularly over the formation of larger
basinal landforms (0.5–>3 km diameter) which
Thorp (1967a) found to be determined by inliers of
Basement Complex rocks or by the configuration
of the ring structures and their dyke intrusions.
Studies of granite terrain have tended to give more
attention to the positive landforms than to negative

areas (Twidale, 1982), though granite embayments
and basins are very widespread features in all
climatic zones and are strikingly developed in the
Scottish Highlands, where differential pre-glacial
weathering is a major reason for their survival in
the present landscape (Linton, 1951; Hall, 1987,
1991).

In Koidu, Sierra Leone, the duricrusted Sula
Schist Belt attains a relief of 350 m above the
basement plateau and nearly everywhere the major
topographic break marks the outcrop. There are
innumerable examples of these relationships on
Precambrian basement terrain, and such a situation
was analysed by Kroonenberg and Melitz (1983, and
Figure 11.11) for Surinam. Moreover, Hack (1979)
also established similar relationships for parts of the
Appalachian Mountains. In an interesting study of
pediment evolution in the Tibesti Mountains,
Busche (1976, p.21) concluded that, 'in the process
of downwearing differences in rock resistance to
weathering become increasingly more important,
until the extent of the youngest pediments was
almost perfectly governed by the different
erodibility of schist and sandstone.'

The lowering of landscapes by low energy
chemical and mechanical transfers of material is
accomplished by the action of water flow within the
weathering profile, as soil throughflow, and as
surface and subsurface flows in small valleys, and
even on pediments (Dohrenwend *et al.*, 1987).

However, although vertical lowering and bedrock
control have been stressed in this section, lateral
movement is self-evident in most landscapes, and
substantial space has been devoted to accounts of
sediment transfer on slopes and the accumulation
of colluvial layers with the formation of stone-lines.
None of this is negated by the foregoing account,
but the different strands of argument still need to
be brought together.

THE BEHAVIOUR OF SLOPES IN THE TROPICAL LANDSCAPE

Arguments about vertical lowering of landscape
cannot in themselves be used to construe a
particular mode of slope development. However,
if Hack's (1960) views of dynamic equilibrium in

the landscape apply widely, then any given slope will have a constant value during periods for which equilibrium is maintained. But if slopes become armoured by coarse clasts that accumulate as a lag deposit on the surface, then they may steepen as a consequence.

On the other hand, geomorphologists have been concerned to account for the formation of erosional plains, variously described as peneplains, pediplains, panplains or etchplains. From the time of Davis (1899a,b) the condition of flatness in the landscape has been associated with the development of a thick, clay-rich regolith or saprolite. The steepest slopes (vertical or overhanging) clearly occur in areas of bare rock, most obviously in granitoid or carbonate (limestone, dolomite) terrain, but also where a quasi-horizontal cap rock occurs, and is undermined at its base.

Very tall rock landforms must eventually become subject to forces that lead to spalling and collapse, and rockfalls probably constitute the major process of slope failure on bare rock cliffs, as around the margins of some inselbergs (Thomas, 1965a). However, Terzaghi (1962) established that few rock cliffs attain their *critical height* according to calculations of mechanical strength, using his formula:

$$Hc = Sc/W \qquad (11.1)$$

where Hc is the critical height at which failure will occur, Sc is the compressive strength of the rock, and W is the unit weight of the rock. Typical calculations indicate that for many rocks such as granite, vertical cliffs of >1000 m can stand without collapse. Such calculations neglect the presence of rock fractures, however, and it is the frequency and distribution of these through the rock mass that leads to premature collapse. Certainly, very high, unbroken rock faces occur on deep-seated Precambrian granite domes (inselbergs and bornhardts). Some of these may exceed 600 m, as at Idanre in southern Nigeria or in the Serra do Mar mountains of eastern Brazil, and although some spalling (or sheeting) occurs, the absence of major talus accumulation often indicates that it is a slow process. King (1962) thought that such rock faces

would undergo parallel slope retreat, but this model probably does not apply in all cases.

Slope retreat concepts

Studies of slope evolution in bedded sedimentary rocks in the American southwest have generally agreed that slope retreat takes place by rockfalls mainly along pre-existing fracture planes. However, the continuation of this process will depend on the weathering and removal of the coarse talus that accumulates against the rock slope (Koons, 1955; Bradley, 1963; Schumm & Chorley, 1966). Chorley *et al.* (1984) illustrated the retreat of cliffs under the influence of different types of cap rock in an arid climate.

If uniform bedrock occurs beneath the entire slope, the manner of slope development will reflect the dominant process. Where this is soil creep within a colluvial layer, Gilbert (1909) showed that convexity would develop and Schumm (1956) was able to demonstrate that slopes will decline in steepness with time, but simultaneously undergo a degree of retreat. Under rainwash which was dominant in another case, parallel slope retreat occurred. This long-held distinction between slopes modelled by mass movement and those subject to water flow appears to remain valid for both field studies and for mathematical models (Kirkby, 1971; Ahnert, 1976). King (1962) also held that the retreat of slopes was largely induced by water flow and cited the advance of gullyheads as the main process bringing about the retreat of major escarpments such as the Drakensberg.

Of these writers, most have chosen to work in semi-arid environments, and many of them in temperate environments; studies of cliff retreat in humid tropical conditions are hard to find. There is little reason to suppose that the principles will vary according to climate, though the balance of process will. It could be argued that talus weathering will advance more rapidly in the tropical climate, thus permitting slope retreat to occur more quickly. On the other hand, rock faces may emerge from beneath a regolith mantle less often and only under the most extreme conditions of uplift, faulting and dissection. When all factors have been considered, it seems probable that lithology and structure act as the most important passive factors, while depth and rate of dissection along stream channels will often determine where rock faces will emerge. Wahrhaftig (1965) has suggested that random events led to exposure of bedrock steps during uplift of the front ranges of the American western cordillera, and his model also encompasses the deep weathering of the step treads.

In humid tropical areas, however, it has already been established that mass movement frequently dominates the erosional system, with debris flows and shallow translational slides occurring widely after prolonged, intense rain, and deep-seated slump-earthflow movements taking place more slowly, where deep saprolite occurs and has been dissected to form slopes typically between 28° and 38°. This process will continue, so long as the landslide debris is removed from the base of the slope by weathering and erosion of fines by running water or other means.

The role of active weathering at the base of slopes has been the subject of some discussion, ranging from the preferential decay of rocks close to the hillslope/piedmont junction (piedmont angle), to the penetration of weathering into the lower hillslope itself. The accumulation of talus undoubtedly promotes subsurface water flow and the rotting of rock below the talus wedge (Figure 10.11), and this may be an important contributory process leading to scarp retreat. Examples of this relationship are widely found in granite terrain, and in South Africa, where the Table Mountain Sandstone overlies basement granites (Thomas, 1978, and Figure 11.6).

Ruxton and Berry (1961) advanced models for hillslope–piedmont weathering and slope retreat under semi-arid and humid conditions, based on studies in Sudan and Hong Kong respectively (Figure 11.13). The situation in the semi-arid environments of the Sudan was based on a study of the behaviour of coarse regolith under occasional heavy storms (Ruxton, 1958). These produced strong throughflow within the coarse hillslope debris, leading to leaching and to eluviation of clay which was deposited on the upper pediments and then later washed across the gentle piedmont slope (Figure 11.14).

In more humid areas, thick weathering profiles are found below major hillslopes (Figures 11.13A and 3.10), their development being favoured by the water entering the lower hillslope zone from above. According to Ruxton and Berry (1961), this situation will favour the penetration of weathering laterally into the lower hillslope, promoting rock falls from the upper slope and retreat of the entire hillslope. The author's study of field situations in the dissected terrain of Natal (Thomas, 1978) indicated that the hillfoot zone displays an alternation of rock buttresses and weathered talus slopes, such that rock domes could be isolated along the scarp front as the hillslope retreated (Chapter 10, Figure 10.20).

Hillslopes and rock pediments

Most writers who have examined granite pediments, allude to the occurrence of a weathered mantle, and often to the role of alternate mantling and stripping of the pediments which must lower the pediment surface (Mabbutt, 1966; Busche, 1976; Mabbutt, 1978; Whitaker, 1978; Oberlander, 1972, 1974, 1989). In humid areas, pediment landforms are not consistently developed in the context of major hillslopes; very often the deeply weathered lower landscape has been dissected into the multi-convex relief that has been discussed in other contexts. This was fully recognised by Ruxton and Berry (1957, 1961), and is a feature of the faulted margins of eastern Brazil.

In seasonal climates, moisture is retained longer towards the base of hillslopes and under the upper pediment; this situation is promoted by accumulations of weathered talus beneath a rock face. This, in turn, creates conditions for the preferential rotting of the rock beneath the mantle of the lower hillslope, and in a similar manner weathering penetration will be favoured beneath the upper pediment (or other piedmont surface). Once prepared by weathering in this manner, dissection and removal of regolith from the hillfoot and upper piedmont zones will lead to the steepening of the hillslope, promoting further rockfalls or landslides.

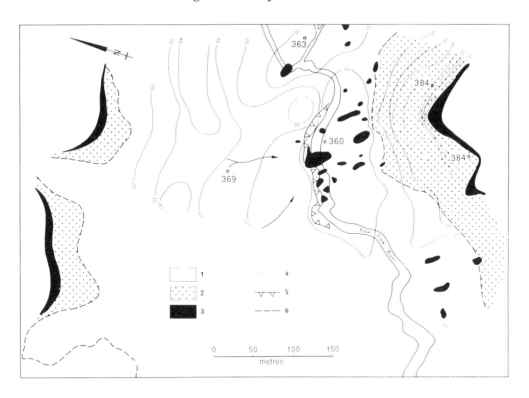

FIGURE 11.12 Sub-talus weathering in Precambrian granite, near Shaki, Nigeria. 1, Soil cover; 2, talus and other boulder-controlled slopes; 3, outcrops of fresh rock; 4, contours of weathering depth at 3 m intervals; 5, convex breaks of slope; 6, slope break at the foot of the talus slope. Compiled from geophysical survey, air photo interpretation and field mapping. (Reproduced by permission of the Macmillan Press Ltd from Thomas, 1974a)

Such a sequence can clearly lead to slope retreat in crystalline rocks, but it depends on the retention of talus against the lower hillslope, and the continuous or episodic removal of the regolith. In many highly seasonal climates only rather small amounts of regolith occur against rock hillslopes, while in humid, forested areas a regolith mantle may clothe the slope from top to bottom, and the frequent rainfall supplies moisture to the entire slope for long periods of the year. It is also clear that some rock hillslopes have been exposed by the removal of former saprolitic mantles (Mabbutt, 1961, 1966, 1978; Thomas, 1974a,b), so that both the gross form and some detailed weathering features may have been modelled within this sub-soil environment.

The base of a slope may also be lowered more or less continuously, or episodically, and this changes the locus of footpoint weathering, reducing its effectiveness as an agent of slope retreat at any one level. According to Twidale (1980, 1981a,b), only narrow fringing pediments (< 100 m wide), flared slopes and minor etch forms such as *tafoni*, have resulted from lateral weathering attack on massive granites and sandstones in Australia, and similar conclusions have been reached by Büdel (1957, 1982).

Both Mabbutt (1978) and Whitaker (1978) addressed these problems in the context of semi-arid environments in Australia, and appear to concur that a composite theory is required to account for the complexity of forms and deposits on piedmont slopes. In the East Kimberleys of Australia, Whitaker (1978) thought that prior development by erosion and stripping of weathered compartments had produced a field of boulder-controlled hills and

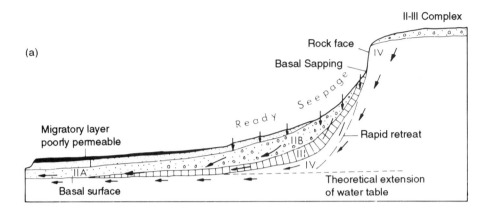

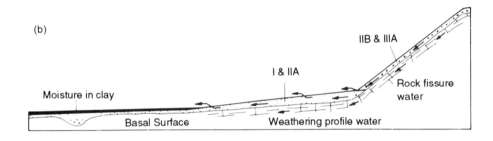

FIGURE 11.13   Weathering processes at the piedmont angle in humid (a) tropical and (b) semi-arid regions. Based on studies from Hong Kong and Sudan by Ruxton and Berry (1961)

inselbergs. Pediments had developed around these and had been extended to some degree by hillslope weathering and retreat. The dissection and retreat was more profound than that implied by the *slope pedimentation* model of Rohdenburg (1969/1970) which corresponds with the notion of *glacis* formation discussed in Chapter 8. The forms described are rock-cut (valley bottom) pediments.

Ahnert (1982) has analysed hillslopes and pediments in connection with inselbergs in the Machakos region of Kenya, and concluded that the hills developed by slope retreat and extension of true pediments, but at the same time he established for 60 hill–pediment complexes, 'a highly significant relationship between the total height of inselberg plus pediment and the width of the pediment' (p.2). The pediments in this area do not carry a deeply weathered regolith, and Ahnert (1982) concluded that they were not shaped by the double planation

process today. He deduced from their relationships with surrounding Miocene volcanic rocks that they had formed by dissection into a pre-Miocene or early Miocene surface (deep weathering was probably present beneath this surface). Within their current phase of development within a dry climate, retreat was estimated at the rate of $100–200 \text{ mm ka}^{-1}$, with downwearing of inselberg summits taking place at 50 mm ka^{-1}. This rate of retreat, at 200 m Ma^{-1}, could well be overtaken by periods of uplift and drainage incision before extensive planation had been achieved.

Oberlander (1989) takes the view that slope retreat by the extension of pediments is not characteristic of the present arid zone climate, but will take place in a sub-humid, open savanna environment, and depends critically on the presence and preservation of a saprolite mantle which may be quite thin. This view is almost identical to that of Büdel (1982) concerning such environments. The radiometric ages of basalts on the upper pediments in the Mojave desert led Oberlander (1989) to calculate a rate of slope retreat of $20\,m\,Ma^{-1}$ ($0.02\,mm\,year^{-1}$). This is very slow and an order of magnitude less than that calculated for the Machakos Hills by Ahnert (1982); possibly two orders of magnitude less than some higher estimates for scarp retreat in layered sedimentary rocks in similar environments. This figure is identical with that advanced by Pavich (1985) for the rate of groundsurface lowering for the Appalachian piedmont. Oberlander (1989) quoted figures of $0.5–6.7\,mm\,year^{-1}$ for the Mesa Verde sandstone in northern Arizona (after Schmidt, 1989), but many other measurements range down to $0.16\,mm\,year^{-1}$. Except where geological evidence is confirmatory, the extrapolation of figures based on short-term measurements is probably hazardous, because of the many factors that can influence erosion rates over longer time periods.

Mabbutt (1978) has, however, stressed two unhelpful constraints commonly accepted when discussing pediment landforms; one being the insistence on a linkage between pediment formation and hillslope retreat, and the other being the assumption that some common set of headward extending and re-grading processes was involved. In fact, pediments, 'are erosional footslopes graded to the requirements of water flow and sediment transport from a hill foot' (Mabbutt, 1978, p.331), and these requirements may vary from instance to instance.

Some other pertinent questions are often avoided. One of these must be the greater erosional energy exerted on steeper slopes, where mass movement operates as well as creep and wash processes. Most pediment studies have taken place in semi-arid areas (temperate as well as tropical) and lie outside the scope of this discussion, but slope retreat in humid tropical areas must take place as landslides recur on hillslopes. The difference is that footslopes are frequently deeply weathered (Figures 11.12 and 11.13) and strong and effective sheetflow processes do not take place under the natural vegetation of humid and sub-humid areas. There is therefore no direct link between hillslope retreat and pediment formation. It has been observed, however, that periods of drier climate have led to the formation of slope pediments or glacis which truncate the saprolite and some weaker rock materials.

In summary, the reality of slope retreat is seen in the landscape and also can be predicted from the operation of surface water flows and mass movement. According to King (1957), once fully developed, the hillslope profile will be maintained by parallel slope retreat, and, as the hillslope retreats away from the lines of channel erosion, the process will become independent of stream incision. Some of the limitations on this model are quite restrictive, and include the following.

(1) On 'soft' rocks such as shales and also saprolites there is no rock face and the dominant process is usually soil creep which leads to declining slope values, unless episodic landsliding occurs.
(2) In humid areas with high stream frequencies, opposing slopes may intersect along sharp ridge crests, leading to the lowering of the entire landscape (assuming stream incision) in dynamic equilibrium.
(3) The decay of cliffed slopes may lead to the build-up of the talus slope until the entire profile is mantled by regolith.
(4) Long-maintained stable base levels, associated with mantled landscapes and soft rocks, can lead to progressive flattening of the landscape (etchplanation).
(5) In contrast, steepening of slopes will occur if stream erosion is renewed (rejuvenation).
(6) As dissection lowers the base level of hillslope erosion, it must continually shift the locus of basal attack on the slope, so that, over long time periods, the hillslope becomes notched at different levels, while its base may continue to lie close to some controlling structure or outcrop.
(7) Strong lithological control over many residual hills indicates that they withstand planation processes, whether vertical or lateral, and actually 'grow' in relative relief above the surrounding surface as this is lowered.

Beyond these points, it is worth remembering that slope evolution on the timescale of the

10^6–10^8 years required for planation must also embrace the climatic oscillations that lead to complex hillslope stratigraphies (explored in Chapter 8) on a timescale of 10^3–10^4 years, and commonly periods of uplift, tilting and dissection which must redirect erosional energy towards vertical incision and lowering of sharp divides.

CONCLUDING REMARKS

Studies of the longer timescales of landscape evolution are, today, often considered less important in geomorphology than those that deal with process measurement over experimental periods or with records of accurately dated landforming events, usually of the later Quaternary. But this can be a false dichotomy, for the landscape is a mosaic of forms and materials of different origins and ages of formation, and is also a palimpsest, in which each component has been overprinted, where it has not been destroyed, by the events that have have occurred since its formation (Starkel, 1987).

Schumm (1985) has argued that consideration of vast spans of time and large areas of the earth's surface set geology and geomorphology apart from the other physical sciences (except for astronomy and cosmology), and that to encompass this canvas the geomorphologist should experience as many conditions as possible and must use analogy as the basis for explanation. Such views are not echoed by other polemicists such as Yatsu (1992), but they underlie much of what has been written in this book.

Gupta (1993) has criticised what he calls 'traditional tropical geomorphology' for its preoccupation with deep weathering beneath stable plains, and for its adherence to the principles of climato-genetic geomorphology as outlined by Büdel (1977, 1982). More important, Gupta (1993) regrets the way in which tropical geomorphology has been seen as a separate entity, isolated from the mainstream of geomorphological research. This is certainly an important issue, but it springs as much from the way in which geomorphologists have sought normative models for the behaviour of

geomorphic systems from the perspective of temperate and high latitudes, as from the (many) deficiencies in our understanding or approach to geomorphology in the tropics.

But what is the 'mainstream' of geomorphology? Baker and Twidale (1991) have criticised many of the ideological bandwagons of recent decades and called both for greater creativity and for a closer link to the inspiration of field conditions. This study has tried to avoid isolating geomorphology in the tropics from landform studies in other areas, and findings from the temperate zone have been freely incorporated. It is a salutary thought that in studies of saprolitic landscapes, new ideas of common interest have come from detailed studies of the southern Appalachian region. The extension of warm moist conditions into such temperate areas was signposted at the outset, and it is clear that there should be no climatic boundary to the study of weathering and landform development.

In the reassessment of earlier concepts, such as that of etchplanation, it would be unwise to dismiss their premises or their arguments because they were worked out on cratonic landscapes or came with climato-genetic 'baggage'. In the humid tropics, especially, the role of biogeochemical processes in denudation is proving to be more, not less, important than we thought, and the understanding of deep saprolitic weathering remains central to the concerns of engineers. The nature and importance of chemical denudation remains inadequately researched and our understanding of this term in most landforming systems is negligible.

The fact that much of the inter-tropical zone coincides with the ancient Gondwana cratons has undoubtedly biased the study of geomorphology in the tropics. Wirthmann (1981, 1985) has called attention to the special characteristics of the Gondwana continents. In such areas, Brunsden (1980) has argued for form convergence irrespective of climate and climate history, while Ruxton (1967) claimed that differentiation of forms according to climate was opposed by uplift and deep dissection. If both these contentions are correct, this leaves little room for the appearance of climato-genetic forms. However, it can be argued that, in tectonically quiescent terranes, the landscape as a whole becomes more complex as it evolves. The particular complexities of tropical and sub-tropical landscapes commonly centre on the nature of the superficial deposits, which have received considerable emphasis in this study because they are both thick and extensive and they control much of the sediment delivered to the oceans.

A tropical perspective in geomorphology therefore views the evolution of the weathered mantle through time as one central theme. From this mantle are derived nearly all the sediments deposited or in transport, many mineral resources and construction materials, and it is a source of fragility in many tropical landscapes. Sustainable use of tropical lands therefore requires a full understanding of these materials and of the landforms with which they are associated.

The uncertainties that surround explanation and extrapolation (prediction) in geomorphology have been identified by Schumm (1985) as including scale, location, convergence, divergence, singularity, sensitivity and complexity. To tackle these problems, Schumm (1985, p. 15) urged that: 'perhaps the most important consideration is that landforms have a history. They change through time, and they respond to climatic fluctuations and tectonic influences. Therefore any historical information on prior behaviour is of value'. This information has to be added to the studies of processes and properties of materials, seen by Yatsu (1992) as the 'main stream' of geomorphology, if it is to become possible to offer some limited predictions, in space as well as time, concerning the future behaviour of the earth's surface and its sensitive mantle of superficial deposits.

Scale problems in geomorphology underlie many apparent contradictions amongst hypotheses of landform origin. For example, views about regional planation often appear to conflict with studies of lithological and structural control over landscape morphology, but it should be possible to contain both within a broader frame of reasoning about uplift, scarp retreat and differential erosion. Similarly, arguments about pediments need to be separated from debates about pediplanation and, as the latter term appears to imply a particular mode of landscape evolution involving pediment

extension, it should be used very sparingly, if at all. Brunsden (1993) has recently returned to these themes to examine the persistence of landforms, a problem that exercised Twidale (1976) and also the present author (Thomas, 1980) in relation to the relief of part of the west African craton.

One conclusion put forward here is that the last 100 My (very approximately) have seen the progressive differentiation of relief and the partial destruction of many ancient landforms, at least on the passive margins of continents. Planation within this time frame has been local and controlled geologically. On the other hand, in tectonically mobile areas, this period has seen the construction of new landforms with strong tectonic components. But, even on the passive margins of continents, domal and plateau uplifts have re-activated the weathering systems, leading to syngenetic bauxites and the deepening of weathering profiles, except where rapid stripping of unprotected saprolite has produced stripped etchsurfaces.

A second conclusion might be that the climate changes of the last 5–6 My, embracing cooling and aridification during the Pliocene as well as the Pleistocene, have been responsible for many of the mesoscale erosional landforms and sedimentary covers into which the smaller scale features produced by the Quaternary oscillations of climate have become inset. The evolution of this final patchwork has been accompanied by the evolution of *Homo sapiens*. We have yet to learn in detail how the natural systems interacted with early cultural developments in the tropics, but the study of this relationship will provide important clues concerning the stability of landsurfaces, in the future as well as the past.

The poor data base from which any generalisations about tropical geomorphology must be projected is certainly a restriction (Ollier, 1983a; Gupta, 1993). But the starting point for any future departures in geomorphology must be a full appraisal of where we are now. This study has attempted to make a modest contribution to this appraisal, specifically for tropical environments.

References

Ab'Saber, A. N. (1957). Conhecimentos sôbre as fluctuaçoes climaticas do Quaternario no Brasil. *Boletim do S. B. G.*, **6**, 41–48, São Paulo, Brazil.

Ab'Saber, A. N. (1982). The paleoclimate and paleoecology of Brazilian Amazonia. In *Biological Diversification in the Tropics* (ed. G. T. Prance), pp. 1–59, Columbia, New York.

Ab'Saber, A. N. (1988). A Serra do Mar na Regiao de Cubatao: avalanches de Janeiro de 1985. In *A Ruptura do Equilibrio Ecologico Na Serra De Paranapiacaba E A Poliçao Industrial* (ed. A. N. Ab'Saber), 74–116, Brazil.

Ab'Saber, A. N. (1989a). Palaeoecological variations with altitude in Brazil. In *Campos De Altitude* (ed. J. Bandeira), pp. 22–37. Editora:Index, São Paulo, Brazil.

Ab'Saber, A. N. (1989b). Paleo-climas Quaternarios e Pré-Historia da América tropical. *Dédalo*, **1**, 9–25, São Paulo.

Absy, M. L. (1982). Quaternary palynological studies in the Amazon Basin. In *Biological Diversification in the Tropics* (ed. G. T. Prance), pp. 67–73. Columbia, New York.

Absy, M. L., Van Der Hammen, T., Soubiès, F., Suguio, K., Martin, L., Fournier, M. & Turcq, B. (1989). Data on the history of vegetation and climate in Carajas, eastern Amazonia. In *International Symposium on Global Changes in South America During the Quaternary*, São Paulo, 1989, Special Publication, Vol. 1, pp. 129–131, São Paulo, Brazil.

Absy, M. L., Cleef, A., Fournier, M., Martin, L., Servant, M., Siffedine, A., Ferreira da Silva, M., Soubiès, F., Suguio, K., Turcq, B. & Van Der Hammen, T. (1991). Mise en évidence de quatre phases d'ouverture de la forêt dense dans le sud-est de l'Amazonie au cours des 60 000 dernières années. Première comparaison avec d'autres régions tropicales. *Comptes Rendus Academie des Sciences, Paris, Série II*, **312**, 673–678.

Ackermann, E. (1936). Dambos in Nordrhodesien. *Wissenschaft Veröff. Dt. Mus. Länderkde*, Leipzig, N.F., **4**, 147–157.

Acworth, R. I. (1987). The development of crystalline basement aquifers in a tropical environment. *Quarterly Journal of Engineering Geology*, **20**, 265–272.

Adam, D. P., Byrne, R. & Luther, E. (1981). A late Pleistocene and Holocene pollen record from laguna de las Tancas, northern coastal Santa Cruz County, California. *Madrono*, **28**, 255–272.

Adams, G. (ed.) (1975). *Planation Surfaces*. Benchmark Papers in Geology, vol. 22, Dowden, Hutchinson and Ross, Pennsylvania.

Adamson, D. A., Gasse, F., Street, F. A. & Williams, M. A. J. (1980). Late Quaternary history of the Nile. *Nature*, **287**, 50–55.

Afifi, A. A. & Bricker, O. P. (1983). Weathering reactions, water chemistry and denudation rates in drainage basins of different bedrock types: 1—sandstone and shale. In *Dissolved Loads of Rivers and Surface Water Quantity/Quality Relationships*, pp. 193–203. International Association of Hydrological Sciences Publication No. 141.

Agassiz, L. (1865). Journal in Brazil. *American Journal of Science, 2nd Series*, **XI**, 390.

Ahnert, F. (1970). Functional relationships between denudation relief and uplift in large, mid-latitude drainage basins. *American Journal of Science*, **268**, 243–263.

Ahnert, F. (1976). A brief description of a comprehensive three dimensional process-response model of landform development. *Zeitschrift für Geomorphologie, N.F., Supplementband*, **25**, 29–49.

Ahnert, F. (1982). Investigations on the morphoclimate and on the morphology of the inselberg region of Machakos, Kenia. *Beiträge zur Geomorphologie der Tropen (Ostafrika, Brasilien, Zentral- und Westafrika), Catena Supplement*, **2**, 1–72.

Ahnert, F. (1987). An approach to the identification of morphoclimates. In *International Geomorphology*

1986, Proc. 1st International Conference on Geomorphology, Part II (ed. V. G. Gardiner), pp. 159–188. Wiley, Chichester.

Alderton, D. H. M. & Rankin, A. H. (1983). The character and evolution of hydrothermal fluids associated with the kaolinised St. Austell granite, SW England. *Journal of the Geological Society of London*, **140**, 297–309.

Aleva, G. J. J. (1973). Aspects of the historical and physical geology of the Sunda Shelf essential to the exploration of submarine tin placers. *Geologie en Mijnbouw*, **52**, 79–91.

Aleva, G. J. J. (1979). Bauxite and other duricrusts in Suriname: a review. *Geologie en Mijnbouw*, **58**, 321–336.

Aleva, G. J. J. (1981). Bauxites and other duricrusts on the Guiana Shield, South America. In *Lateritisation Processes*, Proceedings International Seminar on Lateritisation Processes, Trivandrum, 1979, pp. 261–269. Balkema, Rotterdam.

Aleva, G. J. J. (1983). On weathering and denudation of humid tropical interfluves and their triple planation surfaces. *Geologie en Mijnbouw*, **62**, 383–388.

Aleva, G. J. J. (1984). Lateritisation, bauxitisation and cyclic landscape development in the Guiana Shield. In *Bauxite*, Proceedings 1984 Symposium, Los Angeles (ed. L. Jacob Jr.), pp. 111–151. Society of Mining Engineers, American Institute of Mining, Metallurgical & Petroleum Engineers, New York.

Aleva, G. J. J. (1985). Indonesian fluvial cassiterite placers and their genetic environments. *Journal of the Geological Society, London*, **142**, 815–836.

Aleva, G. J. J. (1991). Tropical weathering, denudation and mineral accumulation. *Geologie en Mijnbouw*, **70**, 35–38.

Alexander, L. T. & Cady, J. G. (1962). *Genesis and Hardening of Laterite in Soils*. Technical Bulletin, No. 1282, pp. 90, US Department of Agriculture, Soil Conservation Service.

Alexandre J. & Symoens, J. J. (eds.) (1989). Stone-Lines. Journée d'Étude, Bruxelles, 24 mars 1987. *Académie Royale des Sciences D'Outre-Mer, Geo-Eco-Trop*, **11**, 239.

Alger, C. S. & Ellen, S. D. (1987). Zero order basins shaped by debris flows, Sunol, California, USA. In *Erosion and Sedimentation in the Pacific Rim* (ed. R. L. Beschta, T. Blinn, C. E. Grant, F. J. Swanson & G. G. Ice), pp. 111–119. IAHS Publication No. 165.

Allison, R. J. (1991). Slopes and slope processes. *Progress in Physical Geography*, **15**, 423–437.

Aloni, J. (1975). Le sol et l'évolution morphologiques des termitières géantes du Haut-Shaba (Rép. Zaire). *Pedologie*, **25**, 25–39.

Alpers, C. N. & Brimhall, G. H. (1988). Middle Miocene climatic change in the Atacama Desert, Northern Chile: evidence from supergene mineralisation at La Escondida. *Geological Society of America Bulletin*, **100**, 1640–1656.

Alpers, C. N. & Brimhall, G. H. (1989). Paleohydrologic evolution and geochemical dynamics of cumulative supergene metal enrichment at La Escondida, Atacama Desert, northern Chile. *Economic Geology*, **84**, 229–255.

Amajor, L. C. (1989). Grain size characteristics and geologic controls on the bedload sediments: a case study from the Imo River in southeastern Nigeria. *Journal of African Earth Sciences*, **9**, 507–515.

Anderson, M. G. & Richards, K. S. (1987). *Slope Stability*. Wiley, Chichester.

Andriesse, J. (1968). A study of the environment and characteristics of tropical podzols in Sarawak. *Geoderma*, **2**, 201–227.

Andriesse, J. (1970). The development of podzol morphology in the tropical lowlands of Sarawak. *Geoderma*, **3**, 261–279.

Arakel, A. V., Jacvobson, G., Salehi, M. & Hill, C. M. (1989). Silification of calcrete in palaeodrainage basins of the Australian arid zone. *Australian Journal of Earth Sciences*, **36**, 73–89.

Archer, D. R. & Mäckel, R. (1973). Calcrete deposits on the Lusaka Plateau of Zambia. *Catena*, **1**, 19–30.

Atkinson, T. C. & Smith, D. I. (1975). The erosion of limestones. In *The Science of Speleology* (ed. T. D. Ford & C. H. D. Cullingford), pp. 151–177. Academic Press, London.

Aubreville, A. (1962). Savanisation tropicale et glaciations quaternaires. *Adansonia*, N.S., **2**, 16–84.

Avenard, J.-M. (1973). Évolution géomorphologique au quaternaire dans le centre-ouest de la Côte D'Ivoire. *Revue Géomorphologie Dynamique*, **22**, 145–160.

Avenard, J.-M. & Michel, P. (1985). Aspects of present-day processes in the seasonally wet tropics of West Africa. In *Environmental Change and Tropical Geomorphology* (ed. I. Douglas & T. Spencer), pp. 75–92. Allen and Unwin, London.

Ayoade, J. O. (1976). A preliminary study of the magnitude, frequency and distribution of intense rainfall in Nigeria. *Hydrological Sciences Bulletin*, **21**, 419–429.

Baes, C. F., Jr. & Mesmer, R. E. (1976). *The Hydrolysis of Cations*. Wiley, New York.

Baillie, I. C. (1975). Piping as an erosional process in the uplands of Sarawak. *Journal of Tropical Geography*, **41**, 9–15.

Baker, V. R. (1978). Adjustment of fluvial systems to climate and source terrain in tropical and sub-tropical environments. In *Fluvial Sedimentology* (ed. A. D. Miall), pp. 211–230. Canadian Society of Petroleum Geologists, Memoir 5.

Baker, V. R. (1983a). Late-Pleistocene fluvial systems. In *Late Quaternary environments of the United States Vol. I, The Late Pleistocene* (ed. H. E. Wright), pp. 115–129. University Minnesota Press, Minneapolis.

Baker, V. R. (1983b). Large-scale fluvial palaeohydrology. In *Background to Palaeohydrology* (ed. K. J. Gregory), pp. 453–478. Wiley, Chichester.

Baker, V. R. (1988). Flood geomorphology and palaeohydrology of bedrock rivers. In *Geomorphological Studies in Southern Africa* (ed. G. F. Dardis & B. P. Moon), pp. 473–486. Balkema, Rotterdam.

Baker, V. R. & Twidale, C. R. (1991). The reenchantment of geomorphology. *Geomorphology*, **4**, 73–100.

Baker, W. E. (1978). The role of humic acid in the transport of gold. *Geochimica et Cosmochimica Acta*, **42**, 645–649.

Bakker, J. P. (1960). Some observations in connection with recent Dutch investigations about granite weathering in different climates and climatic changes. *Zeitschrift für Geomorphologie, N.F., Supplementband*, **1**, 69–92.

Bakker, J. P. (1965). A forgotten factor in the development of glacial stairways. *Zeitschrift für Geomorphologie, N.F.*, **9**, 18–34.

Bakker, J. P. (1967). Weathering of granites in different climates. In *L'Evolution des Versants* (ed. P. Macar). *Congrès et Colloque Université Liége*, **40**, 51–68.

Bakker, J. P. & Levelt, Th. W. M. (1964). An inquiry into the probability of a polyclimatic development of peneplains and pediments (etchplains) in Europe during the Senonian and Tertiary Period. *Publication Service Carte Géologique, Luxembourg*, **14**, 27–75.

Balazs, D. (1973). Relief types of tropical karst areas. In *Symposium on Karst-morphogenesis* (ed. L. Jacucs), pp. 16–32. Attila Joseph University, Szeged.

Balek, J. (1977). Hydrology and water resources in Tropical Africa. *Developments in Water Science*, Amsterdam, **8**.

Balek, J. & Perry, J. E. (1973). Hydrology of seasonally inundated African headwater swamps. *Journal of Hydrology*, **19**, 227–249.

Balfour Beatty, & Company Limited (1961). Niger Dams Report, Geological Supplement. Niger Dams Authority, Federal Government of Nigeria.

Baltzer, F. (1991). Late Pleistocene and recent detrital sedimentation in the deep parts of northern Lake Tanganyika (East African Rift). International Association of Sedimentologists, Special Publication No. 13, 147–173, Blackwell, Oxford.

Bannerman, R. R. (1973). Problems associated with development of groundwater in igneous and metamorphic rocks—a case study from Ghana. *Groundwater*, **11**, 31–34.

Barber, A. J. (1985). The relationship between the tectonic evolution of southeast Asia and hydrocarbon occurrences. In *Tectono-Stratigraphic Terranes of the Circum-Pacific Region* (ed. D. G. Howell), pp. 523–528, Earth Science Series, Vol. 1. Circum-Pacific Council for Energy and Mineral Resources.

Barbier, R. (1957). Aménagements hydroélectriques dans le sud du Brésil. *Comptes rendus et Sommaires, Bulletin Société Géologique de la France*, **6**, 877–892.

Barbier, R. (1967). Nouvelles réflexions sure le problème des 'pains de sucre' à propos d'observations dans le Tassili N'Ajjer (Algerie). *Travaux Laboratorie Geologique, Genoble*, **43**, 15–21.

Bardossy, J. & Aleva, G. J. J. (1990). Lateritic Bauxites. *Developments in Economic Geology*, **27**, pp. 624. Elsevier, Amsterdam.

Barnola, J. M., Raynaud, D., Korotkevich, Y. S. & Lorius, C. (1987). Vostok ice core provides 160,000-year record of atmospheric CO_2. *Nature*, **329**, 408–414.

Barrett, P. J., Elston, D. P., Harwood, D. M., McKelvey, B. C. & Webb, P. N. (1987). Mid-Cenozoic record of glaciation and sea level change on the margin of the Victoria Land Basin, Antarctica. *Geology*, **15**, 634–637.

Batchelor, B. C. (1979a). Geological characteristics of certain coastal and offshore placers as essential guides for tin exploration in Sundaland, Southeast Asia. *Geological Society of Malaysia Bulletin*, **11**, 283–313.

Batchelor, B. C. (1979b). Discontinuously rising late Cainozoic eustatic sea levels, with special reference to Sundaland, southeast Asia. *Geologie en Mijnbouw*, **58**, 1–20.

Batchelor, D. A. F. (1988). Dating of Malaysian fluvial tin placers. *Journal of Southeast Asian Earth Sciences*, **2**, 3–14.

Battiau-Queney, Y. (1984). The pre-glacial evolution of Wales. *Earth Surface Processes and Landforms*, **9**, 229–252.

Baumhauer, R. (1991). Paleolakes of the south central Sahara—problems of paleoclimatic interpretation. *Hydrobiologia*, **214**, 347–357.

Baynes, F. J. & Dearman, W. R. (1978). The microfabric of a chemically weathered granite. *Bulletin of the International Association of Engineering Geologists*, **18**, 91–100.

Beauvais, B. & Tardy, Y. (1991). Formation et degradation des cuirasses ferrugineuses sous climat tropical humide, a la lisiere de la forêt equatoriale (degradation features of iron duricrusts under tropical humid climate at the edge of the tropical rainforest). *Comptes Rendues, Acadamie des Sciences, Série II*, **313**, 1539–1545.

Becker, G. F. (1895). A reconnaissance of the goldfields of the southern Appalachians. *US Geological Survey, 16th Annual Report*, **3**, 251–331.

Benda, L. & Dunne, T. (1987). Sediment routing by debris flow. In *Erosion and Sedimentation in the Pacific Rim* (ed. R. L. Beschta, T. Blinn, C. E. Grant, F. J. Swanson & G. G. Ice), pp. 213–223. IAHS Publication No. 165. Wallingford.

Bennett, P. C., Melcer, M. E., Siegal, D. I. & Hassett, J. P. (1988). The dissolution of quartz in dilute aqueous solutions of organic acids at 25 °C. *Geochimica et Cosmochimica Acta*, **52**, 1521–1530.

Berger, W. H., Killingly, J. S., Metzler, C. V. & Vincent, E. (1985). Two-step deglaciation: ^{14}C-dated high-resolution $\delta^{18}O$ records from the tropical Atlantic Ocean. *Quaternary Research*, **23**, 258–271.

Berry, L. (1970). Some erosional features due to piping and subsurface wash with special reference to the Sudan. *Geografiska Annaler*, **52A**, 113–119.

Berry, L. & Ruxton, B. P. (1960). The evolution of Hong Kong harbour basin. *Zeitschrift für Geomorphologie, N.F.*, **4**, 97–115.

Beuf, S., Biju-Duval, B., de Charpal, O., Rognon, P., Gabriel, O. & Bennacef, A. (1971). Le grés du Paléozoique inférieur au Sahara. *L'Institute Français de Pétrole*, pp. 460, Paris.

Bhaskara Rao, A. (1987a). Lateritised gravel bed: a new guide horizon for lateritic gold? *Chemical Geology*, **60**, 287–291.

Bhaskara Rao, A. (1987b). Guide horizons for gold mineralisation in lateritic crusts. *Chemical Geology*, **60**, 293–298.

Bigarella, J. J. & Ab'Saber, A. N. (1964). Paläogeographische und paläoklimatische aspekte des Känozoikums in Südbrazilien. *Zeitschrift für Geomorphologie N.F.*, **8**, 286–312.

Bigarella, J. J. & Andrade, G. O. de (1965). Contribution to the study of the Brazilian Quaternary. *Geological Society of America Special Paper*, **84**, 435–451.

Bigarella, J. J. & Andrade-Lima, D. de (1982). Palaeoenvironmental changes in Brazil. In *Biological Diversification in the Tropics* (ed. G. T. Prance), pp. 27–39. Columbia, New York.

Bigarella, J. J. & Becker, R. D. (eds.) (1975). International symposium on the Quaternary (Southern Brazil, 15–31 July, 1975). *Boletim Paranaense de Geociências*, **33**.

Bigarella, J. J. & Mousinho, M. R. (1966). Slope development in southeastern and southern Brazil. *Zeitschrift für Geomorphologie, N.F.*, **10**, 150–160.

Bigarella, J. J. & Salamuni, R. (1958). Consideraçoes sôbre o Paleoclima da bacia de Curitiba. *Boletim do Instituto de Historia Natural, Curitiba, Parana, Brasil. Geologia*, **1**, 1–10.

Bigarella, J. J. & Salamuni, R. (1962). Caracteres texturais dos sedimentos da bacia de Curitiba. *Boletim da Universidade do Parana. Geologia*, **7**, 1–164 (+ Anexos 1–14).

Bigarella, J. J., Mousinho, M. G. & Da Silva, J. X. (1965). Processes and environments of the Brazilian Quaternary. *Symposium on Cold Climate Processes and Environments, Fairbanks, Alaska, VII INQUA Congr., 1965*, pp. 71. University Parana, Curitiba, Brazil.

Bird, M. I. & Chivas, A. R. (1988a). Stable-isotope evidence for low-temperature kaolinitic weathering and post-formational hydrogen-isotope exchange in Permean kaolinites. *Chemical Geology*, **72**, 249–265.

Bird, M. I. & Chivas, A. R. (1988b). Oxygen isotope dating of the Australian regolith. *Nature*, **331**, 513–516.

Bird, M. J., Chivas, A. R. & Andrew, A. S. (1989). A stable isotope study of lateritic bauxites. *Geochimica & Cosmochimica Acta*, **53**, 1411–1420.

Birkeland, P. W. (1984). *Soils and Geomorphology*. Oxford University Press, New York.

Birot, P. (1958). Les dômes crystallines. *Memoires et Documents, CNRS*, **6**, 8–34.

Birot, P. (1968). *The Cycle of Erosion in Different Climates*, Batsford, London (originally published in Rio de Janeiro as, *Le Cycle d'Érosion sous les Différents Climats*, University of Brazil, 1960).

Birot, P. & Dresch, J. (1966). Pédiments et glacis dans l'ouest des Etats-Unis. *Annals de Géographie*, **75**, 513–522.

Birot, P., Godard, A., Petit, M. & Ters, M. (1974). Contribution a l'étude des surfaces d'aplanissement et de l'érosion différentielle dans le Transvaal septentrional et oriental (Afrique du Sud). *Revue de Géographie Physique et de Géologie Dynamique*, **16**, 421–454.

Bishop, P. (1987). Geomorphic history of the Yom River floodplain, north central Thailand, and its implications for floodplain evolution. *Zeitschrift für Geomorphologie N.F.*, **31**, 195–211.

Bishop, W. W. (1966). Stratigraphic geomorphology: a review of some East African landforms. In *Essays in Geomorphology* (ed. G. H. Dury), pp. 139–176. Heinemann, London.

Black, C. A. (1957). *Soil–plant Relationships*. Wiley, New York.

Blain, C. F. & Andrew, R. L. (1977). Sulphide weathering and the evaluation of gossans in mineral exploration. *Minerals Science and Engineering*, **9**, 119–149.

Blanck, H. R. (1978). Fossil laterite on bedrock in Brooklyn, New York. *Geology*, **6**, 21–24.

Bleackley, D. (1964). Bauxites and laterites of British Guiana. *Geological Survey of British Guiana, Bulletin*, **34**, 1–154.

Bleackley, D. & Kahn, E. J. A. (1963). Observations on white sand areas of the Berbice Formation, British Guiana. *Journal of Soil Science*, **14**, 44–51.

Blight, G. E. (1990). Keynote paper: Construction in tropical soils. In *Geomechanics in Tropical Soils, Proceedings of the Second International Conference on Geomechanics in Tropical Soils, Singapore, 1988*, Vol. 2, pp. 449–467. Balkema, Rotterdam.

Blong, R. J. (1985). Gully sidewall development in New South Wales, Australia. In *Soil Erosion and Conservation* (ed. S. A. El-Swaify, W. C. Moldenhauer & A. Lo), pp. 574–584. Soil Conservation Society of America, Iowa, USA.

Bloom, A. L., Broecker, W. S., Chappell, J., Matthews, R. S. & Mesolella, K. J. (1974). Quaternary sea-level fluctuations on a tectonic coast: new ^{230}Th/^{234}U dates from the Huon Peninsula, New Guinea. *Quaternary Research*, **4**, 185–205.

Blümel, W. D. (1982). Calcretes in Namibia and SE Spain — relations to substratum, soil formation, and geomorphic factors. *Catena Supplement*, **1**, 67–95.

Boast, R. (1990). Dambos: a review. *Progress in Physical Geography*, **14**, 153–177.

Bocco, G. (1991). Gully erosion: processes and models. *Progress in Physical Geography*, **15**, 392–406.

Bonatti, E. & Gartner, S. (1973). Caribbean climate during Pleistocene ice ages. *Nature*, **244**, 563–565.

Bond, G. (1964). Pleistocene environments in southern Africa. In *African Ecology and Human Evolution* (ed. F. C. Howell & F. Boulière), pp. 308–314. Methuen, London.

Bond, G. (1967). River valley morphology, stratigraphy and palaeoclimatology in southern Africa. In *Background to Evolution in Africa* (ed. W. W. Bishop & J. D. Clark), pp. 303–312. Chicago.

Bonell, M. & Gilmour, D. A. (1978). The development of overland flow in a tropical rainforest catchment. *Journal of Hydrology*, **39**, 365–382.

Bonell, M., Gilmour, D. A. & Cassells, D. S. (1983). Runoff generation in tropical rainforests of northeast Queensland, Australia, and the implications for land use management. In *Hydrology of Humid Tropical Regions* (ed. E. Keller), pp. 287–297. IAHS Publication No. 140, Wallingford.

Bonnefille, R. (1983). Evidence for a cooler and drier climate in the Ethiopian uplands towards 2.5 Myr ago. *Nature*, **303**, 487–491.

Bonnefille, R. & Riollet, G. (1988). The Kashiru pollen sequence (Burundi) palaeoclimatic implications for the last 40 000 yr BP in Tropical Africa. *Quaternary Research*, **30**, 19–35.

Bonnefille, R., Roeland, J. C. & Guiot, J. (1990). Temperature and rainfall estimates for the past 40 000 years in equatorial Africa. *Nature*, **346**, 347–349.

Bornhardt, W. (1900). *Zur Oberflächengestaltung und Geologie Deutsch-Afrikas*. Berlin.

Botha, G. A., De Villiers, J. M. & Vogel, J. C. (1990). Cyclicity of erosion, colluvial sedimentation and palaeosol formation in Quaternary hillslope deposits from northern Natal, South Africa. *Palaeoecology of Africa*, **21**, 195–210.

Bouchard, M. (1989). Characteristics and significance of pre Wisconsinan saprolites in the northern Appalachians. *Zeitschrift für Geomorphologie, N.F., Supplementband* **72**, 125–137.

Boulangé, B. (1983). *Les formations bauxitiques latéritiques de Côte d'Ivoire*. Thèse, Université de Paris.

Boulangé, B. & Millot, G. (1988). La distribution des bauxites sure le craton Ouest-Africain. *Sciences Géologique, Bulletin*, Strasbourg, **41**, Latérites Périatlantiques, 113–123.

Boulet, R., Chauvel, A. & Lucas, Y. (1984). *Les Systèmes de Transformation en Pédologie*, 167–179, Association Française pour l'Étude du Sol, Livre Jubilaire du Cinquantenaire, Paris.

Boulvert, Y. (1985). Aplanissements en Afrique Centrale, relations avec le cuirassement, la tectonique, le bioclimat. Problemes posés, progrés des connaissances. *Bulletin Association Géographie Française*, **4**, 299–309.

Bourman, R. P. (1993). Modes of ferricrete genesis: evidence from southeastern Australia. *Zeitschrift für Geomorphologie, N.F.*, **37**, 77–101.

Bourman, R. P. (1993) Perennial problems in the study of laterite. A review. *Australian Journal of Earth Science*, **40**, 387–401.

Bowden, D. J. (1980). Sub-laterite cave systems and other pseudokarst phenomena in the humid tropics; the example of the Kasewe Hills, Sierra Leone. *Zeitschrift für Geomorphologie*, **24**, 77–90.

Bowden, D. J. (1987). On the composition and fabric of the footslope laterites (duricrust) of Sierra Leone, West Africa, and their geomorphological significance. *Zeitschrift für Geomorphologie, N.F., Supplementband*, **64**, 39–53.

Bowler, J. M. (1978a). Aridity in Australia: age origins and expression in aeolian landforms and sediments. *Earth Science Reviews*, **12**, 279–230.

Bowler, J. M. (1978b). Quaternary climate and tectonics in the evolution of the Riverine Plain, southeastern Australia. In *Landform Evolution in Australasia* (ed. J. L. Davies & M. A. J. Williams), 70–122. ANU Press, Canberra.

Bowler, J. M., Hope, G. S., Jennings, J. N., Singh, G. & Walker, D. (1976). Late Quaternary climates of Australia and New Guinea. *Quaternary Research*, **6**, 359–394.

Boye, M. & Fritsch, P. (1973). Dégagement artificiel d'un dôme crystallin au Sud-Cameroun. *Travaux et Documents de Géographie Tropicale* (Bordeaux), **8**, 31–62.

Brabb, E. E. (1985). Innovative approaches to landslide hazard and risk mapping. *IVth International Symposium on Landslides*, Toronto, 1984, vol. 1, 307–323.

Brabant, P. (1987). La répartition des podzols à Kalimantan. In *Podzols et Podzolisation* (ed. D. Righi & A. Chauvel), pp. 13–24. AFES and INRA, Plaisir et Paris.

Bradbury, J. P., Leyden, B., Salgado-labouriau, M. L., Lewis, W. M., Schubert, C., Binford, M. W., Frey, D. G., Whitehead, D. R. & Weibezahn, F. H. (1981). Late Quaternary environmental history of Lake Valencia, Venezuela. *Science*, **214**, 1199–1205.

Bradley, W. C. (1963). Large scale exfoliation in massive sandstones of the Colorado Plateau. *Bulletin of the Geological Society of America*, **74**, 519–528.

Brakenridge, G. R. (1980). Widespread episodes of stream erosion during the Holocene and their climatic cause. *Nature*, **283**, 655–656.

Brand, E. W. (1985). Landslides in Southeast Asia: a state of the art report. In *IV International Symposium on Landslides*, Toronto, 1984, Vol. 1, pp. 17–59.

Brand, E. W., Premchitt, J. & Phillipson, H. B. (1985). Relationship between rainfall and landslides in Hong Kong. In *IV International Symposium on Landslides*, Toronto, 1984, Vol. 1, pp. 377–384.

Branner, J. C. (1896). Decomposition of rocks in Brazil. *Bulletin of the Geological Society of America*, 7, 255–314.

Branner, J. C. (1911). Aggraded limestone plains of the interior of Brazil and the climatic changes suggested by them. *Bulletin Geological Society of America*, 22, 187–206.

Bravard, S. & Righi, D. (1990). Podzols in Amazonia. *Catena*, 17, 461–475.

Brazilian Society for Soil Mechanics (1985). *First International Conference on Geomechanics in Tropical Lateritic and Saprolitic Soils*, Brasilia, 1985, Vol. 1, Technical Sessions. ABMS, São Paulo.

Bremer, H. (1965). Ayers Rock, ein Beispiel für klimagenetische Morphologie. *Zeitschrift für Geomorphologie, N.F.*, 9, 249–284.

Bremer, H. (1971). Flüsse, Flächen- und Stufenbildung in den feuchten Tropen. *Wuzburger Geographische Arbeiten*, 35, pp. 194.

Bremer, H. (1981). Reliefformen und reliefbildende Prozesse in Sri Lanka. In *Relief Boden Paläoklima* (Zur Morphogenese in den feuchten Tropen. Verwitterung und Reliefbildung am Beispiel von Sri Lanka), Vol. 1, pp. 7–183. Borntraeger, Berlin.

Bremer, H. (1985a). Randschwellen: a link between plate tectonics and climatic geomorphology. *Zeitschrift für Geomorphologie, N.F., Supplementband*, 54, 11–21.

Bremer, H. (1985b). Soil and slope development in the wet zone of Sri Lanka. In *Environmental Change and Tropical Geomorphology* (ed. I. Douglas & T. Spencer), pp. 295–302. Allen and Unwin, London.

Briceño, H. O. & Schubert, C. (1990). Geomorphology of the Gran Sabana, Guayana Shield, southeastern Venezuela. *Geomorphology*, 3, 125–141.

Brink, A. B. A. & Partridge, T. C. (1967). Kayalami land system: an example of physiographic classification for the storage of terrain data. In *Proceedings 4th Regional Conference for Africa on Soil Mechanics and Foundation Engineering*, Vol. 1, Cape Town, pp. 9–14.

Brink, A. B. A., Partridge, T. C. & Matthews, G. B. (1970). Air photo interpretation in terrain evaluation. *Photointerpretation*, 5, 15–30.

Brinkman, R. (1970). Ferrolysis, a hydromorphic soil forming process. *Geoderma*, 3, 199–206.

Brinkman, R. (1979). Ferrolysis, a soil forming process in hydromorphic conditions. *Agricultural Research Reports*, 887. Wageningen, Holland.

Brinkman, R., Jongmans, A. G., Miedema, R. & Maskaut, P. (1973). Clay decomposition in seasonally wet, acid soils: micromorphological, chemical and mineralogical evidence from individual argillans. *Geoderma*, 10, 259–270.

Bristow, C. M. (1977). A review of the evidence for the origin of kaolin deposits in S.W. England. In *Proceedings of the 8th Kaolin Symposium, Madrid/Rome*. Vol. K-2, pp. 1–19, Rome.

British Standards Institution (1981). *BS 5930. Code of Practice for Site Investigations*. British Standards Institution, London.

Broecker, W. S. (1986). Oxygen isotope constraints on surface ocean temperatures. *Quaternary Research*, 26, 121–134.

Broecker, W. S. & Denton, G. H. (1990). The role of ocean-atmosphere reorganisations in glacial cycles. *Quaternary Science Reviews*, 9, 305–341.

Broecker, W. S., Bond, G., Klas, M., Clark, E. & McManus, J. (1992). Origin of the northern Atlantic Heinrich events. *Climate Dynamics*, 6, 265–273.

Brook, G. A. (1978). A new approach to the study of inselberg landscapes. *Zeitschrift für Geomorphologie, N.F., Supplementband*, 31, 138–160.

Brook, G. A. (1982). Stratigraphic evidence of Quaternary climatic change at Echo Cave, Transvaal, and a paleoclimatic record for Botswana and northeastern South Africa. *Catena*, 9, 343–351.

Brosch, A. & Gerson, R. (1975). Hill-top depressions on folded metasediments under a subhumid equatorial climate. *Geografiska Annaler*, 57A, 135–142.

Browne, R. W., Rust, D. J., Summerfield, M. A., Gleadow, A. J. W. & De Witt, M. C. J. (1990). An early Cretaceous phase of accelerated erosion on the southwest margin of Africa: evidence from apatite fission track analysis and the offshore sedimentary record. *Nuclear Tracks and Radiation Measurements, International Journal of Radiation Applications and Instrumentation*, Part D, 17, 339–350.

Brückner, W. D. (1955). The mantle rock (laterite) of the Gold Coast and its origin. *Geologische Rundschau*, 43, 307–327.

Brückner, W. D. (1957). Laterite and bauxite profiles of west Africa as an index of rhythmical climatic variations in the tropical belt. *Géologia, Ecologicae Helvetiae*, 50, 239–256.

Bruijnzeel, L. A. (1983). Evaluation of runoff sources in a forested basin in a wet monsoonal environment: a combined hydrological and hydrochemical approach. In *Hydrology of Humid Tropical Regions* (ed. E. Keller), pp. 165–174. IAHS Publication No. 140.

Brünig, E. F. (1975). Tropical ecosystems: state and targets of research into the ecology of humid tropical ecosystems. *Plant Research and Development*, 1, 22–38.

Brunsden, D. (1980). Applicable models of long term landform evolution. *Zeitshrift für Geomorphologie, Supplementband*, 36, 16–26.

Brunsden, D. (1984). Mudslides. In *Land Instability* (ed. D. Brunsden & D. B. Prior), pp. 363–418. Wiley, Chichester.

Brunsden, D. (1993). The persistence of landforms. *Zeitschrift für Geomorphologie, N.F., Suppl*, **93**, 13–28.

Brunsden, D. & Jium-Chuan, L. (1991). The concept of topographic equilibrium in neotectonic terrains. In *Neotectonics and Resources* (ed. J. Cosgrove & M. Jones), pp. 120–143. Belhaven, London.

Brunsden, D. & Prior, D. B. (ed.) (1984). *Land Instability*. Wiley, Chichester.

Brunsden, D. & Thornes, J. B. (1977). *Geomorphology and Time*. Methuen, London.

Brunsden, D. & Thornes, J. B. (1979). Landscape sensitivity and change. *Institute of British Geographers, Transactions*, NS4, 463–484.

Brunsden, D., Doornkamp, J. C., Fookes, P. G., Jones, D. K. C. & Kelly, J. M. N. (1975). Large scale geomorphological mapping and highway engineering design. *Quarterly Journal of Engineering Geology*, **8**, 227–253.

Buckley, D. K. & Zeil, P. (1984). The character of fractured rock aquifers in eastern Botswana. In *Challenges in African Hydrology and Water Resources* (Proceedings of the Harare Symposium, 1984), pp. 25–36. IASH Publication, No. 144. Wallingford.

Büdel, J. (1948). Das system de klimatischen morphologie. *Deutscher Geographen-Tag, Munchen*, 65–100.

Büdel, J. (1957). Die 'doppelten Einebnungsflächen' in den feuchten Tropen. *Zeitschrift für Geomorphologie, N.F.*, **1**, 201–288.

Büdel, J. (1965). Die Relieftypen de Flächenspül-Zone Sud-Indiens am Ostabfall Dekans gegen madras. *Colloque Geographie, Bonn* **8**.

Büdel, J. (1968). Geomorphology principles. In *Encyclopaedia of Geomorphology* (ed. R. W. Fairbridge), 416–422. Rheinhold, New York.

Büdel, J. (1970). Pedimente, Rumpflächen und Rückland-Steilhänge. *Zeitschrift für Geomorphologie, N.F.*, **14**, 1–57.

Büdel, J. (1977). *Klima-Geomorphologie*. Borntraeger, Stuttgart.

Büdel, J. (1980). Climatic and climatomorphic geomorphology. *Zeitschrift für Geomorphologie, N.F., Supplementband*, **36**, 1–8.

Büdel, J. (1982). *Climatic Geomorphology* (transl. L. Fischer & D. Busche). Princeton University Press, Princeton.

Budyko, M. I. (1963). *Atlas of the Heat Balance of the Earth*. Izd. Geofizichekogo Komiteta.

Bull, W. B. (1991). Conceptual models for changing landscapes. In *Geomorphic Responses to Climatic Change* (ed. W. B. Bull), pp. 3–31. Oxford, New York.

Burke, K. & Dewey, K. F. (1973). Plume-generated triple junctions: key indicators in applying plate tectonics to old rocks. *Journal of Geology*, **81**, 406–433.

Burke, K. & Durotoye, B. (1971). Geomorphology and superficial deposits related to late Quaternary climatic variation in southwestern Nigeria. *Zeitschrift für Geomorphologie, N.F.*, **15**, 430–444.

Busche, D. (1972). Untersuchungen zur Pedimententwicklung im Tibesti-gebirge. *Zeitschrift für Geomorphologie, N.F., Supplementband*, **15**, 21–38.

Busche, D. (1976). Pediments and climate. In *Palaeoecology of Africa, the Surrounding Islands and Antarctica*, Vol. 11 (ed. E. M. Van Zinderen Bakker Sr), pp. 20–24. Balkema, Cape Town.

Busche, D. (1989). Geomorphic effects of subterranean sandstone karst in arid eastern Niger. Abstracts of Papers, Second International Conference on Geomorphology, Frankfurt/Main, 1989. *Geöko-plus*, **1**, 46.

Busche, D. & Erbe, W. (1987). Silicate karst landforms of the southern Sahara (north-eastern Niger and southern Libya). *Zeitschrift für Geomorphologie, N.F., Supplementband*, **64**, 55–72.

Bush, M. B., Colinvaux, P., Weimann, M. C., Piperno, D. R. & Lui, K-B. (1990). Late Pleistocene temperature depression and vegetation change in Ecuadorian Amazonia. *Quaternary Research*, **34**, 300–345.

Bushinsky, G. I. (1975). *Geology of Bauxites. 2nd Edn.* Izd. Nedra, Moscow.

Butler, B. E. (1959). *Periodic Phenomena in Landscapes as a basis for Soil Studies*. CSIRO Australia, Soil Publication, **14**.

Butler, B. E. (1967). Soil periodicity in relation to landform development in southeastern Australia. In *Landform Studies from Australia and New Guinea* (ed. J. N. Jennings & J. A. Mabbutt), pp. 231–255. Cambridge University Press, London.

Butt, C. R. M. (1981). Some aspects of geochemical exploration in lateritic terrains in Australia. In *Lateritisation Processes, Proceedings of the International Seminar on Lateritisation Processes*, Tivandrum, India, 1979, pp. 369–380. Balkema, Rotterdam.

Butt, C. R. M. (1985). Granite weathering and silcrete formation on the Yilgarn Block, Western Australia. *Australian Journal of Earth Science*, **32**, 415–433.

Butt, C. R. M. (1987). A basis for geochemical exploration models for tropical terrains. *Chemical Geology*, **60**, 5–16.

Butty D. L. & Chapallaz, C. A. (1984). Bauxite genesis. In *Bauxite. Proceedings 1984 Symposium*, Los Angeles, 1984 (ed. L. Jacob, Jr.), pp. 111–151. Society of Mining Engineers, American Institute for Mining, Metallurgical and Petroleum Engineers, New York.

Butzer, K. W. (1980). Pleistocene history of the Nile Valley in Egypt and Lower Nubia. In *The Sahara and the Nile* (ed. M. A. J. Williams & H. Faure), pp. 253–280. Balkema, Rotterdam.

Butzer, K. W. (1984). Late Quaternary environments in

South Africa. In *Late cainozoic Palaeoclimates of the Southern Hemisphere*, SASQUA Symposium, 1983 (ed. J. C. Vogel), pp. 235–264. Balkema, Rotterdam.

Butzer, K. W., Isaac, G. L., Richardson, J. L. & Washburn-Kamau, C. (1972). Radiocarbon dating of east African lake levels. *Science*, **175**, 1069–1076.

Buurman, J. & Subagjo (1978). Soil formation on granodiorites near Pontianak, West Kalimantan. *Proc. 1st., Nat. Soil Workshop,* Bogor, Indonesia, 1978, pp. 107–117, Bogor.

Byers, B. R. (1974). Iron transport in gram-positive and acid fast bacilli. In *Microbial Iron Metabolism — A Comprehensive Treatise* (ed. Neilands). Academic Press, New York.

Caine, N. (1980). The rainfall intensity-duration control of shallow landslides and debris flows. *Geografiska Annaler*, **62A**, 23–27.

Campbell, J. M. (1917). Laterite. *Mineralogical Magazine*, **17**, 67–77, 120–128, 171–179, 220–229.

Cannon, S. H. & Ellen, S. D. (1988). Rainfall that resulted in abundant debris-flow activity. In *Landslides, Floods, and Marine Effects of the Storm of January 3–5, 1982, in the San Francisco Bay Region, California* (ed. S. D. Ellen & G. F. Wieczorek), pp. 27–34. US Geological Survey Professional Paper No. 1434, Washington.

Caratini, C. & Tissot, C. (1988). Paleogeographical evolution of the Mahakam Delta in Kalimantan, Indonesia, during the Quaternary and late Pliocene. *Review of Paleobotany and Palynology*, **55**, 217–228.

Carrara, A. (1983). Multivariate models for landslide hazard identification. *Mathematical Geology*, **15**, 403–426.

Carroll, D. (1970). *Rock Weathering*. Plenum, New York.

Carson, M. A. (1976). Mass wasting, slope development and climate. In *Geomorphology and Climate* (ed. E. Derbyshire), p. 101–136. Wiley, New York.

Carson, M. A. & Kirkby, M. J. (1972). *Hillslope Form and Process*. Cambridge University Press, London.

Chakela, Q. K. (1981). Soil erosion and reservoir sedimentation in Lesotho. *Uppsala Universitets Naturgeografiska Institutionen Rapport*, **54**.

Chalmers, R. (1898). The pre-glacial decay of rocks in eastern Canada. *American Journal of Science*, 4th Series, **5**, 273–282.

Chan, R. A. (1988). Regolith terrain mapping for mineral exploration in Western Australia. *Zeitschrift für Geomorphologie, N.F., Supplementband*, **68**, 205–221.

Chan, R. A., Craig, M. A. & Gibson, D. L. (1986). *The Regolith Terrain Map of Australia 1:5,000,000*. Bureau of Mineral Resources, Records 1986/27, Canberra.

Chappell, J. (1974) Geology of coral terraces, Huon Peninsula, New Guinea: a study of Quaternary tectonic movements and sea level changes. *Geological Society of America, Bulletin*, **85**, 555–570.

Chappell, J. (1983). A revised sea-level record for the last 300,000 years from Papua New Guinea. *Search*, **14**, 99–101.

Chappell, J. (1991). Late Quaternary environmental changes in eastern and central Australia and their climatic interpretation. *Quaternary Science Reviews*, **10**, 377–390.

Chatterjea, K. (1989). Surface wash: the dominant geomorphic process in the surviving rainforest of Singapore. *Singapore Journal of Tropical Geography*, **10**, 95–109.

Chauvel, A. (1977). Recherches sur la transformation des sols ferrallitiques dans la zone tropicale à saisons contrastées. *Travaux et Documents de L'ORSTOM, Paris*, **62**, p. 532.

Chauvel, A., Lucas, Y. & Boulet, R. (1987). On the genesis of the soil mantle of the region of Manaus, Central Amazonia, Brazil. *Experientia*, **43**, 234–241.

Chen, X. Y., Prescott, J. R. & Hutton, J. T. (1990). Thermoluminescence dating on gypseous dunes of Lake Amadeus, central Australia. *Australian Journal of Earth Sciences*, **37**, 93–102.

Chilton, P. J. & Smith-Carington, A. K. (1984). Characteristics of the weathered basement aquifer in Malawi in relation to rural water supplies. In *Challenges in African Hydrology and Water Resources* (Proceedings of the Harare Symposium, 1984), pp. 57–72. IASH Publication No. 144.

Chivas, A. R. (1983). The climatic conditions during regolith formation: oxygen- and hydrogen-isotope evidence. In *Regolith in Australia: Genesis and Economic Significance* (ed. G. E. Wilford), Australia Bureau of Mineral Resources: Geology and Geophysics. Record 1983/27, pp. 42–47.

Chorley, R. J., Schumm, S. A. & Sugden, D. E. (1984). *Geomorphology*. Methuen, London.

Christian, C. S. & Stewart, G. A. (1968). Methodology of integrated surveys. In *Aerial Surveys and Integrated Studies*, Natural Resources Research, Vol. 6, pp. 233–280. UNESCO, Paris.

Chukrov, F. V. (1981). On transformations of iron oxides in chemogenic eluvium in tropical and sub-tropical regions. In *Proc. International Seminar on Lateritisation Processes, Trivandrum*, 1979, pp. 11–14. Oxford IBH, Delhi.

Churchward, H. M. (1969). Erosional modification of a lateritised landscape over sedimentary rocks. Its effects on soil distribution. *Australian Journal of Soil Research*, **8**, 1–19.

Churchward, H. M. & Gunn, R. H. (1983). Stripping of deep weathered mantles and its significance to soil patterns. In *Soils from an Australian Viewpoint*, pp. 73–81. Division of Soils CSIRO, Academic Press, Melbourne.

Clapperton, C. M. (1993a). *Quaternary Geology and Geomorphology of South America*. Elsevier, Amsterdam.

Clapperton, C. M. (1993b). Nature of environmental changes in South America at the Last Glacial Maximum. *Palaeogeography, Palaeoclimatology and Palaeoecology*, **101**, 189–208.

Clark, D. A. (1974). *The Kankoma Clay Deposit*. Economic Report No. 49, Geological Survey, Zambia.

Clark, L. (1985). Groundwater abstraction from basement complex areas of Africa. *Quarterly Journal of Engineering Geology*, **18**, 25–34.

Clayton, R. W. (1956). Linear depressions (Bergfüssneiderungen) in savanna landscapes. *Geographical Studies*, **3**, 102–126.

Cleaves, E. T. (1974). Petrologic and chemical investigation of chemical weathering in mafic rocks, eastern piedmont of Maryland. *Maryland Geological Survey, Report No. 25*.

Cleaves, E. T. (1983). Chemical weathering in a humid temperate environment. *Sciences Géologiques, Mémoire*, **72**, 47–55.

Cleaves, E. T. (1989). Appalachian piedmont landscapes from the Permian to the Holocene. *Geomorphology*, **2**, 159–179.

Cleaves, E. T. & Costa, J. E. (1979). *Equilibrium, Cyclicity and Problems of Scale — Maryland's Piedmont Landscape*, pp. 32, Maryland Geological Survey Department of Natural Resources, Information Circular, **29**, Johns Hopkins University, Baltimore.

Cleaves, E. T., Fisher, D. W., & Bricker, O. P. (1974). Chemical weathering of serpentonite in the eastern Piedmont of Maryland. *Geological Society of America Bulletin*, **85**, 437–444.

Cleaves, E. T., Godfrey, A. E. & Bricker, O. P. (1970). Geochemical balance of a small watershed and its geomorphic implications. *Geological Society of America Bulletin*, **81**, 3015–3032.

CLIMAP Project Members (1976). The surface of the ice-age earth. *Science*, **191**, 1131–1138.

CLIMAP Project Members (1981). Seasonal reconstruction of the earth's surface at the last glacial maximum. *Geological Society of America, Map and Chart Series No. 36*.

Coates, D. R. (1977). Landslide perspectives. In *Landslide Perspectives* (ed. D. R. Coates), pp. 3–23. Geological Society of America, Reviews in Engineering Geology, Vol. III.

Coetzee, J. A. (1967). Pollen analytical studies in east and southern Africa. *Palaeoecology of Africa*, **3**, 1–146.

Coetzee, J. A. & Van Zinderen Bakker, E. M. (1989). Palaeoclimatology of East Africa during the last glacial maximum: a review of changing theories. In *Quaternary and Environmental Research on East African Mountains* (ed. W. C. Mahaney), pp. 189–198. Balkema, Rotterdam.

Colinvaux, P. A. (1987). Environmental history of the Amazon Basin. In *Journal of The Quaternary of South America and Antarctic Peninsula*, **5**, 223–237.

Colinvaux, P. (1991). A commentary on: palaeoecological background: neotropics. *Climate Change* (Special Issue: Tropical Forests and Climate), **19**, 49–51.

Collins, K. (1985). Towards characterisation of tropical soil microstructure. In *First International Conference on Geomechanics in Tropical Lateritic and Saprolitic Soils*, Brasilia, 1985, Vol. 1, pp. 85–96, Brasilia.

Colman, S. M. & Dethier, S. P. (eds.) (1986). *Rates of Chemical Weathering of Rocks and Minerals*, pp. 603. Academic Press, Orlando & London.

Colman, S. M. & Dethier, S. P. (1986). An overview of rates of chemical weathering. In *Rates of Chemical Weathering of Rocks and Minerals* (eds. S. M. Colman & D. P. Dethier), pp. 1–18. Academic Press, Orlando & London.

Coltrinari, L. (1993). Global Quaternary changes in South America. *Global and Planetary Change*, **7**, 11–23.

Coltrinari, L. & Nogueira, F. (1989). Dambo-like landforms in southeastern Brazil. *Abstracts of Papers, Second International Conference on Geomorphology, Frankfurt/Main, 1989*. Geöko-plus, **1**, 58.

Colyn, M., Gautier-Hion, A. & Verheyen, W. (1991). A re-appraisal of paleoenvironmental history in central Africa: evidence for a major fluvial refuge in the Zaire Basin. *Journal of Biogeography*, **18**, 403–407.

Conrad, G. (1969). L'évolution continentale post-hercynienne du Sahara algérien. *CNRS, Centre de Recherche du Zones Arides, Paris*, **10**, 527.

Cooke, H. B. S. (1958). *Observations Relating to Quaternary Environments in East and Southern Africa*. Alex Du Toit Memorial Lecture No 5, Geological Society of South Africa, Bulletin, **LX**, Annexe.

Cooke, H. J. (1983). The evidence from northern Botswana of late Quaternary climatic change. In *Late Cainozoic Climates of the Southern Hemisphere* (Proceedings of SASQUA International Symposium, Swaziland, 1983) (ed. J. C. Vogel), pp. 265–278. Balkema, Rotterdam.

Cooke, H. J. & Verstappen, H. Th. (1984). The landforms of the western Makgadikgadi basin in northern Botswana, with a consideration of the chronology of the evolution of Lake Palaeo-Makgadikgadi. *Zeitschrift für Geomorphologie, N.F.*, **28**, 1–19.

Cooke, R. U. & Doornkamp, J. C. (1990). *Geomorphology in Environmental Management*, 2nd edn. Oxford University Press, Oxford.

Cooke, R. U., Warren, A. & Goudie, A. (1993). *Desert Geomorphology*. UCL Press, London

Corbel, J. (1957). L'érosion climatiques des granites et silicates sous climats chauds. *Revue Géomorphologie Dynamique*, **8**, 4–8.

Corbel, J. (1959). Vitesse de l'érosion. *Zeitschrift für Geomorphologie, N.F.*, **3**, 1–28.

Cordery, I. & Pilgrim, D. H. (1983). On the lack of dependence of losses from flood runoff on soil and cover characteristics. In *Hydrology of Humid Tropical Regions* (ed. E. Keller), pp. 187–195. IAHS Publication, No. 140.

Costa, J. E. & Cleaves, E. T. (1984). The Piedmont landscape of Maryland: a new look at an old problem. *Earth Surface Processes and Landforms*, **9**, 59–74.

Cotton, C. A. (1917). Block mountains in New Zealand. *American Journal of Science*, **44**, 249–293.

Cotton, C. A. (1942). *Climatic Accidents in Landscape Making*. Whitcomb & Tombs, Wellington.

Cotton, C. A. (1961). Theory of savanna planation. *Geography*, **46**, 89–96.

Cotton, C. A. (1962). Plains and inselbergs of the humid tropics. *Royal Society of New Zealand, Transactions (Geology)*, **1**, 269–277.

Craig, D. C. & Loughnan, F. C. (1964). Chemical and mineralogical transformations accompanying weathering of basic rocks from New South Wales. *Australian Journal of Soil Research*, **2**, 218–234.

Crickmay, C. H. (1933) The late stages of the cycle of erosion. *Geological Magazine*, **70**, 337–347.

Crickmay, C. H. (1974). *The Work of the River*. MacMillan, London.

Crickmay, C. H. (1975). The hypothesis of unequal activity. In *Theories of Landform Evolution* (ed. W. N. Melhorn & R. C. Flemal), pp. 103–109. George, Allen and Unwin, London.

Crofts, R. W. (1968). Unpublished data prepared for the National Diamond Mining Company, Sierra Leone.

Crowley, T. J. & North G. R. (1991). *Paleoclimatology*. Oxford, New York.

Crowther, J. (1984). Soil carbon dioxide and weathering potentials in tropical karst terrain, peninsular Malaysia: a preliminary model. *Earth Surface Processes and Landforms*, **9**, 397–407.

Crozier, M. J. (1986). *Landslides: Causes, Consequences and Environment*. Croom Helm, London.

Cunningham, F. F. (1969). The Crow Tors, Laramie Mountains, Wyoming, USA. *Zeitschrift für Geomorphologie, N.F.*, **13**, 56–74.

Cunningham, F. F. (1971). The Silent City of Rocks, a bornhardt landscape in the Cottrell range, South Idaho, USA. *Zeitschrift für Geomorphologie, N.F.*, **15**, 404–429.

Curtis, C. D. (1976). Chemistry of rock weathering: fundamental reactions and controls. In *Geomorphology and Climate* (ed. E. Derbyshire), pp. 25–57. Wiley, New York.

Damuth, J. E. (1977). Late Quaternary sedimentation in the western equatorial Atlantic. *Geological Society of America Bulletin*, **88**, 695–710.

Damuth, J. E. & Fairbridge, R. W. (1970). Equatorial Atlantic deep-sea arkosic sands and Ice-age aridity in tropical South America. *Bulletin of the Geological Society of America*, **81**, 189–206.

Daniels, R. B., Gamble, E. E. & Holzhey, C. S. (1975). Thick Bh horizons in the North Carolina coastal plain: I. Morphology and relation to texture and soil ground water. *Soil Science Society of America, Proceedings*, **39**, 1177–1181.

Dardis, G. F., Beckedahl, H. R., Bowyer-Bower, T. A. S. & Hanvey, P. M. (1988). Soil erosion forms in southern Africa. In *Geomorphological Studies in Southern Africa* (ed. G. F. Dardis & B. P. Moon), pp. 187–213. Balkema, Rotterdam.

Darwin, C. (1876). *Geological Observations*, 2nd edn. London.

Darwin, C. (1881). *The Formation of Vegetable Mould Through the Action of Worms*. Appleton, New York.

Davidson, D. A. (ed.). (1986). *Land Evaluation*, pp. 373. Van Nostrand Reinhold, New York.

Davies, J. A. & Robinson, P. J. (1969). A simple energy balance approach to the moisture balance climatology of Africa. In *Environment and Land Use in Africa* (ed. M. F. Thomas & G. Whittington), pp. 23–56. Methuen, London.

Davis, J. L. & Williams, M. A. J. (ed.) (1978). *Landform Evolution in Australasia*. ANU Press, Canberra.

Davis, W. M. (1899a). The geographical cycle. *Geographical Journal*, **14**, 481–504.

Davis, W. M. (1899b). The peneplain. *American Geologist*, **22b**, 207–239. Also in Johnson D. W. (ed.) (1954). *Geographical Essays*, Dover Publications, New York; and in Adams, G. (ed.) (1975) *Planation Surfaces*, Benchmark Papers in Geology, Vol. 22, pp. 61–91, Dowden, Hutchinson & Ross, Pennsylvania.

Day, M. J. (1979). The hydrology of polygonal karst depressions in northern Jamaica. *Zeitschrift für Geomorphologie, N.F., Supplementband*, **32**, 25–34.

Day, M. J. (1982). Landslides in the Gunung Mulu National Park. *Geographical Journal*, **146**, 7–13.

Deacon, J. L. & Lancaster, N. (1988). *Late Quaternary Palaeoenvironments of Southern Africa*, Clarendon, Oxford.

Dearman, W. R., Baynes, F. J. & Irfan, T. Y. (1978). Engineering grading of weathered granite. *Engineering Geology*, **12**, 345–374.

D'Costa, D. M., Edney, P., Kershaw, A. P. & De Decker, P. (1989). Late Quaternary palaeoecology of Tower Hill, Victoria, Australia. *Journal of Biogeography*, **16**, 461–482.

Debaveye, J. & De Dapper, M. (1987). Laterite, soil and landform development in Kedah, Peninsular Malaysia. *Zeitschrift für Geomorphologie N.F., Supplementband*, **64**, 145–161.

Debaveye, J., De Dapper, M., De Paepe, P. & Gijbels, R. (1986). Quaternary volcanic ash deposits in the Padang Terap District, Kedah, Peninsular Malaysia. GEOSEA V Proc. 1. *Geological Society of Malaysia Bulletin*, **19**, 533–549.

De Broodt, M. D. & Gabriels, D. (ed.) (1981). *Assessment of Erosion*. Wiley, Chichester.

De Dapper, M. (1978). Couvertures limono-sableuses, stone-line, induations ferrugineuses et action des termites sur le plateau de La Manika (Kolwezi, Shaba, Zaire). *Géo-Eco-Trop*, **2**, 265–278.

De Dapper, M. (1988). Geomorphology of the sand-covered plateaux in southern Shaba, Zaire. In *Geomorphological Studies in Southern Africa* (ed. G. F. Dardis & B. P. Moon), pp. 115–135. Balkema, Rotterdam.

De Dapper, M. (1989). Pedisediments and stone-line complexes in Peninsular Malaysia. *Stone-Lines. Journée d'Étude, Bruxelles, 24 March 1987* (ed. J. Alexandre & J. J. Symoens). Académie Royale des Sciences D'Outre-Mer, Geo-Eco-Trop, **11**, 37–59.

De Dapper, M. & Debaveye, J. (1986). Geomorphology and soils of the Padang Terap District, Kedah, Peninsular Malaysia. *Bulletin Geological Society Malaysia*, **20**, 765–790.

Deere, D. U. & Patton, F. D. (1971). Slope stability in residual soils. In *Proceedings of the 4th Panamerican Conference on Soil Mechanics and Foundation Engineering*, Vol. 1, Caracas, June 1971, pp. 87–170. American Society of Civil Engineers, New York.

De Heinzelin, J. (1952). Sols, paléosols et désertifications anciennes dans le secteur oriental du bassin du Congo. *Publication de l'INEAC*, Colloque **4**, Brussels.

De Heinzelin, J. (1955). Observations sur la génèse de nappes de gravats dans les sols tropicaux. *Publication de l'INEAC*, série Scientifique, **64**, Brussels.

De Hoore, J. L. (1964). *Soil Map of Africa, Scale 1:5 000 000, Explanatory Monograph*. CCTA Publication No. 93, Lagos.

Delwaulle, J. C. (1973). Résultats de six années d'observations sur l'érosion au Niger. *Bois et Forêts des Tropiques*, **150**, 15–37.

Demangeot, J. (1975). Recherches géomorphologique en Inde du Sud. *Zeitschrift für Geomorphologie, N.F.*, **19**, 229–272.

Demangeot, J. (1976). *Les Espaces Naturels Tropicaux*. Masson, Paris.

Demangeot, J. (1978). Discussion. In (ed. J. Dresch), Les reliefs résiduels. *Bulletin Association Géographie Française*, **4**, 97–103.

Demangeot, J. (1985). Montagnes et cascades de Haute-Guyane Vénézuéliennes. *Bulletin Association Géographie Française*, **4**, 243–253.

De Meis, M. R. & Da Silva, J. X. (1968). Mouvements de masse récentes a Rio de Janeiro: une étude de geomorphologie dynamique. *Revue de Géomorphologie Dynamique*, **18**, 145–151.

De Meis, M. R. M. & Machado, M. B. (1978). A morfologia de rampas e terracos no Planalto do Sudeste do Brasil. *Finisterra*, **13**, 201–219.

De Meis, M. R. M. & Monteiro, A. M. F. (1979). Upper Quaternary 'rampas': Doce River valley, Southeastern Brazilian plateau. *Zeitschrift für Geomorphologie, N.F.*, **23**, 132 151.

De Meis, M. R. M. & De Moura, J. R. da Silva (1984). Upper Quaternary sedimentation and hillslope evolution: southeastern Brazilian plateau. *American Journal of Science*, **284**, 241–254.

Demek, J. (ed.) (1972). *Manual of Detailed Geomorphological Mapping*. Academia, Prague.

Demek, J. & Embleton, C. (eds.). (1978). *Guide to Medium Scale Geomorphological Mapping*. IGU, Stuttgart.

De Mello, V. F. B. (1972). Thoughts on soil engineering applicable to residual soils. In *Proceedings 3rd Southeast Asia Conference on Soil Engineering*, Hong Kong, pp. 5–34, Hong Kong.

De Moura, J. R. da S. & Mello, C. L. (1991). Classificaçao aloestratigrafica do Quaternario superior na regiao de Bananal (SP/RJ). *Revista Brasileira Geociencias (Sociedade Brasileira de Geologia, Sao Paulo)*, **21**, 236–254.

De Moura, J. R. da S., Da Silva, T. M., Mello, C. L., Peixoto, M. N. de O., Santos, A. A. de M. & Esteves, A. A. (1989). Roteiro geomorphologico-estratigrafico da regiao de Bananal (São Paulo). *Publicaçao Especial No. 2. 2° Congresso da Associaçao Brasileira de Estudos do Quaternario*, 1989. Universidade Federal do Rio De Janeiro, Brazil.

Dengler, L., Lehre, A. K. & Wilson, C. J. (1987). Bedrock geometry of unchannelised valleys. In *Erosion and Sedimentation in the Pacific Rim* (ed. R. L. Beschta, T. Blinn, C. E. Grant, F. J. Swanson & G. G. Ice), pp. 81–90. IAHS Publication No. 165.

Dennen, W. H. & Norton, H. A. (1977). Geology and geochemistry of bauxite deposits in the lower Amazon basin. *Economic Geology*, **72**, 82–89.

De Oliveira, M. A. T. (1990). Slope geometry and gully erosion development: Bananal, São Paulo, Brazil. *Zeitschrift für Geomorphology N.F.*, **34**, 423–434.

De Ploey, J. (1964). Stone-lines and clayey-sandy mantles in Lower Congo: their formation and the effect of termites. In *Etudes sur les Termites Africains* (ed. A. Bouillon), pp. 399–414. Université de Louvanium, Leopoldville.

De Ploey, J. (1965). Position géomorphologique, génèse et chronologie de certains dépots superficiels au Congo occidentale. *Quaternaria*, **7**, 131–154.

De Ploey, J. (1968). Quaternary phenomena in the western Congo. In *Means of Correlation of Quaternary Succession*, Vol. 8, pp. 501–515. Proceedings VII Congress INQUA, University of Utah Press.

De Ploey, J. (1974). Mechanical properties of hillslopes and their relation to gullying in central semi-arid Tunisia. *Zeitschrift für Geomorphologie, N.F.*, **21**, 177–190.

De Ploey, J. & Cruz, O. (1979). Landslides in the Serra do Mar, Brazil. *Catena*, **6**, 111–122.

De Ploey, J. & Poesen, J. (1989). Experimental geomorphology and the interpretation of stone-lines. In *Stone-Lines, Journée d'Étude*, Bruxelles, 24 March 1987 (ed. J. Alexandre & J. J. Symoens). Académie Royale des Sciences D'Outre-Mer, Geo-Eco-Trop, **11**, 75–82.

De Swardt, A. M. J. (1964). Lateritisation and landscape development in equatorial Africa. *Zeitschrift für Geomorphologie, N.F.*, **8**, 313–333.

De Swardt, A. M. J. & Bennett, G. (1974). Structural and physiographic development of Natal since the late Jurassic. *Transactions Geological Society of South Africa*, **77**, 309–322.

De Swardt, A. M. J. & Trendall, A. F. (1969). The physiographic development of Uganda. *Overseas Geology and Mineral Resources*, **10**, 241–288.

Dethier, D. P. (1986). Weathering rates and the chemical flux from catchments in the Pacific northwest USA. In *Rates of Chemical Weathering of Rocks and Minerals* (eds. S. M. Colman & D. P. Dethier), 503–530. Academic Press, Orlando & London.

De Vleeschauwer, D., Lal, R. & De Boodt, M. (1978). Comparison of detachability indices in relation to soil erodibility for some important Nigerian soils. *Pedologie*, **28**, 5–20.

Didier, P., Nahon, D., Fritz, B. & Tardy, Y. (1983). Activity of water as a chemical controlling factor in ferricretes. A thermodynamic model in the system: kaolinite–Fe–Al–oxyhydroxides. In *Proceedings International Colloquium Petrology Weathering Soils*, pp. 35–44.

Diester-Haas, L. (1976). Late Quaternary climatic variations in N.W. Africa deduced from east Atlantic sediment cores. *Quaternary Research*, **6**, 299–314.

Dietrich, W. E. & Dorn, R. (1984). Significance of thick deposits of colluvium on hillslopes: a case study involving the use of pollen analysis in the coastal mountains of California. *Journal of Geology*, **92**, 147–158.

Dietrich, W. E. & Dunne, T. (1978). Sediment budget for a small catchment in mountainous terrain. *Zeitschrift für Geomorphologie, N.F., Supplementband*, **29**, 191–206.

Dietrich, W. E., Wilson, C. J. & Reneau, S. L. (1986). Hollows, colluvium and landslides in soil-mantled slopes. In *Hillslope Processes* (The Binghamton Symposia in Geomorphology: International Series, no 16) (ed. A. D. Abrahams), pp. 361–388. Allen & Unwin, Boston & London.

Dietrich, W. E., Reneau, S. L. & Wilson, C. J. (1987). Overview: 'zero order basins' and problems of drainage density, sediment transport and hillslope morphology. In *Erosion and Sedimentation in the Pacific Rim* (ed. R. L. Beschta, T. Blinn, C. E. Grant, F. J. Swanson & G. G. Ice), pp. 27–37. IAHS Publication No. 165.

Dijkerman, J. C. & Miedema, R. (1988a). An ustult–aquult–tropept catena in Sierra Leone, West Africa, I. Characteristics, genesis and classification. *Geoderma*, **42**, 1–27.

Dijkerman, J. C. & Miedema, R. (1988b). An ustult–aquult–tropept catena in Sierra Leone, West Africa, II. Land qualities and land evaluation. *Geoderma*, **42**, 29–49.

Dingle, R. V. (1982). Continental margin subsidence: a comparison between the east and west coasts of Africa. In *Dynamics of Passive Margins* (ed. R. A. Scrutton), pp. 59–71. Geodynamics Series, Vol. 6, American Geophysical Union.

Dissanayake, C. B. (1987). Metals in a lateritic peat deposit—a case study from Sri Lanka. *Chemical Geology*, **60**, 137–143.

Dixey, F. (1922). The physiography of Sierra Leone. *Geographical Journal*, **60**, 41–65.

Dixon, J. C. & Young, R. W. (1981). Character and origin of deep arenaceous weathering mantles on the Bega batholith, southeastern Australia. *Catena*, **8**, 87–109.

Dohrenwend, J. C., Wells, S. G., Mcfadden, L. D. & Turrin, B. D. (1987). Pediment dome evolution in the eastern Mojave Desert, California. In *International Geomorphology 1986*, Part II (ed. V. Gardiner), pp. 1047–1062. Wiley, Chichester.

Doornkamp, J. C. (1968). The role of inselbergs in the geomorphology of southern Uganda. *Transactions of the Institute of British Geographers*, **44**, 151–162.

Doornkamp, J. C. (1970). *The Geomorphology of the Mbarara Area—Sheet SA-36-1*. Geological Survey and Mines Department, Uganda (also published as Geomorphological Report No. 1, Department of Geography, University of Nottingham, 1970).

Doornkamp, J. C. (1986). Climate and weathering. In *A Handbook of Engineering Geomorphology* (ed. P. G. Fookes & P. R. Vaughan), pp. 10–24. Surrey University Press, London & Chapman & Hall, New York.

Doornkamp, J. C. & King, C. A. M. (1971). *Numerical Analysis in Geomorphology*. Arnold, London.

Douglas, I. (1967a). Natural and man-made erosion in the humid tropics of Australia, Malaysia and Singapore. *IAHS Publication*, **75**, 17–25.

Douglas, I. (1967b). Man, vegetation and the sediment yield of rivers. *Nature*, **215**, 925–928.

Douglas, I. (1973). *Rates of Denudation in Selected Small Catchments in Eastern Australia*. University of Hull Occasional Papers in Geography, **21**, pp. 128. Hull.

Douglas. I. (1976). Erosion rates and climate: geomorphological implications. In *Geomorphology and Climate* (ed. E. Derbyshire), pp. 269–287. Wiley, New York.

Douglas, I. (1978a). Denudation of silicate rocks in the humid tropics. In *Landform Evolution in Australasia* (ed. J. L. Davies & M. A. J. Williams), pp. 216–237. ANU Press, Canberra.

Douglas, I. (1978b). Tropical geomorphology: present problems and future prospects. In *Geomorphology—Present Problems and Future Prospects* (ed. C. Embleton, D. Brunsden & D. K. C. Jones), pp. 162–184. Oxford University Press, Oxford.

Douglas, I. (1980). Climatic geomorphology—present-day processes and landform evolution, problems of interpretation. *Zeitschrift für Geomorphologie, N.F., Supplementband*, **36**, 27–47.

Douglas, I. & Spencer, T. (1985a). Present-day processes as a key to the effects of environmental change. In *Environmental Change and Tropical Geomorphology* (ed. I. Douglas & T. Spencer), pp. 39–73. Allen and Unwin, London.

Douglas, I. & Spencer, T. (eds.) (1985b). *Environmental Change and Tropical Geomorphology*. Allen and Unwin, London.

Douglas, R. G. & Woodruff, F. (1981). Deep sea benthic foraminifera. In *The Sea*, vol. 7 (ed. C. Emiliani), pp. 1233–1327. Wiley-Interscience, New York.

Dowling, J. W. F. (1968). Land evaluation for engineering purposes in Northern Nigeria. In *Land Evaluation* (ed. G. A. Stewart), pp. 147–159. Macmillan of Australia, Canberra.

Downing, B. H. (1968). Subsurface erosion as a geomorphological agent in Natal. *Transactions of the Geological Society of South Africa*, **81**, 131–134.

Dresch, J. (1978). Les reliefs résiduels. *Bulletin, Association Géographie Française*, **450**, 97–103.

D'Souza, V. P. C. & Morgan, R. P. C. (1976). A laboratory study of the effect of slope steepness and curvature on soil erosion. *Journal of Agricultural Engineering Research*, **21**, 21–31.

Dubreuil, P. L. (1985). Review of field observations of runoff generation in the tropics. *Journal of Hydrology*, **80**, 237–264.

Dubroeucq, D. & Blancaneaux, P. (1987). Les podzols du haut Rio Negro, region de Maroa, Venezuela. Environnement et relations lithologiques. In *Podzols et Podzolisation* (ed. D. Righi & A. Chauvel), pp. 37–52. Comptes Rendus de la Table Ronde Internationale, Association Française pour l'Etude du Sol, ORSTOM/INRA.

Duchaufour, P. (1982). *Pedology, Pedogenesis and Classification* (Trans. T. R. Paton). George Allen and Unwin, London.

Duncan, R. A. & Pyle, D. G. (1988). Rapid eruption of the Deccan flood basalts at the Cretaceous/Tertiary boundary. *Nature*, **333**, 841–843.

Dunne, T. (1979). Sediment yield and land use in tropical catchments. *Journal of Hydrology*, **42**, 281–300.

Dunne, T., Dietrich, W. E. & Brunengo, M. J. (1978). Recent and past erosion rates in semi-arid Kenya. *Zeitschrift für Geomorphologie, N.F., Supplementband*, **29**, 130–140.

Durgin, P. B. (1977). Landslides and the weathering of granite rocks. In *Landslide Perspectives* (ed. D. R. Coates), pp. 127–131. Reviews in Engineering Geology, Vol. III, Geological Society of America.

Dury, G. H. (1969). Rational descriptive classification of duricrusts. *Earth Science Journal*, **3**, 77–86.

Dury, G. H. (1971). Relict deep weathering and duricrusting in relation to the palaeoenvironments of middle latitudes. *Geographical Journal*, **137**, 511–522.

Eden, M. J. (1971). Some aspects of weathering and landforms in Guyana. *Zeitschrift fur Geomorphologie, N.F.*, **15**, 181–198.

Eggler, D. H., Larson, E. E. & Bradley, W. C. (1969). Granites, grusses and the Sherman erosion surface, southern Laramie Range, Colorado–Wyoming. *American Journal of Science*, **267**, 510–522.

Ehlen, J. & Zen, E.-An. (1986). Petrographic factors affecting jointing in the banded series, Stillwater Complex, Montana. *Journal of Geology*, **94**, 575–584.

Ellen, S. D. (1988). Description and mechanics of soil slip/debris flows in the storm. In *Landslides, Floods, and Marine Effects of the Storm of January 3–5, 1982, in the San Francisco Bay Region, California* (ed. S. D. Ellen & G. F. Wieczorek), US Geological Survey Professional Paper No. 1434, Washington.

Ellen, S. D. & Wieczorek, G. F. (ed.) (1988). *Landslides, Floods, and Marine Effects of the Storm of January 3–5, 1982, in the San Francisco Bay Region, California.* US Geological Survey Professional Paper, No. 1434, Washington.

Ellen, S. D., Cannon, S. H. & Reneau, S. (1988). Distribution of debris flows in Marin County. In *Landslides, Floods, and Marine Effects of the Storm of January 3–5, 1982, in the San Francisco Bay Region, California.* (ed. S. D. Ellen & G. F. Wieczorek), pp. 113–132. US Geological Survey Professional Paper No. 1434, Washington.

El-Swaify, S. A., Dangler, E. W. & Armstrong, C. L. (1982). *Soil Erosion by Water in the Tropics.* College of Tropical Agriculture and Human Resources, University of Hawaii.

Elvhage, C. & Lidmar-Bergström, K. (1987). Some working hypotheses on the geomorphology of Sweden in the light of a new relief map. *Geografiska Annaler*, **69A**, 343–358.

Elwell, H. A. (1971). Erosion Research Programmes, Rhodesia. Department of Conservation Extension, Salisbury (Harare), Zimbabwe.

Elwell, H. A. & Stocking, M. A. (1973). Rainfall parameters for soil loss estimation in a sub-tropical climate. *Journal of Agricultural Engineering Research*, **18**, 169–177.

Elwell, H. A. & Stocking, M. A. (1975). Parameters for estimating annual runoff and soil loss from agricultural lands in Rhodesia. *Water Resources Research*, **11**, 601–605.

Elwell, H. A. & Stocking, M. A. (1976). Vegetal cover to estimate soil erosion hazard in Rhodesia. *Geoderma*, **15**, 61–70

Elwell, H. A. & Stocking, M. A. (1982). Developing a simple yet practical method of soil-loss estimation. *Tropical Agriculture (Trinidad)*, **59**, 43–48.

Embrechts, J. & De Dapper, M. (1987). Morphology and genesis of hillslope pediments in the Febe area (South-Cameroon). *Catena*, **14**, 31–43.

Emiliani, C. (1955). Pleistocene temperatures. *Journal of Geology*, **63**, 538–578.

Enslin, J. F. (1961). Secondary aquifers in South Africa and the scientific selection of boring sites in them. *International Conference on Hydrology*, Nairobi, CCTA Publication **66**, Section 4, Groundwater Hydrology.

Erhart, H. (1955). Biostasie et rhexistasie: esquise d'une théorie sur le rôle de la pédongenèse en tant que phénomène géologique. *Comptes Rendues Academie des Sciences Française*, **241**, 1218–1220.

Esson, J. (1983). Geochemistry of a nickeliferous laterite profile, Liberdade, Brazil. In *Residual Deposits: Surface Related Weathering Processes and Materials* (ed. R. C. L. Wilson), pp. 91–99. Geological Society Special Publication, No. 11, Blackwell, Oxford.

Esson, J. & Surcan dos Santos, L. C. (1978). Chemistry and mineralogy of section through lateritic nickel deposit at Liberdade, Brazil. *Transactions Institute of Mining and Metallurgy*, **87 Sec. B**, B53–B60.

Eswaran, H. & Stoops, G. (1979). Surface texures of quartz in tropical soils. *Soil Science Society of America Journal*, **43**, 420–424.

Eswaran, H., Sys, C. & Sousa, E. C. (1975). Plasma infusions — A pedological process of significance in the humid tropics. *Edafologia y Agrobiologia*, **34**, 665–674.

Evans, D. L. (1982). Lateritisation as a possible contributor to gold placers. *Engineering and Mining Journal*, **1**, 86–91.

Evans, I. S. (1971). Salt crystallisation and rock weathering, a review. *Revue de Géomorphologie Dynamique*, **19**, 153–171.

Eybergen, F. A. & Imeson, A. C. (1989). Geomorphological processes and climatic change. *Catena*, **16**, 307–319.

Eyles, R. J. (1969). Depth of dissection of the West Malaysian landscape. *Journal of Tropical Geography*, **28**, 23–31.

Eyles, R. J. (1971). A classification of West Malaysian drainage basins. *Annals of the Association of American Geographers*, **61**, 460–467.

Eyles, R. J. & Ho, R. (1970). Soil creep on a humid tropical slope. *Journal of Tropical Geography*, **31**, 40–42.

Fabre, J. & Petit-Maire, N. (1988). Holocene climatic evolution at 22–23°N from two paleolakes in the Taoudeni area (northern Mali). *Palaeogeography, Palaeoclimatology, Palaeoecology*, **65**, 133–148.

Faillace, C. (1973). Location of groundwater and the determination of the optimum depth of wells in metamorphic rocks of Karamoja, Uganda. *2nd Conversazione Internationale Sulle Aque Sotterranee, Palermo.*

Fairbridge, R. W. (1961). Eustatic changes in sea level. In *Physics and Chemistry of the Earth* (ed. L. H. Ahrens, F. Press, K. Rankama & S. K. Runcorn), **4**, 99–185, Pergamon, London.

Fairbridge, R. W. (1964). African ice-age aridity. In *Problems of Palaeoclimatology* (ed. A. E. M. Nairn), pp. 356–360. Wiley-Interscience, New York.

Fairbridge, R. W. (1976). Effects of Holocene climatic change on some tropical geomorphic processes. *Quaternary Research*, **6**, 529–556.

Fairbridge, R. W. & Finkl, C. W. (1978). Geomorphic analysis of the rifted cratonic margins of Western Australia. *Zeitschrift für Geomorphologie, N.F.*, **22**, 369–389.

Fairbridge, R. W. & Finkl, C. W. Jr. (1980). Cratonic erosional unconformities and peneplains. *Journal of Geology*, **88**, 69–86.

Fairbridge, R. W. & Finkl, C. W. Jr. (1984). Tropical stone lines and podzolised sand plains as palaeoclimatic indicators for weathered cratons. *Quaternary Science Reviews*, **3**, 41–72.

Falconer, J. D. (1911). *The Geology and Geography of Northern Nigeria*. Macmillan, London.

Falconer, J. D. (1912). The origin of kopjes and inselbergs. *British Association for the Advancement of Science, Transactions Section C*, p. 476.

Fan, C.-H., Allison, R. J. & Jones, M. E. (1994). Effects of tropical weathering on the physical characteristics of argillaceous rocks. In *Rocks, Rock Weathering and Landform Evolution* (ed. D. Robinson & R. Williams). Wiley, Chichester.

Faniran, A. (1974). The extent, profile and significance of deep weathering in Nigeria. *Journal of Tropical Geography*, **38**, 19–30.

Faniran, A. & Jeje, L. K. (1983). *Humid Tropical Geomorphology*. Longman, Harlow.

FAO (1985). *Soil Map of the World 1:5 000 000, Revised Legend*. Food and Agriculture Organisation of the United Nations, Paris.

Feininger, T. (1969). Pseudo-karst on quartz diorite, Colombia. *Zeitschrift für Geomorphologie, N.F.*, **13**, 287–296.

Feininger, T. (1971). Chemical weathering and glacial erosion of crystalline rocks and the origin of till. *U. S. Geological Survey Professional Paper*, **750-C**, C65–C81.

Ferreira, R. C. & Monteiro, L. B. (1985). Identification and evaluation of collapsibility of colluvial soils that occur in the São Paulo State. In *First International Conference on Geomechanics in Tropical Lateritic and Saprolitic Soils*, Brasilia, 1985, Vol. 1, pp. 269–280, Brasilia.

Filizola, H. F. & Boulet, R. (1993). Une évaluation de la vitesse de l'érosion géochimique à partir de l'étude de dépressions fermées sur roches sédimentaires quartzo-kaoliniques au Brésil. *Comptes Rendues, Academie Sciences, Paris, Série II*, **316**, 693–700.

Finkl, C. W. Jr. (1979). Stripped (etched) landsurfaces in southern Western Australia. *Australian Geographical Studies*, **17**, 33–52.

Finkl, C. W, Jr. (1980). Stratigraphic principles and practices as related to soil mantles. *Catena*, **7**, 169–194.

Finkl, C. W. Jr. (1981). Catastrophic erosion events on cratons. *Notas Geomorfologia, Campinas*, **21**, 27–40.

Finkl, C. W. Jr. (1982). On the geomorphic stability of cratonic planation surfaces. *Zeitschrift für Geomorphologie, N.F.*, **26**, 137–150.

Finkl, C. W. Jr. (1984). Evaluation of relative pedostratigraphic dating methods, with special reference to Quaternary successions overlying weathered platform materials. In *Quaternary Dating Methods* (ed. W. C. Mahaney), pp. 323–353. Elsevier, Amsterdam.

Finkl, C. W. Jr. & Churchward, H. M. (1973). The etched surfaces of southwestern Australia. *Journal of the Geological Society of Australia*, **20**, 295–307.

Finkl, C. W. Jr. & Churchward, H. M. (1976). Soil stratigraphy in a deeply weathered shield landscape in south-western Australia. *Australian Journal of Soil Research*, **14**, 109–120.

Fitzpatrick, E. A. (1963). Deeply weathered rock in northeast Scotland, its occurrence, age and contribution to the soils. *Journal of Soil Science*, **14**, 33–42.

Flenley, J. R. (1979). *The Equatorial Rainforest: A Geological History*. Butterworth, London.

Flenley, J. R. (1985). Relevance of quaternary palynology to geomorphology in the tropics and subtropics. In *Environmental Change and Tropical Geomorphology* (ed. I. Douglas & T. Spencer), pp. 153–164. Allen and Unwin, London.

Flint, R. F. (1959a). On the basis of Pleistocene correlation in East Africa. *Geological Magazine*, **96**, 265–284.

Flint, R. F. (1959b). Pleistocene climates in eastern and southern Africa. *Bulletin Geological Society of America*, **70**, 342–374.

Flohn, H. & Nicholson, S. E. (1980). Climatic fluctuations in the arid belt of the 'Old World' since the last glacial maximum: possible causes and future implications. *Palaeoecology of Africa*, **12**, 3–21.

Fölster, H. (1969). Slope development in SW Nigeria during the late Pleistocene and Holocene. *Göttinger Bodenkundliche Berichte*, **10**, 3–56.

Fölster, H. (1992). Holocene autochthonous forest degradation in south eastern Venezuela. In *Tropical Forests in Transition* (ed. J. G. Goldammer). Birkenhauer.

Fölster, H., Moshrefi, N. & Ojenuga, A. G. (1971a). Ferralitic pedogenesis on metamorphic rocks, SW Nigeria. *Pedologie*, **21**, 95–124.

Fölster, H., Kalk, E. & Moshrefi, N. (1971b). Complex pedogenesis of ferrallitic savanna soils in south Sudan. *Geoderma*, **6**, 135–149.

Fookes, P. G., Dearman, W. R. & Franklin, J. A. (1971). Some engineering aspects of rock weathering with field examples from Dartmoor and elsewhere. *Quarterly Journal of Engineering Geology, London*, **4**, 139–185.

Fookes, P. G. & Vaughan, P. R. (1986). *A Handbook of Engineering Geomorphology*. Surrey University Press, and Blackie, Glasgow.

Ford, D. C. & Williams, P. W. (1989). *Karst Geomorphology and Hydrology*. Unwin Hyman, London.

Fournier, F. (1960). *Climat et Érosion: La Relation entre l'Érosion du Sol par l'Eau et les Précipitations Atmosphériques*. Presses Universitaires, Paris.

Fournier, F. (1962). *Carte du Danger d'Érosion en Afrique au Sud du Sahara*. CEE-CCTA, Presses Universitaires, Paris.

Frakes, L. A. (1979). *Climates Through Geologic Time*. Elsevier, Amsterdam.

Franks, C. A. M. & Woods, N. W. (1993). Engineering geology of North Lantau, Hong Kong. *Quarterly Journal of Engineering Geology, London*, **26**, 81–98.

Franzinelli, E. & Potter, P. E. (1983). Petrology, chemistry and texture of modern river sands, Amazon River system. *Journal of Geology*, **91**, 23–39.

Freise, F. W. (1932). Erscheinungen des Erdfliessens im Tropenurwalde. *Zeitschrift für Geomorphologie*, **IX**, 85–98.

Freise, F. W. (1938). Inselberge und Inselberglandschaften im granit und gneissgebeite, Brasiliens. *Zeitschrift für Geomorphologie*, **X**, 137–168.

Fritz, B. & Tardy, Y. (1973). Étude thermodynamique du système gibbsite, quartz, kaolinite, gaz carbonique. Application à la genèse des podzols et des bauxites. *Sciences Géologiques* (Strasbourg), **26**, 339–367.

Fukuoka, M. (1980). Landslides associated with rainfall. *Geotechnical Engineering*, **11**, 1–29.

Furch, K. (1984). Water chemistry of the Amazon basin: the distribution of chemical elements amongst freshwaters. In *The Amazon* (ed. H. Sioli), pp. 167–199. Junk, Dordrecht.

Furley, P. A., Proctor, J. & Ratter, J. A. (eds) (1992). *Nature and Dynamics of Forest-Savanna Boundaries*. Chapman & Hall, London.

Gac, J. Y. (1979). *Géochemie du bassin du lac Tchad*. Thèse, Université de Strasbourg.

Gale, S. J. (1992). Long-term landscape evolution in Australia. *Earth Surface Processes and Landforms*, **17**, 323–343.

Gardner, W. R. (1977). Water entry and movement in relation to erosion. In *Soil Conservation and Management in the Humid Tropics* (ed. D. J. Greenland & R. Lai), pp. 25–31. Wiley, Chichester.

Gardner, L. R. (1992). Long-term isovolumetric leaching of aluminium from rocks during weathering: implications for the genesis of saprolite. *Catena*, **19**, 521–537.

Garner, H. F. (1974). *The Origin of Landscapes*. Oxford University Press, New York.

Garrels, R. M. & Christ, C. L. (1965). *Solutions, Minerals and Equilibria*. Harper and Row, New York.

Garwood, N. C., Janos, D. P. & Brokaw, N. (1979). Earthquake-caused landslides: a major disturbance to tropical forests. *Science*, **205**, 997–999.

Gaskin, A. R. G. (1975). Investigation of the residual iron ores of the Tonkilili district, Sierra Leone. *Transactions Institute of Mining and Metallurgy, Section B, Applied Earth Science*, pp. B98–B119.

Gasse, F., Lédée, V., Massault, M. & Fontes, J.-C. (1989). Water level fluctuations of lake Tanganyika in phase with oceanic changes during the last glaciation and deglaciation. *Nature*, **342**, 57–59.

Gaur, R. & Chopra, S. R. K. (1984). Taphonomy, fauna and environment and ecology of Upper Siwaliks (Plio-Pleistocene) near Chandigarh, India. *Nature*, **308**, 353–355.

Gellert, J. F. (1970). Climatomorphology and palaeoclimates of the central European Tertiary. In *Problems of Relief Planation* (ed. M. Pecsi), pp. 107–112. Akadamiai Kaido, Budapest.

Geological Society (1990). Engineering Group Working Party Report: Tropical residual soils. *Quarterly Journal of Engineering Geology*, **23**, 1–101.

Gerrard, A. J. (1988). *Rocks and Landforms*. Unwin Hyman, London.

Ghosh, K. P. & Datta, B. C. (1978). Mineralogy and genesis of Phutkapahar bauxite deposits of eastern Madhya Pradesh, India. *4th International Congress for the Study of Bauxites, Alumina, and Aluminium*, **1**, 35–67, India.

Ghosh, K. P. & McFarlane, M. J. (1984). Some aspects of the geology and geomorphology of the bauxite belt of central India. In (ed. S. Sinha-Roy & S. K. Gosh), *Products and Processes of Rock Weathering, Recent Researches in Geology (India)*, **11**, pp. 8–135, Hindustani Publishing, Delhi.

Giardino, J. R. & Mäckel, R. (1985). Correlative development of dambos and dwalas: plateau regions of Zambia. *Zeitschrift für Geomorphologie Supplementband*, **52**, 187–200.

Gibbs, R. J. (1967). The geochemistry of the Amazon River System, Part I. The factors that control the salinity and composition of the suspended solids. *Geological Society of America Bulletin*, **78**, 1203–1232.

Gidigasu, M. D. (1976). *Laterite Soil Engineering*. Elsevier, Amsterdam.

Gidigasu, M. D. (1988). Potential application of engineering pedology in shallow foundation engineering on tropical residual soils. In *Geomechanics in Tropical Soils, Proceedings of the Second International Conference on Geomechanics in Tropical Soils*, Singapore, 1988, Vol. 1, pp. 17–24. Balkema, Rotterdam.

Gilbert, G. K. (1909). The convexity of hilltops. *Journal of Geology*, **17**, 344–350.

Gilchrist, A. R. & Summerfield, M. A. (1990). Differential denudation and flexural isostasy in formation of rifted-margin upwarps. *Nature*, **346**, 739–742.

Gilchrist, A. R. & Summerfield, M. A. (1991). Denudation, isostasy and landscape evolution. *Earth Surface Processes and Landforms*, **16**, 555–562.

Gilkes, R. J., Scholz, G. & Dimmock, G. M. (1973). Lateritic deep weathering of granite. *Journal of Soil Science*, **24**, 523–536.

Giresse, P. (1978). Le control climatique de la sédimentation marine et continentale en Afrique centrale Atlantique à la fin du Quaternaire. *Palaeogeography, Palaeoclimatology, and Palaeoecology*, **23**, 57–77.

Giresse, P. & Lanfranchi, R. (1984). Les climats et les océans de la région congolaise pendant l'Holocène — Bilans selon les échelles et les méthodes de l'observation. *Palaeoecology of Africa*, **16**, 77–88.

Giresse, P., Bongo-Passi, G., Delibrias, G. & Du Plessy, J. C. (1982). La lithostratigraphie des sediments hemiplagiques du delta profond du fleuve Congo et ses indications sur les paléoclimats de la fin du Quaternaire. *Bulletin Société Géologie française, Série*, **24**, 803–815.

Gjessing, J. (1967). Norways's Paleic surface. *Norsk Geographische Tidschrift*, **21**, 69–132.

Glazovskaya, M. A. (1984). *Soils of the World*, Vol. II. Balkema.

Goddard, A. (1965). *Recherches en Géomorphologie en Écosse de Nord Ouest*. Masson et Cie, Paris.

Goldbery, R. (1979). Sedimentology of the Lower Jurassic flint clay bearing Mishor Formation, Makhtesh Ramon, Israel. *Sedimentology*, **19**, 229–251.

Goldsmith, P. R. & Smith, E. H. (1985). Tunnelling erosion in S. Auckland, New Zealand. *Engineering Geology*, **22**, 1–11.

Gonzalez, M. A. (1981). Evidencias paleoclimaticas en la Salina del Berbedero (San Luis). *Actas 8th Conges Geologie Argentina, San Luis*, **3**, pp. 411–438, San Luis.

Goudie, A. (1973). *Duricrusts in Tropical and Sub-Tropical Landscapes*. Clarendon Press, Oxford.

Goudie, A. S. (1983a). The arid earth. In *Megageomorphology* (ed. R. Gardner & H. Scoging), pp. 152–171. Clarendon Press, Oxford.

Goudie, A. S. (1983b). Calcrete. In *Chemical Sediments and Geomorphology* (ed. A. S. Goudie & K. Pye), pp. 93–131. Academic Press, London.

Goudie, A. S. and Pye, K. (ed.) (1983). *Chemical Sediments and Geomorphology*. Academic Press, London.

Goudie, A. S. & Bull, P. A. (1984). Slope processes and colluvium deposition in Swaziland: an SEM analysis. *Earth Surface Processes and Landforms*, **9**, 289–299.

Govett, G. J. S. (1987). Exploration geochemistry in some low-latitude areas — problems and techniques. *Transactions Institute Mining and Metallurgy (Series B: Applied Earth Sciences)*, **96**, B97–B116.

Grandin, G. (1976). *Aplanissements Cuirassés et Enrichessement des Gisements de Manganèse dans Quelques Régions d'Afrique de l'Ouest. Mémoire de L'ORSTOM*, vol. 82, Paris.

Grandin, G. & Hayward, D. F. (1975). Aplanissements Cuirassés de la Peninsula de Freetown. *Cahiers de L'ORSTOM, Serie Géologie*, **1**, 11–16.

Grant, W. H. (1969). Abrasion pH, and index of weathering. *Clays and Clay Minerals*, **17**, 151–155.

Grant, K. & Aitcheson, G. D. (1970). The engineering significance of silcretes and ferricretes in Australia. *Engineering Geology*, **4**, 93–120.

Grant, K. (1975). *The PUCE Programme for Terrain Evaluation for Engineering Purposes. I. Principles*. CSIRO Australia Division of Soil Mechanics, Technical Paper **15**, pp. 32 (reprinted in D. A. Davidson (ed.) (1986).

Greenland, D. J. (1977a). Soil structure and erosion hazard. In *Soil Conservation and Management in the Humid Tropics* (ed. D. J. Greenland & R. Lal), pp. 17–23. Wiley, Chichester.

Greenland, D. J. & Lal, R. (eds) (1977b). *Soil Conservation and Management in the Humid Tropics.* Wiley, Chichester.

Greenland, D. J. (ed.) (1981). *Characterisation of Soils— Examples from the Humid Tropics.* Clarendon, Oxford.

Greenway, D. R. (1987). Vegetation and slope stability. In *Slope Stability* (ed. M. G. Anderson & K. S. Richards), pp. 187–230. Wiley, Chichester.

Greenway, D. R. C., Anderson, M. G. & Brian-Boys, K. C. (1985). Influence of vegetation on slope stability in Hong Kong. In *IVth International Symposium on Landslides*, Toronto, 1984, Vol. 1, pp. 399–404.

Gregory, K. J. (1983). *Background to Palaeohydrology.* Wiley, Chichester.

Grey, D. R. C. & Cooke, H. J. (1977). Some problems in the Quaternary evolution of the landforms of northern Botswana. *Catena*, **4**, 123–133.

Grillot, J. C., Blavoux, B., Rakotondrainbe, J. H. & Raunet, M. (1989). Dynamique en hautes eaux des aquifères d'altérites sur les hauts plateaux cristallophylliens de madagascar. *Journal of African Earth Sciences*, **9**, 599–607.

Grove, A. T. (1952). *Land Use and Soil Conservation on the Jos Plateau.* Geological Survey of Nigeria, Bulletin No. 22.

Grove, A. T. (1958). The ancient erg of Hausaland and similar formations on the south side of the Sahara. *Geographical Journal*, **124**, 528–533.

Grove, A. T. & Warren, A. (1968). Quaternary landforms and climate on the south side of the Sahara. *Geographical Journal*, **134**, 193–208.

Grubb, P. L. C. (1979). Genesis of bauxite deposits in the lower Amazon Basin and Guianas coastal plain. *Economic Geology*, **74**, 735–750.

Gupta, A. (1988). Large floods as geomorphic events in the humid tropics. In *Flood Geomorphology* (ed. V. R. Baker, R. C. Kochel & P. C. Patton), pp. 301–315. Wiley, New York.

Gupta, A. (1993). The changing geomorphology of the humid tropics. *Geomorphology*, **7**, 165–186.

Gupta, A. & Dutt, A. (1989). The Auranga: description of a tropical monsoon river. *Zeitschrift für Geomorphologie, N.F.*, **33**, 73–92.

Gupta, A., Rahman, A., Poh Poh, W. & Pitts, J. (1987). The Old Alluvium of Singapore and the extinct drainage system to the South China Sea. *Earth Surface Processes and Landforms*, **12**, 259–275.

Haantjens, H. A. & Bleeker, P. (1970). Tropical weathering in the territory of Papua New Guinea. *Australian Journal of Soil Research*, **8**, 157–177.

Haberle, S. G., Hope, G. S. & Defretes, Y. (1991). Environmental change in the Baliem Valley, montane Irian Jaya, Republic of Indonesia. *Journal of Biogeography*, **18**, 25–40.

Hack, J. T. (1960). Interpretation of erosional topography in humid temperate regions. *American Journal of Science*, **258A**, 80–97.

Hack, J. T. (1965). Geomorphology of the Shenandoah Valley, Virginia, and origin of the residual ore deposits. *US Geological Survey, Professional Paper*, **484**, p. 84.

Hack, J. T. (1975). Dynamic equilibrium and landscape evolution. In *Theories of Landform Development* (ed. W. C. Melhorn & R. C. Flemal), pp. 87–102. George, Allen and Unwin.

Hack, J. T. (1979). Rock control and tectonism— their importance in shaping the Appalachian highlands. *US Geological Survey Professional Paper*, **1126-B**.

Hack, J. T. (1982). Physiographic Divisions and Differential Uplift in the Piedmont and Blue Ridge, pp. 49, *US Geological Survey Professional Paper*, **1265**, Washington.

Hack, J. T. & Goodlett, J. C. (1960). Geomorphology and forest ecology of a mountain range in the central Appalachians. *US Geological Survey Professional Paper*, **347**.

Haffer, J. (1969). Speciation in Amazon forest birds. *Science*, **165**, 131–137.

Haffer, J. (1987). Quaternary history of tropical America. In *Biogeography and Quaternary History in Tropical America* (ed. T. C. Whitmore & G. T. Prance), pp. 1–18. Clarendon Press, Oxford.

Haile, N. S. (1971). Quaternary shorelines in West Malaysia and adjacent parts of the Sunda Shelf. *Quaternaria*, **15**, 338–343.

Hails, J. R. (ed.) (1977). *Applied Geomorphology.* Elsevier, Amsterdam.

Haldemann, E. G. (1956). Recent landslide phenomena in the Rungwe volcanic area, Tanganyika. *Tanganyika Notes and Records*, **45**, 3–14.

Hall, A. M. (1985). Cenozoic weathering covers in Buchan, Scotland and their significance. *Nature*, **315**, 392–395.

Hall, A. M. (1986). Deep weathering patterns in north-east Scotland and their geomorphological significance. *Zeitschrift für Geomorphologie, N.F.*, **30**, 407–422.

Hall, A. M. (1987). Weathering and relief development in Buchan, Scotland. In *International Geomorphology 1986*, Part II (ed. V. Gardiner), pp. 991–1005. Wiley, Chichester.

Hall, A. M. (1988). The characteristics and significance of deep weathering in the Gaick area, Grampian Highlands, Scotland. *Geografiska Annaler*, **70A**, 309–314.

Hall, A. M. (1991). Pre-Quaternary landscape evolution in the Scottish Highlands. *Transactions Royal Society of Edinburgh: Earth Sciences*, **82**, 1–26.

Hall, A. M. Mellor, A. M. & Wilson, M. J. (1989). The clay mineralogy and age of deeply weathered rock in north-east Scotland. *Zeitschrift für Geomorphologie, N.F., Supplementband*, **72**, 97–108.

Hall, A. M., Thomas, M. F. & Thorp, M. B. (1985). Late Quaternary alluvial placer development in the humid tropics: the case of the Birim Diamond Placer, Ghana. *Journal of The Geological Society*, **142**, 777–787.

Hall, P. K. (1969). The Diamond Fields of Sierra Leone. *Geological Survey Sierra Leone Bulletin*, **5**, Freetown.

Hamilton, A. C. (1976). The significance of patterns of distribution shown by forest plants and animals in tropical Africa for the reconstruction of the Upper Pleistocene environment: a review. *Palaeoecology of Africa*, **9**, 63–97.

Hamilton, A. C. (1982). *Environmental History of East Africa: A Study of the Quaternary*. Academic Press, New York.

Hamilton, A. C. & Taylor, D. (1991). History of climate and forests in tropical Africa during the last 8 million years. *Climate Change* (Special Issue: Tropical Forests and Climate), **19**, 65–78.

Handley, J. R. F. (1952). The geomorphology of the Nzega area of Tanganyika with special reference to the formation of granite tors. *International Geological Congress, Comptes Rendues*, Algiers, **21e**, 201–210.

Hansen, A. (1984). Engineering geomorphology: the application of an evolutionary model of Hong Kong's terrain. *Zeitschrift für Geomorphologie, N.F. Supplementband*, **51**, 39–50.

Haq, B. U., Hardenbol, J. & Vail, P. R. (1987). Chronology of fluctuating sea levels since the Triassic. *Science*, **235**, 1156–1167.

Hare, F. (1983). Climate on the desert fringe. In *Mega-Geomorphology* (ed. R. Gardner & H. Scoging), pp. 134–151. Clarendon, Oxford.

Harp, E. L., Wilson, R. C. & Wieczorek, G. F. (1981). Landslides from the February 4, 1976, Guatemala Earthquake. *Geological Survey Professional Paper*, **1204-A**.

Harpum, J. R. (1963). The evolution of granite scenery in Tanganyika. *Geological Survey of Tanganyika Records*, **10**, 39–46.

Harrison, J. B. (1934). *The Katamorphism of Igneous Rocks under Humid Tropical Conditions*. Imperial Bureau of Soil Science, Harpenden.

Harrison, S. P. (1993). Late Quaternary lake-level changes and climates of Australia. *Quaternary Science Reviews*, **12**, 211–231.

Hastenrath, S. (1991). *Climate Dynamics of the Tropics*. Kluwer, Dordrecht.

Hasui, Y. & Ponçano, W. L. (1978). Geossuturas e sismicidade no Brasil. *Anais, II Congresso Brasiliero de Geologia de Engenharia*, São Paulo, **1**, 331–338.

Hasui, Y. & Sadowski, G. R. (1976). Evoluçao geologica do Pré Cambriano da regaio sudeste do Estado de São Paulo. *Revisita Brasiliera de Geociéncias*, **6**, 182–200.

Hawthorne, J. B. (1973). Model of a kimberlite pipe. *Physics and Chemistry of the Earth*, **9**, 1–15.

Hays, J. D., Imbrie, J. & Shackleton, N. J. (1976). Variations in the Earth's orbit: pacemaker of the ice ages. *Science*, **194**, 1121–1132.

Heaney, L. R. (1991). A synopsis of climatic and vegetation change in southeast Asia. *Climate Change* (Special Issue: Tropical Forests and Climate), **19**, 53–61.

Heath, W. & Saroso, B. S. (1988). Natural slope problems related to roads in Java, Indonesia. In *Geomechanics in Tropical Soils, Proceedings of the Second International Conference on Geomechanics in Tropical Soils*, Singapore, 1988, Vol. 1, pp. 259–266. Balkema, Rotterdam.

Heede, B. H. (1974). Stages of development of gullies in western United States of America. *Zeitschrift für Geomorphologie, N.F.*, **18**, 260–271.

Heine, K. (1977). Junquartäre pluviale und interpluviale in der Kalahari (südliches Afrika). *Palaeoecology of Africa*, **10/11**, 31–39.

Heine, K. (1978). Radiocarbon chronology of late Quaternary lakes in the Kalahari, southern Africa. *Catena*, **5**, 145–149.

Heine, K. (1979). Reply to Cooke's discussion of K. Heine: Radiocarbon chronology of late Quaternary lakes in the Kalahari, southern Africa. *Catena*, **6**, 259–266.

Heine, K. (1982). The main stages of the late Quaternary evolution of the Kalahari region, southern Africa. *Palaeoecology of Africa*, **15**, 53–76.

Heinrich, H. (1988). Origin and consequences of cyclic ice rafting in the northeast Atlantic Ocean during the past 130,000 years. *Quatern. Res.*, **29**, 142–152.

Helgren, D. M. (1979). *River of Diamonds: An Alluvial History of the Lower Vaal Basin, South Africa*. Department of Geography Research Paper No. 185, University of Chicago.

Hendricks, D. M. & Whittig, L. D. (1968) Andesite weathering II. Geochemical changes from andesite to saprolite. *Journal of Soil Science*, **19**, 147–153.

Hess, R., Bach, R. & Deuel, H. (1960). Models for the reaction between organic and mineral substances in the ground. *Experienta*, **16**, 38–40.

Heusser, C. J. (1984). Late glacial Holocene climate of the lake district of Chile. *Quaternary Research*, **22**, 77–90.

Heusser, C. J. (1990): Ice age vegetation and climate of subtropical Chile. *Palaeogeography, Palaeoclimatology, Palaeoecology*, **80**, 107–127.

Heydeman, M. T., Button, A. M. & Williams, H. D. (1983). Preliminary investigation of micro-organisms occurring in some open blanket lateritic silicate bauxites. In *Lateritisation Processes. Proc. II International Seminar on Lateritisation Processes*, São Paulo, Brazil, 1982 (ed. A. J. Melfi & A. Carvalho), pp. 225–235. University of São Paulo, Brazil.

Heyligers, P. C. (1963). *Vegetation and Soil of a White-Sand Savanna in Suriname*. Verhandelingen der koninkluke Nederlandse, Akademie van

Wetenschappen, AFD Natuurkunde. N. V. Noord-Hollandsche Uitgevers Maatschappij, Amsterdam.

Hill, I. D. & Rackham, I. J. (1978). Indications of mass movement on the Jos Plateau, Nigeria. *Zeitschrift für Geomorphologie, N.F.*, **22**, 258–274.

Hjulström, F. (1935). Studies of the morphological activity of rivers as illustrated by the River Fyries. *Bulletin Geological Institute University of Uppsäla*, **25**, 221–527.

Holeman, J. N. (1968). The sediment yield of the major rivers of the world. *Water Resources Research*, **4**, 737–747.

Holland, H. D. (1978). *The Geochemistry of the Atmosphere and Oceans*. Wiley-Interscience, Somerset, New Jersey.

Holmes, A. & Wray, A. (1913). Mozambique: a geographical study. *Geographical Journal*, **62**, 143–152.

Holmes, A. (1965). *The Principles of Physical Geology*. Nelson, London.

Holzhey, C. S., Daniels, R. B. & Gamble, E. E. (1975). Thick Bh horizons in the North Carolina coastal plain: II. Physical and chemical properties and rates of organic additions from surface sources. *Soil Science Society of America, Proceedings*, **39**, 1182–1187.

Hooghiemstra, H. (1984). *Vegetational and Climatic History of the High Plain of Bogota, Colombia: A Continuous Record of the Last 3.5 Million Years*. Cramer Verlag, Vaduz.

Hooghiemstra, H. (1989). Variations of the African trade wind regime during the last 140,000 years in pollen flux evidenced by marine sediment records. In *Paleoclimatology and Paleometeorology: Modern and Past patterns of Global Atmospheric Transport* (eds. M. Lienen & M. Sarnthein), pp. 733–770. Kluwer, Amsterdam.

Hooghiemstra, H. & Agwu, C. O. C. (1989). Changes in the vegetation and trade winds in equatorial northwest Africa 140,000–70,000 yr B.P. as deduced from two marine pollen records. *Palaeogeography, Palaeoclimatology & Palaeoecology*, **66**, 173–213.

Hori, N. (1977). Weathering and cuirasses in the Yaoundé area and geographical variation of weathering profile and superficial deposits. In *Geomorphological Studies in the Forest and Savanna Areas of Cameroon* (ed. H. Kadomura), pp. 61–71.

Hori, N. (1986). Geographical variation of superficial deposits and its significance in the late-Quaternary environmental changes in the forested south Cameroon. In *Geomorphology and Environmental Changes in Tropical Africa: Case Studies in Cameroon and Kenya* (ed. H. Kadomura), pp. 31–43. Hokkaido University, Sapporo.

Horton, R. E. (1933). The role of infiltration in the hydrologic cycle. *Transactions of American Geophysical Union*, **14**, 446–460.

Hsü, K. J. (1975). Catastrophic debris streams (Sturzstroms) generated by rockfalls. *Geological Society of America Bulletin*, **86**, 129–140.

Huang, W. H. & Keller, W. D. (1970). Dissolution of rock-forming silicate minerals in dilute organic acids. *American Mineralogist*, **55**, 2076–2094.

Hubbard, F. H. (1983). The Phanerozoic cover sequences preserved as xenoliths in the Kimberlite of eastern Sierra Leone. *Geological Magazine*, **120**, 67–71.

Hudson, N. W. (1971). *Soil Conservation*, Batsford, London (2nd edn, 1981).

Hutchinson, J. N. (1977). Assessment of the effectiveness of corrective measures in relation to geological conditions and types of slope movement. *International Association of Engineering Geologists Bulletin*, **16**, 131–155.

Idnurm, M. & Schmidt, P. W. (1978). Palaeomagnetic dating of late Cretaceous and Tertiary weathered profiles in the Eromanga Basin, Queensland. *Palaeogeography, Palaeoclimatology & Palaeoecology*, **24**, 263–277.

Iida, T. & Okunishi, K. (1983). Development of hillslopes due to landslides. *Zeitschrift für Geomorphologie, N.F.*, **Supplementband**, **46**, 67–77.

IITA (1975). *1975 Annual Report*. International Institute for Tropical Agriculture, Ibadan, Nigeria.

Imbrie, J. (1985). A theoretical framework for the Pleistocene ice ages. *Journal of the Geological Society of London*, **42**, 2170–2188.

Imeson, A. C. & Kwaad, F. J. P. M. (1980). Gully types and gully prediction. *Geografisch Tidschrift*, **5**, 430–441.

Irfan, T. Y. (1988). Fabric variability and index testing of a granitic saprolite. In *Geomechanics in Tropical Soils, Proceedings of the Second International Conference on Geomechanics in Tropical Soils*, Singapore, 1988, Vol. 1, pp. 25–35. Balkema, Rotterdam.

Irfan, T. Y. & Dearman, W. R. (1978). Engineering petrography of a weathered granite in Cornwall, England. *Quarterly Journal of Engineering Geology*, **11**, 233–344.

Irfan, T. Y. & Woods, N. W. (1988). The influence of relict discontinuities on slope stability in saprolitic soils. In *Geomechanics in Tropical Soils, Proceedings of the Second International Conference on Geomechanics in Tropical Soils*, Singapore, 1988, Vol. 1, pp. 267–276. Balkema, Rotterdam.

Isaac, K. P. (1983). Tertiary lateritic weathering in Devon, England, and the Palaeogene continental environment of south-west England. *Proceedings Geologists' Association*, **94**, 105–114.

ISSMFE (1985). *First International Conference on Geomechanics in Tropical Lateritic and Saprolitic Soils, Brasilia, 1985*, Brazilian Society for Soil Mechanics & International Society for Soil mechanics and Foundation Engineering (ISSMFE), Brasilia.

ISSMFE (1988/1990). *Geomechanics in Tropical Soils*, Proceedings of the Second International Conference on Geomechanics in Tropical Soils, Singapore, 1988, SE Asian Geotechnical Society & International Society for Soil mechanics and Foundation Engineering (ISSMFE), Vol. 1, 1988; Vol. 2, 1990. Balkema, Rotterdam.

Jackson, I. J. (1989). *Climate Water and Agriculture in the Tropics*, 2nd edn. Longman, Harlow.

Jacob, L. Jr. (ed.) (1984). *Bauxite*. Proceedings of the 1984 Symposium, Los Angeles, 1984, Society of Mining Engineers, American Institute of Mining, Metallurgical, Petroleum Engineers, Inc. New York.

Jansen, J. H. F. & Van Iperan, J. M. (1991). A 222,000 year climatic record for the east equatorial Atlantic Ocean and equatorial Africa: evidence from diatoms and opal phytoliths in the Zaire (Congo) deep sea fan. *Paleoceanography*, **6**, 573–591.

Jansen, J. H. F., Van Weering, T. C. E., Gieles, R. & Van Iperen, J. (1984). Middle and late Quaternary oceanography and climatology of the Zaire-Congo fan and the adjacent eastern Angola Basin. *Netherlands Journal of Sea Research*, **17**, 201–249.

Jeje, L. K. (1970). *Some Aspects of the Geomorphology of Southwestern Nigeria*. Unpublished PhD Thesis, University of Ibadan, Nigeria.

Jeje, L. K. (1972). Landform development at the boundary of sedimentary and crystalline rocks in south western Nigeria. *Journal of Tropical Geography*, **34**, 25–33.

Jeje, L. K. (1987). Runoff and soil loss from erosion plots in Ife area of southwestern Nigeria. In *International Geomorphology 1986. Proceedings First International Conference on Geomorphology* (ed. V. G. Gardiner), Part II, pp. 459–481. Wiley, Chichester.

Jennings, J. N. (1972). The character of tropical humid karst. *Zeitschrift für Geomorphologie, N.F.*, **16**, 336–341.

Jennings, J. N. (1976). A test of the importance of cliff-foot caves in tower karst development. *Zeitschrift für Geomorphologie, N.F., Supplementband*, **26**, 92–97.

Jennings, J. N. (1983). Sandstone pseudo karst or karst? In *Aspects of Australian Sandstone Landscapes* (ed. R. W. Young & G. C. Nanson), pp. 21–30. Australia and New Zealand Geomorphological Group, Special Publication, No. 1.

Jennings, J. N. (1985). *Karst Geomorphology*. Blackwell, Oxford.

Jennings, J. N. & Bik, M. T. (1962). Karst morphology in Australian New Guinea. *Nature*, **194**, 1036–1038.

Jenny, H. (1941). *Factors of Soil Formation*. McGraw-Hill, New York.

Jenny, H. (1980). *The Soil Resource*. Ecological Studies, Vol. 37, Springer-Verlag, New York.

Jessen, O. (1936). *Reisen und Forschungen in Angola*. Reimer, Berlin.

Jessen, O. (1943). Die Randschwellen der Kontinente. *Petermanns Geographische Mittelungen, Ergünzungs-Heft*, **241**, 1–205.

Jibson, R. W. (1989). Debris flows in southern Puerto Rico. In *Landslide Processes of the Eastern United States and Puerto Rico* (ed. A. P. Schultz & R. W. Jibson), pp. 292–55. Geological Society of America, Special Paper No. 236.

Johnson, D. L. (1993). Dynamic denudation evolution of tropical, subtropical and temperate landscapes with three tiered soils: toward a general theory of landscape evolution. *Quaternary International*, **17**, 67–78.

Johnson, D. L. & Watson-Stegner, D. (1987). Evolution model of pedogenesis. *Soil Science*, **143**, 349–366.

Johnson, D. L. & Watson-Stegner, D. (1990). The soil-evolution model as a framework for evaluating pedoturbation in archaeological site formation. *Geological Society of America (Centennial Special Volume)*, **4**, 541–560.

Johnson, D. W. (1931). *Stream Sculpture on the Atlantic Slope, A Study in the Evolution of Appalachian Rivers*. Columbia University Press, New York.

Johnson, D. W., Cole, D. W., Gessel, S. P., Singer, M. J. & Minden, R. V. (1977). Carbonic acid leaching in tropical, temperate, subalpine and northern forest soils. *Arctic and Alpine Research*, **9**, 329–343.

Johnson, K. A. & Sitar, N. (1990). Hydrologic conditions leading to debris-flow initiation. *Canadian Geotechnical Journal*, **27**, 789–801.

Jones, F. O. (1973). Landslides of Rio de Janeiro and the Serra das Araras Escarpment, Brazil. *US Geological Survey Professional Paper*, **697**.

Jones, J. A. A. (1981). *The Nature of Soil Piping — A Review of Research*, British Geomorphological Research Group, Research Monograph No. 3, Geo Books, Norwich.

Jones, J. G. & Veevers J. J. (1982). A Cainozoic history of Australia's southeast highlands. *Journal of the Geological Society of Australia*, **29**, 1–12.

Jones, J. G. & Veevers, J. J. (1983). Mesozoic origins and antecedents of Australia's Eastern Highlands. *Journal of the Geological Society of Australia*, **30**, 305–322.

Jones, M. J. (1985). The weathered zone aquifers of the basement complex areas of Africa. *Quarterly Journal of Engineering Geology*, **18**, 35–46.

Jones, R. G. B. (1975). *Central Nigeria Project: Report on a Soil Conservation Consultancy to Study Soil Erosion Problems on the Jos Plateau*. Land Resource Report, No. 6.

Jones, T. R. (1859). Notes on some granite tors. *Geologist*, **2**, 301–312.

Journaux, M. A. (1975). Géomorphologie des bordures de L'Amazonie Brésilienne: le modelé des versants; essai d'évolution paléoclimatique. *Bulletin, Association Géographie de la France*, **422/423**, 5–19.

Jouzel, J., Lorius, C., Petit, J. R., Genthon, C., Barkov, N. I., Kotlyakov, V. M. & Petrov, V. M. (1987). Vostok ice core: a continuous isotope temperature record over the last climatic cycle (160,000 years). *Nature*, **329**, 403–408.

Jouzel, J., Barkov, N. I., Barnola, J. M., Bender, M., Chappellaz, J., Genthon, C., Kotlyakov, V., Lipenkov, V., Lorius, C., Petit, J. R., Raynaud, D., Raisbeck, G., Ritz, C., Sowers, T., Stievenard, M., Yiou, F. & Yiou, P. (1993). Extending the Vostok ice-core record of palaeoclimate to the penultimate glacial period. *Nature*, **364**, 407–412.

Junner, N. R. (1943). The diamond deposits of the Gold Coast. *Gold Coast Geological Survey, Bulletin*, **12**.

Kadomura, H. (ed.) (1977). *Geomorphological Studies in the Forest and Savanna Areas of Cameroon.* Hokkaido University, Sapporo.

Kadomura, H. (1982). Late Quaternary climatic and environmental changes in tropical Africa: an introduction. In *Geomorphology and Environmental Changes in the Forest and Savanna Cameroon* (ed. H. Kadomura), pp. 1–12. Hokkaido University, Sapporo.

Kadomura, H. (1986a). Late glacial–early Holocene environmental changes in tropical Africa: a comparative analysis with deglaciation history. *Geographical Reports, Tokyo Metropolitan University*, **21**, 1–21.

Kadomura, H. (ed.) (1986b). *Geomorphology and Environmental Changes in Tropical Africa: Case Studies in Cameroon and Kenya.* Hokkaido University, Sapporo.

Kadomura, H. (ed.) (1989). *Savannization Processes in Tropical Africa.* Tokyo Metropolitan University & Zambia Geographical Society, Tokyo.

Kadomura, H. & Hori, N. (1990). Environmental implications of slope deposits in humid tropical Africa: evidence from southern Cameroon and western Kenya. *Geographical Reports, Tokyo Metropolitan University*, **25**, 213–236.

Kaitanen, V. (1969) A geographical study of the morphogenesis of northern Lapland. *Fennia*, **99**, 1–95.

Kaitanen, V. (1985). Problems concerning the origin of inselbergs in Finnish lapland. *Fennia*, **163**, 359–364.

Kamaludin, H., Nakamura, T., Price, D., Woodroffe, C. D. & Fujii, S. (1993). Radiocarbon and thermo-luminescence dating of the Old Alluvium from a coastal site in Perak, Malaysia. *Sedimentary Geology*, **83**, 1–12.

Kaye, C. A. (1967). Kaolinisation of bedrock of the Boston, Massachusetts area. *US Geological Survey Professional Paper*, **575-C**, C165–C172.

Keller, R. (ed.) (1983) *Hydrology of Humid Tropical Regions*, IAHS Publication, **140**. Wallingford.

Keller, W. D. (1957). *The Principles of Chemical Weathering.* Lucas Brothers, Columbia, USA.

Kershaw, A. P. (1976). A Late Pleistocene and Holocene pollen diagram from Lynch's Crater, northeastern Queensland, Australia. *New Phytologist*, **77**, 469–498.

Kershaw, A. P. (1978). Record of last interglacial–glacial cycle from northeastern Queensland. *Nature*, **272**, 159–161.

Kershaw, A. P. (1992). The development of rainforest-savanna boundaries in tropical Australia. In *Nature and Dynamics of Forest–Savanna Boundaries* (ed. P. A. Furley, J. Proctor & J. A. Ratter), pp. 255–271. Chapman & Hall, London.

Kershaw, A. P., D'Costa, D. M., McEwen Mason, J. R. C. & Wagstaff, B. E. (1991). Palynological evidence for Quaternary vegetation and environments of mainland southeastern Australia. *Quaternary Science Reviews*, **10**, 391–404.

Kesel, R. H. (1977). Slope runoff and denudation in the Rupununi savanna, Guyana. *Journal of Tropical Geography*, **44**, 33–42.

Kesel, R. H. (1985). Alluvial fan systems in wet-tropical environment, Costa Rica. *National Geographic Research*, **1**, 450–469.

Khobzi, J. (1972). Érosion chimique et mécanique dans la genèse de dépressions 'pseudo-karstiques' souvent endoréiques. *Revue de Géomorphologie Dynamique*, **21**, 57–70.

Kiewietdejonge, C. J. (1984a). Büdel's Geomorphology I. *Progress in Physical Geography*, **8**, 218–248.

Kiewietdejonge, C. J. (1984b). Büdel's Geomorphology II. *Progress in Physical Geography*, **8**, 365–397.

King, J., Loveday, I. & Schuster, R. L. (1989). The 1985 landslide dam and resulting debris flow, Papua New Guinea. *Quarterly Journal of Engineering Geology*, **22**, 257–270.

King, L. C. (1948). A theory of bornhardts. *Geographical Journal*, **112**, 83–87.

King, L. C. (1949). On the ages of African landsurfaces. *Quarterly Journal of the Geological Society of London*, **104**, 439–453.

King, L. C. (1950a). The study of the world's plainlands. *Quarterly Journal of the Geological Society, London*, **106**, 101–131.

King, L. C. (1950b). The cyclic land-surfaces of Australia. *Journal Royal Society, Victoria*, **62**, 79–95.

King, L. C. (1951). *South African Scenery* (2nd edn.). Oliver & Boyd, Edinburgh.

King, L. C. (1953). Canons of landscape evolution. *Bulletin Geological Society of America*, **64**, 721–752.

King, L. C. (1955). Pediplanation and isostacy: an example from South Africa. *Quarterly Journal of the Geological Society of London*, **111**, 353–359.

King, L. C. (1956). A geomorphological comparison between eastern Brazil and Africa (central and southern). *Quarterly Journal of the Geological Society of London*, **112**, 445–474.

King, L. C. (1957). The uniformitarian nature of hillslopes. *Transactions of the Edinburgh Geological Society*, **17**, 81–102.

King, L. C. (1962). *The Morphology of the Earth*, Oliver and Boyd, Edinburgh (2nd edn 1967).

King, L. C. (1966). The origins of bornhardts. *Zeitschrift für Geomorphologie, N.F.*, **10**, 97,98.

King, L. C. (1972). *The Natal Monocline: Explaining the Origins and Scenery of Natal, South Africa*. University of Natal Press, Durban.

King, L. C. (1975). Bornhardt landforms and what they teach. *Zeitschrift für Geomorphologie, N.F.*, **19**, 299–318.

King, L. C. (1976). Planation remnants upon high lands. *Zeitschrift für Geomorphologie N.F.*, **20**, 133–148.

King, L. C. (1983). *Wandering Continents and Spreading Sea Floors on an Expanding Earth*. Wiley, Chichester.

Kinosita, T. (1983). Runoff and flood characteristics in some humid tropical regions. In *Hydrology of Humid Tropical Regions* (ed. R. Keller) IAHS Publication, vol. 140, pp. 53–62, Wallingford.

Kirkby, M. J. (1971). Hillslope process-response models based on the continuity equation. *Institute of British Geographers, Special Publication*, **3**, 15–30.

Klaer, W. (1957). Karstkegel, Karstinselberg und Poljeboden am Beispiel des Jezeropoljes. *Petermanns Geographische Mitteilungen*, **101**, 108–111.

Klammer, G. (1976). Zur jungquartären Reliefgeschichte de Amazonastales. *Zeitschrift für Geomorphologie N.F.*, **20**, 149–170.

Klammer, G. (1978). Reliefentwicklung im Amazonasbecken und plio-pleistozäne Bewegungen des Meeresspiegels. *Zeitschrift für Geomorphologie N.F.*, **22**, 390–416.

Klammer, G. (1981). Landforms, cyclic erosion and deposition, and Late Cenozoic changes in climate in southern Brazil. *Zeitschrift für Geomorphologie, N.F.*, **25**, 146–165.

Klammer, G. (1982). Die Päleowüste des Pantanal von Mato Grosso und die pleistozäne Klimageschichte der brasilianischen Randtropen. *Zeitschrift für Geomorphologie, N.F.*, **26**, 393–416.

Klein, C. (1974). Tectogenèse et morphogenèse Armoricaines et péri-Armoricaines. *Revue de Géologie Dynamique et de Géographie Physique*, **16**, 87–100.

Klein, C. (1985). La notion de cycle en géomorphologie. *Revue de Géologie Dynamique et de Géographie Physique*, **26**, 95–107.

Klinge, H. (1965). Podzol soils in the Amazon basin. *J. Soil Sci.*, **16**, 5–13.

Knox, J. C. (1972). Valley alluviation in southwestern Wisconsin. *Annals of the Association of American Geographers*, **62**, 401–410.

Knox, J. C. (1975). Concept of the graded stream. In *Theories of Landform Development* (ed.

W. C. Melhorn & R. C. Flemal), pp. 169–198. George, Allen and Unwin, London.

Knox, J. C. (1984). Fluvial responses to small scale climate changes. In *Developments and Applications of Geomorphology* (ed. J. E. Costa & P. J. Fleisher), pp. 318–342. Springer Verlag, Berlin.

Kobilsek, B. & Lucas, Y. (1988). Étude morphologique et pétrographique d'une formation bauxitique d'Amazonie (secteur de Juruti, état du Para, Brésil). *Sciences Géologiques, Bulletin (Strasbourg)*, **41**, 71–84.

Kolla, B., Biscaye, P. E. & Hanley, A. F. (1979). Distribution of quartz in Late Quaternary Atlantic sediments in relation to climate. *Quaternary Research*, **11**, 261–277.

Koo, Y. C. (1982a). Relic joints in completely decomposed granites in Hong Kong. *Canadian Geotechnical Journal*, **19**, 117–123.

Koo, Y. C. (1982b). The mass strength of jointed residual soils. *Canadian Geotechnical Journal*, **19**, 225–231.

Koons, D. (1955). Cliff retreat in the southwestern United States. *American Journal of Science*, **253**, 44–52.

Köppen, W. (1923). *Die Klimate der Erde; Grundriss der Klimakunde*. Griuyter, Berlin.

Kotschubey, B. & Lermos, V. P. (1985). Consideraçaoes sobre a origem e a genese das bauxitas da Serra do Carajas. *Anais do II Symposio de geologia da Amazonia, Belem*, 48–61.

Kotschubey, B. & Trockenbrodt, W. (1981). Evoluçao poligenetica das bauxitas do distrito do Paragominas, Acailandia (estados de Para e mranhao). *Revue Brasileiros de Geociencias*, **11**, 193–202.

Kotschubey, B. & Trockenbrodt, W. (1984). Genèse des bauxites latéritiques du district de Nhamunda-Trombetas, région de la basse Amazone, Brésil. *101e Congrés National de la Sociétie Sav.*, Dijon, **1**, 347–357.

Kowal, J. M. & Kassam, A. H. (1976). Energy load and instantaneous intensity of rainstorms at Samaru, northern Nigeria. *Tropical Agriculture*, **53**, 185–198 (also *Soil Conservation and Management in the Humid Tropics* (ed. D. J. Greenland & R. Lal), pp. 49–56. Wiley, Chichester.

Kraus, E. B. (1973). Comparison between ice age and present general circulation. *Nature*, **245**, 129–133.

Krauskopf, K. B. (1967) *Introduction to Geochemistry*. McGraw-Hill, New York.

Kronberg, B. I. & Melfi, A. J. (1987). The geochemical evolution of lateritic terranes. *Zeitschrift für Geomorphologie N.F., Supplementband*, **64**, 25–32.

Kronberg, B. I., Fyfe, W. S., Leonardos, O. H. Jr. & Santos, A. M. (1979). The chemistry of some Brazilian soils: element mobility during intense weathering. *Chemical Geology*, **24**, 211–229.

Kronberg, B. I., Fyfe, W. S., McKinnon, B. J., Couston, J. F., Filho, B. S. & Nash, R. A. (1982). Model for bauxite formation: Paragominas (Brazil). *Chemical Geology*, **35**, 311–320.

Kronberg, B. I., Nesbitt, H. W. & Lam, W. W. (1986). Upper Pleistocene Amazon deep-sea fan muds reflect intense chemical weathering of their mountainous source lands. *Chemical Geology*, **54**, 283–294.

Kronberg, B. I., Nesbitt, H. W. & Fyfe, W. S. (1987). Mobilities of alkalis, alkaline earths and halogens during weathering. *Chemical Geology*, **60**, 41–49.

Krook, L. (1979). *Sediment Petrographical Studies in Northern Suriname*. Academisch Proefschrift, Drukkerij Elinkwijk BV, Utrecht.

Kroonenberg, S. B. & Melitz, P. J. (1983): Summit levels, bedrock control and the etchplain concept in the basement of Suriname. *Geologie en Mijnbouw*, **62**, 389–399.

Kuete, M. (1986). Palaeoforms and superficial deposits: evolution of footslopes and river valleys in the west-central part of the Cameroon Plateau. In *Geomorphology and Environmental Changes in Tropical Africa: Case Studies in Cameroon and Kenya* (ed. H. Kadomura), Special Publication **4**, pp. 11–30. Graduate School of Environmental Science, Hokkaido University, Sapporo.

Kühnel, R. A. (1987). The role of cationic and anionic scavengers in laterites. *Chemical Geology*, **60**, 31–40.

Kutzbach, J. E. (1976). The nature of climate and climatic variations. *Quaternary Research*, **6**, 471–480.

Laity, J. E. & Malin, M. C. (1985). Sapping processes and the development of theatre-headed valley networks on the Colorado Plateau. *Geological Society of America Bulletin*, **96**, 303–317.

Lal, R. (1977). Analysis of factors affecting rainfall erosivity and soil erodibility. In *Soil Conservation and Management in the Humid Tropics* (ed. D. J. Greenland & R. Lal), pp. 49–56. Wiley, Chichester.

Lal, R. (1981). Deforestation of tropical rainforest and hydrological problems. In *Tropical Agricultural Hydrology* (ed. R. Lal & E. W. Russell), pp. 131–140. Wiley, Chichester.

Lal, R. (1986). Deforestation and soil erosion. In *Land Clearing and Development in the Tropics* (ed. R. Lal, P. A. Sanchez & R. W. Cummings, Jr.), pp. 299–315. Balkema, Rotterdam.

Lal, R. & Russell, E. W. (eds.) (1981). *Tropical Agricultural Hydrology*. Wiley, Chichester.

Lal, R., Lawson, T. L. & Anastase, A. H. (1981). Erosivity of tropical rains. In *Assessment of Erosion* (ed. M. De Broodt & D. Gabriels), pp. 143–151. Wiley, Chichester.

Lancaster, N. (1979). Evidence for a widespread Late Pleistocene humid period in the Kalahari. *Nature*, **279**, 145–146.

Lancaster, N. (1981). Palaeoenvironmental implications of fixed dune systems in southern Africa. *Palaeogeography, Palaeoclimatology, Palaeoecology*, **33**, 327–346.

Lancaster, N. (1989). Late Quaternary palaeoenvironments in the southwestern Kalahari. *Palaeogeography, Palaeoclimatology, Palaeoecology*, **70**, 367–376.

Lautenschlager, M. (1991). Simulation of the ice age atmosphere—January and July means. *Geologische Rundschau*, **80/3**, 513–534.

Laws, J. O. & Parsons, D. A. (1943). The relation of raindrop size to intensity. *Transactions of the American Geophysical Union*, **24**, 452–460.

Le Bas, M. J. (1971). Per-alkaline vulcanism, crustal swelling and rifting. *Nature*, **230**: 85–87.

Le Coeur, C. (1989). La question des altérites profondes dans la région des Hébrides internes (Ecosse occidentale). *Zeitschrift für Geomorphologie, N.F., Supplementband*, **72**, 109–124.

Ledger, D. C. (1964). Some hydrological characteristics of West African rivers. *Transactions Institute British Geographers*, **35**, 73–90.

Ledger, D. C. (1969). Dry season flow characteristics of West African rivers. In *Environment and Land Use in Africa* (ed. M. F. Thomas & G. Whittington), pp. 83–102. Methuen, London.

Ledger, D. C. (1974). The water balance of an exceptionally wet catchment area in west Africa. *Journal of Hydrology*, **24**, 207–214.

Ledru, M.-P. (1993). Late Quaternary environmental and climatic changes in Central Brazil. *Quaternary Research*, **39**, 90–98.

Lee, K. E. & Wood, T. G. (1971). *Termites and Soils*. Academic Press, London & New York.

Lee, S. G. & De Freitas, M. H. (1991). A revision of the description and classification of weathered granite and its application to granites in Korea. *The Quarterly Journal of Engineering Geology*, **22**, 31–48.

Lee, S. W. (1981). Landslides in Taiwan. *Proceedings South East Asian Regional Symposium on Problems of Soil Erosion and Sedimentation*, pp. 195–206, Bangkok.

Lehman, D. S. (1963). Some principles of chelation chemistry. *Soil Science Society of America Proceedings*, **27**, 167–170.

Lehmann, H. (1936). *Morphologische Studien auf Java*. Engelhorn, Stuttgart.

Le Houérou, H. N. (1992). Outline of the biological history of the Sahara. *Journal of Arid Environments*, **22**, 3–20.

Leigh, C. H. (1982). Sediment transport by surface wash and throughflow at the Pasoh Forest Reserve, Negri Sembilan, Peninsular Malaysia. *Geografiska Annaler*, **64A**, 171–180.

Lelong, F. (1966). Régimes des nappes phréatiques contenues dans las formations d'altération tropicale. Conséquences pour la pédogenèse. *Sciences de la Terre*, 11, 205–244.

Lelong, F. (1969). Nature et genèse des produits d'alteration des roches crystallines sous climat tropicale humide. *Mémoire Sciences de la Terre*, pp. 1–14.

Leneuf, N. (1959). *L'altération des granites calco-alcalins et de grano-diorites en Côte d'Ivoire forestière et les sols qui sont dérivé.* Thèse, Université de Paris.

Leneuf, N. & Aubert, G. (1960). Essai d'évaluation de la vitesse de ferrallitisation. *Proceedings 7th International Congress of Soil Science*, 225–228.

Le Roux, J. S. (1991). Is the pediplanation cycle a useful model? *Zeitschrift für Geomorphologie, N.F.*, 35, 175–185.

Lewis, L. A. (1975). Slow slope movement in the dry tropics: La Paguera, Puerto Rico. *Zeitschrift für Geomorphologie, N.F.*, 19, 334–339.

Lewis, L. A. (1976). Soil movement in the tropics—a general model. *Zeitschrift für Geomorphologie, N.F. Supplementband*, 25, 132–144.

Lézine, A.-M. (1988). Les variations de la couvature forestière mésophile d'Afrique occidentale au cours de l'Holocène. *Comptes Rendus, Académie des Sciences, Paris, Série II*, 307, 439–445.

Lézine, A.-M. (1991). West African paleoclimates during the last climatic cycle inferred from an Atlantic deep-sea pollen record. *Quaternary Research*, 35, 456–463.

Lézine, A.-M. & Casanova, J. (1989). Pollen and hydrological evidence for the interpretation of past climates in tropical west Africa during the Holocene. *Quaternary Science Reviews*, 8, 45–55.

Lichte, M. (1989). Arid processes in the SE-Brazilian relief evolution during the last ice-age. *International Symposium on Global Changes in South America During the Quaternary*, Sao Paulo, 1989, Special Publication, 1, pp. 60–64.

Lidmar-Bergström, K. (1982). Pre-Quaternary geomorphological evolution in southern Fennoscandia. *Sveriges Geologiska Undersökning, Series C*, 785, 1–202, Uppsala.

Lidmar-Bergström, K. (1989). Exhumed Cretaceous landforms in south Sweden. *Zeitschrift für Geomorphologie, N.F., Supplementband*, 72, 21–40.

Linton, D. L. (1951). Problems of the Scottish scenery. *Scottish Geographical Magazine*, 67, 65–85.

Linton, D. L. (1955). The problem of tors. *Geographical Journal*, 121, 470–487.

Littmann, T. (1989a). Climatic change in Africa during the Last Glacial: facts and problems. *Palaeoecology of Africa*, 20, 163–179.

Littmann, T. (1989b). Spatial patterns and frequency distribution of Late Quaternary water budget tendencies in Africa. *Catena*, 16, 163–188.

Livingstone, D. A. (1963). Chemical compositions of rivers and lakes. *US Geological Survey Professional Paper*, 440-G, p. 64.

Livingstone, D. A. (1975). Late Quaternary climatic change in Africa. *Annual Review of Ecology & Systematics*, 6, 249–280.

Livingstone, D. A. (1980). Environmental changes in the Nile headwaters. In *The Sahara and The Nile* (ed. M. A. J. Williams & H. Faure), pp. 339–359. Balkema, Rotterdam.

Livingstone, D. A. (1982). Quaternary geography of Africa and the Refuge Theory. In *Biological Diversification in the Tropics* (ed. G. T. Prance), pp. 523–536. Columbia, New York.

Löffler, E. (1974). Piping and pseudokarst features in the tropical lowlands of New Guinea. *Erdkunde*, 28, 13–18.

Löffler, E. (1977). *Geomorphology of Papua New Guinea.* CSIRO/Australian National University Press, Canberra.

Löffler, E. (1978). Karst features in igneous rocks in Papua New Guinea. In *Landform Evolution in Australasia* (ed. J. L. Davies & M. A. J. Williams), pp. 238–249. ANU Press, Canberra.

Loughnan, F. C. (1969). *Chemical Weathering of the Silicate Minerals.* Elsevier, New York.

Louis, H. (1964). Über rumpfflächen—und talbildung in den wechselfeuchten Tropen besonders nach Studien in Tanganyika. *Zeitschrift für Geomorphologie, N.F.*, 8, 43–70.

Lovering, T. S. (1959). Significance of accumulator plants in rock weathering. *Bulletin of the Geological Society of America*, 70, 781–800.

Lowe, D. J. (1986). Controls on the rates of weathering and clay mineral genesis in airfall tephras: a review and New Zealand case study. In *Rates of Chemical Weathering of Rocks and Minerals* (ed. S. M. Colman & D. P. Dethier), pp. 265–330. Academic Press, New York.

Lu, Y. (1987). Karst geomorphological mechanisms and types in China. In *International Geomorphology 1986. Proceedings of 1st International Conference on Geomorphology*, Part II (ed. V. Gardiner), pp. 1077–1093. Wiley, Chichester.

Lucas, Y., Boulet, R., Chauvel, A. & Veillon, L. (1987). Systèmes sols ferrallitiques-podzols en région amazonienne. In *Podzols et Podzolisation* (ed. D. Righi & A. Chauvel), pp. 53–65, Comptes Rendus de la Table Ronde International, 1986, Association Française pour l'Étude du Sol, INRA/ORSTOM, Poitiers.

Lucas, Y., Boulet, R. & Chauvel, A. (1988). Intervention simultanée des phénomènes d'enforcement vertical et de transformation latérale dans la mise en place de systèmes sols ferrallitiques-podzols de l'Amazonie Brésilienne. *C. R. Académie des Sciences Paris, Serie II*, 306, 1395–1400.

Lucas, Y., Luizao, F. J., Chauvel, A., Rouiller, J. & Nahon, D. (1993). The relation between biological activity of the rainforest and mineral composition of soils. *Science*, 260, 521–523.

Lull, H. W. (1959). Soil compaction in forest and range lands. *USDA Miscellaneous Publication*, No. 768.

Lumb, P. (1962). The properties of decomposed granite. *Géotechnique*, **12**, 226–243.

Lumb, P. (1965). The residual soils of Hong Kong. *Géotechnique*, **15**, 180–194.

Lumb, P. (1975). Slope failures in Hong Kong. *Quarterly Journal of Engineering Geology*, **8**, 31–65.

Lumb, P. (1983). Engineering properties of fresh and decomposed igneous rocks from Hong-Kong. *Engineering Geology*, **27**, 287–298.

Lupini, J. F., Skinner, A. E. & Vaughan, P. R. (1981). The drained residual strength of cohesive soils. *Géotechnique*, **31**, 181–213.

Mabbutt, J. A. (1961). A stripped landsurface in Western Australia. *Transactions of the Institute of British Geographers*, **29**, 101–114.

Mabbutt, J. A. (1962). Geomorphology of the Alice Springs area. In *Lands of the Alice Springs Area, Northern Territory, 1956–57* (ed. R. A. Perry *et al.*), pp. 163–178. Land Research Series, Vol. 8, CSIRO Australia.

Mabbutt, J. A. (1965). The weathered landsurface of central Australia. *Zeitschrift für Geomorphologie, N.F.*, **9**, 82–114.

Mabbutt, J. A. (1966). The mantle controlled planation of pediments. *American Journal of Science*, **264**, 78–91.

Mabbutt, J. A. (1967). Denudation chronology of central Australia. In *Landform Studies from Australia and New Guinea* (ed. J. N. Jennings & J. A. Mabbutt), pp. 144–181. ANU Press, Canberra.

Mabbutt, J. A. (1968). Review of concepts of land evaluation. In *Land Evaluation* (ed. G. A. Stewart), pp. 11–28. Macmillan, Melbourne.

Mabbutt, J. A. (1978). Lessons from pediments. In *Landform Evolution in Australasia* (ed. J. L. Davies & M. A. J. Williams), pp. 331–347. ANU Press, Canberra.

Mabbutt, J. A. (1988). Land-surface evolution at the continental time-scale: an example from interior Western Australia. *Earth Science Reviews*, **25**, 457–466.

Mabbutt, J. A. & Scott, R. M. (1966). Periodicity of morphogenesis and soil formation in a savannah landscape near Port Moresby, Papua. *Zeitschrift für Geomorphologie, N.F.*, **10**, 69–89.

Mäckel, R. (1974). Dambos. A study in morphodynamic activity on the plateau regions of Zambia. *Catena*, **1**, 327–365.

Mäckel, R. (1985). Dambos and related landforms in Africa—an example for the ecological approach to tropical geomorphology. In *Dambos: Small Channelless Valleys in the Tropics* (ed. M. F. Thomas & A. S. Goudie), *Zeitschrift für Geomorphologie, N.F., Supplementband*, **52**, 1–23.

Mahaney, W. C. (1988). Late Pleistocene deglaciation of Mount Kenya. *Zeitschrift für Geomorphologie, N.F.*, **32**, 227–230.

Mahaney, W. C. (ed.) (1989). *Quaternary and Environmental Research on East African Mountains*. Balkema, Rotterdam.

Maignien, R. (1958). *Le Cuirassement des Sols en Guinée, Afrique Occidentale*. Mémoire du Service de la Carte Geologique D'Alsace et de Lorraine, **16**, Strasbourg.

Maignien, R. (1966). *Review of Research on Laterites*. UNESCO (Natural Resources Research Vol. 4), Paris.

Mainguet, M. (1972). *Le Modelé de Grès, Problèmes Généraux*. Institut Géographique National, Paris.

Maley, J. (1976). Essai sur le rôle de la zone tropicale dans les changements climatiques; l'example Africain. *Comptes Rendus Academie de Sciences, Paris*, **283D**, 337–340.

Maley, J. (1987). Fragmentation de la forêt dense humide Africaine et extension des biotopes montagnards au Quaternaire récent: nouvelles données polliniques et chronologiques. Implications paléoclimatiques et biogéographiques. *Palaeoecology of Africa*, **18**, 307–334.

Maley, J. (1991). The African rain forest vegetation and palaeoenvironments during Late Quaternary. *Climate Change*, **19**, 79–98 (Special Issue: Tropical Forests and Climate).

Maley, J. & Livingstone, D. A. (1983). Extension d'un élément montagnard dans le sud du Ghana (Afrique de l'Ouest) au Pléistocène supérieur et à l'Holocene inférieur: premières données polliniques. *Comptes Rendus, Académie des Sciences, Paris, Série II*, **296**, 1287–1292.

Mann, A. W. (1983). Hydrogeochemistry and weathering on the Yilgarn block, Western Australia—ferrolysis and heavy metals in continental brines. *Geochimica et Cosmochimica Acta*, **47**, 181–190.

Mann, A. W. (1984). Mobility of gold and silver in lateritic weathering profiles: some observations from Western Australia. *Economic Geology*, **79**, 38–49.

Mann, A. W. & Horwitz, R. C. (1979). Groundwater calcrete deposits in Australia: some observations from Western Australia. *Journal Geological Society of Australia*, **26**, 293–303.

Mann, A. W. & Ollier, C. D. (1985). Chemical diffusion and ferricrete formation. In *Soils and Geomorphology, Catena Supplement* (ed. P. D. Jungerius), **6**, 151–157.

Mark, R. K. & Newman, E. B. (1988). Rainfall totals before and during the storm: distribution and correlation with damaging landslides. In *Landslides, Floods, and Marine Effects of the Storm of January 3–5, 1982, in the San Francisco Bay Region, California* (ed. S. D. Ellen & G. F. Wieczorek), pp. 17–26. US Geological Survey Professional Paper No. 1434, Washington.

Marker, M. E. (1987). Karst areas of southern Africa. In *International Geomorphology 1986. Proceedings of 1st International Conference on Geomorphology, Part II* (ed. V. Gardiner), pp. 1169–1178. Wiley, Chichester.

Markewich, H. W., Pavich, M. J., Mausbach, M. J., Hall, R. L., Johnson, R. G. & Hearn, P. P. (1987).

Age relations between soils and geology in the coastal plain of Maryland and Virginia. *US Geological Survey, Bulletin*, **1589a**.

Markgraf, V. (1989). Palaeoclimates in central and south America since 18000 BP based on pollen and lake-level records. *Quaternary Science Reviews*, **18**, 1–24.

Marmo, V. (1956). On the porphyroblastic granite of central Sierra Leone. *Acta Geographica* (Helsinki), **15**, 1–26.

Marshall, T. R. (1988). The diamondiferous gravel deposits of the Bamboesspruit, southwestern Transvaal, South Africa. In *Geomorphological Studies in Southern Africa* (ed. G. F. Dardis & B. P. Moon), pp. 495–505. Balkema, Rotterdam.

Mayr, E. (1944). Wallace's Line in the light of recent zoogeographic studies. *Quarterly Review of Biology*, **19**, 1–14.

McConnell, R. B. (1968). Planation surfaces in Guyana. *Geographical Journal*, **134**, 506–520.

McDonald, R. Ch. (1976a). Limestone morphology in South Sulawesi, Indonesia. *Zeitschrift für Geomorphologie, N.F., Supplementband*, **26**, 79–91.

McDonald, R. Ch. (1976b). Hillslope base depressions in tower karst topography of Belize. *Zeitschrift für Geomorphologie, N.F., Supplementband*, **26**, 98–103.

McDonald, R. Ch. (1979). Tower karst geomorphology in Belize. *Zeitschrift für Geomorphologie, N.F., Supplementband*, **32**, 35–45.

McFadden, L. D. & Knuepfer, P. L. K. (1990). Soil geomorphology: the linkages of pedology and surficial processes. *Geomorphology*, **3**, 197–205; reprinted in *Soils and Landscape Evolution* (ed. P. L. K. Knuepfer & L. D. McFadden). Elsevier, Amsterdam & New York.

McFarlane, A., Crowe, M. J., Arthurs, J. W. & Wilkinson, A. F. (1981). *The Geology and Mineral Resources of Northern Sierra Leone*. Overseas Memoir No. 7, Institute of Geological Sciences.

McFarlane, M. J. (1969). *Lateritisation and landscape development in parts of Uganda*, PhD Thesis, University of London.

McFarlane, M. J. (1971). Laterisation and landscape development in parts of Uganda. *Quarterly Journal Geological Society, London*, **126**, 501–539.

McFarlane, M. J. (1976). *Laterite and Landscape*. Academic Press, London.

McFarlane, M. J. (1983a). Laterites. In *Chemical Sediments and Geomorphology* (ed. A. S. Goudie & K. Pye), pp. 7–58. Academic Press, London.

McFarlane, M. J. (1983b). The temporal distribution of bauxitisation and its genetic implications. In *Lateritisation Processes* (ed. A. J. Melfi & A. Carvalho), pp. 197–207. Proceedings 2nd International Seminar on Lateritisation Processes, São Paulo, Brazil, 1982, University of São Paulo, Brazil.

McFarlane, M. J. (1984). Bauxites and parent materials. In *Products and Processes of Rock Weathering* (ed. S. Sinha-Roy & S. K. Gosh). *Recent Researches in Geology*, **11**, 82–86.

McFarlane, M. J. (1986). Geomorphological analysis of laterites and its role in prospecting. *Geological Survey of India, Memoirs*, **120**, 29–40.

McFarlane M. J. (1987a). The key role of micro-organisms in the process of bauxitisation. *Modern Geology*, **11**, 325–344.

McFarlane, M. J. (1987b). Dambos—their characteristics and geomorphological evolution in parts of Malawi and Zimbabwe, with particular reference to their role in the hydrogeological regime of surviving areas of African Surface. In *Groundwater Exploration and Development in Crystalline Basement Aquifers, Proceedings*, Zimbabwe, June 1987, Vol. 1, pp. 254–302. Commonwealth Science Council.

McFarlane, M. J. (1991). Some sedimentary aspects of lateritic weathering profile development in the major bioclimatic zones of tropical Africa. *Journal of African Earth Sciences*, **12**, 267–282.

McFarlane, M. J. & Bowden, D. J. (1992). Mobilisation of aluminium in the weathering profiles of the African surface in Malawi. *Earth Surface Processes and Landforms*, **17**, 789–805.

McFarlane, M. J., Chilton, P. J. & Lewis, M. A. (1992). Geomorphological controls on borehole yields: a statistical study in an area of basement rocks in central Malawi. In *Hydrogeology of Crystalline Basement Aquifers in Africa* (ed. W. Burgess & E. P. W. Wright), Geological Society of London, Special Publication, **65**, 131–154.

McFarlane, M. J. & Heydeman, M. T. (1985). Some aspects of kaolinite dissolution by a laterite-indigenous micro-organism. In *Les Processus de Latéritisation. Journée d'Étude, Bruxelles, 22 May 1984* (ed. J. Alexandre & J. J. Symoens). Académie Royale des Sciences D'Outre-Mer, Geo-Eco-Trop, **8**, 73–91.

McFarlane, M. J. & Pollard, S. (1989). Some aspects of stone-lines and dissolution fronts associated with regolith and dambo profiles in parts of Malawi and Zimbabwe. In (ed. J. Alexandre & J. J. Symoens), *Stone-Lines*. Journée D'Étude, Bruxelles, 24 mars 1987. Académie Royale des Sciences D'Outre-Mer, Geo-Eco-Trop, **11**, 23–35.

McFarlane M. J. & Twidale, C. R. (1987). Karstic features associated with tropical weathering profiles. *Zeitschrift für Geomorphologie, N.F., Supplementband*, **64**, 73–95.

McFarlane, M. J. & Whitlow, R. (1990). Key factors affecting the initiation and progress of gullying in dambos in parts of Zimbabwe and Malawi. *Land Degradation & Rehabilitation*, **2**, 215–235.

McFarlane, M. J., Bowden, D. J. & Giusti, L. (1994). The behaviour of chromium in weathering profiles associated with the African surface in parts of Malawi. In *Rocks, Rock Weathering and landform Evolution* (ed. D. Robinson & R. Williams). (In Press.)

McGetchin, T. R., Burke, K. C., Thompson, G. A. & Young, R. A. (1980). Mode and mechanisms of plateau uplifts. In *Dynamics of Plate Interiors* (ed. A. W. Bally,

P. L. Bender, T. R. McGetchin & T. I. Walcott), pp. 99–110. Geodynamic Series, American Geophysical Union, Washington, DC.

McGowran, B. (1979). Comments on early Tertiary tectonism and lateritisation. *Geological Magazine*, **116** (Correspondence), 227–230.

Meadows, M. E. (1983). Past and present environments of the Nyika Plateau, Malawi. *Palaeoecology of Africa*, **16**, 353–390.

Meadows M. E. (1985). Dambos and environmental change in Malawi, central Africa. *Zeitschrift für Geomorphologie, Supplementband*, **52**, 147–169.

Melfi, A. J. & Carvalho, A. (1983). *Lateritisation Processes. Proceedings IInd International Seminar on Lateritisation Processes*, São Paulo, Brazil, 1982. University São Paulo, São Paulo.

Melfi, A. J., Trescases, J. J., Carvalho, A., De Oliveira, S. M. B., Filho, E. R. & Formoso, M. L. L. (1988). The lateritic ore deposits of Brazil. *Sciences Géologiques Bulletin*, **41**, 5–36

Melhorn, W. N. & Edgar, D. E. (1975). The case for episodic, continental-scale erosion surfaces: a tentative geodynamic model. In *Theories of Landform Development* (ed. W. N. Melhorn & R. C. Flemal), pp. 243–276. George Allen and Unwin, London.

Melhorn, W. N. & Flemal, R. C. (eds) (1975). *Theories of Landform Development*. George Allen and Unwin, London.

Menabe, S. & Broccoli, A. J. (1985). The influence of continental ice sheets on the climate of an ice age. *Journal of Geophysical Research*, **90**, 2167–2190.

Menabe, S. & Hahn, D. G. (1977). Simulation of the tropical climate of an ice age. *Journal of Geophysical Research*, **82**, 3889–3911.

Mesolella, K. J., Matthew, R. K., Broecker, W. S. & Thurber, D. L. (1969). The astronomical theory of climatic change: Barbados data. *Journal of Geology*, **77**, 250–274.

Meunier, A., Velde, B., Dudoignon, P. & Beaufort, D. (1983). Identification of weathering and hydrothermal alteration in acidic rocks: petrography and mineralogy of clay minerals. *Sciences Géologiques, Mémoire*, **72**, 93–99.

Meybeck, M. (1976). Total mineral dissolved transport by world major rivers. *International Association of Hydrological Sciences Bulletin*, **21**, 265–284.

Meybeck, M. (1979). Concentrations des eaux fluviales éléments majeurs et apports en solution aux océans. *Revue de Géologie Dynamiques et de Géographie Physique*, **21**, 215–246.

Meyer, L. D. (1965). Mathematical relationships governing soil erosion by water. *Journal of Soil and Water Conservation*, **20**, 149–150.

Meyer L. D. & Wischmeier, W. H. (1969). Mathematical simulation of the process of soil erosion by water. *Transactions of the American Society of Agricultural Engineers*, **12**, 754–758, 762.

Meyerhoff, H. A. (1975). The Penckian model—with modifications. In *Theories of Landform Development* (ed. W. N. Melhorn & R. C. Flemal), pp. 45–68. George, Allen and Unwin, London.

Michel, P. (1973). Les Bassins des fleuves Sénégal et Gambie—étude géomorphologique. *ORSTOM, Mémoire*, **63**, Paris.

Michel, P. (1980). The southwestern Sahara margin: sediments and climatic changes during the recent Quaternary. *Palaeoecology of Africa*, **12**, 297–314.

Milankovitch, M. M. (1941). Canon of insolation and the ice-age problem. *Koniglich Serbische Akademie, Beograd* (Engl. Translation, US Dept. of Commerce & National Science Foundation, Washington).

Milliman, J. D. & Meade, R. H. (1983). World-wide delivery of river sediment to the oceans. *Journal of Geology*, **91**, 1–21.

Milliman, J. D. & Syvitski, P. M. (1992). Geomorphic/tectonic control of sediment discharge to the ocean: the importance of small mountain rivers. *Journal of Geology*, **100**, 525–544.

Milliman, J. D., Qin, Y. S., Ren, M. E. & Saito, Y. (1987). Man's influence on the erosion and transport of sediment by Asian rivers: the Yellow River (Huanghe) example. *Journal of Geology*, **95**, 751–762.

Millington, A. C. (1981) Relationship between three scales of erosion measurement on two small basins in Sierra Leone. In *Erosion and Sediment Transport Measurement*, pp. 485–492. IAHS Publication No. 133.

Millington, A. C. (1982). Establishing soil loss and erosion hazard maps in a developing country: a West African example. *Recent Developments in the Explanantion and Prediction of Erosion and Sediment Yield*, pp. 283–292. IAHS Publication No. 137.

Millot, G. (1964). *Géologie des Argiles*, Masson, Paris.

Millot, G. (1970). *Geology of Clays* (Engl. transl. W. R. Farrand & H. Paquet). Chapman and Hall, London.

Millot, G. (1980). Les grands aplanissements des socles continentaux dans les pays subtropicaux, tropicaux et désertiques. *Mémoire h. Série Société géologique de France*, **10**, 295–305.

Millot, G. (1983). Planation of continents by intertropical weathering and pedogenetic processes. In *Lateritisation Processes. Proceedings IInd International Seminar on Lateritisation Processes*, São Paulo, Brazil, 1982 (ed. A. J. Melfi & Carvalhi), pp. 53–63. University of São Paulo, Brazil.

Minter, W. E. L. (1978). A sedimentological synthesis of placer gold, uranium, and pyrite concentrations in Proterozoic Witwatersrand sediments. In *Fluvial Sedimentology* (ed. A. D. Miall), pp. 801–829. Canadian Society of Petroleum Geology, Memoir No. 5.

Modenesi, M. C. (1983). Weathering and morphogenesis in a tropical plateau. *Catena*, **10**, 237–251.

Modenesi, M. C. (1988). Quaternary mass movements in a tropical plateau (Campos do Jordão, São Paulo, Brazil). *Zeitschrift für Geomorphologie, N.F.*, **32**, 425–440.

Moeyersons, J. (1977). Joint patterns and their influence on the form of granite residuals in NE Nigeria. *Zeitschrift für Geomorphologie, N.F.*, **21**, 14–25.

Moeyersons, J. (1981). Slumping and planar sliding on hillslopes in Rwanda. *Earth Surface Processes and Landforms*, **6**, 265–274.

Moeyersons, J. (1989a). A possible causal relationship between creep and sliding on Rwaza Hill, southern Rwanda. *Earth Surface Processes and Landforms*, **19**, 597–614.

Moeyersons, J. (1989b). The concentration of stones into a stone-line, as a result from subsurface movements in fine and loose soils in the tropics. *Stone-Lines. Journée d'Étude, Bruxelles, 24 mar. 1987* (ed. J. Alexandre & J. J. Symoens), Académie Royale des Sciences D'Outre-Mer, Geo-Eco-Trop, **11**, 11–22.

Moeyersons, J. (1990). Soil loss by rainwash: a case study from Rwanda. *Zeitschrift für Geomorphologie, N.F.* **34**, 385–408.

Mohr, E. C. J. & Van Baren, F. A. (1959). *Tropical Soils*. Interscience Publishers/Royal Tropical Institute, Amsterdam.

Monteith, J. L. (1979). Soil temperature and crop growth in the tropics. In *Soil Physical Properties and Crop Production in the Tropics* (ed. R. Lal & D. J. Greenland), pp. 250–262. Wiley, Chichester.

Montford, H. M. (1970). The terrestrial environment during Upper Cretaceous and Tertiary times. *Proceedings Geologists Association*, **81**, 181–203.

Morgan, R. P. C. (1972). Observations on factors affecting the behaviour of a first order stream. *Transactions of the Institute of British Geographers*, **56**, 171–185.

Morgan, R. P. C. (1974). Estimating regional variations in soil erosion hazard in Peninsular Malaysia. *Malaysian Naturalist Journal*, **28**, 94–106.

Morgan, R. P. C. (1976). The role of climate in the denudation system: a case study from West Malaysia. In *Geomorphology and Climate* (ed. E. Derbyshire), pp. 317–343. Wiley, New York.

Morgan, R. P. C. (1979). *Soil Erosion*. Longman, New York.

Morgan R. P. C. (1986). *Soil Erosion and Conservation* (ed. by D. A. Davidson). Longman, Harlow.

Morgan R. P. C. (1987). Applied geomorphology, land conservation and resource evaluation. In *International Geomorphology 1986. Proceedings 1st International Conference on Geomorphology* (ed. V. Gardiner), Part II, pp. 379–382. Wiley, Chichester.

Morley, R. J. (1981). Development and vegetation dynamics of a lowland ombrogenous peat swamp in Kalimantan Tengah, Indonesia. *Journal of Biogeography*, **8**, 383–404.

Morley, R. J. (1982). A Palaeoecological interpretation of a 10,000 year pollen record from Danau Padang, Central Sumatra, Indonesia. *Journal of Biogeography*, **9**, 151–190.

Morley, R. J. & Flenley, J. R. (1987). Late Cainozoic vegetational and environmental changes in the Malay Archipelago. In *Biogeographical Evolution in the Malay Archipelago* (ed. T. C. Whitmore), Oxford Monograph in Biogeography, Vol. **4**, pp. 40–59, Oxford.

Morrison, R. B. (1978) Quaternary soil stratigraphy — concepts methods and problems. In *Quaternary Soils* (ed. W. C. Mahaney), pp. 77–108. Geobooks, Norwich.

Moss, R. P. (1965). Slope development and soil morphology in a part of south west Nigeria. *Journal of Soil Science*, **16**, 192–209.

Moss, R. P. (1968). Soils, slopes and surfaces in tropical Africa. In *The Soil Resources of Tropical Africa* (ed. R. P. Moss), pp. 29–60. Cambridge University Press, Cambridge.

Moukolo, N. Laraque, A., Olivry, J. C. & Briquet, J. P. (1993). Transport en solution et en suspension par le fleuve Congo (Zaire) et ses principaux affluents de la rive droite. *Hydrological Sciences Journal*, **38**, 133–145.

Mountain, E. D. (1951). The origin of silcrete. *South African Journal of Science*, **48**, 201–204.

Mulcahy, M. J. (1964). Laterite residuals and sandplains. *Australian Journal of Science*, **27**, 54–55.

Murray, L. M. & Olsen, M. T. (1990). Colluvial slopes — A geotechnical and climatic study. In *Geomechanics in Tropical Soils,* Proceedings of the Second International Conference on Geomechanics in Tropical Soils, Singapore, 1988, Vol. 2, pp. 573–579. Balkema, Rotterdam.

Nahon, D. (1977). Time factor in iron crusts genesis. *Catena*, **4**, 249–254.

Nahon, D. (1986). Evolution of iron crusts in tropical landscapes. In *Rates of Chemical Weathering of Rocks and Minerals* (ed. S. M. Colman & D. P. Dethier), pp. 169–191. Academic Press, New York.

Nahon, D. & Bocquier, G. (1983). Petrology of mineral transfers in weathering and soil systems. *Proceedings International Colloquium on Petrology and Weathering of Soils*, 1983, Sciences Géologiques, Strasbourg, Mémoire, **72**, 111–120.

Nahon, D. & Lappartient, J. R. (1977). Time factor and geochemistry in iron crust genesis. *Catena*, **4**, 249–254.

Nair, NG. K., Santosh, M. & Mahadevan, R. (1987). Lateritisation as a possible contributor to gold placers in Nilambur Valley, southwest India. *Chemical Geology*, **60**, 309–315.

Nanson, G. C. (1986). Episodes of vertical accretion and catastrophic stripping: a model of disequilibrium floodplain development. *Geological Society of America Bulletin*, **97**, 1467–1475.

Nanson, G. C. & Croke, J. C. (1992). A genetic classification of floodplains. *Geomorphology*, **4**, 459–486.

Nanson, G. C. & Young, R. W. (1981). Overbank deposition and floodplain formation on small coastal streams of New South Wales. *Zeitschrift für Geomorphologie, N.F.,* **25**, 332–347.

Nanson, G. C. & Young, R. W. (1988). Fluviatile evidence for a period of Late-Quaternary pluvial climate in coastal southeastern Australia. *Palaeogeography, Palaeoclimatology, Palaeoecology,* **66**, 45–61.

Nanson, G. C., Young, R. W., Price D. M. & Rust, B. R. (1988). Stratigraphy, sedimentology and Late-Quaternary chronology of the channel country of western Queensland. In *Fluvial Geomorphology of Australia* (ed. R. F. Warner), pp. 151–175. Academic Press, Sydney.

Nanson, G. C., Price, D. M., Short, S. A., Young, R. W. & Jones, B. G. (1991). Comparative uranium-thorium and thermoluminescence dating of weathered Quaternary alluvium in the tropics of Northern Australia. *Quaternary Research,* **35**, 347–366.

Newill, D. & Dowling, J. W. F. (1968). Laterites in West Malaysia and Northern Nigeria. In *Engineering Properties of Laterite Soils,* Proceedings ICSMFE VII, pp. 133–150.

Nicholson, S. E. & Flohn, H. (1980). African environmental and climatic changes and the general atmospheric circulation in Late Pleistocene and Holocene. *Climatic Change,* **2**, 313–348.

Nikiforoff, C. C. (1949). Weathering and soil evolution. *Soil Science,* **67**, 219–230.

Nikiforoff, C. C. (1959). Reappraisal of the soil. *Science,* **129**, 186–196.

Nilsen, T. H. & Kerr, D. R. (1978). Paleoclimatic and paleogeographic implications of a lower Tertiary laterite (latosol) on the Iceland-Faeroe Ridge, North Atlantic region. *Geological Magazine,* **115**, 153–236.

Nilsen, T. H. & Moore, T. E. (1984). *Bibliography of Fan Deposits,* pp. 96. Geo Books, Norwich.

Nilsen, T. H. & Turner, B. L. (1975). Influence of rainfall and ancient landslide deposits on recent landslides (1950–71) in urban areas of Contra Costa County, California. *Geological Survey Bulletin,* **1388**, 18.

Nilsson, E. (1931). Quaternary glaciations and pluvial lakes in British East Africa. *Geografiska Annaler,* **A13**, 249–349.

Nishimura, S., Nishida, J., Yokoyama, T. & Hehuwat, F. (1986). Neotectonics of the Strait of Sunda, Indonesia. *Journal of South East Asian Earth Sciences,* **1**, 81–91.

Nortcliff, S. & Thornes, J. B. (1981). Seasonal variations in the hydrology of a small forested catchment near Manaus, Amazonas, and the implications for its management. In *Tropical Agricultural Hydrology* (ed. R. Lal & E. W. Russell), pp. 37–57. Wiley, Chichester.

Nortcliff, S., Ross, S. M. & Thornes, J. B. (1990). Soil moisture, runoff and sediment yield from differentially cleared tropical rainforest plots. In *Vegetation and Erosion* (ed. J. B. Thornes), pp. 420–436. Wiley, Chichester.

Nossin, J. J. (1967). Igneous rock weathering on Singapore Island. *Zeitschrift für Geomorphologie, N.F.,* **11**, 14–38.

Nugent, C. (1990). The Zambezi River: tectonism, climatic change and drainage evolution. *Palaeogeography, Palaeoclimatology, Palaeoecology,* **78**, 55–69.

Nye, P. H. (1954). Some soil-forming processes in the humid tropics. Part I: A field study of a catena in the West African forest. *Journal of Soil Science,* **5**, 7–27.

Nye, P. H. (1955). Some soil-forming processes in the humid tropics. Part II: The development of the upper slope member of the catena. *Journal of Soil Science,* **6**, 51–62.

Oades, J. M. & Waters, A. G. (1991). Aggregate hierarchy in soils. *Australian Journal of Soil Research,* **29**, 815–828.

Oberlander, T. M. (1972). Morphogenesis of granitic boulder slopes in the Mojave Desert, California. *Journal of Geology,* **80**, 1–20.

Oberlander, T. M. (1974). Landscape inheritance and the pediment problem in the Mojave Desert of southern California. *American Journal of Science,* **274**, 849–875.

Oberlander, T. M. (1989). Slope and pediment systems. In *Arid Zone Geomorphology* (ed. D. S. G. Thomas), pp. 56–84. Belhaven, London.

Ogunkoya O. O. & Jeje, L. K. (1987). Sediment yield from some third order basins of the basement complex rocks in southwestern Nigeria. *Catena,* **14**, 383–397.

Ogunkoya, O. O., Adejuwon, J. O. & Jeje, L. K. (1984). Runoff response to basin parameters in southwestern Nigeria. *Journal of Hydrology,* **72**, 67–84.

Ojanuga, A. G. & Wirth, K. (1977). Threefold stonelines in southwestern Nigeria; evidence of cyclic soil and landscape development. *Soil Science,* **123**, 249–257.

Oliveira, S. M. B. (1985). Génese de bauxita de Lages — SC. *Boletim Geological Institute, University of Sao Paulo (USP), Série Ceintifica,* **16**, 46–81.

Olivry, J. C. (1977). Transports solides en suspension au Cameroun. In *Erosion and Solid Matter Transport in Inland Waters* (Proceedings of Paris Symposium, July 1977), pp. 134–141. IAHS Publication No. 122.

Ollier, C. D. (1959). A two cycle theory of tropical pedology. *Journal of Soil Science,* **10**, 137–148.

Ollier, C. D. (1960). The inselbergs of Uganda. *Zeitschrift für Geomorphologie, N.F.,* **4**, 43–52.

Ollier, C. D. (1969). *Weathering* (2nd edn., 1984), pp. 270. Longman, London.

Ollier, C. D. (1976). Catenas in different climates. In *Geomorphology and Climate* (ed. E. Derbyshire), pp. 137–169. Wiley, London & New York.

Ollier, C. D. (1977). Terrain classification: methods, applications and principles. In *Applied Geomorphology* (ed. J. R. Hails), pp. 277–316. Elsevier, Amsterdam.

Ollier, C. D. (1978). Inselbergs of the Namib Desert, processes and history. *Zeitschrift für Geomorphologie, N.F., Supplementband,* **31**, 161–176.

Ollier, C. D. (1979). Evolutionary geomorphology of Australia and New Guinea. *Transactions Institute of British Geographers NS,* **4**, 516–539.

Ollier, C. D. (1981). *Tectonics and Landforms*. Longman, London.

Ollier, C. D. (1982). The great Escarpment of eastern Australia: tectonic and geomorphic significance. *Journal of the Geological Society of Australia*, **29**, 13–23.

Ollier, C. D. (1983a). Tropical geomorphology and long-term landform evolution. *Finisterra*, **36**, 203–221.

Ollier, C. D. (1983b). Weathering or hydrothermal alteration. *Catena*, **10**, 57–59.

Ollier, C. D. (1985). Morphotectonics of passive continental margins: Introduction. *Zeitschrift für Geomorphologie, N.F., Supplementband*, **54**, 1–9.

Ollier, C. D. (1988a). Deep weathering, groundwater and climate. *Geografiska Annaler*, **70A**, 285–290.

Ollier, C. D. (1988b). The regolith in Australia. *Earth Science Reviews*, **25**, 355–361.

Ollier, C. D. (ed.) (1991). *Ancient Landforms*. Batsford, London.

Ollier, C. D. (1992). Global change and long-term geomorphology. *Terra Nova*, **4**, 312–319.

Ollier, C. D. & Galloway, R. W. (1990). The laterite profile, ferricrete and unconformity. *Catena*, **17**, 97–109.

Ollier, C. D., Gaunt, G. F. M. & Jurkowski, I. (1988a). The Kimberley Plateau, Western Australia a Precambrian erosion surface. *Zeitschrift für Geomorphologie, N.F.*, **32**, 239–246.

Ollier, C. D., Chan, R. A., Craig, M. A. & Gibson, D. L. (1988b). Aspects of landscape history and regolith in the Kalgoorlie region, Western Australia. *BMR Journal of Geology & Geophysics*, **10**, 309–321.

Ong, H. L. & Swanson, V. E. (1969). Natural organic acids in the transportation, deposition and concentration of gold. *Colorado School of Mines Quarterly*, **64**, 395–425.

Ong, H. L., Swanson, V. E. & Bisque, R. E. (1970). Natural organic acids as agents of chemical weathering. In *Geological Survey Research 1970*, US Geological Survey Professional Paper **700-C**, pp. C130–C137.

Orme, A. R. (1990). Recurrence of debris production under coniferous forest, Cascade foothills, northwest United States. In *Vegetation and Erosion* (ed. J. B. Thornes), pp. 67–84. Wiley, Chichester.

Osborn, G. & Du Toit, C. (1991). Lateral planation by rivers as a geomorphic agent. *Geomorphology*, **4**, 249–260.

Oyagi, N. (1968). Weathering-zone structure and landslides of the area of granitic rocks in Kamo-Daito, Shimane Prefecture. In Studies of the Mechanism and Foreknowledge of Landslides in Weathered Areas of Granite Rocks. *Reports of Cooperative Research for Disaster Prevention*, **14**, 113–127 (English translation, J. Arata and P. Durgin, National Agricultural Library, Washington, DC).

Oyebande, L. (1981). Sediment transport and river basin management in Nigeria. In *Tropical Agricultural Hydrology* (ed. R. Lal & E. W. Russell), pp. 201–225. Wiley, Chichester.

Pachur, H.-J. & Hoelzmann, P. (1991). Paleoclimatic implications of Late Quaternary lacustrine sediments in western Nubia, Sudan. *Quaternary Research*, **36**, 257–276.

Pain, C. F. (1985). Morphotectonics of the continental margins of Australia. *Zeitschrift für Geomorphologie, N.F., Supplementband*, **54**, 23–35.

Pain, C., Chan, R., Craig, M., Hazell, M., Kamprad, J. & Wilford, J. (1991). *RTMAP-BMR Regolith Database Field Handbook*. Bureau of Mineral Resources, Canberra.

Pallister, J. W. (1956). Slope development in Buganda. *Geographical Journal*, **122**, 80–87.

Pardhasaradhi, Y. J. (1976). Etchplain development of Malabar Uplands—a suggestion. *Journal of the Geological Society of India*, **17**, 73–78.

Parker, G. G. (1963). Piping, a geomorphic agent in land development in drylands. *International Association of Scientific Hydrology*, Publication, No. 65, pp. 103–114.

Parker, G. G. & Jenne, E. A. (1967). Structural failure of western US highways caused by piping. *Report US Geological Survey Water Resources Division*, Washington.

Partridge, T. C. & Brink, A. B. A. (1969). Gravels and terraces of the lower Vaal basin. *South African Geographical Journal*, **49**, 21–38.

Partridge, T. C. & Maud, R. R. (1987). Geomorphic evolution of southern Africa since the Mesozoic. *South African Journal of Geology*, **90**, 179–208.

Partridge, T. C. & Maud, R. R. (1988). The geomorphic evolution of southern Africa: a comparative review. In *Geomorphological Studies in Southern Africa* (ed. G. F. Dardis & B. P. Moon), pp. 5–15. Balkema, Rotterdam.

Passarge, S. (1904). Rumpfflächen und inselberge. *Zeitschrift Deutsche Gesselschaft*, **56**, 193–215.

Passarge, S. (1923). Die Inselberglandschaft der Massaisteppe. *Petermanns Geographische Mitteilungen*, **65**, 41–46.

Pastouret, L., Chamley, H., Delibrias, G., Duplessy, J. C. & Theide, J. (1978). Late Quaternary climatic changes in western tropical Africa deduced from deep-sea sedimentation off the Niger Delta. *Oceanologica Acta*, **1**, 217–232.

Pathak, P. C., Pandy, A. N. & Singh, J. S. (1984). Overland flow, sediment output and nutrient loss from certain forested sites in the central Himalaya, India. *Journal of Hydrology*, **71**, 239–251.

Paton, T. R. & Williams, M. A. J. (1972). The concept of laterite. *Annals Association of American Geographers*, **62**, 42–56.

Paulissen, E. (1989). The inundations of the Saharan Nile around 12,500 BP: unique catastrophic floods of an exotic river. *Abstracts of Papers, Second International Conference on Geomorphology, Frankfurt/Main, 1989. Geöko-plus*, **1**, 216.

Pavich, M. J. (1985). Appalachian piedmont morphogenesis: weathering, erosion, and Cenozoic uplift. In *Tectonic Geomorphology* (ed. M. Morisawa & J. T. Hack), pp. 27–51. George, Allen and Unwin, London.

Pavich, M. J. (1986). Processes and rates of saprolite production and erosion on a foliated granitic rock of the Virginia Piedmont. In *Rates of Chemical Weathering of Rocks and Minerals* (ed. S. M. Colman & D. P. Dethier), pp. 552–590. Academic Press, New York.

Pavich, M. J. (1989). Regolith residence time and the concept of surface age of the Piedmont 'peneplain'. *Geomorphology*, 2, 181–196.

Pavich, M. J. & Obermeier, S. F. (1985). Saprolite formation beneath Coastal Plain sediments near Washington, D.C. *Geological Society of America, Bulletin*, 96, 886–900.

Pavich, M. J., Brown, L, Valette-Silver, J. N., Klein, J. & Middleton, R. (1985). ^{10}Be analysis of a Quaternary weathering profile in the Virginia Piedmont. *Geology*, 13, 39–41.

Peart, M. (1991). The Kaiapit Landslide: events and mechanisms. *Quarterly Journal of Engineering Geology*, 24, 399–411.

Pedro, G. (1966). Essai sur la caractéristation géochemique de minéraux argileux dans les altérations et les sols des climats méditerranéens et tropicaux à saisons constrastées. *Mémoire Service Carte Géologique* Alsace Lorraine, 30.

Pedro, G. (1968). Distributions des principaux types d'altération chimique à la surface du globe. *Revue de Géographie Physique et Géologie Dynamique*, X, 457–470.

Pedro, G. (1983). Structuring some basic pedological processes. *Geoderma*, 31, 289–299.

Pedro, G. (1984). La genèse des argiles pédologiques ses implications minérologiques, physico-chimiques et hydriques. *Sciences Géologiques, Bulletin*, 37, 333–347.

Pedro, J. (1991). Personal communication.

Peltier, L. C. (1950). The geographical cycle in periglacial regions as it is related to climatic geomorphology. *Annals Association of American Geographers*, 40, 214–236.

Peltier, L. C. (1962). Area sampling for terrain analysis. *Professional Geographer*, 14, 24–28.

Penck, W. (1924). *Die Morphologische Analyse*. Geographische Abhandlungen.

Peters, M. & Tetzlaff, G. (1990). West African palaeosynoptic patterns at the last glacial maximum. *Theoretical and Applied Climatology*, 42, 67–79.

Petit, M. (1985). Aspects morphologiques du massif central Guyanais. *Bulletin Association Géographie Française, Paris*, 4, 254–267.

Pettijohn, F. J. (1975). *Sedimentary Rocks*. Harper & Row, New York.

Peulvast, J.-P. (1989). Les altérites et l'identification des reliefs préglaciaires dans une montagne de haute latitude: l'example des Scandes. *Zeitschrift für Geomorphologie, N.F., Supplementband*, 72, 55–78.

Peyrot, B. & Lanfranchi, R. (1984). Les oscillations morphologiques récents dans la vallée du Niari (République Populaire du Congo). *Palaeoecology of Africa*, 16, 265–281.

Pflug, R. (1969). Quaternary lakes of eastern Brazil. *Photogrammetria*, 24, 29–35.

Philip, J. R. (1957). The theory of infiltration. I. The infiltration equation and its solution. *Soil Science*, 83, 345–357.

Pickup, G. (1984). Geomorphology of tropical rivers I. Landforms, hydrology and sedimentation in the Fly and Lower Purari, Papua New Guinea. *Catena Supplement*, 5, 1–18.

Pickup, G. (1991). Event frequency and landsurface stability on the floodplain systems of arid central Australia. *Quaternary Science Reviews*, 10, 463–473.

Pickup, G. & Warner, R. F. (1984). Geomorphology of tropical rivers II. Channel adjustment, sediment load and discharge in the Fly and Lower Purari, Papua New Guinea. *Catena Supplement*, 5, 19–41.

Pickup, G., Higgins, R. J. & Warner, R. F. (1980). Erosion and sediment yield in Fly River Basin. International Association of Hydrological Sciences (IAHS), Publication No. 132, pp. 438–456, Wallingford.

Pilgrim, D., Cordery, I. & Baron, B. (1982). Effects of catchment size on runoff relationships. *Journal of Hydrology*, 58, 205–221.

Pillans, B. J. (1977). An early Tertiary age for deep weathering at Bredbo, southern NSW. *Search*, 8, 81–83.

Pissis, M. A. (1842). La Position Géologique des Terrains de la Partie Austral du Brésil. *Mémoire de L'Institute de France*, X, 358.

Pokras, E. M. & Mix, A. C. (1985). Eolian evidence for spatial variability of late Quaternary climates in tropical Africa. *Quaternary Research*, 24, 137–149.

Pollack, H. R. (1983). Land surfaces and lateritisation in Surinam. In *Lateritisation Processes. Proceedings IInd International Seminar on Lateritisation Processes* (ed. A. J. Melfi & A. Carvalhi), pp. 295–308. University of São Paulo, Brazil.

Potter, P. E. (1978). Petrology and chemistry of modern big river sands. *Journal of Geology*, 86, 423–449.

Pouyllau, M. (1985). Les karsts gréseux dans la Gran Sabana (Guyane Vénézuélienne). *Bulletin Association Géographie Française, Paris*, 4, 269–283.

Prance, G. T. (ed.) (1982). *Biological Diversification in the Tropics*. Columbia, New York.

Prell, W. L. & Van Campo, E. (1986). Coherent response of Arabian Sea upwelling and pollen transport to late Quaternary monsoonal winds. *Nature*, 323, 526–528.

Prell, W. L. & Kutzbach, J. E. (1987). Monsoon variability over the past 150,000 years. *Journal of Geophysical Research*, 92 (D7), 8411–8425.

Prescott, J. A. & Pendleton, R. L. (1952). *Laterites and Lateritic Soils*. Commonwealth Bureau of Soil Science, Technical Communication No. 47.

Pretorius, D. A. (1976). The nature of the Witwatersrand gold-uranium deposits. In *Handbook of Strata-bound and Stratiform Ore Deposits* (ed. K. H. Wolf), pp. 29–32. Elsevier, Amsterdam.

Pugh, J. C. (1955). Isostatic readjustment and the theory of pedimentation. *Quarterly Journal Geological Society of London*, **111**, 361–369.

Pugh, J. C. (1956). Finging pediments and marginal depressions in the inselberg landscape of Nigeria. *Transactions of the Institute of British Geographers*, **22**, 15–31.

Pugh, J. C. (1966). Landforms in low latitudes. In *Essays in Geomorphology* (ed. G. H. Dury), pp. 121–138. Heinemann, London.

Pullan, R. A. (1979). Termite hills in Africa: their characteristics and evolution. *Catena*, **6**, 267–291.

Pye, K. (1983): Red beds. In *Chemical Sediments and Geomorphology* (ed. A. S. Goudie & K. Pye), pp. 227–263. Academic Press, London.

Pye, K., Goudie, A. S. & Thomas, D. S. G. (1984). A test of petrological control in the development of bornhardts and koppies on the Matopos batholith, Zimbabwe. *Earth Surface Processes and Landforms*, **9**, 455–467.

Radwan, A. M. (1988). Properties of granite soil in Aswan, Egypt. *Proceedings of the 2nd International Conference on Geomechanics in Tropical Soils, Singapore*, **1**, 203–210.

Ramesh, R. & Subramanian, V. (1986). Mass transport in Krishna River basin, India. In *Drainage Basin Sediment Delivery* (ed. R. F. Hadley), pp. 185–197. IAHS Publication No. 159.

Rao, S. M., Sridharan, A. & Chandrakaran, S. (1988). The role of iron oxide in tropical soil properties. *Geomechanics in Tropical Soils*, Proceedings of the Second International Conference on Geomechanics in Tropical Soils, Singapore, 1988, Vol. 1, pp. 43–48. Balkema, Rotterdam.

Rapp, A. (1975). Some views on the Ordovician palaeoglaciation in Saharan Africa. *Geologiska Föreningens i Stockholm Förhandlingar*, **97**, 142–150.

Rapp, A., Axelsson, V., Berry, L. & Murray-Rust, D. H. (1972). Soil erosion and sediment transport in the Morogorao river catchment, Tanzania. *Geografiska Annaler*, **54-A**, 125–155.

Räsänen, M. E., Salo, J. S. & Kalliola, R. J. (1987). Fluvial perturbance in the western Amazon Basin: regulation by long-term sub-Andean tectonics. *Science*, **238**, 1398–1400.

Raunet, M. (1985). Les bas-fonds en Afrique et à Madagascar. *Zeitschrift für Geomorphologie, N.F., Supplementband*, **52**, 25–62.

Reich, B. M. (1963). Short-term rainfall intensity estimates and other design aids for regions of sparse data. *Journal of Hydrology*, **1**, 3–28.

Reneau, S. L. & Dietrich, W. E. (1987). Size and location of colluvial landslides in a steep forested landscape. In *Erosion and Sedimentation in the Pacific Rim* (ed. R. L. Beschta, T. Blinn, C. E. Grant, F. J. Swanson & G. G. Ice), pp. 39–48. IAHS Publication No. 165.

Reneau, S. L., Dietrich, W. E., Dorn, R. I., Berger, C. R. & Rubin, M. (1986). Geomorphic and palaeoclimatic implications of latest Pleistocene radiocarbon dates from colluvium-mantled hollows, California. *Geology*, **14**, 655–658.

Riccomini, C., Peloggia, A. U. G., Saloni, J. C. L., Kohnke, M. W. & Figueira, R. M. (1989). Neotectonic activity in the Serra do Mar rift system (southeastern Brazil). *Journal of South American Earth Sciences*, **2**, 191–197.

Riezebos, H. Th. (1983). Geomorphology, soils and vegetation differentiation in a tropical rainforest environment in Suriname. *Geologie en Mijnbouw*, **62**, 669–675.

Riezebos, H. Th. (1984). Geomorphology and savannisation in the upper Sipaliwini River basin (S. Suriname). *Zeitschrift für Geomorphologie, N.F.*, **28**, 265–284.

Rind, D. & Peteet, D. (1985). Terrestrial conditions at the last glacial maximum and CLIMAP sea-surface temperature estimates: are they consistent? *Quaternary Research*, **24**, 1–22.

Roberts, D. G., Morton, A. C. & Backmann, J. (1984). Late Paleocene–Eocene volcanic events in the northern North Atlantic Ocean. *Initial Report Deep-Sea Drilling Project*, **81**, 913–923.

Roberts, N. & Barker, P. (1993). Landscape stability and biogeomorphic response to past and future climatic shifts in intertropical Africa. In *Landscape Sensitivity* (ed. D. S. G. Thomas & R. J. Allison). Wiley, Chichester.

Roche, M. A. (1981). Watershed investigations for development of forest resources of the Amazon Region in French Guyana. In *Tropical Agricultural Hydrology* (ed. R. Lal & E. W. Russell), pp. 75–92. Wiley, Chichester.

Rodenburg, J. K. (1984). Geology, genesis and bauxite reserves of West Kalimantan, Indonesia. In *Bauxite*. Proceedings of the 1984 Symposium, Los Angeles, 1984 (ed. L. Jacob, Jr.), pp. 603–618. Society of Mining Engineers, American Institute of Mining, Metallurgical, Petroleum Engineers, New York.

Rodier, J. (1975). Evaluation of runoff in tropical African Sahel. *ORSTOM, Travaux et Documents*, **145**, Paris.

Rodin, L. E. & Basilevic, N. I. (1967). *Production and Mineral Cycling in Terrestrial Vegetation*. Oliver and Boyd, Edinburgh.

Rodolfi, G. (1988). Geomorphological mapping applied to land evaluation and soil conservation in agricultural planning: some examples from Tuscany (Italy). *Zeitschrift für Geomorphologie, N.F., Supplementband*, **68**, 155–174.

Roe, F. W. (1951). *The Geology and Mineral Resources of the Fraser's Hill Area, Selangor, Perak and Pehang, Federation of Malaya*. Memoir **5**. Geological Survey, Federation of Malaya.

Rogers, N. W. & Selby, M. J. (1980). Mechanisms of shallow translational landsliding during summer rainstorms: North Island, New Zealand. *Geografiska Annalaer*, **62A**, 11–27.

Rognon, P. & Williams, M. A. J. (1977). Late Quaternary climatic changes in Australia and North Africa: a preliminary investigation. *Palaeogeography, Palaeoclimatology and Palaeoecology,* 21, 285–327.

Rohdenburg, H. (1969/1970). Hangpedimentation und Klimawechsel als wichstigste Faktoren der Flächen und Stufenbildung in den wechselfeuchten Tropen an Beispielen aus Westafrika, besonders aus dem Schichtstufenland Südost-Nigerias. *Zeitschrift für Geomorphologie, N.F.,* 14, 58–78.

Rohdenburg, H. (1982). Geomorphologisch-bodenstratigraphischer Vergleich zwischen dem Nordostbrazilianischen Treckengebiet und immerfeucht-tropischen Gebieten Südbraziliens mit Ausfürungen zum Problemkreis der Pediplain–Pediment–Terrassentreppen. *Catena,* Supplement 2, 74–122.

Rohdenburg, H. (1983). Beiträge zur allgemeinen geomorphologie der Tropen und Subtropen. *Catena,* 10, 393–438.

Roose, E. J. (1973). *Dix-sept années de mesures éxperimentales de l'érosion et du russellement sur un sol ferrallitique sableux de basse Côte d'Ivoire.* Thèse Docteur Ingénieur, Faculté des Sciences, Université d'Abidjan, Publié ORSTOM, Abidjan.

Roose, E. J. (1977). Application of the universal soil loss equation of Wischmeier and Smith in West Africa. In *Soil Conservation and Management in the Humid Tropics* (ed. D. J. Greenland & R. Lal), pp. 177–187. Wiley, Chichester.

Roose, E. J. (1981). Approach to the definition of rain erosivity and soil erodibility in West Africa. In *Assessment of Erosion* (ed. M. De Broodt & D. Gabriels), pp. 143–151. Wiley, Chichester.

Roose, E. J. (1986). Runoff and erosion before and after clearing depending on the type of crop in western Africa. In *Land Clearing and Development in the Tropics* (ed. R. Lal, P. A. Sanchez & R. W. Cummings, Jr), pp. 317–330. Balkema, Rotterdam.

Rooseboom, A. (1978). Sedimentafvoer in Suider-Afrikaanse Riviere. *Water, South Africa,* 4, 14–17.

Rose, J. (1984a). Alluvial terraces of an equatorial river, Melinau drainage basin, Sarawak. *Zeitschrift für Geomorphologie, N.F.,* 28, 155–177.

Rose, J. (1984b). Contemporary river landforms and sediments in an area of equatorial rainforest, Gulong Mulu National Park, Sarawak. *Transactions Institute British Geographers,* NS 9, 345–363.

Rossi, G. (1976). Karst et dissolution des calcaires en milieu tropical. *Zeitschrift für Geomorphologie, N.F., Supplementband,* 26, 124–152.

Rossignol-Strick, M. (1985). Mediterranean Quaternary sapropels, an immediate response of the African monsoon to variation of insolation. *Palaeogeography, Palaeoclimatology and Palaeoecology,* 49, 237–263.

Rossignol-Strick, M. & Duzer, D. (1979). A late Quaternary continuous climatic record from palynology of three marine cores off Senegal. *Palaeoecology of Africa,* 11, 185–188.

Rossignol-Strick, M., Nesteroff, W., Olive, P. & Vergnaud-Grazzini, C. (1982). After the deluge: Mediterranean stagnation and sapropel formation. *Nature,* 295, 105–110.

Rougerie, G. (1955). Un mode de dégagement probable de certains dômes granitiques. *Comptes Rendues Academie des Sciences,* 246, 327–329.

Rougerie, G. (1960). *Le Façonnement Actuel des Modelés en Côte D'Ivoire Forestière.* Memoire Institut Français Afrique Noire, No. 58, Dakar.

Ruddiman, W. F. & Kutzbach, J. E. (1989). Forcing of late Cenozoic northern hemisphere climate by plateau uplift in southern Asia and the American west. *Journal of Geophysical Research,* 94, (D15), 18 409–18 427.

Ruddock, E. C. (1967). Residual soils of the Kumasi district in Ghana. *Géotechnique,* 17, 359–377.

Ruhe, R. V. (1956). *Landscape Evolution in the High-Ituri, Belgian Congo.* Publication INEAC, Série Scientifique, Vol. 66.

Ruhe, R. V. (1959). Stonelines in soils. *Soil Science,* 87, 223–231.

Ruhe, R. V. (1960). Elements of the soil landscape. *7th International Congress of Soil Science, Madison, Transactions,* 4, 165–170.

Runge, J. (1993). Geomorphological observations concerning palaeoenvironmental conditions in eastern Zaire. *Zeitschrift für Geomorphologie, N.F., Supplementband,* 91, 109–122.

Rushton, K. R. & Weller, J. (1985). Response to pumping of a weathered-fractured granite aquifer. *Journal of Hydrology,* 80, 299–309.

Ruxton, B. P. (1958). Weathering and sub-surface erosion in granite at the piedmont angle, Balos Sudan. *Geological Magazine,* 95, 353–377.

Ruxton, B. P. (1967). Slopewash under mature primary rainforest in northern Papua. In *Landform Studies from Australia and New Guinea* (ed. J. N. Jennings & J. A. Mabbutt), pp. 85–94. Cambridge University Press, London.

Ruxton, B. P. (1968a). Order and disorder in landform. In *Land Evaluation* (ed. G. A. Stewart), pp. 29–39. Macmillan, Melbourne.

Ruxton, B. P. (1968b). Measures of the degree of chemical weathering of rocks. *Journal of Geology,* 76, 518–527.

Ruxton, B. P. & Berry, L. (1957). Weathering of granite and associated erosional features in Hong Kong. *Geological Society of America Bulletin,* 68, 1263–1292.

Ruxton, B. P. & Berry L. (1961). Weathering profiles and geomorphic position on granite in two tropical regions. *Revue de Géomorphologie Dynamique,* 12, 16–31.

Sanchez, P. (1976). *Properties and Management of Soils in the Tropics.* Wiley, New York.

Sandroni, S. S. (1985). Sampling and testing of residual soils in Brazil. In *Sampling and Testing of Residual*

Soils (ed. E. W. Brand & H. B. Phillipson), pp. 31–50. Scorpion Press, Hong Kong.

Sapper, K. (1935). *Geomorphologie der Feuchten Tropen.* Teubner, Leipzig.

Sarin, M. M., Krishnaswami, S., Dilli, K., Somayajulu, B. L. K. & Moore, W. S. (1989). Major ion chemistry of the Ganga–Brahmaputra river system: weathering processes and fluxes to the Bay of Bengal. *Geochimica et Cosmochimica Acta,* **53,** 997–1009.

Sarnthein, M. (1978). Sand deserts during glacial maximum and climatic optimum. *Nature,* **272,** 43–46.

Saunders, I. & Young, A. (1983). Rates of surface processes on slopes, slope retreat and denudation. *Earth Surface Processes and Landforms,* **8,** 473–501.

Savage, K. M., De Cesero, P. & Potter, P. E. (1988). Mineralogic maturity of modern sand along a high energy tropical coast: Baixada de Jacarepaguá, Rio de Janeiro, Brazil. *Journal of South American Earth Sciences,* **1,** 317–328.

Savat, J. (1975). Some morphological and hydraulic characteristics of river patterns in the Zaire Basin. *Catena,* **2,** 161–180.

Savat, J. (1977). The hydraulics of sheet flow on a smooth surface and the effect of simulated rainfall. *Earth Surface Processes,* **2,** 125–140.

Savigear, R. A. G. (1960). Slopes and hills in West Africa. *Zeitschrift für Geomorphologie, N.F., Supplementband,* **1,** 156–171.

Savigear, R. A. G. (1965). A technique of morphological mapping. *Annals of the Association of American Geographers,* **55,** 513–538.

Schalascha, E. B., Appelt, H. & Schatz, A. (1967). Chelation as a weathering mechanism — 1. Effect of complexing agents on the solubilisation of iron from minerals and granodiorite. *Geochimica et Cosmochimica Acta,* **31,** 587–596.

Schellmann, W. (1978). Behaviour of nickel, cobalt and chromium in ferruginous lateritic nickel ores. *Bulletin Bureau Recherche Géologique et Minière (BRGM), Section II,* 275–282.

Schellmann, W. (1981). Considerations on the definition and classification of laterites. In *Lateritisation Processes,* Proceedings International Seminar on Lateritisation Processes, Trivandrum, 1979, pp. 1–10. Balkema, Rotterdam.

Schellmann, W. (1983). A new definition of laterite. *Natural Resources and Development,* **18,** 7–21. Tubingen.

Schmidt, K-H. (1989). The significance of scarp retreat for Cenozoic landform evolution on the Colorado Plateau. *Earth Surface Processes and Landforms,* **14,** 93–105.

Schmidt, P. W. & Ollier, C. D. (1988). Paleomagnetic dating of Late Cretaceous to Early Tertiary weathering in New England, N.S.W., Australia. *Earth Science Reviews,* **25,** 363–371.

Schnütgen, A. & Bremer, H. (1985). Die entstehung von Decksanden im oberen Rio Negro-Gebiet. *Zeitschrift*

für Geomorphologie, N.F., Supplementband, **56,** 55–67.

Scholz, C. A. & Rosendahl, B. R. (1988). Low lake stands in Lakes Malawi and Tanganyika, East Africa, delineated with multifold seismic data. *Science,* **240,** 1645–1648.

Schubert, C. (1980). Contribution to the paleolimnology of Lake Valencia, Venezuela: seismic stratigraphy. *Catena,* **7,** 275–292.

Schubert, C. (1988). Climatic changes during the last glacial maximum in northern South America and the Caribbean: a review. *Interciencia,* **13,** 128–137.

Schubert, C. & Hubert, O. (1990). *Le Gran Sabana.* Cuadernos Lagoven, Caracas.

Schubert, C. & Valastro, S. Jr. (1980). Quaternay Esnujaque Formation, Venezuelan Andes: preliminary alluvial chronology in a tropical mountain range. *Zeitschrift für Deutsche Geologie, Gesselschaft,* **131,** 927–947.

Schumm, S. A. (1956). The role of creep and rainwash on the retreat of badland slopes. *American Journal of Science,* **254,** 693–706.

Schumm, S. A. (1965). Quaternary paleohydrology. In *The Quaternary of the United States* (ed. H. E. Wright & D. G. Frey), pp. 783–794. Princeton University Press, Princeton.

Schumm, S. A. (1968). Speculations concerning paleohydrologic controls of terrestrial sedimentation. *Bulletin Geological Society of America,* **79,** 1573–1588.

Schumm, S. A. (1975). Episodic erosion: a modification of the geomorphic cycle. In *Theories of Landform Development* (ed. W. N. Melhorn & R. C. Flemal), pp. 69–86. George, Allen and Unwin.

Schumm S. A. (1977). *The Fluvial System.* Wiley, Chichester.

Schumm, S. A. (1979). Geomorphic thresholds: the concept and its applications. *Transactions Institute of British Geographers,* **NS 4,** 485–515.

Schumm S. A. (1981). Evolution and response of the fluvial system, sedimentologic implications. *SEPM Special Publication,* **31,** 19–29.

Schumm, S. A. (1985). Explanation and extrapolation in geomorphology: seven reasons for geologic uncertainty. *Transactions of the Japanese Geomorphological Union,* **6,** 1–18.

Schumm, S. A. & Chorley, R. J. (1966). Talus weathering and scarp recession in the Colorado Plateau. *Zeitschrift für Geomorphologie, N.F.,* **10,** 11–36.

Schumm, S. A. & Lichty, R. W. (1965). Time, space and causality in geomorphology. *American Journal of Science,* **263,** 110–119.

Schwartz, D. (1988). Some podzols on Bateke Sands and their origins, People's Republic of Congo. *Geoderma,* **43,** 229–247.

Sciences Géologiques (1988). Latérites périatlantiques. *Sciences Géologiques, Bulletin,* **41** (1), Strasbourg.

Scott, L. (1989). Climatic conditions in southern Africa since the last glacial maximum, inferred from pollen analysis. *Palaeogeography, Palaeoclimatology, Palaeoecology*, **70**, 345–353.

Scrivenor, J. B. (1931). *The Geology of Malaya*. Macmillan, London.

Segalen, P. (1971). Metallic oxides and hydroxides in soils of the warm and humid areas of the world; formation, identification evolution. In *Soils and Tropical Weathering Processes, Proceedings of the Bandung Symposium*, 1969, pp. 29–50. UNESCO.

Selby, M. J. (1977a). On the origin of sheeting and laminae in granitic rocks: evidence from Antarctica, the Namib Desert and the central Sahara. *Madoqua*, **10**, 171–179.

Selby, M. J. (1977b). Bornhardts of the Namib Desert. *Zeitschrift für Geomorphologie, N.F.*, **21**, 1–13.

Selby, M. J. (1982). *Hillslope Materials and Processes*. Oxford University Press, Oxford.

Senior, B. B. & Mabbutt, J. A. (1979). A proposed method of defining deeply weathered rock units based on regional geological mapping in southwest Queensland. *Journal of the Geological Society of Australia*, **26**, 237–254.

Servant, M. (1973). *Séquences continentales et variations climatiques: évolution du Bassin du Tchad au Cénozoique Supérieur*. Thèse, ORSTOM, Paris.

Servant, M. & Servant-Vildary, S. (1980). L'environment Quaternaire du bassin du Tchad au Cénozoic Supérieur. In *The Sahara and the Nile* (ed. M. A. J. Williams & H. Faure), pp. 133–162. Balkema, Rotterdam.

Servant, M., Soubiés, F., Suguio, K., Turcq, B. & Fournier, M. (1989). Alluvial fans in southeastern Brazil as an evidence for early Holocene dry climatic period. In *International Symposium on Global Changes in South America During the Quaternary, S*ão Paulo, 1989, Special Publication, Vol 1, pp. 75–77, São Paulo.

Seshagiri, D. N., Badrinarayananan, S., Upendran, R., Lakshimikantham, C. B. & Srinivasan, V. (1982). *The Nilgiri Landslides*. Geological Survey of India, Miscellaneous Publication, No. 57, Delhi.

Shackleton, N. J. (1987). Oxygen isotopes, ice volume and sea level. *Quaternary Science Reviews*, **6**, 183–190.

Shackleton, N. J. & Kennett, J. P. (1975). Paleo-temperature history of the Cenozoic and the initiation of Antarctic glaciation: oxygen and carbon isotope analysis in the DSDP sites 277, 279 and 281. In *Initial Report Deep-Sea Drilling Project*, Vol. 29 (ed. J. P. Kennett *et al.*), pp. 743–755. US Government Printing Office, Washington.

Shackleton, N. J. & Opdyke, N. D. (1973). Oxygen isotope and palaeomagnetic stratigraphy of equatorial Pacific core V28-238: oxygen isotope temperatures and ice volumes on a 10^5-year time scale. *Quaternary Research*, **3**, 39–55.

Shackleton, N. J. & Opdyke, N. D. (1976). Oxygen isotope and palaeomagnetic stratigraphy of equatorial Pacific core V28-239 late Pliocene to latest Pleistocene. In *Investigation of late Quaternary Paleoceanography and Paleoclimatology* (ed. R. M. Cline & J. D. Hays), Geological Society of America, Memoir, pp. 449–464, Boulder.

Sharma, K. D. & Correia, J. F. (1989). Upland erosion in a mixed pine forest ecosystem. *Proceedings of Indian Academy of Sciences (Earth and Planetary Science)*, **98**, 133–145.

Sharpe, C. F. S. (1938). *Landslides and Related Phenomena*, pp. 137, Columbia University Press, New York (reprinted 1960, Pageant Books, New Jersey).

Shaw, P. A. & Cooke, H. J. (1986). Geomorphic evidence for the Late Quaternary palaeoclimates of the middle Kalahari of northern Botswana. *Catena*, **13**, 349–359.

Shaw, P. A., Thomas, D. S. G. & Nash, D. J. (1992). Late Quaternary fluvial activity in the dry valleys (Mekacha) of the middle and southern Kalahari. *Journal of Quaternary Science*, **7**, 273–281.

Sheffield, C. (1977). *Man on Earth*. Sidgwick and Jackson, London.

Sheppard, S. M. F. (1977). The Cornubian batholith, SW England: D/H and $^{18}O/^{16}O$) studies of kaolinite and other alteration minerals. *Journal of the Geological Society of London*, **133**, 573–591.

Sherman, G. D. (1952). The genesis and morphology of the alumina-rich laterite clays. In *Clay and Laterite Genesis*, pp. 154–161. Institute of Mining and Metallurgy, 154–161.

Shroder, Jr. J. F. (1976). Mass movements on the Nyika Plateau, Malawi. *Zeitschrift für Geomorphologie, N.F.*, **20**, 56–77.

Sieffermann, R. G. (1988). Le système des grandes tourbières equatoriales. *Annales de Géographie*, **544**, 642–666.

Simon, A., Larsen, M. C. & Hupp, C. R. (1990). The role of soil processes in determining mechanisms of slope failure and hillslope development in a humid-tropical forest, eastern Puerto Rico. *Geomorphology*, **3**, 263–286.

Simonett, D. S. (1967). Landslide distribution and earthquakes in the Bewanni and Toricelli Mountains, New Guinea, a statistical analysis. In *Landform Studies from Australia and New Guinea* (ed. J. N. Jennings & J. A. Mabbutt), pp. 64–84. Cambridge University Press, London.

Singh, J., Wapakala, W. W. & Chebosi, P. K. (1984). Estimating groundwater recharge based on infiltration characteristics of layered soil. In *Challenges in African Hydrology and Water Resources*, Proceedings of the Harare Symposium, 1984, pp. 37–45. IASH Publication, No. 144.

Sirocko, F., Sarnthein, M., Erlenkeuser, H., Lange, H., Arnold, M. & Duplessy, J. C. (1993). Century-scale events in monsoonal climate over the past 24,000 years. *Nature*, **364**, 322–324.

Sivarajasingham, S. (1968). *Soil and Land Use Survey in the Eastern Province*. Report to Government of Sierra Leone, UNDP/FAO; TA 2584.

Sivarajasingham, S., Alexander, L. T., Cady, J. G. & Cline, M. G. (1962). Laterite. *Advances in Agronomy*, **14**, 1–60.

Skempton, W. A. & Delory, F. A. (1957). Stability of natural slopes in London Clay. In *Proceedings 4th International Conference on Soil Mechanics*, Vol. 2, London, pp. 378–381.

Skempton, W. A. & Hutchinson, J. N. (1969). Stability of natural slopes and embankment sections. In *Proceedings of the 7th International Conference on Soil Mechanics and Foundation Engineering*, State of the Art Volume, pp. 291–340.

Slaymaker, O. (1978). Sediment yield in unglacierised Alpine and Tropical low to moderate relief rainforest environments. In *Géomorphologie Dynamique dans les Régions Interpropicale* (ed. J. Alexandre), *Géo-Eco-Trop*, **2**, 113–130.

Slingerland, R. & Smith, N. D. (1986). Occurrence and formation of water-laid placers. *Annual Review of Earth and Planetary Sciences*, **14**, 113–147.

Smart, P. (1973). Structure of a red clay soil from Nyeri, Kenya. *Quarterly Journal of Engineering Geology*, **6**, 129–140.

Smart, P. & Tovey, N. K. (1982). *Electron Microscopy of Soils and Sediments — Techniques*. Oxford University Press, Oxford.

Smith, A. M. (1992). Holocene paleoclimatic trends from palaeoflood analysis. *Palaeogeography, Palaeoclimatology, Palaeoecology, (Global and Planetary Change Section)*, **97**, 235–240.

Smith, B. J. (1982). Effects of climate and land use on gully development: an example from northern Nigeria. *Zeitschrift für Geomorphologie, N.F., Supplementband*, **44**, 33–51.

Smith, B. J. & de S. Sanchez, B. A. (1992). Erosion hazards in a Brazilian suburb. *Geography Review*, **6**, 37–41.

Smith, B. J. & McAlister, J. J. (1987). Tertiary weathering environment and products in northeast Ireland. In *International Geomorphology 1986, Proceedings 1st International Conference on Geomorphology*, Part II (ed. V. Gardiner), 1007–1031.

Smith, D. I. (1971). The residual hypothesis for the formation of Jamaican bauxite — a consideration of the rate of limestone erosion. *Journal of the Geological Society of Jamaica*, **11**, 3–12.

Smith, D. I. & Atkinson, T. C. (1976). Process, landforms and climate in limestone regions. In *Geomorphology and Climate* (ed. E. Derbyshire), pp. 369–409. Wiley, London & New York.

Smith, T. C. (1988). A method for mapping relative susceptibility to debris flows, with an example from San Mateo County. In *Landslides, Floods and Marine Effects of the Storm of January 3–5, 1982, in the San Francisco Bay Region, California* (ed. S. D. Ellen &

G. F. Wieczorek), pp. 185–194. US Geological Survey Professional Paper No. 1434, Washington.

So, C. L. (1971). Mass movements associated with the rainstorm of June 1966 in Hong Kong. *Transactions Institute British Geographers*, **53**, 55–66.

Söderman, G. (1985). Planation and weathering in eastern Fennoscandia. *Fennia*, **163**, 347–352.

Soil Survey Staff (1975). *Soil Taxonomy, A Basic System of Soil Classification for Making and Interpreting Soil Surveys*. Soil Conservation Service, US Department of Agriculture, Agricultural Handbook, pp. 486, Washington, DC.

Sombroek, W. G. (1971). Ancient levels of plinthisation in N.W. Nigeria. In *Paleopedology* (ed. D. H. Yaalon), pp. 329–336. ISSS & Israel University Press, Jerusalem.

Soubiès, F., Suguio, K., Martin, L., Leprun, J. C., Servant, M., Turcq, B., Fournier, M., Delaume, M. & Siffedine, A. (1989). The Quaternary lacustrine deposits of the Serra dos Carajas (State of Para, Brazil): ages and other preliminary results. In *International Symposium on Global Changes in South America During the Quaternary*, São Paulo, 1989, Special Publication No. 1, pp. 125–128, São Paulo.

Soyer, J. (1989). Rôle des termites dans la formation du complexe de la stone-line. In *Stone-Lines*. Journée D'Étude, Bruxelles, 24 mars 1987 (ed. J. Alexandre & J. J. Symoens), pp. 97–108. Académie Royale des Sciences D'Outre-Mer, Geo-Eco-Trop, 11.

Späth, H. (1981). Bodenbildung und Reliefentwicklung in Sri Lanka. *Relief Boden Paläoklima*, **1** (Zur Morphogenese in den feuchten Tropen. Verwitterung und Reliefbildung am Beispiel von Sri Lanka) 1, pp. 185–238. Borntraeger, Berlin.

Späth, H. (1987). Landform development and laterites in northwestern Australia. *Zeitschrift für Geomorphologie, N.F., Supplementband*, **64**, 25–32.

Speight, J. G. (1965). Flow and channel characteristics of the Angabunga River, Papua. *Journal of Hydrology*, **3**, 16–36.

Speight, J. G. (1971). Log-normality of slope distribution. *Zeitschrift für Geomorphologie, N.F.*, **14**, 290–311.

Spencer, T., Douglas, I., Greer, T. & Sinun, W. (1990). Vegetation and fluvial geomorphic processes in south-east Asian tropical rainforests. In *Vegetation and Erosion* (ed. J. B. Thornes), pp. 420–436. Wiley, Chichester.

Stallard, R. F. & Edmond, J. M. (1981). Geochemistry of the Amazon: 1. Precipitation chemistry and the marine contribution to the dissolved load at the time of peak discharge. *Journal of Geophysical Research*, **86** (C10), 9844–9858.

Stallard, R. F. & Edmond, J. M. (1983). Geochemistry of the Amazon: 2. The influence of geology and weathering environment on the dissolved load. *Journal of Geophysical Research*, **88** (C14), 9671–9688.

Stallard, R. F. & Edmond, J. M. (1987). Geochimstry of the Amazon: 3. Weathering chemistry amd limits to dissolved inputs. *Journal of Geophysical Research*, **92** (C8), 8293–8302.

Staples, R. R. (1936). *Run-off and Soil Erosion Tests in Semi-arid Tanganyika Territory* (Second Report). Annual Report, Department of Veterinary Science and Animal Husbandry, pp. 134–141.

Stark, N. M. & Jordan, C. F. (1978). Nutrient retention by the root mat of an Amazonian rainforest. *Ecology*, **59**, 434–437.

Starkel, L. (1972). The role of catastrophic rainfall in shaping the relief of the Lower Himalaya (Darjeeling Hills). *Geographica Polonica*, **21**, 103–147.

Starkel, L. (1976). The role of extreme (catastrophic) meteorological events in contemporary evolution of slopes. In *Geomorphology and Climate* (ed. E. Derbyshire), pp. 203–246. Wiley, New York.

Starkel, L. (1983). The reflection of hydrologic changes in the fluvial environment of the temperate zone during the last 15,000 years. In *Background to Palaeohydrology* (ed. K. J. Gregory), pp. 213–236. Wiley, Chichester.

Starkel, L. (1987). The role of the inherited forms in the present-day relief of the Polish Carpathians. In *International Geomorphology 1986, Proceedings 1st International Conference on Geomorphology* (ed. V. Gardiner), Part II, 1033–1045.

Stauffer, P. H. (1973). Cenozoic. In *Geology of the Malay Peninsula* (ed. D. J. Gobbett & C. S. Hutchison), pp. 43–176. Wiley, New York.

Stein, R. & Sarnthein, M. (1984). Late Neogene events of atmospheric and oceanic circulation off-shore northwest Africa: high resolution record from deep sea sediments. *Palaeoecology of Africa*, **16**, 9–36.

Stephens, C. G. (1971). Laterite and silcrete in Australia: a study of the genetic relationship of laterite and silcrete and their companion materials, and their collective significance in the weathered mantle, soils, relief and drainage of the Australian continent. *Geoderma*, **5**, 5–52.

Sternberg, H. O'Reilly (1949). Floods and landslides in the Paraiba valley, December, 1948. Influence of destructive exploitation of land. *International Geographical Congress*, **3**, 335–364.

Stewart, A. J., Blake, D. H. & Ollier, C. D. (1986). Cambrian river terraces and ridgetops in central Australia: oldest persisting landforms? *Science*, **233**, 758–761.

Stocking, M. A. (1978a). Interpretation of stone-lines. *The South African Geographical Journal*, **60**, 121–134.

Stocking, M. A. (1978b). The measurement, use and relevance of rainfall energy in investigation into erosion. *Zeitschrift für Geomorphologie, N.F., Supplementband*, **29**, 141–150.

Stocking, M. A. (1981a). A model of piping in soils. *Transactions, Japanese Geomorphological Union*, **2**, 263–278.

Stocking, M. A. (1981b). Causes and prediction of the advance of gullies. In *Proceedings of the South-east Asian Regional Symposium on Problems of Soil Erosion and Sedimentation*, Bangkok, Thailand, 1981, pp. 37–47.

Stocking, M. A. (1984). Rates of erosion and sediment yield in the African environment. In *Challenges in African Hydrology and Water Resources*, Proceedings of the Harare Symposium, 1984, pp. 285–293. IASH Publication No. 144.

Stoddart, D. R. (1969). Climatic geomorphology: review and assessment. *Progress in Physical Geography*, **1**, 160–222.

Stoops, G. (1989a). Relict properties in soils of humid tropical regions with special reference to central Africa. In *Paleopedology* (ed. A. Bronger & J. A. Catt). *Catena, Supplement*, **16**, 95–106.

Stoops, G. (1989b). Contribution of *in situ* transformations to the formation of stone-layer complexes in Central Africa. In (ed. J. Alexandre & J. J. Symoens), *Stone-Lines. Journée D'Étude, Bruxelles, 24 mars 1987*. Académie Royale des Sciences D'Outre-Mer, Geo-Eco-Trop, **11**, 139–150.

Strakhov, N. M. (1967). *Principles of Lithogenesis*. Vol. 1. Oliver and Boyd, Edinburgh.

Street, F. A. (1980). The relative importance of climate and hydrogeological factors in influencing lake-level fluctuations. *Palaeoecology of Africa*, **12**, 137–158.

Street, F. A. (1981). Tropical palaeoenvironments. *Progress in Physical Geography*, **5**, 157–185.

Street, F. A. & Grove, A. T. (1976). Environmental and climatic implications of late Quaternary lake level fluctuation in Africa. *Nature*, **261**, 385–390.

Street, F. A. & Grove, A. T. (1979). Global maps of lake level fluctuations since 30,000 BP. *Quaternary Research*, **12**, 83–118.

Street-Perrott, F. A. & Perrott, R. A. (1990). Abrupt climatic fluctuations in the tropics: the influence of Atlantic Ocean circulation. *Nature*, **343**, 607–612.

Street-Perrott, F. A., Roberts, N. & Metcalfe, S. (1985). Geomorphic implications of late Quaternary hydrological and climatic changes in the Northern Hemisphere tropics. In *Environmental Change and Tropical Geomorphology* (ed. I. Douglas & T. Spencer), pp. 164–183. Allen and Unwin, London.

Stromquist, L. (1981). Recent studies on soil erosion, sediment transport and reservoir sedimentation in central Tanzania. In *Tropical Agricultural Hydrology* (ed. R. Lal & E. W. Russell), pp. 189–200. Wiley, Chichester.

Stuijts, I., Newsome, J. C. & Flenley J. R. (1988). Evidence for Late Quaternary vegetational change in the Sumatran and Javan highlands. *Review of Palaeobotany and Palynology*, **55**, 207–216.

Styles, K. A., Hansen, A., Dale, M. J. & Burnett, A. D. (1985). Terrain classification for development planning and geotechnical appraisal: A Hong Kong case. *IVth International Symposium on Landslides*, Toronto, 1984, Vol. 1, 561–568.

Sueoka, T. (1988). Identification and classification of granitic residual soils using chemical weathering index. In *Geomechanics in Tropical Soils*, Proceedings of the

Second International Conference on Geomechanics in Tropical Soils, Singapore, 1988, Vol. 1, pp. 55–61. Balkema, Rotterdam.

Sucoka, T., Lee, I. K., Muramatsu, M. & Imamura, S. (1985). Geomechanical properties and engineering classification for decomposed granite soils in Kaduna District, Nigeria. In *First International Conference on Geomechanics in Tropical Lateritic and Saprolitic Soils*, Brasilia, 1985, Vol. 1, pp. 175–186, Brasilia.

Sugden, D. (1984). The laterite–bauxite deposits of the Trombetas and Paragominas regions of the Amazon Basin. In *Bauxite*. Proc. 1984 Symposium, Los Angeles, 1984 (ed. L. Jacob Jr.), Society of Mining Engineers, American Institute Mining, Metallurgical & Petroleum Engineers, Inc. New York, pp. 141–151, New York.

Sugden, D. E. (1968). The selectivity of glacial erosion in the Cairngorm Mountains, Scotland. *Transactions Institute of British Geographers*, **45**, 79–92.

Sugden, D. E. (1976). A case against deep erosion of shields by ice sheets. *Geology*, **4**, 580–582.

Summerfield, M. A. (1983a). Silcrete. In *Chemical Sediments and Geomorphology* (ed. A. S. Goudie & K. Pye), pp. 59–91. Academic Press, London.

Summerfield, M. A. (1983b). Geochemistry of weathering profile silcretes, southern Cape Province, South Africa. In *Residual Deposits: Surface Related Weathering Processes and Materials* (ed. R. C. L. Wilson), pp. 167–178. Geological Society Special Publication, No. 11, Blackwell, Oxford.

Summerfield, M. A. (1985). Plate tectonics and landscape development of the African continent. In *Tectonic Geomorphology* (ed. M. Morisawa & J. T. Hack), pp. 27–51. George, Allen and Unwin, London.

Summerfield, M. A. (1988). Global tectonics and landform development. *Progress in Physical Geography*, **12**, 389–404.

Summerfield, M. A. (1991a). Sub-aerial denudation of passive margins: regional elevation versus local relief models. *Earth and Planetary Science Letters*, **102**, 460–469.

Summerfield, M. A. (1991b). *Global Geomorphology*. Longman, Harlow.

Sundquist, E. T. (1987). Ice core links CO_2 to climate. *Nature*, **329**, 389–390.

Sutherland, D. G. (1982). The transport and sorting of diamonds by fluvial and marine processes. *Economic Geology*, **77**, 1613–1620.

Sutherland, D. G. (1984). Geomorphology and mineral exploration: some examples from exploration for diamondiferous placer deposits. *Zeitschrift für Geomorphologie, N.F., Supplementband*, **51**, 95–108.

Sutherland, D. G. (1985). Geomorphological controls on the distribution of placer deposits. *Journal of The Geological Society*, **142**, 727–737.

Sweeting, M. M. (1968). Karst. In *Encyclopaedia of Geomorphology* (ed. R. W. Fairbridge), pp. 582–587. Reinhold, New York.

Sweeting, M. M. (1972). *Karst Landforms*. Macmillan, London.

Talbot, M. R. (1982). Holocene chronostratigraphy of tropical Africa. *Striae*, **16**, 17–20.

Talbot, M. R. & Delibrias, F. (1980). A new late Pleistocene—Holocene water-level curve for lake Bosumtwi, Ghana. *Earth and Planetary Science Letters*, **47**, 336–344.

Talbot, M. R. & Johannessen, T. (1992). A high resolution palaeoclimatic record of the last 27,500 years in tropical West Africa from the carbon and nitrogen isotopic composition of lacustrine organic matter. *Earth and Planetary Science Letters*, **110**, 23–37.

Talbot, M. R., Livingstone, D. A., Palmer, P. G., Maley, J., Melack, J. M., Delibrias, G. & Gulliksen, S. (1984). Preliminary results from sediment cores from lake Bosumtwi, Ghana. *Palaeoecology of Africa*, **16**, 173–192.

Talma, A. S. & Vogel, J. C. (1992). Late Quaternary paleotemperatures derived from a speleothem from Cango caves, Cape Province, South Africa. *Quaternary Research*, **37**, 203–213.

Tamura, T. (1982). Recent morphogenetic changes as revealed in the slope form and regolith characteristics of the West Cameroon Highlands. In *Geomorphology and Environmental Change in the Forest and Savanna Cameroon* (ed. H. Kadomura). Special Publication No. 2, pp. 67–78. Graduate School of Environmental Science, Hokkaido University, Sapporo.

Tamura, T. (1986). Regolith-stratigraphic study of Late Quaternary environmental history in the West Cameroon Highlands and the Adamaoua Plateau. In *Geomorphology and Environmental Changes in Tropical Africa: Case Studies in Cameroon and Kenya*, Special Publication No. 4 (ed. H. Kadomura), pp. 69–93. Graduate School of Environmental Science, Hokkaido University, Sapporo.

Tamura, T. (1991). Termite's role in changing the surface of tropical lands. *Transactions of the Japanese Geomorphological Union*, **12**, 203–218.

Tardy, Y. (1971). Characterisation of the principal weathering types by the geochemistry of waters from some European and African crystalline massifs. *Chemical Geology*, **7**, 253–271.

Tardy, Y. & Nahon, D. (1985). Geochemistry of laterites, stability of Al-goethite, Al-hematite, and Fe^{3+}-kaolinite in bauxites and ferricretes: an approach to the mechanism of concretion formation. *American Journal of Science*, **285**, 865–903.

Tardy, Y., Bocquier, G., Paquet, H. & Millot, G. (1973). Formation of clay from granite and its distribution in relation to climate and topography. *Geoderma*, **10**, 271–284.

Taylor, D. M. (1990). Late Quaternary pollen records from two Ugandan mires: evidence for environmental change in the Rukiga Highlands of southwest Uganda. *Palaeogeography, Palaeoclimatology & Palaeoecology*, **80**, 283–300.

Taylor, G. R. (1983). Landscape evolution of the northern Monaro, NSW, and its implications for the uplift history of the southern Australian highlands. *Geological Society of Australia, Abstracts*, **9**, 114.

Taylor, G. R. & Ruxton, B. P. (1987) A duricrust catena in South-east Australia. *Zeitschrift für Geomorphologie, N.F.*, **31**, 385–410.

Taylor, G. R., Eggleton, R. A., Holzhauer, C. C., Maconachie, L. A., Gordon, M., Brown, M. C. & McQueen, K. G. (1992). Cool climate lateritic and bauxitic weathering. *Journal of Geology*, **100**, 669–677.

Taylor, H. P. (1977). Water/rock interactions and the origin of H_2O in granitic batholiths. *Journal of the Geological Society of London*, **133**, 509–558.

Teeuw, R. M. (1986). *Geomorphology and Surficial Geology of the Koidu Area, Sierra Leone*. Unpublished PhD thesis, University of Stirling.

Teeuw, R. M. (1989). Variations in the composition of gravel layers across the landscape. Examples from Sierra Leone. In *Stone-Lines*, Journée D'Étude, Bruxelles, 24 mars 1987. (ed. J. Alexandre & J. J. Symoens), Académie Royale des Sciences D'Outre-Mer, Geo-Eco-Trop, **11**, pp. 151–169.

Teeuw, R. M. (1991a). A catenary approach to the study of gravel layers and tropical landscape morphodynamics. *Catena*, **18**, 71–89.

Teeuw, R. M. (1991b). Comparative studies of adjacent drainage basins in Sierra Leone: some insights into tropical landscape evolution. *Zeitschrift für Geomorphologie, N.F.*, **35**, 257–268.

Teeuw, R. M., Thomas, M. F. & Thorp, M. B. (1991). Geomorphology applied to exploration for tropical placer deposits. In *Alluvial Mining* (Institute of Mining and Metallurgy), pp. 458–480. Elsevier Applied Science, London.

Temple, P. H. (1972). Measurements of runoff and soil erosion at an erosion plot scale with particular reference to Tanzania. *Geografiska Annaler*, **54**, 203–220.

Temple, P. H. & Rapp, A. (1972). Landslides in the Mgeta area, western Uluguru Mountains, Tanzania. *Geografiska Annaler*, **54A**, 157–193.

Temple, P. H. & Sundborg, A. (1972). The Rufiji River, Tanzania—hydrology and sediment transport. *Geografiska Annaler*, **54A**, 345–368.

Terzaghi, K. (1950). Mechanism of landslides. In *Application of Geology for Engineering Practice* (ed. S. Paige), pp. 83–213. Geological Society of America, New York.

Terzaghi, K. (1962). Stability of steep slopes on hard unweathered rocks. *Geotechnique*, **12**, 251–270.

Thomas, D. S. G. (1984). Ancient ergs of the former arid zones of Zimbabwe, Zambia and Angola. *Transactions, Institute of British Geographers*, **NS9**, 75–88.

Thomas, D. S. G. (1989). Reconstructing ancient arid environments. *Arid Zone Geomorphology* (ed. D. S. G. Thomas), pp. 311–334. Belhaven, London.

Thomas, D. S. G. & Shaw, P. A. (1993). The evolution and characteristics of the Kalahari, southern Africa. *Journal of Arid Environments*, **25**, 97–108.

Thomas, M. F. (1965a). An approach to some problems of landform analysis in tropical climates. In *Essays in Geography for Austin Miller* (ed. J. B. Whittow & P. D. Wood), pp. 118–144. University of Reading, Reading.

Thomas, M. F. (1965b). Some aspects of the geomorphology of domes and tors in Nigeria. *Zeitschrift für Geomorphologie, N.F.*, **9**, 63–81.

Thomas, M. F. (1966a). Some geomorphological implications of deep weathering patterns in crystalline rocks in Nigeria. *Transactions of the Institute of British Geographers*, **40**, 173–193 (reprinted in Adams, 1975).

Thomas, M. F. (1966b). The origin of bornhardts. *Zeitschrift für Geomorphologie, N.F.*, **10**, 478–480.

Thomas, M. F. (1967a). A bornhardt dome in the plains near Oyo, Western Nigeria. *Zeitschrift für Geomorphologie, N.F.*, **11**, 239–261.

Thomas, M. F. (1967b). *A contribution to the study of inselberg landscapes in Nigeria*. PhD Thesis, University of London.

Thomas, M. F. (1968). Etchplain. In *Encyclopaedia of Geomorphology* (ed. R. W. Fairbridge), pp. 331–333. Reinhold, New York.

Thomas, M. F. (1969). Geomorphology and land classification in tropical Africa. In *Environment and Land Use in Africa* (ed. M. F. Thomas and G. W. Whittington), pp. 103–145. Methuen, London.

Thomas, M. F. (1974a). *Tropical Geomorphology*. Macmillan, London & Basingstoke.

Thomas, M. F. (1974b). Granite landforms: a review of some recurrent problems of interpretation. In *Institute British Geographers, Special Publication* (ed. E. H. Brown & R. S. Waters), No. 7, 13–37.

Thomas, M. F. (1976). Denudation in the tropics and the interpretation of the tropical legacy in higher latitudes—a view of the British experience. In *Geomorphology—Present Problems and Future Prospects* (ed. C. Embleton, D. Brunsden & D. K. C. Jones), pp. 185–202. Oxford University Press, Oxford.

Thomas, M. F. (1978a). The study of inselbergs. *Zeitschrift für Geomorphologie, Supplementband*, **31**, 1–41.

Thomas, M. F. (1978b). Chemical denudation, lateritisation and landform development in Sierra Leone. In *Géomorphologie Dynamique dans les Régions Intertropicale* (ed. J. Alexandre). *Géo-Eco-Trop*, **2**, 243–264.

Thomas, M. F. (1980). Timescales of landform development on tropical shields—a study from Sierra Leone. In *Timescales in Geomorphology* (ed. R. A. Cullingford, D. A. Davidson & J. Lewin), pp. 333–354. Wiley, Chichester.

Thomas, M. F. (1983). Contemporary denudation systems and the effects of climatic change in the humid tropics—some problems from Sierra Leone. In *Studies in Quaternary Geomorphology* (ed. D. J. Briggs & R. S. Waters), pp. 195–214. Geo Books, Norwich.

Thomas, M. F. (1986). Savanna. In *Handbook of Engineering Geomorphology* (ed. P. G. Fookes & P. R. Vaughan), pp. 125–136. Surrey University Press, Guildford.

Thomas, M. F. (1988). Superficial deposits as resources for development—some implications for applied geomorphology. *Scottish Geographical Magazine*, 104, 72–83.

Thomas, M. F. (1989a). The role of etch processes in landform development—I. Etching concepts and their applications. *Zeitschrift für Geomorphologie, N.F.*, 33, 129–142.

Thomas, M. F. (1989b). The role of etch processes in landform development—II. Etching and the formation of relief. *Zeitschrift für Geomorphologie, N.F.*, 33, 257–274.

Thomas, M. F. (1994). Ages and relationships of saprolite mantles. In *Rocks, Rock Weathering and Landform Evolution* (ed. D. Robinson & R. Williams), Wiley, Chichester (in press).

Thomas, M. F. & Goudie, A. S. (ed.) (1985). Dambos: small channelless valleys in the tropics. *Zeitschrift für Geomorphologie, Supplementband*, 52, pp. 222.

Thomas, M. F. & Summerfield, M. J. (1987). Long-term landform evolution: key themes and research problems. In *International Geomorphology 1986*, Part II (ed. V. Gardiner), pp. 935–956. Wiley, Chichester.

Thomas, M. F. & Thorp, M. B. (1980). Some aspects of the geomorphological interpretation of Quaternary alluvial sediments in Sierra Leone. *Zeitschrift für Geomorphologie, N.F., Supplementband*, 36, 140–161.

Thomas, M. F. & Thorp, M. B. (1985). Environmental change and episodic etchplanation in the humid tropics of Sierra Leone: the Koidu etchplain. In *Environmental Change and Tropical Geomorphology* (ed. I. Douglas & T. Spencer), pp. 239–267. Allen and Unwin, London.

Thomas, M. F. & Thorp, M. B. (1992). Landscape dynamics and surface deposits arising from late Quaternary fluctuations in the forest–savanna boundary. In *Nature and Dynamics of Forest–Savanna Boundaries* (ed. P. A. Furley, J. Proctor & J. A. Ratter), pp. 215–253. Chapman & Hall, London.

Thomas, M. F. & Thorp, M. B. (1993). The geomorphology of some Quaternary placer deposits. *Zeitschrift für Geomorphologie Supplementband*, 87, 183–194.

Thomas, M. F., Thorp, M. B. & Teeuw, R. M. (1985). Palaeogeomorphology and the occurrence of diamondiferous placer deposits in Koidu, Sierra Leone. *Journal of the Geological Society*, 142, 789–802.

Thornbury, W. D. (1954). *Principles of Geomorphology*. Wiley, New York.

Thornes J. B. & Brunsden, D. (1979). *Geomorphology and Time*. Methuen, London.

Thorp, M. B. (1967a). Closed basins in the Younger Granite massifs, northern Nigeria. *Zeitschrift für Geomorphologie, N.F.*, 11, 459–480.

Thorp, M. B. (1967b). Jointing patterns and the evolution of landforms in the Jarawa granite massif, northern Nigeria. In *Essays in Geography* (ed. R. Lawton & R. W. Steel), pp. 65–83. Longmans, London.

Thorp, M. B. (1975a). Geomorphic evolution in the Liruei Younger Granite hills, Nigeria. *Savanna*, 4, 139–154.

Thorp, M. B. (1975b). Geomorphology. In *An Advanced Geography of Africa* (ed. J. I. Clarke & M. B. Thorp), pp. 21–73. Hulton, London.

Thorp, M. B. & Thomas, M. F. (1992). The timing of alluvial sedimentation and floodplain formation in the lowland humid tropics of Ghana, Sierra Leone, and western Kalimantan (Indonesian Borneo). *Geomorphology*, 4, 409–422.

Thorp, M. B., Thomas, M. F., Martin, T. & Whalley, W. B. (1990). Late Pleistocene sedimentation and landform development in western Kalimanatan (Indonesian Borneo). *Geologie en Mijnbouw*, 69, 133–150.

Toh, E. S. C. (1978). Comparison of exploration for alluvial tin and gold. *Proceedings 11th Commonwealth Mining and Metallurgy Congress*, Hong Kong, pp. 269–278. Institute of Mining and Metallurgy, London.

Tomlinson, R. W. (1974). Preliminary biogeographical studies on the Inyanga Mountains, Rhodesia. *South African Geographical Journal*, 56, 15–26.

Touraine, F. (1972). Erosion et planation. *Revue de Géographie Alpine*, 60, 101–121.

Trendall, A. F. (1962). The formation of 'apparent peneplains' by a process of combined lateritisation and surface wash. *Zeitschrift für Geomorphologie, N.F.*, 6, 183–197.

Trescases, J.-J. (1973). Weathering and geochemical behaviour of the elements of ultramafic rocks in New Caledonia. *Bureau of Mineral Resources, Geology and Geophysics*, Canberra, 141, 149–161.

Trescases, J.-J. (1983). Nickeliferous laterites: a review of the contributions of the last ten years. In *Lateritisation Processes*, Proceedings IInd International Seminar on Lateritisation Processes, São Paulo, Brazil, 1982 (ed. A. J. Melfi & A. Carvalhi), University of São Paulo, Brazil.

Trescases, J.-J., Melfi, A. J. & De Olivera Sonia, M. B. (1981). Nickeliferous 'laterites' of Brazil. In *Lateritisation Processes*, Proceedings International Seminar on Lateritisation Processes, Trivandrum, India, 1979, pp. 170–184, Balkema, Rotterdam.

Trewartha, G. T. (1961). *The Earth's Problem Climates*. University of Wisconsin Press, Madison.

Tricart, J. (1965). *The Landforms of the Humid Tropics, Forests and Savannas* (transl. C. J. KiewietdeJonge, 1972), Longmans, London.

Tricart, J. (1974). Existence de périodes sèches au Quaternaire en Amazonie et dans les régions voisines. *Revue de Géomorphologie Dynamique*, 23, 145–158.

Tricart, J. (1975). Influence des oscillations climatiques récentes sur le modelé en Amazonie Orientale (région de Santarém) d'après des images de radar latéral. *Zeitschrift für Geomorphologie, N.F.*, 19, 140–163.

Tricart, J. (1977). Types des fleuvieux en Amazonie brésilienne. *Annals de Géographie*, **473**, 1–54.

Tricart, J. (1985). Evidence of Upper Pleistocene dry climates in northern South America. In *Environmental Change and Tropical Geomorphology* (ed. I. Douglas & T. Spencer), pp. 197–217. Allen and Unwin, London.

Tricart, J. & Cailleux, A. (1965). *Introduction à la Géomorphologie Climatique*. SEDES, Paris.

Tricart, J. & KieweitdeJonge, C. (1992). *Ecogeography and Rural Management*. Longman, Harlow.

Trivandrum (1981). *Lateritisation Processes*, Proceedings International Seminar on Lateritisation Processes, Trivandrum, India, 1979, Balkema, Rotterdam.

Truckenbrodt, W., Kotschoubey, B. & Schellmann, W. (1991). Composition and origins of the clay cover on north Brazilian laterites. *Geologische Rundschau*, **80**, 591–610.

Trudgill, S. (1985). *Limestone Geomorphology*. Longman, London.

Tsukamoto, Y. & Kusakabe, O. (1984). Vegetative influences on debris slide occurrences on steep slopes in Japan. In *Proceedings, Symposium on Effects of Forest Land Use on Erosion and Slope Stability*. Environment and Policy Institute, Honolulu, Hawaii.

Tsukamoto, Y. & Minematsu, H. (1987). Hydrogeomorphological characteristics of a zero order basin. In *Erosion and Sedimentation in the Pacific Rim* (ed. R. L. Beschta, T. Blinn, C. E. Grant, F. J. Swanson & G. G. Ice), pp. 61–70. IAHS Publication No. 165, Wallingford.

Tuncer, R. E. & Lohnes, R. A. (1977). An engineering classification of certain basalt-derived lateritic soils. *Engineering Geology*, **11**, 319–339.

Twidale, C. R. (1962). Steepened margins of inselbergs from north-western Eyre Peninsula, South Australia. *Zeitschrift für Geomorphologie, N.F.*, **6**, 51–69.

Twidale, C. R. (1964). A contribution to the general theory of domed inselbergs. *Transactions of the Institute of British Geographers*, **34**, 91–113.

Twidale, C. R. (1967). Hillslopes and pediments in the Flinders Ranges, South Australia. In *Landform Studies from Australia and New Guinea* (ed. J. N. Jennings & J. A. Mabbutt) 95–117. University Press, Cambridge.

Twidale, C. R. (1976). On the survival of palaeoforms. *American Journal of Science*, **276**, 77–94.

Twidale, C. R. (1978). Granite platforms and the pediment problem. In *Landform Evolution in Australasia* (ed. J. L. Davies & M. A. J. Williams), pp. 288–304. ANU Press, Canberra.

Twidale, C. R. (1980). The origin of bornhardts. *Journal of the Geological Society of Australia*, **27**, 195–208.

Twidale, C. R. (1981a). Origins and environments of pediments. *Journal of the Geological Society of Australia*, **28**, 423–434.

Twidale, C. R. (1981b). Inselbergs—exhumed and exposed. *Zeitschrift für Geomorphologie, N.F.*, **25**, 219–221.

Twidale, C. R. (1982). *Granite Landforms*. Elsevier, Amsterdam.

Twidale, C. R. (1983). Pediments, peneplains and ultiplains. *Revue de Géomorphologie Dynamique*, **32**, 1–35.

Twidale, C. R. (1987). Etch and intracutaneous landforms and their implications. *Australian Journal of Earth Science*, **34**, 367–386.

Twidale, C. R. (1988). The missing link: planation surfaces and etch forms in southern Africa. In *Geomorphological Studies in Southern Africa* (ed. G. F. Dardis & B. P. Moon), pp. 31–46. Balkema, Rotterdam.

Twidale, C. R. (1990). The origin and implications of some erosional landforms. *Journal of Geology*, **98**, 343–364.

Twidale, C. R. (1991). A model of landscape evolution involving increased and increasing relief amplitude. *Zeitschrift für Geomorphologie, N.F.*, **35**, 85–109.

Twidale, C. R. & Bourne, J. A. (1975). Episodic exposure of inselbergs. *Geological Society of America Bulletin*, **86**, 1473–1481.

Twidale, C. R., Bourne, J. A. & Smith, D. M. (1974). Reinforcement and stabilisation mechanisms in landform development. *Revue de Géomorphologie Dynamique*, **23**, 115–125.

Ugolini, F. C. & Sletten, R. S. (1991). The role of proton donors in pedogenesis as revealed by soil solution studies. *Soil Science*, **151**, 59–75.

UNESCO (1978). *World Water Balance and the Water Resources of the Earth*. Studies and Reports in Hydrology, Vol. 25, UNESCO, Paris.

Valeton, I. (1972). *Bauxites*, Elsevier, Amsterdam.

Valeton, I. (1983). Palaeoenvironment of lateritic bauxites with vertical and lateral differentiation. In *Residual Deposits: Surface Related Weathering Processes and Materials* (ed. R. C. L. Wilson), pp. 77–90. Geological Society Special Publication, No. 11, Blackwell, Oxford.

Valeton, I., Beissner, H. & Carvalho, A. (1991). *The Tertiary Bauxite Belt on Tectonic Uplift Areas in the Serra da Mantiqueira, South East Brazil*. Contributions to Sedimentology, **17**. Schweizerbart, Stuttgart.

Van, Asch, Th. W. J., Deimel, M. S., Haak, W. J. C. & Simon, J. (1989). The viscous creep component in shallow clayey soil and the influence of the load on creep rates. *Earth Surface Processes and Landforms*, **14** (6 & 7), 557–560.

Van Campo, E. (1986). Monsoon fluctuations in two 20,000 Yr BP oxygen-isotope pollen records off southwest India. *Quaternary Research*, **26**, 376–388.

Van de Graaff, W. J. E. (1983). Silcrete in Western Australia: geomorphological settings, textures, structures, and their genetic implications. In *Residual*

Deposits: Surface Related Weathering Processes and Materials (ed. R. C. L. Wilson), pp. 159–166. Geological Society Special Publication No. 11, Blackwell, Oxford.

Van der Hammen, T. (1972). Changes in vegetation and climate in the Amazon basin and surrounding areas during the Pleistocene. *Geology en Mijnbouw*, **51**, 641–643.

Van der Hammen, T. (1974). The Pleistocene changes of vegetation and climate in tropical South America. *Journal of Biogeography*, **1**, 3–26.

Van der Hammen, T. (1991). Palaeoecological background: neotropics. *Climate Change* (Special Issue: Tropical Forests and Climate), **19**, 37–47.

Van der Hammen, T. & Gonzales, E. (1960). Upper Pleistocene and Holocene climate and vegetation of the Sabana de Bogota (Colombia South America). *Leidse Geologische Mededelingen*, **26**, 261–315.

Van der Hammen, T., Duivenvoorden, J. F., Lips, J. M., Urrego, L. E. & Espejo, N. (1992a). Late Quaternary of the middle Caquetá River area (Colombian Amazonia). *Journal of Quaternary Science*, **7**, 45–55.

Van der Hammen, T., Urrego, L. E., Espejo, N., Duivenvoorden, J. F. & Lips, J. M. (1992b). Late-glacial and Holocene sedimentation and fluctuations of river water level in the Caquetá River area (Colombian Amazonia). *Journal of Quaternary Science*, **7**, 57–67.

Van Wambeke, A. R. (1962). Criteria for classifying tropical soils by age. *Journal of Soil Science*, **13**, 124–132.

Van Zinderen Bakker, E. M. (1967). Upper Pleistocene and Holocene stratigraphy and ecology on the basis of vegetation changes in Sub-Saharan Africa. In *Background to Evolution in Africa* (ed. W. W. Bishop & J. D. Clark), pp. 125–147. Chicago, USA.

Van Zinderen Bakker, E. M. & Mercer, J. H. (1986). Major late Cainozoic climatic events and palaeoenvironmental changes in Africa viewed in a world wide context. *Palaeogeography, Palaeoclimatology and Palaeoecology*, **56**, 217–235.

Vargas, M. (1990). Special lecture: Collapsible and expansive soils in Brazil. In *Geomechanics in Tropical Soils*, Proceedings of the Second International Conference on Geomechanics in Tropical Soils, Singapore, 1988, Vol. 2, pp. 489–492. Balkema, Rotterdam.

Varnes, D. J. (1958). Landslide types and processes. In *Landslides and Engineering Practice* (ed. E. B. Eckel), pp. 20–47. Highway Research Board, Special Report No. 29, National Academy of Sciences–National Research Council, Publication No. 544, Washington, DC.

Varnes, D. J. (1978). Slope movement types and processes. In *Landslide Analysis and Control* (ed. R. L. Schuster & R. J. Krizek), pp. 11–33. National Academy of Sciences, Special Report No. 176.

Vaughan, P. R. & Kwan, C. W. (1984). Weathering, structure and in situ stress in residual soils. *Géotechnique*, **34**, 42–59.

Veevers, J. J. (1981). Morphotectonics of rifted continental margins in embryo (East Africa), youth (Africa–Arabia), and maturity (Australia). *Journal of Geology*, **89**, 57–82.

Velbel, M. A. (1985). Geochemical mass balances and weathering rates in forested watersheds of the southern Blue Ridge. *American Journal of Science*, **285**, 904–930.

Velbel, M. A. (1986). The mathematical basis for determining rates of geochemical and geomorphic processes in small forested watersheds by mass balance: examples and implications. In *Rates of Chemical Weathering of Rocks and Minerals* (ed. S. M. Colman & D. P. Dethier), pp. 431–451. Academic Press, New York.

Verstappen, H. Th. (1960). Some observations on karst development in the Malay Archipelago. *Journal of Tropical Geography*, **14**, 1–10.

Verstappen, H. Th. (1975). On palaeoclimates and landform development in Malaysia. In *Modern Quaternary Research in SE Asia* (ed. G. Bartstra & W. A. Casparie), **1**, 3–35. Balkema, Rotterdam.

Verstappen, H. Th. (1980). Quaternary climatic changes and natural environment in SE Asia. *Geo Journal*, **4**, 45–54.

Vincens, A. (1991). Late Quaternary vegetation of the southern Tanganyika Basin. Climatic implications for southern central Africa. *Palaeogeography, Palaeoclimatology and Palaeoecology*, **86**, 207–226.

Vincens, A., Chalié, F., Bonnefille, R., Guiot, J. & Tiercelin, J-J. (1993). Pollen-derived rainfall and temperature estimates from Lake Tanganyika and their implication for late Pleistocene water levels. *Quaternary Research*, **40**, 343–350.

Vincent, C. E., Davies, T. D., Brimblecombe, P. & Beresfors, A. K. C. (1989). Lake levels and glaciers: indicators of the changing rainfall in the mountains of East Africa. In *Quaternary and Environmental Research on East African Mountains* (ed. W. C. Mahaney), pp. 199–216. Balkema, Rotterdam.

Vincent, E. (1972). Climatic change at the Pleistocene–Holocene boundary in the southwestern Indian Ocean. *Palaeoecology of Africa*, **6**, 45–54.

Vincent, P. L. (1966). Les formations meubles superficielles au sud du Congo et au Gabon. *Bulletin du Bureau de Recherches Géologiques et Minières*, **4**, 53–111.

Virgo, K. J. & Munro, R. N. (1978). Soil and erosion features of the Central Plateau region of Tigrai, Ethiopia. *Geoderma*, **20**, 131–157.

Vogel, J. C. (1989). Evidence of past climatic change in the Namib Desert. *Palaeogeography, Palaeoclimatology, Palaeoecology*, **70**, 355–366.

Vogt, J. (1966). Le complexe de la stone-line. Mise au point. *Bulletin du Bureau de Recherches Géologiques et Minières*, **4**, 3–51.

Wahrhaftig, C. (1965). Stepped topography of the southern Sierra Nevada, California. *Bulletin Geological Society of America*, 76, 1165–1190.

Walker, D. (1982). Speculations on the origin and evolution of Sunda-Sahul rain forests. In *Biological Diversification in the Tropics* (ed. G. T. Prance), pp. 554–575. Columbia, New York.

Walling, D. E. (1984). The sediment yields of African rivers. In *Challenges in African Hydrology and Water Resources*, Proceedings of the Harare Symposium, 1984, pp. 265–283. IASH Publication No. 144.

Walling, D. E. & Webb, B. W. (1986). Solutes in river systems. In *Solute Processes* (ed. S. T. Trudgill), pp. 251–327. Wiley, Chichester.

Wang, C. & Ross, G. J. (1989). Granite saprolites: their characteristics, identification and influence on soil properties in the Appalachian region of Canada. *Zeitschrift für Geomorphologie, N.F., Supplementband*, 72, 150–161.

Ward, P. R. B. (1980). Sediment transport and a reservoir siltation formula for Zimbabwe-Rhodesia. *The Civil Engineer in South Africa*, Jan., 9–15.

Watkins, J. R. (1967). The relationship between climate and the development of landforms in the Cainozoic rocks of Queensland. *Journal Geological Society of Australia*, 14, 153–168.

Watson, A., Price-Williams D. & Goudie A. S. (1983). Palaeoenvironmental interpretation of colluvial sediments and palaeosols of the Late Pleistocene hypothermal in southern Africa. *Palaeogeography, Palaeoclimatology and Palaeoecology*, 45, 225–250.

Watson, J. P. (1974). Termites in relation to soil formation, groundwater, and geochemical prospecting. *Soils and Fertilizers*, 37, 111–114.

Wayland, E. J. (1929). African pluvial periods. *Nature*, 123, 607.

Wayland, E. J. (1933). Peneplains and some other erosional platforms. *Annual Report and Bulletin*, Protectorate of Uganda Geological Survey, Department of Mines, *Note 1*, 77–79.

Wayland, E. J. (1952). The study of past climates in tropical Africa. *Proceedings First Pan African Congress on Prehistory* (1947), pp. 59–66. Blackwell, Oxford.

Wellman, H. W. & Wilson, A. T. (1965). Salt weathering, a neglected erosive agent in coastal and arid environments. *Nature*, 205, 1097–1098.

Wellman, P. (1979). On the Cainozoic uplift of the south-eastern Australian highland. *Journal Geological Society of Australia*, 26, 1–9.

Wellman, P. & McDougall, I. (1974). Potassium-argon ages on the Cainozoic volcanic rocks of New South Wales. *Journal of the Geological Society of Australia*, 21, 247–272.

Wells, N. A., Andriamihaja, B. & Rakotovololona, H. F. S. (1990). Stonelines and landscape development on the laterised craton of Madagascar. *Geological Society of America Bulletin*, 102, 615–627.

Wells, N. A., Andriamihaja, B. & Rakotovololona, H. F. S. (1991). Patterns of development of lavaka, Madagascar's unusual gullies. *Earth Surface Processes and Landforms*, 16, 189–206.

Wentworth, C. K. (1943). Soil avalanches on Oahu, Hawaii. *Bulletin Geological Society America*, 54, 53–64.

West, G. & Dumbleton, J. J. (1970). The minerology of tropical weathering illustrated by some West Malysian soils. *Quarterly Journal of Engineering Geology*, 3, 25–40.

Whipkey, R. Z. & Kirkby, M. J. (1978). Flow within the soil. In *Hillslope Hydrology* (ed. M. J. Kirkby), pp. 121–144. Wiley, Chichester.

Whitaker, C. R. (1978). Pediment form and evolution on granite in the East Kimberleys. In *Landform Evolution in Australasia* (ed. J. L. Davies & M. A. J. Williams), pp. 305–330. ANU Press, Canberra.

White, S. (1949). Processes of erosion on steep slopes of Oahu, Hawaii. *American Journal of Science*, 247, 168–186.

White, W. B. (1984). Rate processes: chemical kinetics and karst landform development. In *Groundwater as a Geomorphic Agent* (ed. R. G. La Fleur), pp. 227–248. Allen and Unwin, London.

White, W. B., Jefferson, G. L. & Haman, R. J. (1966). Quartzitic karst in southeastern Venezuela. *International Journal of Speleology*, 2, 309–316.

Whitlow, J. R. (1979). Bornhardt terrain on granite rocks in Zimbabwe: a preliminary assessment. *Zambian Geographical Journal*, 33/34, 75–94.

Whitlow, J. R. (1988a). Potential versus actual erosion in Zimbabwe. *Applied Geography*, 8, 87–100.

Whitlow, J. R. (1988b). *Land Degradation in Zimbabwe: A Geographical Study*. Department of Natural Resources, Harare.

Whitmore, T. C. (ed.) (1987). *Biogeographical Evolution in the Malay Archipelago*, Oxford Monograph in Biogeography, Vol. 4.

Whitmore, T. C. & Prance G. T. (eds.) (1987). *Biogeography and Quaternary History in Tropical America*. Clarendon, Oxford.

Wigwe, G. A. (1966). *Drainage Composition and Valley Forms in Parts of Northern and Western Nigeria*. Unpublished PhD Thesis, University of Ibadan.

Wijmstra, T. A. & Van der Hammen, T. (1966). Palynological data on the history of tropical savannas in northern South America. *Leidse Geologische Mededelingen*, 38, 71–90.

Wilford, G. E. (1991). Exposure of landsurfaces, drainage age and erosion rates. In *Ancient Landforms* (ed. C. D. Ollier), pp. 91–103. Belhaven, London.

Wilhelmy, H. (1958). *Klimamorphologie der Massengesteine*, Westermann, Brunswick.

Wilkinson, G. E. & Aina, P. O. (1976). Infiltration of water into two Nigerian soils under secondary forest and subsequent arable cropping. *Geoderma*, **15**, 51–59.

Williams, M. A. J. (1968). Termites and soil development near Brocks Creek, Northern Territory. *Australian Journal of Science*, **31**, 135–154.

Williams, M. A. J. (1973). The efficacy of creep and slope wash in tropical and temperate Australia. *Australian Geographical Studies*, **11**, 62–78.

Williams, M. A. J. (1978). Termites, soils and landscape equilibrium in the Northern Territory of Australia. In *Landform Evolution in Australasia* (ed. J. L. Davis & M. A. J. Williams), pp. 128–141. ANU Press, Canberra.

Williams, M. A. J. & Faure, H. (eds). (1980). *The Sahara and the Nile*. Balkema, Rotterdam.

Williams, M. A. J. (1985). Pleistocene aridity in tropical Africa, Australia and Asia. In *Environmental Change and Tropical Geomorphology* (ed. I. Douglas & T. Spencer), pp. 219–233. Allen and Unwin, London.

Williams, P. W. (1972). Morphometric analysis of polygonal karst in New Guinea. *Bulletin of the Geological Society of America*, **83**, 761–796.

Williams, P. W. (1978). Interpretation of Australasian karsts. In *Landform Evolution in Australasia* (ed. J. L. Davies & M. A. J. Williams), pp. 259–286, ANU Press, Canberra.

Williams, P. W. (1987). Geomorphic inheritance and the development of tower karst. *Earth Surface Processes and Landforms*, **12**, 453–465.

Willis, B. (1936). *East African Plateaus and Rift Valleys — Studies in Comparative Seismology*. Carnegie Institute, Washington, Publication, **470**.

Wilson, A. F. (1984). Origin of quartz-free gold nuggets and supergene gold found in laterites and soils — a review and some new observations. *Australian Journal of Earth Science*, **31**, 303–316.

Wilson, C. J. & Dietrich, W. E. (1987). Contribution of bedrock groundwater flow to storm runoff and high pore pressure development in hollows. In *Erosion and Sedimentation in the Pacific Rim* (ed. R. L. Beschta, T. Blinn, C. E. Grant, F. J. Swanson & G. G. Ice), IAHS Publication No. 165, 49–59.

Wilson, L. (1973a). Relationships between geomorphic processes and modern climates as a method in paleoclimatology. In *Climatic Geomorphology* (ed. E. Derbyshire), pp. 269–284. Macmillan, London.

Wilson, L. (1973b). Variations in mean annual sediment yield as a function of mean annual precipitation. *American Journal of Science*, **273**, 335–349.

Wilson, N. W. & Marmo, V. (1958). *Geology, Geomorphology and Mineral Resources of the Sula Mountains*. Geological Survey of Sierra Leone, Bulletin No. 1.

Wilson, R. C. L. (ed.) (1983). *Residual Deposits: Surface Related Weathering Processes and Materials*, Geological Society Special Publication, No. 11. Blackwell, Oxford.

Wirthmann, A. (1987). *Geomorphologie der Tropen — Erträge der Forschung*. Wissenschaftl. Buchgeselschaft, Darmstadt.

Wischmeier, W. H. & Smith, D. D. (1958). Rainfall energy and its relationship to soil loss. *Transactions of The American Geophysical Union*, **39**, 285–291.

Wise, D. U. (1972). Freeboard of continents through time. *Geological Society of America*, Memoir, **132**, 87–100.

Wise, D. U. (1974). Continental margins, freeboard and the volumes of continents through time. In *The Geology of Continental Margins* (ed. C. A. Burke & C. L. Drake), pp. 45–68. Springer, Berlin.

WMO (World Meteorological Organisation) (1983). Operational hydrology in the humid tropical regions. In *Hydrology of Humid Tropical Regions with Particular Reference to the Hydrological Effects of Agriculture and Forestry Practice*, pp. 3–26. IAHS Publication No. 140.

Wolfe, J. A. (1977). Large, Holocene low angle landslide, Samar Island, Philippines. In *Landslide Perspectives* (ed. D. R. Coates), pp. 149–153. Reviews in Engineering Geology, Vol. III, Geological Society of America.

Wolman, M. G. & Leopold, L. B. (1957). River flood plains: some observations on their formation. *US Geological Survey Professional Paper*, **282-C**, 88–107.

Wolman, M. G. & Miller J. P. (1960). Magnitude and frequency of forces in geomorphological processes. *Journal of Geology*, **68**, 54–74.

Woolnough, W. G. (1918). Physiographic significance of laterite in Western Australia. *Geological Magazine, NS*, **6**, 385–393.

Woolnough, W. G. (1927). The duricrust of Australia. *Journal Proceedings of the Royal Society of New South Wales*, **61**, 25–53.

Woolnough, W. G. (1930). Influence of climate and topography on the products of weathering. *Geological Magazine*, **67**, 123–132.

Wopfner, H. (1974). Post-Eocene history and stratigraphy of northeastern South Australia. *Transactions of the Royal Society of South Australia*, **98**, 1–12.

Wopfner, H. (1978). Silcretes of northern South Australia and adjacent regions. In *Silcrete in Australia* (ed. T. Langford-Smith), pp. 93–141. University of New England Press, Australia.

Wopfner, H. (1983). Environments of silcrete formation: a comparison of examples from Australia and the Cologne embayment, West Germany. In *Residual Deposits: Surface Related Weathering Processes and Materials* (ed. R. C. L. Wilson), pp. 151–158. Geological Society Special Publication, No. 11, Blackwell, Oxford.

Yatsu, E. (1992). To make geomorphology more scientific. *Transactions of the Japanese Geomorphological Union*, **13**, 87–124.

Young, A. (1972). *Slopes*. Oliver & Boyd, Edinburgh.

Young, A. (1976). *Tropical Soils and Soil Survey*. Cambridge University Press, London.

Young, A. & Saunders, I. (1986). Rates of surface processes and denudation. In *Hillslope Processes*, (ed. A. D. Abrahams), pp. 3–27. Allen & Unwin, Boston & New York.

Young, R. W. (1981). Denudational history of the south central uplands of New South Wales. *Australian Geographer*, **15**, 17–88.

Young, R. W. (1983). The tempo of geomorphological change: evidence from southeastern Australia. *Journal of Geology*, **91**, 221–230.

Young, R. W. (1987a). Tower karst in sandstone: Bungle Bungle massif, northwestern Australia. *Zeitschrift für Geomorphologie, N.F.*, **30**, 189–202.

Young, R. W. (1987b). Sandstone landforms of the tropical East Kimberley region, northwestern Australia. *Journal of Geology*, **95**, 205–218.

Young, R. W. (1988). Quartz etching and sandstone karst: examples from the East Kimberleys, northwestern Australia. *Zeitschrift für Geomorphologie, N.F.*, **32**, 409–423.

Yuan, D. (1987). New observations on tower karst. In *International Geomorphology 1986*. Proceedings of 1st International Conference on Geomorphology, Part II (ed. V. Gardiner), pp. 1109–1123. Wiley, Chichester.

Zeese, R. (1983). Reliefentwicklung in Nordost-Nigeria reliefgenerationen oder morphogenetische sequenzen. *Zeitschrift für Geomorphologie, N.F., Supplementband*, **48**, 225–234.

Zeese, R. (1991a). Paleosols of different age in central and northeast Nigeria. *Journal of African Earth Sciences*, **12**, 311–318.

Zeese, R. (1991b). Fluviale geomorphodynamik im Quartar Zentral- und Nordostnigerias. *Freiburger Geographische Hefte*, **33**, 199–208.

Zeuner, F. E. (1959). *The Pleistocene Period*. Hutchinson, London.

Ziemer, R. R. & Albright, J. S. (1987). Subsurface pipeflow dynamics on north-coastal California swale systems. In *Erosion and Sedimentation in the Pacific Rim* (ed. R. L. Beschta, T. Blinn, C. E. Grant, F. J. Swanson & G. G. Ice), pp. 71–80. IAHS Publication No. 165.

Zonneveld, J. I. S. (1972). Sulas and sula complexes. *Göttinger Geographische Abhandlungen (Hans-Poser-Festschrift)*, **60**, 93–101.

Zonneveld, J. I. S. (1982). Summit levels in Surinam. *ITC Journal*, **3**, 237–242.

Zonneveld, J. I. S. (1993). Planation and summit levels in Suriname (S. America). *Zeitschrift für Geomorphologie, N.F., Supplementband*, **93**, 29–46.

Author Index

(Where a work has more than two authors, only the first named appears in the Index. Numbers in *italics* refer to Figures in the text)

Subject Index

(Place-name entries are grouped under countries except where referring to trans-national features—e.g. Amazon, Kalahari—Numbers in **bold** refer to Figures in the text)